Disease Processes in Marine Bivalve Molluscs

Funding for the publication of this book was provided by

Horn Point Environmental Laboratories

Center for Environmental and Estuarine Studies
University of Maryland, USA

New Jersey Agricultural Experiment Station

Cook College, Rutgers University
Fisheries and Aquaculture Technology Extension Center of the
New Jersey Commission on Science and Technology, USA

Virginia Institute of Marine Science

School of Marine Science
The College of William and Mary, USA

Sea Grant College Program

Texas A&M University, USA

Consejo Superior de Investigaciones Científicas

Instituto de Investigaciones Marinas
Vigo, Spain

Disease Processes in Marine Bivalve Molluscs

Edited by

William S. Fisher

American Fisheries Society
Special Publication 18

Bethesda, Maryland
1988

The American Fisheries Society Special Publication series is a registered serial. Suggested citation formats follow.

Entire book

Fisher, W. S., editor. 1988. Disease processes in marine bivalve molluscs. American Fisheries Society Special Publication 18.

Article within the book

Comps, M. 1988. Epizootic diseases of oysters associated with viral infections. American Fisheries Society Special Publication 18:23–37.

Library of Congress Catalog Card Number: 88-83123

ISSN 0097-0638   ISBN 0-913235-52-0

.

Address orders to

American Fisheries Society
5410 Grosvenor Lane, Suite 110
Bethesda, Maryland 20814, USA

Contents

Preface

Epizootic diseases of marine bivalves, or at least descriptions of these diseases, have increased dramatically over the last 35 years. Most have been recorded in oyster populations of Europe and North America. It is possible that we are more acutely aware of oyster diseases now than in the past because the fisheries have been harvested beyond their production capacity and remaining stocks are very valuable. But, more likely, the spread of parasites has been facilitated by the transport of oysters from one region to another. Parasites associated with imported oysters can proliferate in naive hosts and cause a disease of epizootic proportions. Only one such case has been documented so far (and that, retroactively), but oyster growers and government institutions, with the sound intention of maintaining a viable industry, have transported oysters all over the world; the potential for disease introductions is great.

During this time, scientists have concentrated on recognition and diagnosis of diseases, description of parasites and their life cycles, and determination of epizootic patterns related to host, season, and environmental conditions. These studies have led to management guidelines and practices directed toward reducing the impact on a fishery and preventing further spread of diseases. A common theme in these recommendations has been to limit the movement of stocks. Indeed, rigorous criteria for the transport of aquatic organisms into nonendigenous regions have been developed and adopted by the International Council for the Exploration of the Sea. Hopefully, reasonable regulation and common sense will protect fisheries from introduced pathogens in the future.

The role of the scientist should now extend toward a deeper understanding of host and parasite interactions. There are highly dynamic aspects of bivalve health and disease (such as parasite changes within a host, sublethal parasitism, environmental influence on hosts and parasites, physiological and ecological responses to parasitism, coevolution, and others) that can be pursued, Knowledge from vertebrate and mammalian systems can be useful templates for these studies insofar as parallels are demonstrated and not merely presumed. Certainly the technology developed for human medicine will prove invaluable. Advanced cellular and biochemical technology can be applied to host–parasite interactions, parasite life cycles, and host defense mechanisms, and can even be merged with bivalve hatchery technology to test and manipulate stocks for disease-resistant or -tolerant traits. In this respect, there is great opportunity for liaison between commercial operations and "fundamental" science.

The study of marine bivalve diseases is at a juncture; the published works of pioneers in the field, advanced technology, and hatchery production are all available. It is no coincidence then, that contributions from the many scientists in this volume reflect a strong dependence on past knowledge and a keen interest in the dynamic mechanisms that underly pathologies.

I wish to thank the Assisting Editors, Jay D. Andrews, Antonio J. Figueras, and Susan E. Ford for their conscientious efforts during various stages of this publication. I wish also to thank the authors of contributed articles; the high quality of their work made this an extremely educational and relatively simple editorial experience. We are all indebted to the more than 50 scientific referees who provided clear insight and valuable feedback on the articles. Their hidden contributions are greatly appreciated by the editors and authors. I would also like to thank Managing Editor Robert Kendall, Production Manager Sally Kendall, and the entire American Fisheries Society Publication Staff for their excellent work in preparing and publishing this volume. My family and my colleagues at the University of Maryland and University of Texas Medical Branch are due a note of thanks for their patience and enthusiasm. Finally, I extend my appreciation and thanks to the institutional sponsors of this publication, as noted opposite the title page. The administrators of these institutions are sufficiently dedicated and attuned to the problems of marine bivalve diseases to recognize the need for increased communication in the field. Thank you all.

WILLIAM S. FISHER[1]
University of Maryland
Horn Point Environmental Laboratories
Center for Environmental and
  Estuarine Studies
Cambridge, Maryland 21613, USA

[1]Present address: Marine Biomedical Institute, University of Texas Medical Branch, 200 University Boulevard, Galveston, Texas 77550, USA.

Common and Scientific Names of Molluscs

The colloquial names of many native and introduced mollusc species in North America have been standardized in *Common and Scientific Names of Aquatic Invertebrates from the United States and Canada: Mollusks*, 1988, American Fisheries Society Special Publication 16. Within each paper of the present volume, species covered by Special Publication 16 are cited by both their common and scientific names when they are first mentioned, and by their common names thereafter. The list below collates all such species that are mentioned in this volume. Former scientific or other frequently used common names are given in parentheses.

Bivalvia

Adula californiensis .... California datemussel
Anodonta oregonensis ...... Oregon floater
Arctica islandica .... ocean quahog
Argopecten irradians .... bay scallop
Crassostrea gigas .... Pacific oyster (Japanese oyster)
Crassostrea virginica ...... eastern oyster (American oyster)
Dendostrea frons .... frond oyster (*Ostrea frons*)
Macoma balthica .... Baltic macoma
Macoma calcarea .... chalky macoma
Macoma inquinata .... stained macoma
Macoma nasuta .... bent-nose macoma
Mercenaria mercenaria ..... northern quahog
Modiolus modiolus .... northern horsemussel
Mya arenaria .... softshell (softshell clam)
Mya truncata .... truncate softshell
Mytilus californianus .... California mussel
Mytilus edulis .... blue mussel
Ostrea edulis .... edible oyster (European flat oyster)
Ostreola conchaphila .... Olympia oyster (*Ostrea lurida*)
Ostreola equestris .... crested oyster (*Ostrea equestris*)
Protothaca staminea .... Pacific littleneck
Pseudochama radians .... Atlantic jewelbox
Saxidomus giganteus .... butter clam
Tapes philippinarum .... Japanese littleneck
Teredo navalis .... naval shipworm

Gastropoda

Achatina fulica .... giant African snail
Aplysia californica .... California seahare
Biomphalaria glabrata ...... bloodfluke planorb
Boonea impressa .... impressed odostome
Cepaea nemoralis .... grovesnail
Crepidula fornicata .... common Atlantic slippersnail
Helix aspersa .... brown gardensnail
Helix pomatia .... escargot
Ilyanassa obsoleta .... eastern mudsnail
Limax flavus .... yellow gardenslug
Lymnaea stagnalis .... swamp lymnaea
Urosalpinx cinerea .... Atlantic oyster drill

Cephalopoda

Octopus vulgaris .... common octopus

American Fisheries Society Special Publication 18:1–4, 1988

MAJOR PARASITIC AND PATHOLOGICAL CONDITIONS

Bonamiasis: A Model Study of Diseases in Marine Molluscs

HENRI GRIZEL, ERIC MIALHE, DOMINIQUE CHAGOT,
VIVIANE BOULO, AND EVELYNE BACHÈRE

Laboratoire de Pathologie et de Génétique des Invertébrés Marins
Boîte Postale 133, 17390 La Tremblade, France

Abstract.—Research on bonamiasis, an epizootic disease of the edible oyster *Ostrea edulis* caused by the protozoan *Bonamia ostreae,* is discussed in relation to oyster farming, research technology, epizootiology, and management. Morphological and infectious characteristics of the parasite are described. Recent progress in isolation and purification of the parasite have permitted investigations into host defense mechanisms, parasite infectivity, and the development of a mollusc–pathogen model.

Since 1955, several important infectious diseases have been observed in molluscs, but, with few exceptions, no effective measures have been developed for eliminating them. This deficiency is a consequence of (1) a lack of knowledge concerning the host's anatomy, biology, and physiology; the infectious agent's metabolism, life cycle, etc.; and the mechanisms that control the host–parasite interrelationships; (2) the lack of a global view from which to study a disease; (3) a lack of coordination among research teams working in complementary disciplines; (4) insufficient contact and flow of information between research institutions, growers, and government agencies; and (5) a lack of training to recognize factors that induce or favor development of pathogenic agents.

The economic impact of the disease caused by *Bonamia ostreae,* a parasite of the edible oyster *Ostrea edulis* (Grizel 1983; Meuriot and Grizel 1984), led to research studies that have yielded original and substantial advances in mollusc pathology.

The Pathogen: Morphology, Ultrastructure, and Life Cycle

The tinctorial affinities of the cytoplasm of *Bonamia ostreae* reveal two different and distinct cellular types. Highly basophilic cells are seen most frequently. They are spheroid in shape, measure 2–3 μm in diameter, and contain a nucleus surrounded by a pale halo. Slightly basophilic cells are seen less frequently. They are somewhat elongated and measure 3–5 μm. In addition to these two types, binucleated cells and certain multinucleate plasmodial forms can also be ob-served, especially in postmortem oysters (Brehelin et al. 1982). The plasmodia can be as large as 6 μm in diameter and contain from 3 to 5 nuclei. Identical forms have also been reported by Dinamani et al. (1987) in *Bonamia* sp., a parasite of the New Zealand oyster *Tiostrea lutaria.* These different forms of *Bonamia ostreae* have been described from observations made with the electron microscope (Pichot et al. 1980; Brehelin et al. 1982; Comps 1983; Grizel 1985).

The cytoplasm of the highly basophilic cells contain many ribosomes 2.5 nm in diameter and mitochondria as large as 1.8 μm. In addition, the cells also contain two spheroid electron-dense inclusions and other dense particles that have the same structure as the haplosporosomes described by Perkins (1971) in several genera of the Ascetospora. The nucleus is 1 μm in diameter and contains a finely granular and electron-dense nucleoplasm. The cytoplasm of the slightly basophilic cells contains mitochondria with numerous tubular cristae, Golgi bodies composed of stacked saccules and small vesicles, and haplosporosomes. Spheroid inclusions are also present. The nucleus consists of a granular nucleoplasm and contains a dense nucleolus located peripherally. The cytoplasm of the plasmodial forms is similar to that of the highly basophilic cells.

The developmental cycle of *Bonamia ostreae* is unknown. Schizogony occurs in the host, but these divisions are probably rapid and are rarely observed. The same is likely true of the final multiplication of the parasite in the postmortem phase from which the plasmodial forms have been described. Perkins (1988, this volume) discusses the probability that *B. ostreae* is one of the Haplosporidia.

The Disease

Gross Pathology

The most distinctive gross symptoms of bonamiasis are nonspecific gill lesions on one or more lamellae. According to Tigé et al. (1980), these lesions appear as perforations and indentations surrounded by characteristic yellowish bands. The principal microscopic findings are (1) dense cellular accumulations due to hemocytic infiltration and (2) *B. ostreae* present in the hemocytes or in the extracellular tissue spaces of the host oyster.

Descriptive Epidemiology

B. ostreae was discovered in Europe in edible oysters raised on Tudy Island (Brittany, France) in beds affected by high rates of mortality (Comps et al. 1980). This protozoan, previously undetected in molluscs cultivated along the French coast, was soon found in most areas of Brittany where the edible oyster was cultivated (Tigé et al. 1980). Only a few natural breeding areas off the coast of Brittany remain unaffected by the disease. Areas in which there was no breeding of edible oysters (such as Vendée) remain free of disease, as has the Mediterranean coast even after several transfers of infected oysters. The prevalence of the parasitic infection varies from one site to another and also from one age-class of edible oyster to another (Grizel 1985). This parasite has been found in several other European countries, including Spain, The Netherlands, England, and Ireland. Most likely, the parasite was introduced into France following the importation of spat from west coast hatcheries of the USA. The identity of *B. ostreae* and "microcells" imported from California in the 1960s (Katkansky et al. 1969) has been proposed on evidence from electron microscopy (Elston et al. 1986). This diagnosis has been confirmed with monoclonal antibodies specific for *B. ostreae* (E. Mialhe, unpublished data).

Analytic Epidemiology

Determination of the period of infection.—Experiments carried out in situ revealed that the disease is transmissible throughout the year (Tigé and Grizel 1984). Tigé and Grizel (1984) observed that different groups of presumably healthy edible oysters consistently became infected with the parasite 3–4 months after they were immersed in infested waters. The highest infection rates occurred during the summer.

Influence of environmental factors.—Temperature is not a major factor affecting transmission of the disease. In contrast to *Marteilia refringens*, which requires at least 17°C, infections of *B. ostreae* can be obtained at ambient temperatures as low as 4–5°C. The specific influence of salinity on infectivity has not been tested. However, infections have been observed in tanks where the salinity was 39‰. In general, there is insufficient evidence to describe the role of the environment in the expression of the disease or transmission of the parasite. Environment may have played a role in cases where the disease could not be transmitted in the laboratory.

Host resistance.—Susceptibility to *B. ostreae* varies with host genus and species. Tests conducted with several genera of molluscs living in a *Bonamia*-contaminated habitat did not result in infections (*Crassostrea gigas, Mytilus edulis, Ruditapes decussatus, R. philippinarum,* and *Cardium edule*). In New Zealand, however, *Bonamia* sp. has been detected in the genus *Haliotis* by M. Hine (Fisheries Research Center, Wellington, personal communication). Furthermore, Grizel et al. (1983) and Bougrier et al. (1986) observed during acclimatization tests that *Ostrea chilensis* and *Ostrea angasi* were susceptible to *B. ostreae*. Tests of resistance conducted with edible oysters collected from natural environments or hatcheries in different geographical areas revealed that all tested stocks contracted the disease.

Experimental Pathogenesis

Studies of experimental pathogenesis depend on the standardized purification and production of large quantities of the pathogenic agent. Recent progress made in the purification of certain molluscan protozoa (Mialhe et al. 1985) has enabled us to undertake such research.

In Vivo Pathogenesis

Poder et al. (1982) and Grizel (1985) conducted tests of pathogenicity by using either the proximity method or the method of inoculating infected hemocytes, but they did not quantify the parasite dose in either case. The second method resulted in a more rapid and uniform onset of the disease than the first. The time it takes for the infection to become evident varies with the method (Bachère et al. 1982; Poder et al. 1982) and can range from 15 d to 4 months. However, the rate of infection never exceeded 50% during the first 4 months of any experiment. Recently, experimental infections were initiated by injecting solutions of 50 µL of seawater containing 1, 100, 10,000, or 1,000,000 *B. ostreae*-infected cells. The percentages of successful infections increased as the concentration of the inoculum was increased, 27% for 100 injected cells to 70% for 10^5 cells. Tests replicated five times with 30 injected edible oysters per concentration yielded similar results (Mialhe and others, unpublished data). At the lowest concentration, only one case of infection was detected in 150 edible oysters that were examined. These results showed the limits of diagnostic sensitivity and the need to increase the observation time to obtain the maximum response. The results also suggested high variability in individual responses. Improvement of the experimental technique should result in a useful and repeatable experimental model of parasitism in a marine mollusc.

In Vitro Pathogenesis

The availability of techniques to purify *B. ostreae* and the ability to maintain primary in vitro cultures of oyster cells, especially hemocytes, make it possible to apply the long-established techniques of in vitro mammal–parasite systems. Such research could characterize the mechanisms of cellular recognition, penetration of the host cell by the parasite, and parasite survival in the host cell. Fisher (1988) found that *B. ostreae* adhered more to *Crassostrea gigas* hemocytes than to *O. edulis* hemocytes in vitro, but latex beads adhered equally to both kinds of hemocytes. He suggested that hemocyte recognition of the parasite may play a role in the different susceptibilities to *B. ostreae* of these two oyster species. In our laboratory, we introduced *B. ostreae* into test vials containing hemocytes from *O. edulis* and *C. gigas*. One-half hour later, the parasites had adhered to the host cells, and after 2 h, *B. ostreae* entered the blood cells of both species of oysters, infecting the hyalinocytes as well as the granulocytes. In the latter cell type, the parasite was always enclosed in a parasitophorous vacuole (D. Chagot, unpublished data). Additional experiments are being conducted to answer other questions, such as (1) Is there simple cellular adhesion or a true process of recognition and binding? (2) Does *B. ostreae* actively enter the host cell, or is it phagocytized? (3) If phagocytized, how does *B. ostreae* survive in the phagocyte of *O. edulis?* (4) What happens to the parasite in the hemocyte of *C. gigas,* in which neither natural nor experimental infections have been observed? These important fundamental studies are intermediate steps that will lead to new paths of research to be pursued in genetics and therapy. Such research should broaden our present means to combat epizootic bonamiasis in marine molluscs.

Prophylaxis

After the discovery of *B. ostreae* in Europe, preventive measures were put into practice. These measures included destruction of the infectious areas by harvesting the edible oysters from them and prohibition of transfers of edible oysters from contaminated to disease-free areas. Van Banning (1987) studied eradication of *B. ostreae* during 1982–1986. He introduced test lots into previously contaminated areas and monitored them for 1 year. Until 1985, he detected the parasite at low levels (5%). The important question now is whether a large-scale reintroduction of edible oysters will again result in epizootic disease.

In France, the application of these disease control measures, together with health management practices, has made it possible to continue deepwater breeding of the edible oyster in the bays of Mont St. Michel (Cancale) and St. Brieuc (Binic) (Grizel et al. 1987). Excellent results have been obtained by breeding and growout to market size at the same site and by reducing spat density from 5 to 2 tonnes per hectare. Furthermore, data acquired in the course of epizootiological studies have been useful in developing a plan for safeguarding the edible oyster (Grizel et al. 1987). The absence of infection in juveniles made it possible to continue gathering spat even in infested bays.

Tests of disease resistance are being conducted on edible oysters that have shown individual resistance within a breeding population (Elston et al. 1987) or after experimental inoculations. We expect other experiments conducted in our laboratory to provide information about the genetic nature and transmissibility of this resistance.

Lastly, we have developed an immunodiagnostic procedure based on monoclonal antibodies and an enzyme substrate reaction. This procedure is more specific and sensitive than the commonly practiced histological techniques and provides a rapid diagnosis. This tool should facilitate sanitary checks and should also serve research in immunopathology.

Conclusions

The study of bonamiasis is an important example of the recent research developments in molluscan pathology and should continue to provide meaningful information. Our studies of bonamiasis, conducted both in the field (descriptive and analytical epizootiology) and in the laboratory (experimental pathogenesis), have resulted in the following developments: (1) progress in conceptual approaches in the domain of marine molluscan pathology, (2) establishment of in vivo and in vitro experimental models, (3) proposals to check the spread of the disease and to permit continued breeding despite the presence of the parasite, (4) increased awareness within the edible oyster farming industry and government agencies of the need for measures to prevent the spread of disease, and (5) cooperation between international research teams concerned with epizootic diseases of molluscs. One work–study group has already met in France in April 1987.

References

Bachère, E., J. L. Durand, and G. Tigé. 1982. *Bonamia ostreae* (Pichot et coll., 1979) parasite de l'huître plate: comparaison de deux méthodes de diagnostic. Conseil International pour l'Exploration de la Mer, C. M. 1982/F:28, Copenhagen.

Bougrier, S., G. Tigé, E. Bachère, and H. Grizel. 1986. *Ostrea angasi*, acclimatization to French coasts. Aquaculture 58:151–154.

Brehelin, M., J. R. Bonami, F. Cousserans, and C. P. Vivares. 1982. Existence de formes plasmodiales vraies chez *Bonamia ostreae* parasite de l'huître plate *Ostrea edulis*. Compte Rendus Hebdomadaires des Séances de l'Académie des Sciences, Série D, Sciences Naturelles 295:45–48.

Comps, M. 1983. Recherches histologiques et cytologiques sur les infections intracellulaires des mollusques bivalves marins. Thèse de Doctorat. Université des Sciences et Techniques du Languedoc, Montpellier, France.

Comps, M., G. Tigé, and H. Grizel. 1980. Recherches ultrastructurales sur un Protiste parasite de l'huître plate *Ostrea edulis*. Compte Rendus Hebdomadaires des Séances de l'Académie des Sciences. Série D, Sciences Naturelles 290:383–384.

Dinamani, P., R. Hickman, P. Hine, J. Jones, and H. Cranfield. 1987. Report on investigations into the disease outbreak in Foveaux Strait Oysters, *Tiostrea lutaria*, 1986–87. Ministry of Agriculture and Fisheries, New Zealand.

Elston, R., C. Farley, and M. Kent. 1986. Occurrence and significance of bonamiasis in European flat oyster *Ostrea edulis* in North America. Diseases of Aquatic Organisms 2:49–54.

Elston, R., M. Kent, and M. Wilkinson. 1987. Resistance of *Ostrea edulis* to *Bonamia ostreae* infection. Aquaculture, 64:237–242.

Fisher, W. S. 1988. In vitro binding of parasites (*Bonamia ostreae*) and latex particles by hemocytes of susceptible and insusceptible oysters. Developmental and Comparative Immunology 12:43–53.

Grizel, H. 1983. Impact de *Marteilia refringens* et de *Bonamia ostreae* sur l'ostreiculture bretonne. Conseil International pour l'Exploration de la Mer, C. M. 1983/F:9, Copenhagen.

Grizel, H. 1985. Étude des récentes épizooties de l'huître plate *Ostrea edulis* L. et de leur impact sur l'ostreiculture bretonne. Thèse de Doctorat. Université des Sciences Techniques du Languedoc, Montpellier, France.

Grizel, H., M. Comps, D. Raguennes, Y. Leborgne, G. Tigé, and A. G. Martin. 1983. Bilan des essais d'acclimatation d'*Ostrea chilensis* sur les côtes de Bretagne. Revue des Travaux de l'Institut des Pêches Maritimes 46(3):209–225.

Grizel, H., E. Bachère, E. Mialhe, and G. Tigé. 1987. Solving parasite related problems in cultured molluscs. Pages 301–308 *in* M. J. Howell, editor. Proceedings of the Sixth International Congress of Parasitology. Australian Academy of Science, Brisbane.

Katkansky, S. C., W. A. Dahlstron, and R. W. Warner. 1969. Observations on survival and growth of the European flat oyster, *Ostrea edulis*, in California. California Fish and Game 55(1):69–74.

Meuriot, E., and H. Grizel. 1984. Note sur l'impact économique des maladies de l'huître plate en Bretagne. Rapports Techniques de l'Institut Scientifique et Technique des Pêches Maritimes 12:1–20.

Mialhe, E., E. Bachère, C. Le Bec, and H. Grizel. 1985. Isolement et purification de *Marteilia* (Protozoa, Ascetospora) parasite de bivalves marins. Compte Rendus Hebdomadaires des Séances de l'Académie des Sciences, Série D, Sciences Naturelles 301: 137–141.

Perkins, F. O. 1971. Sporulation in the trematode hyperparasite *Urosporidium crescens*, Deturk, 1940, (Haplosporida: Haplosporidiidae). An electron microscope study. Journal of Parasitology 57:9–23.

Perkins, F. O. 1988. Structure of protistan parasites found in bivalve molluscs. American Fisheries Society Special Publication 18:93–111.

Pichot, Y., M. Comps, G. Tigé, H. Grizel, and M. A. Rabouin. 1980. Recherche sur *Bonamia ostreae* gen. n., sp. n., parasite nouveau de l'huître plate *Ostrea edulis* L.. Revue des Travaux de l'Institut des Pêches Maritimes 43(1):131–140.

Poder, M., A. Cahour, and G. Balouet. 1982. Hemocytic parasitosis in European oyster *Ostrea edulis* L.: pathology and contamination. Proceeding of the 15th Annual Meeting of the Society for Invertebrate Pathology, Brighton, United Kingdom, 254–257.

Tigé, G., H. Grizel, and M. Comps. 1980. Données sur le nouveau parasite de l'huître plate. Situation épidémiologique. Conseil International pour l'Exploration de la Mer, C. M. 1980/F:39, Copenhagen.

Tigé, G., and H. Grizel. 1984. Essai de contamination d'*Ostrea edulis* L. par *Bonamia ostreae* (Pichot et al., 1979) en rivière de Crach (Morbihan). Revue des Travaux de l'Institut des Pêches Maritimes 46(4):1–8.

van Banning, P. 1987. Further results of the *Bonamia ostreae* challenge tests in the Dutch oyster culture. Aquaculture 67:191–194.

American Fisheries Society Special Publication 18:5–22, 1988
© Copyright by the American Fisheries Society 1988

Uncertainties and Speculations about the Life Cycle of the Eastern Oyster Pathogen *Haplosporidium nelsoni* (MSX)

HAROLD H. HASKIN

Rutgers Shellfish Research Laboratory, Port Norris, New Jersey 08349 USA

JAY D. ANDREWS

*Virginia Institute of Marine Science, School of Marine Science, College of William and Mary
Gloucester Point, Virginia 23062 USA*

Abstract.—For 30 years, the pathogen *Haplosporidium nelsoni* (MSX) has been causing serious mortalities of eastern oysters *Crassostrea virginica* in the Delaware and Chesapeake bays of the eastern USA. Its life cycle is largely unknown, and methods for control are wanting. Breeding of resistant eastern oyster strains, at this time, offers the best hope for some degree of control of the disease. Although haplosporidians are known by their spores, controlled transmission, with one possible exception, has not been achieved in any of the 30 recognized species. *Haplosporidium nelsoni* rarely sporulates in eastern oysters, and this and other observations led to early speculation that another host probably exists. Based on apparent effects of environment on *H. nelsoni* abundance, particularly in Delaware Bay, we are attempting to profile a hypothetical alternate host. Recent increases in abundance and activity of MSX in Chesapeake and Delaware bays and in Virginia rivers are associated with droughts, but this is not true elsewhere. Distribution of *H. nelsoni* along the Atlantic coast appears to have spread slowly southward from Chesapeake and Delaware bays, but is not identified with eastern oyster mortalities in southern localities. To the north of the original epizootics, *H. nelsoni* has been scattered along the Long Island, Connecticut, and Massachusetts coasts for 30 years or more, usually without occurrence of serious mortalities of eastern oysters. Foci of increasing *H. nelsoni* activity with mortality, over the past 5 years, are of great concern in these areas. Careful examination of changes in these northern areas may contribute significantly to our understanding of the relationships between *H. nelsoni* and its environment.

In spring 1957, the disease of the eastern oyster *Crassostrea virginica* caused by *Haplosporidium nelsoni* (commonly called MSX) appeared without warning in Delaware Bay, USA (Haskin et al. 1966; Ford and Haskin 1982). It spread rapidly throughout the bay, killing 90–95% of all eastern oysters on the planting grounds and 60% on the seed beds within 3 years (Figures 1, 2). Two years later, in spring 1959, *H. nelsoni* was associated with serious eastern oyster mortalities in the high-salinity areas of Chesapeake Bay (Andrews and Wood 1967); 30–50% of the 122 million kg of eastern oysters in Mobjack Bay–Egg Island areas were destroyed. Within the next 2–3 years, total mortalities reached 90–95% in the lower bay; however, losses on the James River seedbeds were negligible (Andrews 1964). The disease has not abated in virulence or intensity since that time. It spread far up the Chesapeake Bay during periods of drought and is now causing mortalities in Long Island and New England waters. Resistance to mortality has developed in Delaware Bay native eastern oysters (Haskin and Ford 1979) and in wild

stocks in the lower Chesapeake Bay (Andrews 1968; Farley 1975), but, during years of intensive activity of the pathogen, native resistant eastern oysters in both estuaries have been overwhelmed. Current understanding of survival mechanisms is reviewed by Ford (1988, this volume).

Haplosporidium nelsoni has many peculiarities (Andrews 1982, 1984b). Its life cycle is not known, although patterns of infection and mortality are well known as to timing and duration (Andrews 1966; Ford and Haskin 1982). The pathogen has not been cultured, and controlled transmission of infection has not been achieved. Furthermore, the spore stage is known (Couch et al. 1966; Perkins 1968) but is rarely found in eastern oysters (Andrews 1979; Ford and Haskin 1982), and the source of infective stages remains unknown. This lack of information has led to much speculation about other hosts (Ford and Haskin 1982; Andrews 1984a, 1984b; Burreson 1988, this volume), but none has been found. The wide dispersion of infective materials throughout the high-salinity areas of both bays, whether large

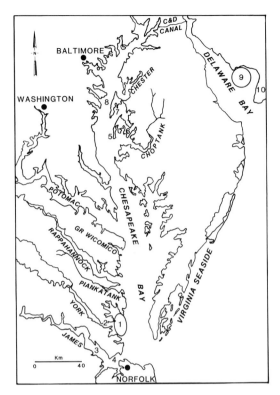

FIGURE 1.—Map of Chesapeake and Delaware bays.
1 = Mobjack Bay–Egg Island area of 1959 MSX outbreak; 2 = Virginia Institute of Marine Science Laboratory; 3 = James River seed beds; 4 = Hampton Bar;
5 = Tilghman, Maryland; 6 = Oxford Laboratory; 7 =
Eastern Bay; 8 = Bay Bridge; 9 = MSX outbreak area
in 1957; 10 = Rutgers Cape Shore Laboratory.

eastern oyster populations are present or absent,
and the apparent lack of contagious transfer among
eastern oysters strengthen the speculations about
the existence of an alternate or reservoir host for *H.
nelsoni*. Alternate hosts have not been demonstrated for any haplosporidian species (Andrews
1984a, 1984b), nor has any controlled transmission
of a haplosporidian parasite disease been reported,
with one possible exception (Barrow 1965).

The annual cycle of infection and mortality is
basically the same in both Chesapeake and Delaware bays, although there are some differences in
timing and detail of interest to the eastern oyster
industry. Some of these differences are also of
interest because they lead to questions of *H. nelsoni* dosage and of influences of temperature and
salinity on host–parasite reactions. In any event,
they lead to speculations that can be explored.

Data accumulated over the past 30 years have
documented epizootiological factors and opinions

on life cycle. Our purpose is not to review the
enormous literature on *H. nelsoni* disease, but to
review some of the data in general, emphasize
several important gaps in our knowledge, and
speculate on explanations of some of the un-
knowns and uncertainties. When interpretations
and data analyses differ for Delaware and Chesapeake bays, alternative positions will be given.

Recent History of *H. nelsoni*

The pressure of *H. nelsoni* on the eastern oyster
resources of the U.S. east coast has not only
continued but has intensified since the early outbreaks. In the early outbreaks and rapid spread,
the pathogen did not kill eastern oysters on the
uppermost beds in both Delaware and Chesapeake bays and in tributary rivers and creeks.
Low salinities were probably setting the limits for
H. nelsoni penetration. Since then, extensive
studies of infection and mortality patterns along
the salinity gradients in Delaware Bay, Chesapeake Bay, and the James River have established
that: (a) above a salinity of 20‰, *H. nelsoni* is not
inhibited in its activities; (b) below 15‰, infections are generally rare, and development is inhibited; and (c) below 10‰, *H. nelsoni* cannot survive in eastern oysters (Andrews 1964, 1983;
Haskin and Ford 1982; Ford 1985; Ford and
Haskin 1988). Therefore, in examining areas of
changing *H. nelsoni* activity, we should ask first if
there has been a corresponding change in salinity
regimes. In the drought of the mid 1960s, *H.
nelsoni* invaded Maryland eastern oyster beds and
moved up Chesapeake Bay as far as Tilghman,
Maryland (Farley 1975). For the first time, it killed
oysters on the James River seedbeds and was
active in Delaware Bay on the uppermost productive seedbed (Figures 1, 2). All three of these
examples of expanding range were predictable
because increased salinities occurred in areas
where low salinities previously had restricted the
parasite. At the end of the drought, *H. nelsoni*
retreated to its earlier boundaries coincident with
the return of normal river flows. Occurrences of
intensified *H. nelsoni* activity that cannot be explained by increased salinity are puzzling.

Increased *H. nelsoni* activity, culminating in
increased eastern oyster mortality, may be influenced by factors such as temperatures favorable
for the development of critical stages in the *H.
nelsoni* life cycle, for host response (Ford 1988),
or for the release of infective stages. The influence
may be on the magnitude and the timing of infective dosages available to eastern oysters. Many

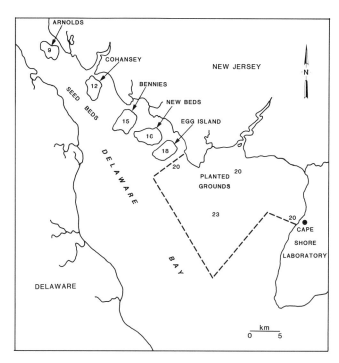

FIGURE 2.—Delaware Bay showing major eastern oyster seedbeds, planting grounds, and location of Rutgers University Cape Shore Laboratory. Numbers shown indicate bottom salinities (‰) at mean river flows.

factors that stress eastern oysters directly may also result in *H. nelsoni*-related mortality when *H. nelsoni* pressure alone would not be fatal. These stresses may include shell repair after damage by enemies such as boring sponges, scavenging crabs, and *Polydora* spp. Noxious algal blooms and failures in phytoplankton production that can reduce eastern oyster condition may make the animals susceptible to disease. Diversion of energy reserves into gametes may also leave less reserves available to meet the immediate energetic demands imposed by *H. nelsoni* disease. These factors underlie proposals to produce triploid east coast eastern oysters to reduce losses due to disease.

Two enzootic areas of recently increased *H. nelsoni* activity not attributable to salinity changes have been closely linked in long-term studies that provide excellent data for assessing *H. nelsoni* activity. One area is on tidal flats in lower Delaware Bay off the Rutgers Cape Shore Laboratory and the other is in the York River near the Virginia Institute of Marine Science Laboratory (VIMS) at Gloucester Point, Virginia (Figure 1). Since spring 1959, we have placed seed of susceptible James River eastern oysters in trays at these two locations in advance of the early sum-

mer infection period for *H. nelsoni* (late May–June). Salinities in the two locations are quite similar: Cape Shore, 18–22‰ and up to 26‰ in drought; VIMS, 17–24‰ and up to 25‰ in drought. Live and dead eastern oysters have been carefully monitored for *H. nelsoni* prevalence and intensity. The James River eastern oysters were not previously exposed to *H. nelsoni* and were transplanted as uninfected individuals to both test locations. At Cape Shore and at VIMS, James River imports also served as the highly *H. nelsoni*-susceptible constant controls against which the performance of *H. nelsoni*-resistant strains being developed at these locations could be compared. Details of tray handling and calculation of cumulative mortalities for Cape Shore and VIMS groups were presented elsewhere (Andrews 1968; Haskin and Ford 1979).

The mortality patterns at VIMS for susceptible eastern oysters exposed to *H. nelsoni* in early summer (May and June) are illustrated in Figure 3. Infections usually appeared in July as shown by increased prevalences. Mortality usually began about the first of August, peaked in late August and September, and declined sharply in October and November. Typically, there was a small end-of-winter kill in March or April of eastern oysters

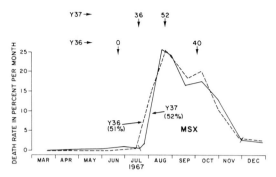

FIGURE 3.—Mortality of, and MSX prevalence in, duplicate lots (trays Y36 and Y37) of MSX-susceptible eastern oysters transplanted from the James River to Gloucester Point in March 1967, before the early-summer infection period for MSX. Percent prevalences of MSX in samples of 25 live eastern oysters are shown above the vertical arrows, which indicate sampling dates. Total MSX-related mortalities of eastern oysters in each lot during mid-July to end November (hash marks) are in parentheses.

with advanced infections. The June to December first-season mortality averaged 52% for the two tray lots of eastern oysters.

The mortality pattern for late-summer lots, that is, lots imported after the first of August, is shown in Figure 4 for VIMS. Late-summer infections usually

remained subpatent until spring, but in 1966 they appeared in October. Slight mortality occurred in March and April followed by a June–July peak. Prevalence of the disease was high throughout the winter and spring of 1966 without causing appreciable mortality. The August 1967 mortality shoulder was caused by June 1967 infections; total mortality by the end of the year was nearly 74%.

The Cape Shore mortality patterns for early summer infections were essentially the same as those at VIMS, but the total mortalities were usually greater (Figure 5). Infections in late August or September imports developed more rapidly at Cape Shore than at VIMS and often killed eastern oysters that same fall. At Cape Shore, fall infections (September–October) killed some eastern oysters as early as March, but losses were usually much heavier in June.

First season (June–December) cumulative mortalities at both locations showed both similarities and differences (Figure 5). In the York River, the 3-year peak in 1965–1967 was related to the drought. The decreased mortality in 1972 was attributed to the freshwater input by Hurricane Agnes, and the 1973 low may have indicated a residual salinity effect of that big freshwater discharge.

Cape Shore mortalities were definitely more variable from year to year than those in the York

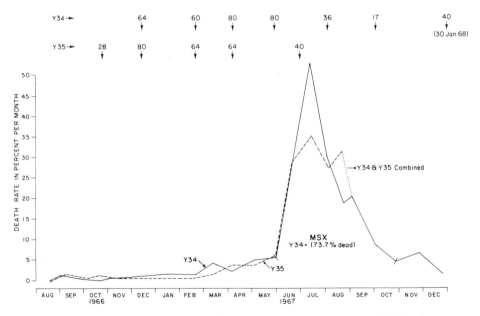

FIGURE 4.—Mortality of, and MSX prevalence in, duplicate lots (trays Y34 and Y35) of eastern oysters transplanted from Deep Water Shoal to Gloucester Point before the late-summer infection period for MSX. Percent prevalences of MSX in samples of 25 live eastern oysters are shown above the vertical arrows, which indicate sampling dates.

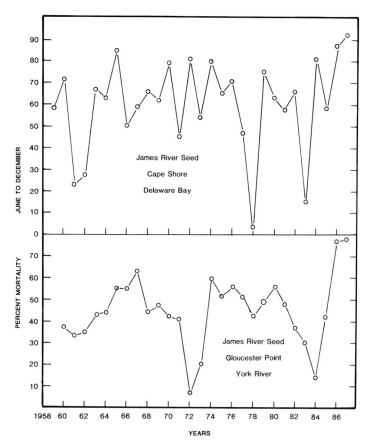

FIGURE 5.—Cumulative mortalities of eastern oysters during June–December at Cape Shore, Delaware Bay, and at Gloucester Point, York River. James River eastern oysters were placed in trays shortly before the MSX infection period.

River. They indicated little or no influence of Delaware River flows, reflecting neither the drought of 1963–1967 nor the eight consecutive years of higher-than-average runoff in 1972–1979. The extremely low mortalities in 1961, 1962, 1971, 1978, and 1983 occurred after unusually cold winters (Ford and Haskin 1982). We speculated that the extremely cold winters influenced *H. nelsoni* activity by killing reservoir hosts from which the infective stages were released (Ford and Haskin 1982).

Low-mortality periods following unusually cold winters have not been apparent in the York River. For the 28 years that records have been kept, the average June–December mortality at VIMS was 42% compared with 58% at Cape Shore. Except for the four cold winters, the Cape Shore mortalities ranged generally from 60 to 80%. Few VIMS mortalities exceeded 60%, and it is clear that selection pressure for survival against *H. nelsoni*

was greater at the Cape Shore. Mortalities caused by *H. nelsoni* have persisted in both locations and have intensified dramatically, especially in the last 2 years.

Except for a few years in the early 1960s, Delaware Bay oystermen have continued to plant eastern oysters brought downbay from the seed-beds in May and June. Every year, the Rutgers Shellfish Laboratory has followed a number of these plantings for determination of *H. nelsoni* prevalences and mortalities from time of planting until harvest (Ford and Haskin 1982). June–December mortalities of these planted native eastern oysters for each year were averaged for comparison with the mortalities of James River imports described previously (Figure 6).

As at the Cape Shore tray station, mortalities on the planting grounds obviously did not reflect the periods of drought or high runoff described above. This conclusion leads to speculation that

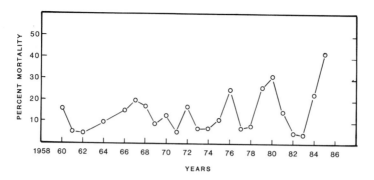

FIGURE 6.—First-season nonpredatory eastern oyster mortality on Delaware Bay planting grounds from 1958 to 1986. Eastern oysters were transplanted from the seedbeds in May and June each year and were monitored on selected grounds until harvested.

there is a variable discharge of infective particles by reservoir hosts that leads to variable mortalities in an ever-favorable salinity regime. Four of the five lowest points on the graph of mortality on planted grounds (Figure 6) coincide with the four lowest points of mortality in the Cape Shore trays (Figure 5), and these points correlate strongly with preceding cold winters. Since 1972, mortality peaks have become more severe after each low-mortality period. The upward trend in Delaware Bay mortalities is remarkably similar to that found for James River seed exposed to *H. nelsoni* on the Virginia seaside (Figure 7).

The highest first-season mortality ever recorded on Delaware Bay planting grounds occurred in 1985 and was related to drought (Figure 6). It was a direct result of intense 1984 *H. nelsoni* infections extending upbay to all seedbeds in the dry fall followed by the lowest winter–spring flows on record in 1984–1985 (Table 1). Normal spring

flows probably purge *H. nelsoni* from seedbeds of eastern oysters when eastern oysters begin to feed in mid to late March. With seedbed salinities 3–4‰ higher than usual, this purging did not occur on the lower seedbeds in the spring of 1985, and, for the first time, New Jersey oystermen planted native seed already heavily infected with *H. nelsoni*. In late May 1985, 65–80% of the eastern oysters on the lower seedbeds were systemically infected (Table 2). They died in large numbers over the next 2 months, before the onset of new summer infections. Reduced river flows, with their corresponding salinity increases, are not sufficient of themselves to permit *H. nelsoni* to move farther upbay. An abundance of *H. nelsoni* infective stages must also be present. For example, in June 1980, a severe drought led to low river

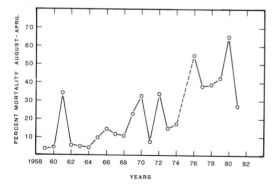

FIGURE 7.—Mortality of eastern oysters due to MSX at Virginia seaside, 1958–1981. James River seed oysters were placed at seaside each year in trays. Mortalities are cumulative for the period August to April.

TABLE 1.—Mean monthly Delaware River flows at Trenton, New Jersey, during the drought of 1984–1985 compared with the long-term means of 1913–1983. (USGS District Engineer 1913–1985.)

Month	Mean flow (m³/s)	
	1984–1985	1913–1983
Aug	168	169
Sep	104	150
Oct	103	189
Nov	100	291
Dec	197	343
Jan	171	343
Feb	191	365
Mar	273	595
Apr	187	638
May	215	395
Jun	171	252
Jul	132	198
Aug	100	
Sep	262	
Oct	273	
Nov	501	

TABLE 2.—*Haplosporidium nelsoni* winter prevalence (%)[a] and cumulative annual nonpredatory mortality (%)[b] of eastern oysters on four Delaware Bay seedbeds, 1980–1986.

Year	Arnolds bed		Cohansey bed		Bennies bed		Egg Island bed	
	Prevalence	Mortality	Prevalence	Mortality	Prevalence	Mortality	Prevalence	Mortality
1980	0	6	10	21	55	26	50	13
1981	10	19	21	14	65	33	45	19
1982	0	7	15	2	25	4	55	5
1983					10[c]	14[c]	25	10
1984	10	11	20	11	65	12	85	26
1985	60	45	50	41	65	46	80	62
1986		13		26	86	37		21

[a] Sample usually taken in December.
[b] July to June.
[c] Sample taken from adjoining bed.

flows that continued below long-term averages through April 1981. River flows through the fall and early winter were substantially lower than in corresponding periods of the 1984–1985 drought when higher *H. nelsoni* prevalences and mortalities occurred on the seedbeds (Table 2). These relationships support the conclusion that an increase in infection pressure is not directly correlated with an increase in salinity.

The Virginia seaside is another monitoring area receiving James River seed imported by VIMS over a period of at least 25 years. The seaside bays have high salinities, and there is no question of restraint of *H. nelsoni* development by low salinity. Over the first 5 or 6 years of import, although a few infections occurred, mortalities associated with *H. nelsoni* were so low (about 5% annually) that a tentative conclusion was reached that salinities of about 30‰ were inhibiting development of *H. nelsoni* (Andrews and Castagna 1978). From about 1969, however, *H. nelsoni*-related mortalities trended upward, though with wide swings, to the 40–60% range (Figure 7). Because a second haplosporidian parasite of eastern oysters, *Haplosporidium costale* (SSO), complicated the seaside picture in spring, the *H. nelsoni*-related mortalities were sorted out for the period from August to April. The upward trend in *H. nelsoni*-related mortality in this area cannot be related to salinity changes. Perhaps *H. nelsoni* was new to the region in late 1958, and, after several years, became abundant enough to cause substantial eastern oyster mortalities. Another possibility is that a reservoir host population was producing and releasing ever-more infective particles.

Expanded Range Along the East Coast of North America

As river runoffs returned to normal levels after the mid-1960s drought, *H. nelsoni* receded from its upbay and uptributary excursions in the Chesapeake and Delaware bays. More recently, two 3-year periods of drought in the mid-Atlantic region (1980–1982 and 1985–1987) have permitted *H. nelsoni* to extend its range even further than in the 1960s and to do great damage to the eastern oyster industries.

Because of the presence of *H. nelsoni*, nearly all eastern oysters in Chesapeake Bay are grown in areas where late-summer salinities do not exceed 18–20‰. Most areas above the mouth of the Rappahannock River (Figure 1) have late spring and early summer salinities less than 15‰, which inhibit or delay *H. nelsoni* infections. As a result, infections are more likely to occur in late summer, after salinities increase. These infections usually do not become patent or serious until spring of the following year. In average or wet years, however, many of these areas have early spring salinities below 10‰, which purge eastern oysters of infections. Successive dry years during 1980–1982 and 1985–1987 have permitted both early- and late-summer infections to proliferate and cause serious mortalities. Late-summer infections are most insidious during dry periods because salinities in the coastal plains estuaries such as the Great Wicomico and Piankatank in Virginia and the Choptank in Maryland (Figure 1) are controlled by the Chesapeake Bay regime and not by local freshwater discharges.

In 1982, an intensive spatfall occurred in Great Wicomico River, but in late summer, native eastern oysters (including the spat) became infected with *H. nelsoni* and died in May–June 1983. This mortality would not have occurred with average salinities, but, in spring 1983, the salinities never fell below 10‰. The winter and spring of 1983 was wetter than average, but three preceding drought years had allowed Chesapeake Bay salinities to

become unusually high, 20‰ at the Bay Bridge in
Maryland in the summer of 1982. Higher bay
salinities were still controlling salinities in the
coastal plain estuaries in spring 1983. In the James
and Rappahannock rivers, fall infections also occurred but were mostly purged by low salinities in
April and May.

The pattern of mortalities caused by *H. nelsoni*
epizootics in Virginia and Maryland has been
essentially repeated in the drought years of 1985–
1987. The losses in all rivers of Virginia have been
severe. In Maryland, heavy losses extended into
the lower Choptank River and Eastern Bay, and
H. nelsoni was also reported in the lower reaches
of the Chester River (Figure 1).

Shortly after *H. nelsoni* appeared in Chesapeake Bay, it was reported in North Carolina
waters (Albemarle Sound). This occurrence was
not surprising because North Carolina planters
had traditionally imported eastern oyster seed
from Virginia. More recently, *H. nelsoni* has been
identified in eastern oysters from South Carolina
and Georgia. Some mortalities have been reported
in these areas, although they have not been extensive, and prevalences remain low. Just a few
years ago, samples from South Carolina and
Georgia were negative. The pattern of positive *H.
nelsoni* samples now indicates that it is gradually
spreading southward but has probably not made
big jumps. In both states, *Perkinsus marinus* is
also present, and attributing mortalities due to *H.
nelsoni* is difficult (C. A. Farley and F. Kern,
National Marine Fisheries Service, Oxford,
Maryland, personal communication).

In 1985, a sample of 30 eastern oysters from the
St. Johns River system near Jacksonville, Florida,
had three eastern oysters with *H. nelsoni*, one of
which had a heavy infection. In spring 1986, four
of 30 eastern oysters from Biscayne Bay, Florida,
had *H. nelsoni* gill infections. A 1987 sample had
none (R. Hillman, Battelle Laboratory, Duxbury,
Massachusetts, personal communication). These
Florida reports mark the known southern limit of
H. nelsoni at this time.

To the north of Delaware Bay along the New
Jersey Coast, the two small eastern oyster producing areas, Great Egg Harbor and Great Bay,
have been plagued with *H. nelsoni* since the late
1950s (Figure 8). Test samples, wherever taken in
the back bays, have been positive. A surviving
eastern oyster population in the Navesink River,
relict from an industry that ended shortly after
World War I, was free of *H. nelsoni* until 1980
when a substantial kill occurred. A small eastern

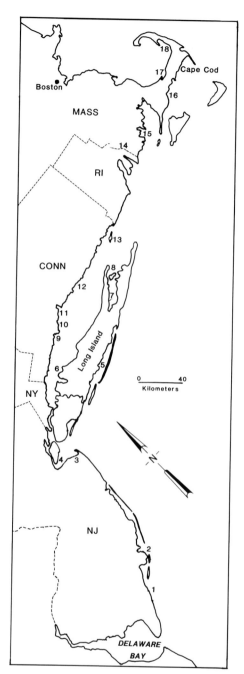

FIGURE 8.—Map of the east coast of the USA from
Delaware Bay to Cape Cod. With the exception of
Fishers Island, MSX has been found in each identified
location. 1 = Great Egg Harbor; 2 = Great Bay; 3 =
Navesink River; 4 = Raritan Bay; 5 = Great South Bay;
6 = Oyster Bay; 7 = Peconic Bay; 8 = Gardiners Bay;
9 = Bridgeport; 10 = Milford; 11 = New Haven; 12 =
Hammonasset River; 13 = Fishers Island; 14 =
Swansea; 15 = West River; 16 = Cotuit Harbor; 17 =
Barnstable; 18 = Wellfleet Harbor.

oyster population in Raritan Bay, sampled irregularly between 1979 and 1988, averaged about 10–30% *H. nelsoni* prevalence (Farley, personal communication; Rutgers Shellfish Laboratory, unpublished data).

In the Long Island area, Great South Bay eastern oyster grounds (Figure 8) were sampled in 1965, and three grounds had *H. nelsoni* with prevalences ranging from 25 to 54%. A principal grower in that area at that time estimated annual mortalities in hatchery-reared stock at 15% (Rutgers Shellfish Laboratory, unpublished data).

Along the Connecticut shore of Long Island Sound, *H. nelsoni* prevalence of 40% was found in eastern oysters in Milford harbor in 1960 and 12% prevalence in 1974. New Haven harbor eastern oysters were positive for *H. nelsoni* when examined in 1966 and 1967. In 1985, prevalences of 18 and 32% were reported in Bridgeport harbor and the Hammonasset River, respectively (C. A. Farley and F. Kern, personal communication).

In Massachusetts, *H. nelsoni* was first found in eastern oysters from Wellfleet harbor in 1967 and then in follow-up samples in 1969 and 1970 (Krantz et al. 1972). Krantz et al. (1972) suggested that *H. nelsoni* was introduced by importation of eastern oysters from an epizootic area. They also noted that its presence in native eastern oysters in 1969 and 1970 indicated that it had become enzootic in Wellfleet harbor. *Haplosporidium nelsoni* persisted at Wellfleet with prevalences of 12 and 28% in 1975 and 1978, respectively (C. A. Farley and F. Kern, personal communication). Major eastern oyster mortalities were reported there in 1982, 1985, and 1987 (B. Chapman, Shellfish Constable of Wellfleet, Massachusetts, personal communication). Presently, substantial mortalities are reported at Barnstable and Swansea in addition to Wellfleet, although they have not yet been linked to *H. nelsoni* by histological examination (F. Germano, Division of Marine Fisheries, Sandwich, Massachusetts, personal communication). In 1985, prevalences of 48 and 93% were reported in the West River near New Bedford harbor and in Cotuit harbor, respectively (C. A. Farley and F. Kern, personal communication).

A nagging question over the past 20 years has been, "What is preventing massive epizootics, like those in Chesapeake and Delaware bays, from occurring in the eastern oyster areas extending from Great South Bay, Long Island, to the tip of Cape Cod?" Since 1983, there seem to be some shifts in this situation, as illustrated by two well-documented examples of serious eastern oyster mortalities associated with *H. nelsoni,* one in Long Island and a second on Cape Cod. A third occurrence of mortality in eastern bays of Long Island is not so clearly due to *H. nelsoni*.

In the Flower Hatchery at Bayville, Long Island, spat are set on fragments of shell and then rafted in trays for 6 weeks or longer. The clusters of spat are then planted directly on hard bottom grounds in Oyster Bay that have been previously suction-dredged to remove drills, starfish, and other predators. Within 2.5 years, they usually reach marketable size. Over the years, returns by number have averaged about 30% but range from 15% in bad years to 50% in good years. In any one season, there are usually three year classes on the grounds, and in summer 1983, when *H. nelsoni* appeared, these were year classes 1981, 1982, and 1983. Unusually heavy mortalities continued through 1984 and into 1985. On final tally of harvests, approximately 90% of all three year classes had been lost (J. Zahtila, Franklin B. Flowers & Sons, Bayville, New York, personal communication). Histological examination of samples of eastern oysters by one of us (H. H. H.) and by the Oxford Laboratory (C. A. Farley and F. Kern) established involvement of *H. nelsoni* in these losses.

The Cape Cod eastern oyster operation in Cotuit harbor traditionally depended totally on Connecticut eastern oyster growers for wild seed. In recent years, part of the seed has been produced in a small eastern oyster hatchery of the Ocean Pond Corporation, Fishers Island, New York. In 1984, wild eastern oyster seed of several year classes from the Hammonasset River in Connecticut were planted in Cotuit harbor over the summer and early fall. Some of these eastern oysters were probably infected with *H. nelsoni*. A sample of Hammonasset seed taken from Cotuit harbor in November, had 6% prevalence of *H. nelsoni* (F. Perkins, Virginia Institute of Marine Science, Gloucester Point, personal communication). Hatchery-reared seed that had been rafted on Fishers Island was also imported at the same time. Over the winter and spring the Hammonasset eastern oysters were harvested. A sample of 50 eastern oysters taken in the spring, presumably the Fishers Island seed, had 15 *H. nelsoni* infections. Losses in the summer were heavy, about 40%, and a sample examined after the mortality had a 92% prevalence.

Before the 1985 mortality, eastern oysters from Ocean Pond on Fishers Island had been examined in 1968, 1972, 1976, and 1978, and all were negative for *H. nelsoni*. Two additional samples in January and September 1985 were also negative. Eastern oysters from Hammonasset River were negative in 1978 and 1983, but in December 1985, 32% of a sample had *H. nelsoni* (C. A. Farley and F. Kern, personal communication).

In 1986, experimental trays of Fishers Island eastern oysters were placed in Cotuit harbor monthly and sampled monthly from April 1 to November 1. A preliminary report on the timing and early development of *H. nelsoni* infections in these eastern oysters indicated close conformance with the earlier findings of investigators in New Jersey, Maryland and Virginia (L. Leibovitz, Marine Biological Laboratory, Woods Hole, Massachusetts, personal communication).

All evidence indicated that seed brought from Ocean Pond in 1986 was not infected with *H. nelsoni*. Therefore, the tray studies demonstrated that *H. nelsoni* was established in Cotuit harbor and able to infect imported susceptible eastern oysters. Probably, *H. nelsoni* was introduced into Cotuit harbor with the Hammonasset eastern oysters or possibly, infective stages entered the harbor in the coastal flow in the summers of 1984, 1985, and 1986. This speculation is based on the unpredictable occurrences from the 1960s to 1985 of *H. nelsoni* in so many scattered locations spread along the coast of Connecticut and the entire southern coast of Massachusetts.

Mortality of eastern oysters in Peconic and Gardiners bays in late summer and fall of 1985 was massive—a loss of 80% of a planting of market-size eastern oysters valued at $1–1.5 million—but the role of *H. nelsoni* in that mortality is not clear. Most of the seed eastern oysters came from wild stock in New Haven harbor and Norwalk, Connecticut, with smaller additions of hatchery-reared seed from Maine and the Shinnecock Indian hatchery on Long Island. Brown tide, a bloom of a recently identified small phytoplankter *Aureococcus anorexefferens,* appeared in 1985 in eastern Long Island bays. Brown tide is considered responsible for widespread mortality of the bay scallop *Argopecten irradians* and the destruction of eel grass by shading (Cosper et al. 1987). During the period of bay scallop mortality, the eastern oyster plantings were checked weekly without evidence of deaths. The owner thought his eastern oysters had escaped damage from the algal bloom, but as the bloom was dissipating, the eastern oysters "died within a period of 1–2 weeks." After the mortality, in mid-October 1985,

two samples of survivors were sent to the Rutgers Shellfish Research Laboratory. Samples containing 20 eastern oysters each from a Cedar Beach ground and a Long Beach ground had 3 and 2 lightly infected eastern oysters, respectively. Because survivors of an *H. nelsoni*-caused epizootic are usually highly infected (Ford and Haskin 1982), we do not believe that *H. nelsoni* was the primary cause of the massive kill. However, light *H. nelsoni* stress, added to the burden of the noxious alga, might have been enough to trigger the high mortality.

Resistance with Increasing Disease Pressure

After the early epizootic mortalities in the lower Chesapeake Bay in 1959 and 1960, Virginia planters retreated from grounds in Mobjack Bay and Hampton Roads to areas of lower salinity in tributary rivers. Because no major planting areas of low salinity were available in Delaware Bay, the industry largely confined its planting from the seedbeds to the upper edge of the traditional planting grounds. By the mid 1960s, the natural seed—offspring of survivors of the heavy *H. nelsoni* pressure of 1957–1959—had about three times more resistance to mortality caused by *H. nelsoni* than the preepizootic population. With this level of resistance, the use of larger seed of eastern oysters from the lower seedbeds, and a shorter planting time (harvesting spring plants in the following fall and winter), the Delaware Bay industry managed to survive, although it was greatly stressed and suffered reduced production (Haskin and Ford 1983). The generally upward trend in mortalities from about 1972 (Figure 6), capped by the peak in 1986, has discouraged further planting of natural seed and has raised questions as to why the resistant native seed was overwhelmed by *H. nelsoni*. Parallel experiences with native eastern oysters from the surviving populations in Mobjack Bay and Hampton Roads led to the same questions for the Chesapeake Bay.

We use the term "resistance" to mean resistance to death from disease caused by *H. nelsoni*. It does not imply resistance to infection but rather the capacity to restrict parasite numbers to tolerated levels (Ford 1988). If two stocks of eastern oysters are exposed for a time to infection with *H. nelsoni,* and one has 10% survival and the other 30%, we conclude that the second group is three times more resistant to mortality than the first. Resistance in eastern oysters is not absolute; rather, it may differ among stocks or individuals.

TABLE 3.—Mortality of seven Cape Shore (Delaware Bay) tray stocks of eastern oysters, June–December 1986.

Stock	Mortality (%)
Control—1986 imports of James River seed	88
Offspring of James River seed[a]	93
Control—1986 imports from Maryland	91
Offspring of Maryland import seed[a]	98
Cape Shore wild seed[a]	41
Resistant yearlings, 6th generation[a]	5
Resistant yearlings, 6th generation[a]	26

[a] 1985 year class.

TABLE 4.—MSX-related mortality of six 1978 year-class groups of eastern oysters at Cape Shore (Delaware Bay) over a 3-year test, October 1978–July 1981.

Stock	Mortality (%)
Virginia control, susceptible	97
Navesink control, susceptible	98
F_6 Virginia, resistant[a]	60
F_4 Long Island, resistant[b]	35
F_4 and F_5 Long Island, Delaware Bay, resistant[b]	39
Wild Cape Shore set, resistant	54

[a] Produced at Virginia Institute of Marine Science Laboratory.
[b] Produced at Rutgers Cape Shore Laboratory.

The level of resistance to *H. nelsoni* mortality in an eastern oyster population or in an inbred line may be expected to reflect the rigor with which it has been selected through generations of exposure to *H. nelsoni* infections. Surviving older wild eastern oysters in areas of high salinity in both Chesapeake and Delaware bays are being rigorously selected against *H. nelsoni* disease. However, newly set spat in these locations may derive partly from eastern oysters outside the immediate area. In all low-salinity creek and river tributaries of Delaware Bay and on the seedbeds, eastern oysters are under little or no pressure due to *H. nelsoni*. As the larvae of these eastern oysters swim, they mix with all other eastern oyster larvae in the estuary, including those produced by survivors of heavy selective pressure in the lower bay. The natural set, wherever it occurs, comes from this mix (Haskin and Ford 1982). In contrast, laboratory-reared resistant eastern oysters have been held in the lower bay (Cape Shore) where selection against *H. nelsoni* is very intense. Each generation in a resistant line was exposed for at least 2.5 years before the survivors of that generation became parents for the next. Broodstock was spawned, and larvae were raised and set in the laboratory, without mixing with wild larvae from the native bay eastern oysters. The hardiest of these lines after six generations are about three times as resistant as the present Delaware Bay wild seed, and they are about ten times more resistant to *H. nelsoni*-caused mortality than the original Delaware Bay wild stock. Mortalities of several stocks in trays on the Cape Shore flats over the 1986 June–December test period illustrate this point (Table 3).

In 1986, *H. nelsoni* pressure in lower Delaware Bay, judged by mortality in susceptible control stocks, was higher than in any earlier year (Figure 5). In an average year, the first year June–November mortalities in the James River seed imports would have been about 60%; in the Cape Shore wild set, about 15–25%. Losses of susceptible controls over that same 5-month period in 1986 approached the long-term average of 93% for a 33-month period (Table 3). In the presence of this intense disease pressure, mortality for the sixth generation of one of the best Delaware Bay lines was 5%, for another line (mixed Delaware Bay and Long Island stock) it was 26%. Thus, in contrast to four susceptible controls, the rigorously selected resistant lines were not overwhelmed.

In earlier years, the authors developed *H. nelsoni*-resistant lines and exchanged groups of eastern oysters with each other for tests at Cape Shore and at VIMS. In May 1978, a fifth generation resistant line, with a record of about 10% annual mortality over 2 years at VIMS, was sent to Cape Shore for breeding and testing. In that summer, *H. nelsoni* pressure at Cape Shore was very slight (Figure 5), and testing of the yearling spat of the Virginia resistant line extended through the 1981 season. Results of these tests were compared with results from tests with offspring of two imported susceptible controls, with two resistant lines under development at the Cape Shore, and with Cape Shore wild set of the 1978 year class (Table 4). The offspring of the line selected under the test conditions at VIMS had the highest mortality of the four resistant groups including the wild Cape Shore set.

The severe eastern oyster losses in both Chesapeake and Delaware bays in recent years are not an indictment of disease-resistant oysters. They simply indicate that the disease pressure is now higher than we have seen before and certainly higher than the infection pressure against which those eastern oysters have been selected. We are encouraged that some of the Rutgers selectively bred resistant lines were able to withstand the increased pressure.

Even the best stocks, held under continuing *H. nelsoni* attack for several years, will begin to have substantial mortalities caused by *H. nelsoni*, and, after 5–6 years, most stocks will have died as a result of advanced infections (Ford and Haskin 1987). Growers of eastern oysters can be encouraged that the most resistant eastern oysters can effectively localize and tolerate infections until they reach market size. With *H. nelsoni* under control, they can feed, grow, reproduce, and develop as a high quality eastern oyster.

Environmental Effects on *H. nelsoni* Disease

Variation in *H. nelsoni* activity along the east coast is difficult to understand but no more so than in the bays in which *H. nelsoni* first appeared 30 years ago. Eastern oyster mortalities caused by *H. nelsoni* in James River seed are substantially higher in Delaware Bay at the Cape Shore than in the York River at Gloucester Point (VIMS), even though both are enzootic waters with very similar salinity and temperature regimes. This observation indicates that other factors controlling *H. nelsoni* activity may be different in the two areas. The Cape Shore intertidal area consistently receives the heaviest setting of eastern oysters in the bay. Eastern oyster growth rates there are high and correlate well with phytoplankton populations (measured by total chlorophyll) that are the highest in the bay (W. Canzonier, Rutgers Shellfish Laboratory, Port Norris, New Jersey, personal communication). The hydrographic system that concentrates larvae and phytoplankton at the Cape Shore may also concentrate the infective stage of *H. nelsoni,* and this system may not have a counterpart in the York River at Gloucester Point.

Although salinity and temperature data indicate somewhat similar habitats in these two locations, there are faunal indicators that are probably more sensitive than our physical measurements. For example, the eastern oyster pathogen *Perkinsus marinus* was not found north of the lower Chesapeake Bay until imported to the Delaware Bay with seed eastern oysters from Virginia and Maryland in the early 1950s. With an embargo on all eastern oysters for commercial imports and exports after the 1957–1959 *H. nelsoni* epizootic, *P. marinus* died out in Delaware Bay, and the northern boundary of this pest is currently reestablished in the Chesapeake Bay. This boundary probably indicates a sensitivity to temperature regimes that are milder in the Chesapeake Bay area. Andrews (1988, this volume) discusses tem-

perature relative to *P. marinus*. Ford and Haskin (1982) also suggested that the winter temperatures of the lower Chesapeake Bay never become low enough to reduce *H. nelsoni* activity in a cyclic pattern such as that in Delaware Bay.

The exceptionally high simultaneous mortalities of 1986 and 1987 in susceptible stocks at Cape Shore and VIMS (Figure 5) suggest changes in a common controlling factor that is probably climatological. The prime factor could be the drought commencing in the fall of 1984 and extending through 1987. Because salinities in both locations are well within the range for optimal parasite activity under normal conditions, the infective dosage may have peaked during 1986 and 1987. Furthermore, drought may have permitted an increase in supply of infective particles at both locations.

An earlier deduction based on *H. nelsoni* prevalence patterns in lower Delaware Bay (Ford and Haskin 1982) led to the conclusion that the source of infective stages was downbay and that the stages were diluted with increasing distance upbay from this source. An extension of this hypothesis is that a reservoir host was held downbay by unfavorable salinities. A direct result of persistent drought and increasing salinities would be to permit upbay migration of such a host. This migration would increase the concentration (dosage) of infective stages, released by this host, throughout the bay. This scenario, assuming that migration of the reservoir host requires one or more reproductive seasons, would explain observations that the increased *H. nelsoni* activity upbay does not occur until the second or third year of drought in Chesapeake Bay.

The upbay movement of the host would not only extend the range for infective stages but in all probability would increase their abundance within the area of release. The quick increase in *H. nelsoni* activity in Delaware Bay, compared to the Chesapeake Bay, in times of drought may reflect its smaller size and its consequent reduced reserve of freshened water.

The complex role of temperature in eastern oyster–*H. nelsoni* interactions is also not clearly established. Earlier experimental work by Myhre (1973) and Douglass (1977) in the Rutgers Laboratory has been discussed (Ford 1988). Douglass (1977) demonstrated in mortality-resistant stocks of eastern oysters that *H. nelsoni,* typically restricted to gill epithelia, becomes systemic as fall temperatures drop below 18–20°C. As temperatures rise in spring, infections in resistant oysters may be suppressed or eliminated. Shifts in temperature regimes, such as an early fall or a late winter with delayed spring warming, could influence *H. nelsoni* infections, causing mortality rates to rise or fall.

Two further cold winter relationships to *H. nelsoni* levels are possible. (1) In comparing mortalities at Cape Shore and VIMS (Figure 5), the 1972 low mortality at VIMS was attributed to the influence of Hurricane Agnes. A similar low in 1984 has not been explained. Both the Gloucester Point lows of 1972 and 1984 follow cold winter lows, at the Cape Shore. The other two Cape Shore cold winter lows, 1961–1962 and 1978, coincide with smaller but distinct drops in mortality at Gloucester Point. These are probably coincidental but perhaps a look at Virginia coastal climatological data would be justified. (2) In considering coastal relationships of *H. nelsoni*, Ford and Haskin (1982) pointed out that the parasite had been present in several locations in Long Island and southern New England since at least the 1960s without substantial eastern oyster mortalities. Arguing from the cold winter syndrome of *H. nelsoni* cycles in Delaware Bay, they suggested that the colder winters to the north might prevent full-fledged disease development. The outbreaks of *H. nelsoni* in Long Island and Cape Cod described above overlap in the years 1983–1986 and, therefore, invite a study of temperature trends in that area.

Life Cycle Clues to Recent Range Expansions of *H. nelsoni*

We are certain that the *H. nelsoni* stage that infects oysters is waterborne and can be spread through the waterways. The longer the infective stage remains viable, the greater the distances it can travel by this route. Spreading would be expected to continue until boundaries set by temperature, salinity, or some other factor are reached. If this stage is released by infected eastern oysters, the spread within these boundaries would be most quickly accomplished by the movement of infected eastern oysters to the new areas.

On the other hand, if there is an obligate reservoir or intermediate host species that supplies the *H. nelsoni* life cycle stage infective for eastern oysters, that species must be established within the range of the eastern oysters. If that reservoir host species is not resident in the area, transport of infected eastern oysters alone to virgin territory would then not permit transmission and establishment of *H. nelsoni*. Presumably *H. nelsoni*-in-

fected eastern oysters that are moved to virgin territory might also carry with them the reservoir or alternate host which would then have an opportunity to become established in that territory. Moreover, *H. nelsoni* may not be species specific for its presumed alternate or reservoir host, that is, more than one species might host the infective stages of *H. nelsoni*.

The *H. nelsoni*-caused mortality of 1985 and 1986 in Cotuit harbor followed an importation of infected eastern oysters from Connecticut in 1984. In 1985, the mortality included both the Connecticut imported seed and the disease-free seed brought in from Ocean Pond in 1984. In 1986, Ocean Pond eastern oysters that were brought in experimentally at monthly intervals also became infected and had heavy mortality. The *H. nelsoni*-caused mortality of the Ocean Pond eastern oysters established that transmission of *H. nelsoni* occurred within Cotuit harbor. Based on *H. nelsoni* activity in other areas, we think it very unlikely that it was directly transmitted from one eastern oyster to another. The presence of *H. nelsoni* in Ocean Pond eastern oysters in spring 1985 indicates infections existed in fall of 1984. This would mean very quick cycling from eastern oyster to alternate host to eastern oyster if the alternate host required exposure to parasitized eastern oysters before it could become infected. Probably other species dredged from Hammonasset River accompanied the eastern oysters to Cotuit. Among these may have been the host species already carrying *H. nelsoni* infective stages that were released upon arrival to infect previously unexposed Ocean Pond eastern oysters. The Cotuit oyster planter indicated that he had brought eastern oyster seed from the Hammonasset River for several years before the 1984 import. Two earlier samples from the Hammonasset River (1979 and 1983) were negative for *H. nelsoni*. One sample in October 1985 had a prevalence of 32%. Another possibility is that alternate or reservoir hosts from the Hammonasset River had been established in Cotuit harbor with the earlier imports. Other possibilities are considered at the end of this article.

The most important advance yet to be made in our understanding of *H. nelsoni* biology is to work out its complete life cycle. That knowledge may point the way to control of *H. nelsoni*. Many of the gaps in our information were indicated in the introduction. Our frequent reference to a hypothetical reservoir host also emphasizes that we really do not know the source of *H. nelsoni* infectivity for eastern oysters.

Although most haplosporidians commonly undergo sporulation in their hosts, *H. nelsoni* rarely achieves this stage in eastern oysters. Because of the scarcity of spores and the consequent uncertainty of its affinities, *H. nelsoni* was not named until 8 years after its discovery. Andrews (1979) reported 44 cases of sporulation in 170,000 slide preparations of stained tissues from living and recently dead eastern oysters over a period of 16 years. When only infected oysters were considered, the sporulation rate was less than one case per 2,000. These cases of sporulation were scattered throughout the year, although those in June and July were most common. In the Delaware Bay area, "fewer than a dozen cases. . .among the many thousands of tissue slides and fresh smears" were reported (Ford and Haskin 1982). Most of these cases were in yearling eastern oysters and invariably in the epithelia of the digestive tubules. Sprague (1965) hypothesized that *H. nelsoni* "may normally occur within some associated organisms living in the vicinity of oyster populations and may sporulate regularly to provide forms infective to both oysters and the other host."

During the drought of the mid-1960s, *H. nelsoni* moved far up Chesapeake Bay and attacked highly susceptible eastern oysters with a small increase in the abundance of sporulation stages. Whether the ratio of sporulation to numbers of eastern oysters examined increased is not clear. By selecting sick and moribund specimens from thousands of young susceptible eastern oysters, Couch et al. (1966) were able to recognize and describe sporulation of *H. nelsoni* after which Farley (1967) proposed a tentative life cycle.

Unlike most haplosporidians where sporulation occurs in all tissues, *H. nelsoni* confines sporulation to the epithelia of the digestive tubules. Therefore, all prespore stages either migrate to the digestive diverticula or develop in them (Farley 1967; Andrews 1979). Multinucleate plasmodia enlarge and undergo nuclear division, and the chromatin material acquires a punctate appearance before 50 or more spores are formed in the sporocysts. This enlargement of sporonts occurs between epithelial cells and forms protrusions into the tubule lumina. Such restricted sporulation limits the quantity of spores that can be produced and may facilitate release of spores from live eastern oysters sporadically as the epithelia are destroyed by the bulging sporonts.

Eastern oysters may live several months after sporulation.

Sporulation by *H. costale* (SSO), the seaside organism, is more typical of haplosporidians in general. All plasmodia enlarge when the punctate stage is reached, and sporulation occurs in all tissues, including mantle and gill. Sporulation in *H. costale* occurs regularly in late May and early June each year. Eastern oysters with massive numbers of sporocysts die rapidly, often before spores become mature. Distinctive sporulation sites for the two species indicate that different biochemical or physiological processes may be occurring in spore formation.

The rarity of sporulation by *H. nelsoni* has led to some speculation that *H. nelsoni* is not really adapted to parasitism of the eastern oyster and that the eastern oyster is an accidental host. Such opinions are refuted by an interesting account of a massive sporulation in eastern oysters in Virginia in 1976 (Andrews 1979). Of thousands of highly susceptible, hatchery-raised young (25 mm) spat in a single tray, 39% had *H. nelsoni* sporulation. They were set in mid-May, held in a disease-free pond for early growth, and then transferred to the York River enzootic area on 8 July 1976. Nine weeks later, on 21 September, 40% had died, 88% of the survivors were infected with *H. nelsoni*, and 39% were in sporulation. A second group of the same brood was transplanted to the York River on 16 August 1976. In this group, patent infections did not appear until December, and little mortality occurred until May and June 1977 when infections were intensive. In this second lot there was only one case of sporulation in 93 infections diagnosed during 1977.

At the time the first tray was placed in the York River, there were 75 other lots of oysters in the vicinity, some within 50 feet. None of these exhibited sporulation, although *H. nelsoni* activity was intense in several lots of susceptible eastern oysters.

The question raised by these observations is what induced sporulation in this one lot of eastern oysters while others in nearby trays developed only plasmodial infections. Most likely, the differences in the two lots were genetic. Among the 75 lots in the York River, no others were newly set Rappahannock River stock receiving first exposures to *H. nelsoni* infectivity. Even within the Rappahannock gene pool, individual variation in response to *H. nelsoni* challenge would be great.

Such a spectacular sporulation event leads to a certain uneasiness. Could we all have been missing such events occurring in young spat in their first exposure to *H. nelsoni,* events that could have produced infective stages to account for the massive mortalities in our bays? We do not believe so. Surely such events would have been detected by the large-scale monitoring programs with accompanying histological studies in the three states of Virginia, Maryland, and New Jersey. The highly susceptible control stocks, spawned and examined year after year in the disease-resistance breeding programs at VIMS and at the Rutgers Laboratory, would also have been prime candidates for such sporulations, if they were occurring.

Although at least 30 species of *Haplosporidium* are known (La Haye et al. 1984), none of these have been transmitted to their respective hosts under controlled conditions, with the possible exception of *H. pickfordae* in freshwater snails (Barrow 1965). The complete life cycles are not known for any of these species.

We hope that the immunological technique being adapted by our colleagues will enable us to search for the hypothetical reservoir or alternate host(s) of *H. nelsoni* (Burreson 1988). To date, we have avoided a grueling systematic histological search through the hundreds of possible candidates in and around our coastal estuaries. The enzyme-linked antibodies, DNA probes, or both promise to reduce the drudgery and to speed the search for *H. nelsoni.*

It is certainly time to intensify and concert the efforts of investigators along the coast to resolve the many questions remaining about the life cycle of *H. nelsoni.* We need to lift our eyes and perhaps open our imaginations to some of the exciting discoveries in related fields. Sweeney et al. (1985) working with a microsporidian parasite (*Amblyospora* sp.) infecting the Australian encephalitis mosquito vector *Culex annulirostris,* demonstrated stages in an intermediate copepod host necessary to complete the life cycle of the microsporidian. This discovery is cited as the first evidence of alternate host involvement in the life cycles of microsporidia. Another spore is formed in the copepod, and this spore is then infectious to larval mosquitoes.

Andreadis (1985), working in Connecticut, demonstrated that haploid spores from another species of *Amblyospora* parasitic in another mosquito species, are also transmitted to an alternate copepod host. He reported that members of the genus *Amblyospora* have at least three distinct developmental cycles, each producing a different spore.

A discovery of particular relevance, for those who have been frustrated for nearly 30 years in pursuit of the *H. nelsoni* life cycle, concerns the whirling disease of salmonid fish which has been known and studied for 80 years (Wolf and Markiw 1984). The causative agent was recognized as a myxosporean named *Myxosoma cerebralis* which produces an abundance of spores in trout. However, the spores were not infectious to other fish. This myxosporean disease of fish is initiated by an organism known since 1899 as an actinosporean, parasitic in a tubificid oligochaete. Wolf and Markiw (1984) showed conclusively that "instead of being considered as representatives of separate classes in the phylum *Myxozoa,* the myxosporean and actinosporean are alternating life forms of a single organism." They suggest that, if the host worms could be eradicated by selectively lethal chemicals, the whirling disease of trout may be prevented.

We stated earlier that *H. nelsoni* is a poorly adapted parasite in the eastern oyster. Wolf and Markiw (1984) noted that *Myxosoma cerebralis* infections were well tolerated in the parasite's original host fish, the brown trout *Salmo trutta.* In a new host, the rainbow trout *S. gairdneri,* the parasite produces the virulent whirling disease. They note that the rainbow trout was introduced into Europe in the late 1800s and that *Myxosoma cerebralis* was accidently brought to the USA in the 1950s. *Haplosporidium nelsoni* may prove to be nonvirulent in another host or perhaps even in a parallel host, such as the Pacific oyster *Crassostrea gigas.*

Profile of an Alternate or Reservoir Host for *H. nelsoni*

Recent findings in parasite life cycles encourage us to attempt a profile of an alternate or reservoir host of *H. nelsoni.* Based on observations of *H. nelsoni* activity, what deductions can we reasonably make about its source, that is, the hypothetical host that releases the stage infective for eastern oysters?

Observation 1.—Infection intensity (dosage) of *H. nelsoni* in eastern oysters is independent of location or size of eastern oyster populations. Infection pressures actually appear to be increasing in recent years as the eastern oyster populations diminish in our bays.

Deduction 1.—This observation may indicate that the oyster has no obligate role in the life cycle of *H. nelsoni.* That is, it is an accidental host and irrelevant to the cycling of *H. nelsoni* in the bays. This deduction would be in line with Sprague's

(1965) earlier suggestion that an organism in proximity to the eastern oysters is supplying the infective stage both for the eastern oysters and for that other organism.

Observation 2.—In Delaware Bay, there is a pattern of timing of *H. nelsoni* infections in oysters in summer and early fall with first infections in the lower regions of the planting grounds and later infections progressing slowly upbay in a wave. This pattern has not been observed in the lower Chesapeake Bay.

Deduction 2.—The source (host) of the infective stages is in the lower bay, either on or below the lowest planting grounds. This may indicate that the host is restricted to higher salinity areas.

Observation 3.—In times of drought, higher prevalences and mortalities of eastern oysters are delayed in upper Virginia and Maryland sectors of Chesapeake Bay until the second or third year of the drought. In Delaware Bay, the delays are shorter.

Deduction 3.—In drought periods, the salinity-limited host may move upbay as the salinity increases, probably with a jump in population size each reproductive period, especially if there is a pelagic larval stage. A substantial upbay movement of the host would then require 1 or 2 years to establish larger populations. The upbay migration of the host then would increase the concentration of infective material in upbay and tributary areas.

Observation 4.—In Delaware Bay, the years of lowest *H. nelsoni* prevalence and mortality follow unusually cold winters, indicating that the host may be damaged, killed, or inhibited by low temperatures.

Deduction 4.—The host in the lower bay is vulnerable to damage by cold because it inhabits shallow water on shoals, rock jetties, or bases of light houses, or in salt marshes. Prolongation of the cold period may also increase winter casualties, even in deepwater populations.

Observation 5.—The reduction in *H. nelsoni* activity in Delaware Bay does not occur until a full year after the cold winter, for example, the 1983 low was preceded by an unusually cold January in 1982.

Deduction 5.—(a) Perhaps the simplest deduction would be that the infective stages were released in average numbers before, during, or even after the winter damage to the host and persisted until the usual eastern oyster infection period begins in late May or June. Lack of infections during the following summer could be related to the time required to rebuild the host population.

(b) An alternate deduction is one that would require a two-host alternating cycle in addition to the eastern oyster. In this scenario, host A would release infective materials supplying both the eastern oyster and host B. Host A is not cold sensitive and would release its infective materials in the season immediately following the cold winter. The eastern oyster would receive its dosage of *H. nelsoni* particles. But the damaged or decimated host B population would not be able to receive and process its usual dosage. The host B survivors would be producing a reduced amount of infective materials to cycle to host A. The reduced infection of host A would then be reflected in its reduced output of infective materials for the eastern oyster population.

If the eastern oyster is indeed an accidental host for *H. nelsoni,* there is a consequence of practical importance for management. *Haplosporidium nelsoni*-infected oysters by themselves would not be effective in spreading the disease into a new area. Rather, the true host(s) would be the effective carrier(s) of the infective stages. Maintenance of the disease in the new area would require infection of the true host species in that area, if already resident. If not resident, the true host carrier to the new area would have to become established to maintain the *H. nelsoni* population.

However, there is no direct evidence that the eastern oyster is an accidental host for *H. nelsoni.* The speculation is based on the premise that if the eastern oyster is an obligate host, the supply of infectious stages should diminish with the reduction of the high-salinity eastern oyster population. There is no certainty, however, that a small residual eastern oyster population could not be remarkably productive of *H. nelsoni* infectious stages.

Acknowledgments

We are indebted to staff members C. Austin Farley and Fred Kern, U.S. National Marine Fisheries Service Laboratory, Oxford, Maryland, and to Frank Perkins, Director of Virginia Institute of Marine Science (VIMS) for unpublished information on the distribution of *H. nelsoni* along the east coast from New England to Georgia. Robert Hillman, Battelle Laboratories, Duxbury, Massachusetts, extended that range to Florida. Frank Germano, Division of Marine Fisheries, Sandwich, Massachusetts, provided information on eastern oyster mortalities along the southern Massachusetts coast, and Billy Chapman, Shellfish Constable of Wellfleet, Massachusetts, updated the eastern oyster mortalities there. Details of the Cotuit harbor outbreak were provided by George C. Matthiessen, Ocean Pond Corporation, and J. R. Nelson, Cotuit Oyster Company. *Haplosporidium nelsoni* experiences of the Bayville Hatchery of Frank Flower & Sons were detailed by David Relyea and Joseph Zahtila. Jack Mulhall, Long Island Oyster Company, discussed with us the details of his eastern oyster losses in eastern Long Island bays. Recent data on diseases in Virginia were provided by E. M. Burreson, VIMS. George Krantz, Director of the Oxford Laboratory, has released to us the most recent information on *H. nelsoni* distribution in Maryland waters. Searching criticisms by several anonymous reviewers were constructive and much appreciated by the authors.

This is publication F-32504-3-88 supported by state funds of the New Jersey Agricultural Experiment Station and publication 1490 of VIMS.

References

Andreadis, T. G. 1985. Experimental transmission of a microsporidian pathogen from mosquitoes to an alternate copepod host. Proceedings of the National Academy of Sciences of the USA 82:5574–5577.

Andrews, J. D. 1964. Oyster mortality studies in Virginia. IV. MSX in James River public seed beds. Proceedings National Shellfisheries Association 53: 65–84.

Andrews, J. D. 1966. Oyster mortality studies in Virginia. V. Epizootiology of MSX, a protistan pathogen of oysters. Ecology 47:19–31.

Andrews, J. D. 1968. Oyster mortality studies in Virginia. VII. Review of epizootiology and origin of *Minchinia nelsoni*. Proceedings National Shellfisheries Association 58:23–36.

Andrews, J. D. 1979. Oyster diseases in Chesapeake Bay. U.S. National Marine Fisheries Service Marine Fisheries Review 41(1–2):45–53.

Andrews, J. D. 1982. Epizootiology of late summer and fall infections of oysters by *Haplosporidium nelsoni*, and comparison to annual life cycle of *Haplosporidium costalis*, a typical haplosporidian. Journal of Shellfish Research 2:15–23.

Andrews, J. D. 1983. *Minchinia nelsoni* (MSX) infections in the James River seed-oyster area and their expulsion in spring. Estuarine, Coastal and Shelf Science 16:255–269.

Andrews, J. D. 1984a. Epizootiology of diseases of oysters (*Crassostrea virginica*), and parasites of associated organisms in eastern North America. Helgolaender Meeresuntersuchungen 37:149–166.

Andrews, J. D. 1984b. Epizootiology of haplosporidian diseases affecting oysters. Comparative Pathobiology 7:243–269 (Plenum, New York).

Andrews, J. D. 1988. Epizootiology of the disease caused by the oyster pathogen *Perkinsus marinus* and its effects on the oyster industry. American Fisheries Society Special Publication 18:47–63.

Andrews, J. D., and M. Castagna. 1978. Epizootiology of *Minchinia costalis* in susceptible oysters in seaside bays of Virginia's eastern shore, 1959–1976. Journal of Invertebrate Pathology 32:124–138.

Andrews, J. D., and J. L. Wood. 1967. Oyster mortality studies in Virginia. VI. History and distribution of *Minchinia nelsoni*, a pathogen of oysters in Virginia. Chesapeake Science 8:1–13.

Barrow, J. H., Jr. 1965. Observations on *Minchinia pickfordae* (Barrow 1961) found in snails of the Great Lakes region. Transactions of the American Microscopical Society 80:319–329.

Burreson, E. M. 1988. Use of immunoassays in haplosporidian life cycle studies. American Fisheries Special Publication 18:298–303.

Cosper, E. M., and seven coauthors. 1987. Recurrent and persistent brown tide blooms perturb coastal marine ecosystems. Estuaries 10:284–290.

Couch, J. A., C. A. Farley, and A. Rosenfield. 1966. Sporulation of *Minichinia nelsoni* (Haplosporida, Haplosporidiidae) in *Crassostrea virginica* (Gmelin), Science (Washington, D.C.) 153:1529–1531.

Douglass, W. R. 1977. *Minchinia nelsoni* disease development, host defense reactions and hemolymph enzyme alterations in stocks of oysters (*Crassostrea virginica*) resistant and susceptible to *Minchinia nelsoni*-caused mortality. Doctoral dissertation. Rutgers University, New Brunswick, New Jersey.

Farley, C. A. 1967. A proposed life cycle of *Minchinia nelsoni* (Haplosporida, Haplosporidiidae) in the American oyster *Crassostrea virginica*. Journal of Protozoology 14:616–625.

Farley, C. A. 1975. Epizootic and enzootic aspects of *Minchinia nelsoni* (Haplosporida) disease in Maryland oysters. Journal of Protozoology 22:418–427.

Ford, S. E. 1985. Effects of salinity on survival of the MSX parasite *Haplosporidium nelsoni* (Haskin, Stauber, and Mackin) in oysters. Journal of Shellfish Research 5:85–90.

Ford, S. E. 1988. Host–parasite interactions in eastern oysters selected for resistance to *Haplosporidium nelsoni* (MSX) disease: survival mechanisms against a natural pathogen. American Fisheries Society Special Publication 18:206–224.

Ford, S. E., and H. H. Haskin. 1982. History and epizootiology of *Haplosporidium nelsoni* (MSX), an oyster pathogen, in Delaware Bay, 1957–1980. Journal of Invertebrate Pathology 40:118–141.

Ford, S. E., and H. H. Haskin. 1987. Infection and mortality patterns in strains of oysters *Crassostrea virginica* selected for resistance to the parasite *Haplosporidium nelsoni* (MSX). Journal of Parasitology 73:368–376.

Ford, S. E., and H. H. Haskin. 1988. Comparison of in vitro salinity tolerance of the oyster parasite *Haplosporidium nelsoni* (MSX) and hemocytes from the host *Crassostrea virginica*. Comparative Biochemistry and Physiology A, Comparative Physiology 90:183–187.

Haskin, H. H., and S. E. Ford. 1979. Development of resistance to *Minchinia nelsoni* (MSX) mortality in laboratory-reared and native oyster stocks in Delaware Bay. U.S. National Marine Fisheries Service Marine Fisheries Review 41(1–2):54–63.

Haskin, H. H., and S. E. Ford. 1982. *Haplosporidium nelsoni* (MSX) on Delaware Bay seed oyster beds: a host–parasite relationship along a salinity gradient. Journal of Invertebrate Pathology 40:388–405.

Haskin, H. H., and S. E. Ford. 1983. Quantitative effects of MSX disease *(Haplosporidium nelsoni)* on production of the New Jersey oyster beds in Delaware Bay, USA. International Council for the Exploration of the Sea, C.M. 1983/E:56, Copenhagen.

Haskin, H. H., L. A. Stauber, and J. A. Mackin. 1966. *Minchinia nelsoni* n. sp. (Haplosporida, Haplosporidiidae): causative agent of the Delaware Bay oyster epizootic. Science (Washington, D.C.) 153:1414–1416.

Krantz, E. L., L. R. Buchanan, C. A. Farley, and A. H. Carr. 1972. *Minchinia nelsoni* in oysters from Massachusetts waters. Proceedings National Shellfisheries Association 62:83–88.

La Haye, C. A., N. D. Holland, and N. McLean. 1984. Electron microscopic study of *Haplosporidium comatulae* n. sp. (Phylum Ascetospora: Class Stellatosporea), a haplosporidian endoparasite of an Australian crinoid, *Oligometra serripinna* (Phylum Echinodermata). Protistologica 20:507–515.

Myhre, J. L. 1973. Levels of infection in spat of *Crassostrea virginica* and mechanisms of resistance to the haplosporidian parasite *Minchinia nelsoni*. Master's thesis. Rutgers University, New Brunswick, New Jersey.

Perkins, F. O. 1968. Fine structure of the oyster pathogen, *Minchinia nelsoni* (Haplosporida, Haplosporidiidae). Journal of Invertebrate Pathology 10:287–307.

Sprague, V. 1965. Comments on the life cycle, host parasite relationships and epizootiology of MSX. University of Maryland, Natural Resources Institute 65-13. Chesapeake Biological Laboratory, Solomons, Maryland. (Not seen; cited in Farley 1967.)

Sweeney, A. W., E. I. Hazard, and M. F. Graham. 1985. Intermediate host for an *Amblyospora* sp. (Microspora) infecting the mosquito, *Culex annulirostris*. Journal of Invertebrate Pathology 46:98–102.

USGS (U.S. Geological Survey) District Engineer. 1913–1985. [Water resources reports for New Jersey.] 1913–1928: New Jersey Department of Conservation and Development, Division of Waters Report, 1929. 1929–1960: irregular reports of the relevant New Jersey water supply agency. 1961–1985: Water resources data, New Jersey; USGS Water Resources Division, West Trenton, New Jersey.

Wolf, K., and M. E. Markiw. 1984. Biology contravenes taxonomy in the Myxozoa: new discoveries show alternation of invertebrate and vertebrate hosts. Science (Washington, D.C.) 225:1449–1452.

American Fisheries Society Special Publication 18:23–37, 1988

Epizootic Diseases of Oysters Associated with Viral Infections

Michel Comps

Institut Francais de Recherche pour l'Exploitation de la Mer
rue Jean Vilar, 34200 Sete, France

Abstract.—Virus infections have been associated with major diseases of oysters of the genus *Crassostrea*. These infections include gill necrosis virus and hemocytic infection virus diseases of the Portuguese oyster *C. angulata* and more recently the oyster velar virus disease affecting larval, hatchery-reared Pacific oysters *C. gigas*. This report presents histo- and cytopathological characteristics of these infections in which large icosahedral cytoplasmic deoxyriboviruses, similar to the Iridoviridae, are implicated. The gill necrosis virus causes, principally in the Portuguese oyster, an evolutive ulceration of the gills, including cellular hypertrophy and severe inflammation. Mortalities have been observed in the most serious cases. Hemocytic infection virus has caused mass mortalities of Portuguese oysters in Europe (1970–1974); it induces cytoplasmic lesions in the hemocytes and causes severe injuries to interstitial tissues. Similar infections were observed in 1977 in Pacific oysters. The oyster velar virus develops in the cytoplasm of epithelial cells of the velum and causes tissue lesions. It is the apparent cause of some hatchery mortalities.

Until recently, much of our knowledge of epizootic mortalities of molluscs associated with microbial disease agents has been restricted to bacterial, mycotic, and protozoan diseases, and little attention has been given to viral diseases. During the past decade, important virological studies have led to the discoveries of viral diseases in molluscs, particularly molluscs of economic importance.

Although a virus has been suspected as the etiologic agent of the epizootic Malpeque Bay disease of the eastern oyster *Crassostrea virginica* (Rosenfield 1969), the actual occurrence of a viral infection in molluscs was first demonstrated by Rungger et al. (1971) in the common octopus *Octopus vulgaris*. In the same year, Devauchelle and Vago (1971) noted the presence of viruslike particles, similar to reoviruses in the cuttlefish *Sepia officinalis*.

The first case of viral infection found in marine bivalve molluscs was described in the eastern oyster by Farley et al. (1972), who reported hexagonal, single-enveloped, herpeslike virus particles in tissues of eastern oysters subjected to elevated water temperatures. The viral particles were empty or possessed dense nucleoids, and the enveloped virions (200–250 nm) were observed in the cytoplasm of infected cells (Farley 1978). Subsequently, from electron microscope studies, viruslike particles were recognized in several species of oysters. Ribonucleic acid-like, enveloped viral particles, 100–110 nm, were noted by Farley (1976) in the epithelial cells of the digestive diverticula of *C. virginica*. This author also described

virus particles (53 nm) associated with marked hypertrophy of gametogenic epithelial cells of *C. virginica* and similar lesions in tissues of *C. gigas* (Farley 1976). He later observed, in the hemocytes of the Olympia oyster *Ostrea lurida*, viruslike ribonucleic acid particles budding through the plasma membranes of the cells and enveloped virions (50 nm) and conforming to the morphological description of Togaviridae.

While these cases of viral infections, which are rare and of little pathological importance, were reported, other new viroses developing in epizootic form were observed during major disease outbreaks of oysters: gill necrosis virus disease and hemocytic infection virus disease (HIVD) associated with mass mortalities of the Portuguese oyster *C. angulata* (Comps and Duthoit 1976; Comps et al. 1976) and HIVD observed in the Pacific oyster *C. gigas* during summer mortalities in the Bay of Arcachon (France) in 1977 (Comps and Bonami 1977). The pathogenic agents of these diseases have properties resembling iridoviruses.

A third type of iridovirus, the oyster velar virus (OVV), has been found in the velar epithelial cells of hatchery-reared, Pacific oyster larvae, which were undergoing serious mortalities (Elston 1979).

Gill Necrosis Virus and Hemocytic Infection Virus Diseases

Both gill necrosis and hemocytic infection viruses (GNV, HIV) have been implicated in epizootic diseases of the Portuguese oyster and have very similar characteristics including size, shape, and intracytoplasmic morphogenesis; however, they have different pathogenic effects.

Hosts

Although GNV infection was observed more commonly in the Portuguese oyster, a few cases of this infection have been observed in the diseased gills of Pacific oysters cultured in the region of Marennes–Oléron, France (Comps 1969; Comps 1970; Comps and Duthoit 1976).

Hemocytic infection virus caused serious infection and mass mortalities of Portuguese oysters in France from 1970 to 1973 (Comps et al. 1976). A similar type of virus infection has been reported in Pacific oysters in the Bay of Arcachon long after the disappearance of the Portuguese oyster (Comps and Bonami 1977).

Viruses

Morphology.—Mature particles (380 nm in diameter) scattered through the cytoplasm of infected cells have an icosahedral symmetry, with five-, three-, and twofold axes of rotational symmetry to their sections. The shells of the virions appear to consist of two trilaminar layers (Figure 1A, B). However, some sections show that the outer layer is composed of subunits, corresponding to an icosahedral lattice. A fringe of knobbed fibrils is attached to the subunits (Figure 1B).

The electron-opaque central core, 250 nm in diameter, is limited by a three-layered fringe of definite width and surrounded by a layer of dense material. Occasionally, sections of the core reveal an ordered stacking of the fine filaments, which may represent the nucleic acid arrangement. These features of the shell conform to the model established by Stolz (1973) for icosahedral cytoplasmic deoxyriboviruses (ICDV).

Morphogenesis.—The morphogenesis of these viruses takes place in the cytoplasm of oyster hemocytes or gill cells (Comps 1983). Several distinctive features occur during this process.

At the periphery of the virogenic stroma, a trilaminar element develops similar to the inner membrane which constitutes the inner component of the shell (Figure 2A). In juxtaposition to the unit membrane, in association with the development of the shell, a layer of poorly defined subunits appears; the subunits induce a sharp outline to the incomplete particles due to the increased density of the virogenic granular material (Figure 2B). Then the immature particle becomes detached from the viroplasmic matrix (Figures 1D, 2C); a dense nucleoid with two layers is contained

within the shell of the developing particle. The external layer is formed of subunits, while the inner layer corresponds to a unit membrane.

The maturation stages of the particles are visible during the pathogenesis of gill disease. They are observed during the development of the electron-dense nucleoid, which retracts from the inner surface of the shell (Figure 1D).

Nucleic acid.—The Portuguese oyster viruses were not isolated from the oysters during the course of the disease, nor were they characterized chemically. The presence of a deoxyribonucleic acid (DNA) virus was demonstrated by histochemical techniques. Fluorescent microscopy of histologic sections stained with acridine orange at pH 3.8 revealed large greenish-yellow areas that fluoresced in the cytoplasm of cells infected by GNV. The intensification of the emission of fluorescence indicated DNA concentrations compatible with the replication of the observed virus. Similar observations that were made with the fluorochrome 4′,6-diamidino-2-phenylindole (DAPI) confirmed the virus of gill disease to be a DNA virus (Comps and Masso 1978). The same techniques also confirmed HIV as a DNA virus.

Pathologic Lesions

Gill disease (GNV infection).—Portuguese oysters infected with GNV exhibit various stages of ulcerations of the gill and labial palps (Comps 1970). Three stages of lesion development have been described according to the extent of tissue damage. The first detectable gross lesions are small perforations in the center of yellowish discolored zones of tissue. Further development and extension of the lesions result in larger and deeper ulcerations with jagged edges (Figures 3, 4).

Histologically, the response to the infection is characterized by tissue necrosis and massive hemocytic cellular infiltration around the lesions; the infiltration induces marked visible changes in the structural organization of the gills. The gill filaments, the interfilamentar junctions, and the interlamellar septae can be affected (Figure 5). Marked inflammatory reactions are observed in the connective tissue of ulcerated labial palps (Figure 5D).

The most distinctive cellular lesion associated with the disease is the occurrence of giant polymorphic cells, which may be up to 30 μm in size. The nucleus of the cell is hypertrophied and usually oval; its maximum width is 10 μm and its length is 15 μm. The diameter of the nucleolus is between 2 and 4 μm. The cytoplasm contains

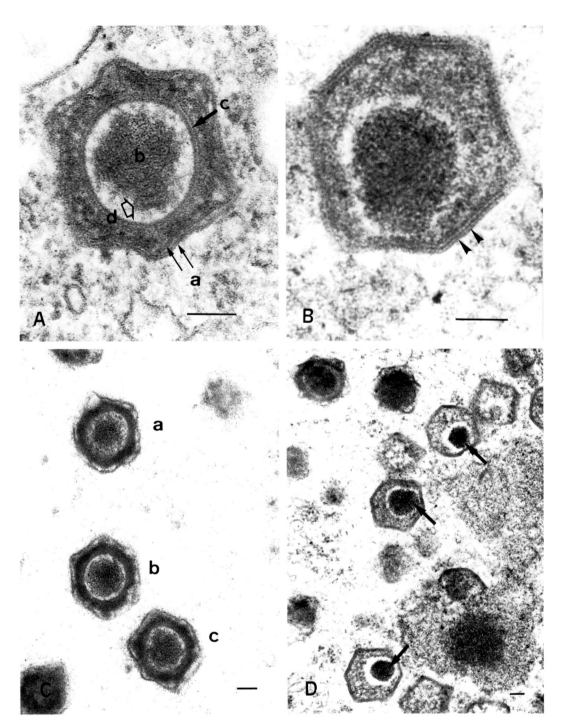

FIGURE 1.—Morphology of the hemocytic infection virus (HIV) and the gill necrosis virus (GNV). **A.** Characteristic HIV mature particle. a = trilaminar elements forming the shell, b = central core, c = three-layered fringe, d = filament, possibly representing nucleic acid arrangement. **B.** Section of GNV immature particle showing the structural subunits of the outer element of the shell (arrows). **C.** HIV particles exhibiting different axes of rotational symmetry. a = fivefold axes, b = threefold axes, c = twofold axes. **D.** Stages of the development of electron-dense nucleoids (arrows) into immature GNV particles. Electron micrographs; bars = 100 nm.

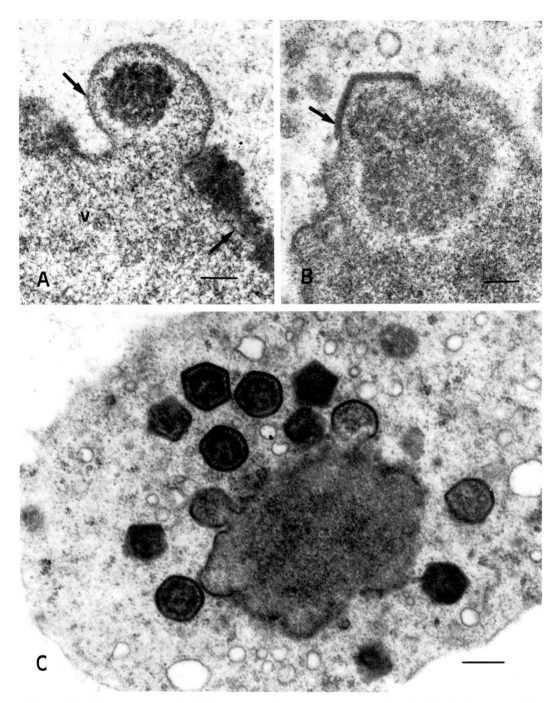

FIGURE 2.—Electron micrographs of the sequence of morphogenetic events involved in the formation of the virions. **A.** Development on the periphery of the virogenic stroma (v) of a trilaminar structure (arrows) typical of a unit membrane. Bar = 100 nm. **B.** Addition of outer layer of subunits (arrow). Bar = 100 nm. **C.** Separating immature hemocytic infection virus particles. Bar = 300 nm.

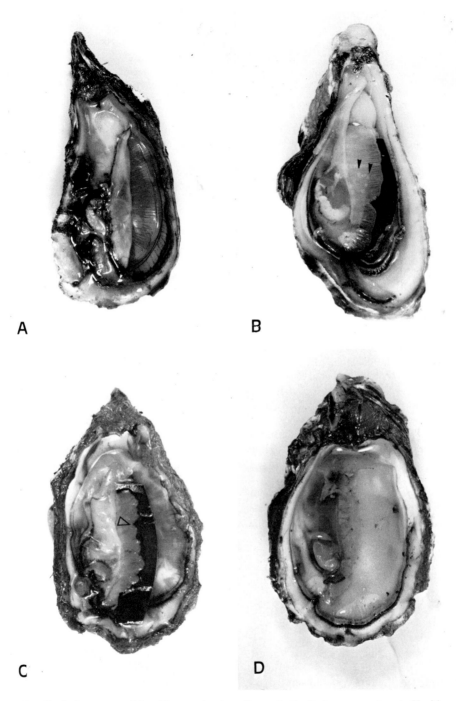

FIGURE 3.—Evolutive stages of the gill necrosis virus disease in the Portuguese oyster. **A.** Healthy oyster. **B.** Microperforations associated with inflammation on the gills (arrows). **C.** Severe ulceration of the gills exhibiting deeply jagged edges (arrow). **D.** Prelethal stage: the gills are nearly destroyed, and the oyster is very thin.

FIGURE 4.—Detail of a typical lesion of the gill necrosis virus disease. The necrosis of the tissues (arrow) induces major alterations in gill morphology. Bar = 1 mm.

large, lightly refringent fuchsinophilic granules, 3 to 5 μm in diameter, that are frequently associated with small basophilic granules and vacuoles (Figure 6). In some giant cells, generally found at the edge of necrotic lesions, a voluminous basophilic inclusion (5–15 μm) occupies the greatest part of the cytoplasm in which finer basophilic grains (0.4–0.5 μm) are also present (Figure 7). The presence of DNA in the inclusion body is indicated by a positive reaction when the inclusion body is stained by the Feulgen method, and the reaction is confirmed by the acridine orange and DAPI reactions, yielding intense yellow-green and blue staining, respectively. In some cases, the same cytochemical properties can be observed in the fine granular inclusions in the cytoplasm.

Electron microscopy of thin sections revealed that these cytoplasmic elements correspond to the intracytoplasmic viral lesion: the inclusions are the viroplasm and the fine granules, the virions. The intense reaction of the finely granular dense viroplasm with DAPI staining demonstrated the high rate of DNA synthesis. The formation of the virions takes place at the periphery of the viroplasm as previously described (Figure 8).

During the course of development of the viral lesions, the GNV induces progressive hypertrophication of infected cells. Other cytopathological changes are observed as the enlargement of

the nucleus combines with a disappearance of the chromatin and the nucleolus. The cytoplasm is also affected by the formation of numerous vacuoles and of dense multilayered spheroidal inclusions, 2–3 μm in diameter (Figure 9).

Mass mortality during the 1970 HIV infection.—Some of the 1970 HIV-affected Portuguese oysters exhibited atrophy and weakness of the adductor muscle; however, no distinctive clinical signs were noted that could be associated with the disease.

The most characteristic histological lesion of the disease is an acute inflammatory response associated with the presence of atypical hemocytes in the connective tissue and an increase in the number of brown cells (Comps 1983). Pycnotic nuclei and basophilic intracytoplasmic inclusion bodies, often with attached groups of fine granules, are found in atypical blood cells. The inclusions (2 or 3 μm in diameter) are round and Feulgen-positive. Acridine orange and DAPI staining demonstrated that the granules and inclusions contain DNA (Figure 10).

Ultrathin sections revealed that these cells were infected with a virus (HIV) (Comps et al. 1976). The inclusions correspond to the virogenic stroma and the granules to the virions. The finely granular viroplasm mass is spheroid in shape. Incomplete virions adhere to the periphery of the viroplasm, whereas detached particles are scattered throughout the cytoplasm. The morphology of the virus was described above and infected cells are shown in Figure 11. The lesion caused by the HIV in the Pacific oyster can be associated with cytoplasmic paracrystalline arrays (Figure 12).

Important Epizootiological and Pathological Aspects

Gill disease (GNV infection).—Gill disease was originally observed by Trochon in the region of La Tremblade in November 1966 (Marteil 1968). The incidence of the disease increased, became epizootic in 1967, and extended to all areas of Portuguese oyster culture in France, including Marennes–Oléron, Bay of Arcachon, Vendée, and some sectors of Brittany. Abnormally high mortalities observed at this time confirmed the virulence of the disease, which reached its highest level in 1967. In most of the affected areas, 70–80% of the Portuguese oysters evidenced gill necrosis. The occurrence of gill disease was also noted in Great Britain, Spain, and Portugal (Ferreira and Dias 1973). After 1968, the disease persisted in an enzootic state in natural stocks of

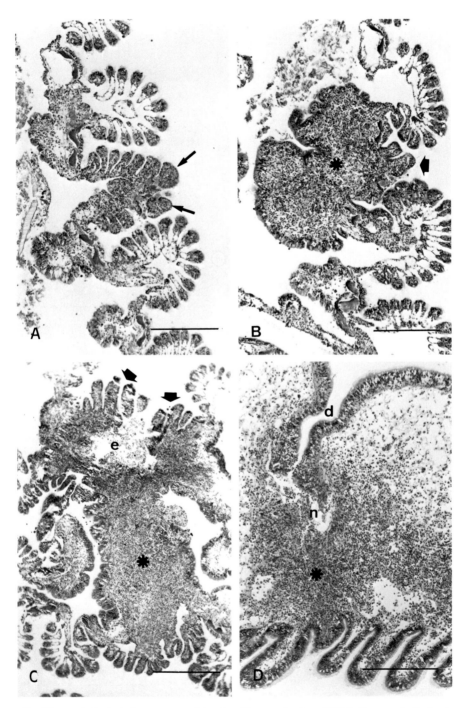

FIGURE 5.—Histopathological changes caused by the gill necrosis virus (GNV) disease. **A.** Early stage of the infection: the inflammatory reaction involves swelling of the gill filaments (arrows). **B.** Advanced stage of the disease: a group of filaments (arrow) is completely destroyed, and the inflammation spreads to the interfilamentar junction (star). **C.** Heavy ulceration of two plicae of filaments (arrows) producing a large excavation (e); in this case, the infection affects also the interlamellar septum (star). **D.** GNV lesion on the labial palp: localized degradation of the epithelium (d) associated with a necrosis of the connective tissue (n); note the intense inflammation around the lesion (star). Bar = 200 μm.

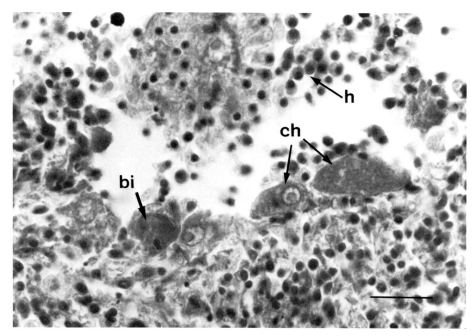

FIGURE 6.—Detail of a gill lesion showing polymorphic hypertrophic cells (ch) and hypertrophic globular cells with basophilic inclusion (bi); hemocytes are scattered around the lesion (h). Azan; bar = 20 μm.

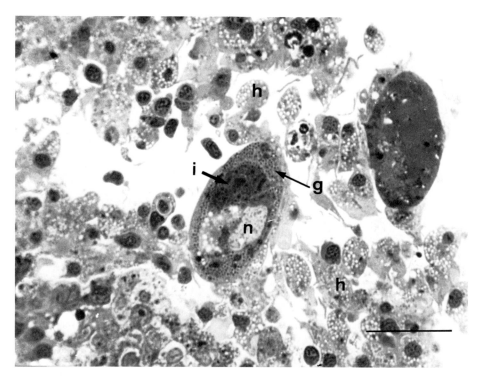

FIGURE 7.—Section of hypertrophic cells infected by gill necrosis virus; n = nucleus; i = basophilic inclusion; g = basophilic granules or virions; h = hemocytes. Toluidine blue; bar = 20 μm.

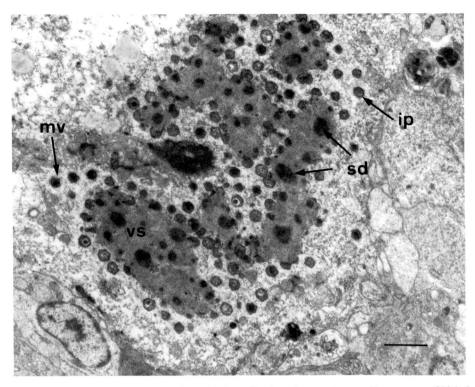

FIGURE 8.—Intracytoplasmic viral lesion in a globular cell; sd = virogenesis according to spots of high density; vs = virogenic stroma; ip = immature particles; mv = mature virions. Electron micrograph; bar = 1 μm.

Portuguese oysters from the Sado River and from the area of Cadiz (Comps and Masso 1978).

Environmental conditions influenced the course of the disease. In the most severe cases, massive necrosis was evident, and destruction of the gills was extensive. Respiratory disorders followed (His 1969), and gametogenesis was interrupted (Marteil 1969). Severe mortality (40%) occurred in Marennes and Bay of Arcachon (Marteil 1968). Surviving Portuguese oysters that were sampled showed healing of gill lesions (Comps 1970).

Historically, some protistans were implicated in necrosis of the gills. Besse (1968) attributed a role to a ciliate of the genus *Trichodina*. Gras (1969) noted the presence of a *Dermocystidium*-like parasite, and Arvy and Franc (1968) ascribed the cause of the disease to a new protistan, *Thanatostrea polymorpha* (Franc and Arvy 1969). However, an ultrastructural study demonstrated that the large cells described as forms of *T. polymorpha* were actually host cells infected with the virus (Comps and Duthoit 1976). The virus was not isolated, and the experimental transmission studies needed to demonstrate its pathogenicity were not undertaken dur-

ing the progress of the disease. However, numerous histological sections of Portuguese oysters were prepared during the disease epizootics, and a clear relationship was established between the gill lesions and the virus infection.

Mass mortality during the 1970 HIV infection.—Commencing in August 1970, mass mortalities of Portuguese oysters occurred among Marennes–Oléron oysters and in several oyster-producing areas in Brittany. In December, mortalities of this species were reported also in the Bay of Arcachon. The disease spread to other Portuguese oyster culture beds in France with the exception of the Étang de Thau on the Mediterranean coast (Comps 1983). In the south of Spain, recurring mortalities of Portuguese oysters associated with HIV were found in the Gadalquivir River in 1974 (M. Guttierez, Instituto de Investigaciones Pesqueras, Cadiz, personal communication). In 1983, long after the disappearance of the Portuguese oysters in French oyster beds, mortalities due to HIV occurred in Portuguese oyster spat from an experimental hatchery study located in Brittany (Bougrier et al. 1986).

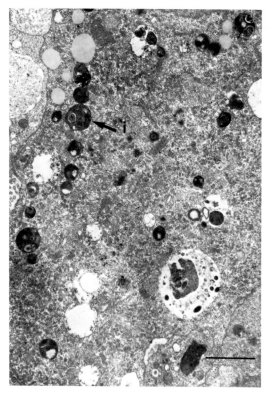

FIGURE 9.—Disorganized cytoplasm of cells infected with gill necrosis virus containing dense multilayered inclusions (i). Electron micrograph; bar = 1 μm.

Oyster Velar Virus Disease (OVVD)

A virus infection of larval Pacific oysters, producing velar lesions and mortalities, has been reported (Leibovitz et al. 1978). Elston and Wilkinson (1985) described the new disease in detail, and they named it oyster velar virus disease. The disease was detected during an 8-year period in the state of Washington in Pacific oysters grown in Wilapa Bay and Puget Sound.

Affected animals were less active and had velar lesions with progressive loss of velar epithelial cells, which then became dissociated, rounded, and detached from the deformed velum. These velar cells contained inclusion bodies and showed viral cellular lesions. DNA was demonstrated by histochemical methods (Feulgen and acridine orange reactions). The intracytoplasmic lesion consisted of a granular, electron-dense viroplasmic inclusion body.

The morphogenesis of the virus particles began with the formation of capsids around the viroplasm, which became irregular in shape. De-

tached, empty particles had an icosahedral structure. The process of morphogenesis of the viral particles was completed with the formation of the dense viral core. The zone separating the core from the capsids became denser during the process of morphogenesis.

The virions, averaging 228 nm from edge to edge, have capsids 20.6 nm in thickness and are composed of two bilayered membranous structures. The dense inner core, 103 nm × 160 nm, is ovoid in section and is surrounded by a less dense zone.

The importance of this viral infection is evident; however, its relationship to the larval disease has not been established experimentally. Moreover, the seasonal transmission of the disease is difficult to explain, especially as to the mode of transmission.

Discussion

Compared with vertebrates and insects, the knowledge of marine invertebrate virology is very limited and is usually restricted to histopathological and epizootiological data. The viral etiology of the major epizootics of Portuguese oysters was demonstrated in 1976, but research was effectively halted when the susceptible host disappeared from France. Because of this disappearance and the lack of research methodology when the diseases first appeared, the HIV and GNV have not been isolated, purified, or analyzed. Without information on the fundamental characteristics such as density and composition of the polypeptides and nucleic acids, it is not possible to make a serious comparison between the two agents. Presently, they are considered different only on the basis of histological and cytopathological analyses.

The study of OVV, although more recent (Elston and Wilkinson 1985), is also limited to morphological considerations. It has not been isolated, experimentally transmitted, or analyzed biochemically. In spite of these problems, a rough comparison is instructive.

At present, the precise taxonomic positions of HIV and GNV are uncertain, yet they resemble the iridoviruses in general morphology, size, intracytoplasmic morphogenesis, and nucleic acid composition. The lymphocystis virus in fish has recently been classified in the genus *Iridovirus* (Matthews 1979), and the virus causing piscine erythrocytic necrosis (PEN) of marine fish (Atlantic cod *Gadus morhua* and others) is also similar. This group is also a member of the genus *Iridovirus* (Reno et al. 1978; Walker and Sherburne 1977). A further comparison can be made

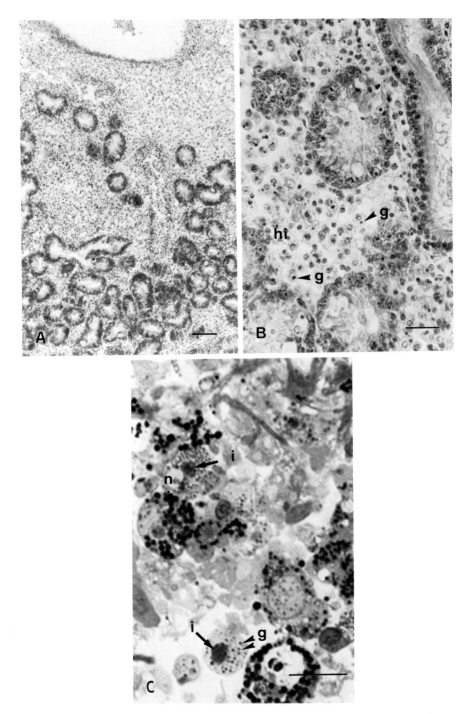

FIGURE 10.—Infection caused by the hemocytic infection virus (HIV). **A.** Characteristic aspect of the connective tissue of an infected Portuguese oyster. Hematoxylin and eosin; bar = 100 μm. **B.** Section of connective tissue showing infected hemocytes (hi) containing basophilic granules (g). Hematoxylin and eosin; bar = 20 μm. **C.** Section through hemocytes with HIV lesion; i = basophilic inclusion or virogenic stroma; g = basophilic granules or virions; n = nucleus showing abnormal aspect. Toluidine blue; bar = 10 μm.

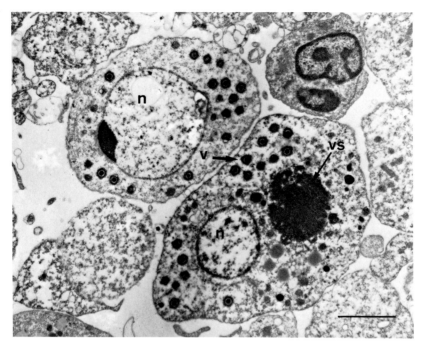

FIGURE 11.—Electron micrograph of cells infected by hemocytic infection virus; n = nucleus; vs = virogenic stroma; v = virions. Bar = 2 μm.

with the virus described by Elston and Wilkinson (1985), which causes OVVD in *C. gigas*. These workers demonstrated similarities in the viral morphogenesis and structures of the virions; however, the viruses differed in some characteristics such as size (228 nm versus 380 nm) and shape of the inner core, which was pararectangular in OVV. Another virus, similar but smaller than OVV (120–140 nm), has been described by Rungger et al. (1971) in the common octopus and classified in the genus *Iridovirus* (Fenner 1976).

Although paracrystalline arrays of virions have been demonstrated in the cytoplasm of some iridovirus-infected cells (Walker and Weissenberg 1965; Devauchelle and Durchon 1973), this characteristic is not a permanent feature of the iridoviruses and has not been observed in the cells of *C. angulata* infected by HIV or GNV. In contrast, the viral lesion caused by HIV, which infects the hemocytes of *C. gigas*, can be associated with paracrystalline inclusions, showing similarities with the inclusion bodies observed during the course of replication of some iridescent viruses (Kelly and Tinsley 1972). Similar inclusions have been reported in the cells infected by an iridovirus causing muscle lesions in *Octopus vulgaris* (Rungger et al. 1971).

If, for the present, the taxonomic position of the GNV and HIV cannot be determined more precisely, both viruses, along with the PEN viruses, the OVV, and some lymphocystis disease viruses, can be considered an original group of marine ICDV.

Another consideration is the host-virus relationship and the susceptibility of both species, *C. angulata* and *C. gigas*, to the pathogen. Other than an exceptional case of HIV and GNV infection found in *C. gigas*, this species, under natural conditions, appears to be highly resistant to these infections. During the summer of 1970 and all of 1971, when Portuguese oysters were greatly affected with mass mortalities, no losses were noted in the Pacific oysters coming from Japan or in the area of Marennes–Oléron. On the contrary, these Pacific oyster-producing areas saw exceptional growth (Comps 1972). Although Pacific and Portuguese oysters are considered the same species by some investigators (Menzel 1974), there is a difference in their susceptibility to HIV.

The ability of the Pacific oyster to resist the 1970 mass mortalities resolved a very serious economic problem. Pacific oyster production was maintained by massive imports of Pacific oyster seed from Japan, Korea, and Canada and by the

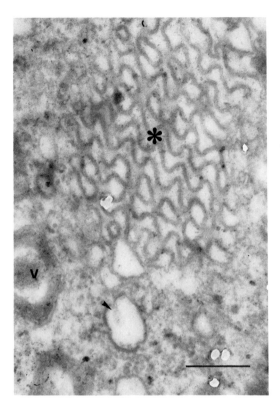

FIGURE 12.—Hemocytic infection virus lesion in Pacific oyster showing paracrystalline inclusion (star) associated with virus particles (v). Electron micrograph; bar = 200 nm.

regeneration of the natural Pacific oyster beds with Pacific oysters from British Columbia, Canada. Now, the culture of the Pacific oyster encompasses all the coasts of Europe.

The initial difficulties in isolation and analysis of these viruses stemmed partly from the relative scarcity of the virions. In the GNVD and OVVD, very few cells are infected, and in all cases, the number of virions is light. Similar conditions characterize the erythrocytic infection caused by the large ICDV in Atlantic cod (Reno and Nicholson 1981) and in Atlantic herring *Clupea harengus harengus* (Reno et al. 1978).

The resolution of these problems would be greatly facilitated by the ability to replicate the viruses in vitro in tissue cultures. Some attempts have been conducted with fish cell lines, but the viruses isolated from several species of oysters were not pathogenic. Hill (1976) described three serotypes of viruses isolated from several marine bivalves, particularly from the Pacific and eastern oysters and the edible oyster *Ostrea edulis*, which

were cultured in bluegill fry (*Lepomis macrochirus*) cell culture (BF 2). The isolates have identical properties to those of a number of different infectious pancreatic necrosis serotypes. The effects of these viruses on molluscan hosts are unknown.

More recently, a reolike virus designated 13p2, was isolated by Meyers (1979) from juvenile eastern oysters and grown in bluegill fry cell cultures (BG 20). The virus replicated in cell cultures of freshwater bluegill, brown bullhead *Ictalurus nebulosus*, Atlantic salmon *Salmo salar*, guppy *Poecilia reticulata*, and walleye *Stizostedion vitreum*. Experimentally, the virus produced lesions and mortalities in bluegills but was not pathogenic to eastern oysters (Meyers 1980). Clearly, obtaining cell lines from oysters or other marine bivalves is of primary importance to future progress in research on marine molluscan viral diseases.

Acknowledgment

I thank Louis Leibovitz, Laboratory for Marine Animal Health, Marine Biological Laboratory, Woods Hole, Massachusetts, for his suggestions and critical review of the manuscript.

References

Arvy, L., and A. Franc. 1968. Sur un protiste nouveau, agent de destruction des branchies et des palpes de l'huître portugaise. Comptes Rendus Hebdomadaires des séances de l'Académie des Sciences, Série D, Sciences Naturelles 267:103–105.

Besse, P. 1968. Résultats de quelques observations sur une affection branchiale des huîtres (*Crassostrea angulata* Lmk). Bulletin Académie Vétérinaire de France 41:87–91.

Bougrier, S., G. Raguenès, E. Bachére, G. Tigé, and H. Grizel. 1986. Essai de réimplantation de *Crassostrea angulata* en France. Résistance au chambrage et comportement des hybrides *C. angulata–C. gigas*. Conseil International pour l'Exploration de la Mer, C.M. 1986/F:38, Copenhagen.

Comps, M. 1969. Observations relatives à l'affection branchiale des huîtres portugaises (*Crassostra angulata* Lmk). Revue des Travaux de l'Institut des Pêches Maritimes 33(2):151–160.

Comps, M. 1970. La maladie des branchies chez les huîtres du genre *Crassostrea*, caractéristiques et évolutions des altérations, processus de cicatrisation. Revue des Travaux de l'Institut des Pêches Maritimes 34(1):23–44.

Comps, M. 1972. Observations sur la résistance des huîtres du genre *Crassostrea* au cours de la mortalité massive de 1970–1971 dans le bassin de Marennes–Oléron. Conseil International pour l'Exploration de la Mer, C.M. 1972/K22, Copenhagen.

Comps, M. 1983. Recherches histologiques et cytologiques sur les infections intracellulaires des mollusques bivalves marins. Doctoral dissertation. Université de Sciences et Techniques du Languedoc, Montpellier, France.

Comps, M., and J. R. Bonami. 1977. Infection virale associée à des mortalités chez l'huître *Crassostrea gigas* Th. Comptes Rendus Hebdomadaires des séances de l'Académie des Sciences, Série D, Sciences Naturelles 285:1139–1140.

Comps, M., J. R. Bonami, C. Vago, and A. Campillo. 1976. Une virose de l'huître portugaise (*Crassostrea angulata*). Comptes Rendus Hebdomadaire des Séance de l'Académie des Sciences, Série D, Sciences Naturelles 282:1991–1993.

Comps, M., and J. L. Duthoit. 1976. Infection virale associée à la "maladie des branchies" de l'huître portugaise *Crassostrea angulata* Lmk. Comptes Rendus Hebdomadaire des Séances de l'Académie des Sciences, Série D, Sciences Naturelles 283: 1595–1596.

Comps, M., and R. M. Masso. 1978. Study with fluorescent technique of the virus infections of the Portuguese oyster *Crassostrea angulata* Lmk. Pages 39–40 *in* Proceedings of the International Colloquium on Invertebrate Pathology. Prague, Czechoslovakia.

Devauchelle, G., and M. Durchon. 1973. Sur la présence d'un virus du type Iridovirus dans les cellules mâles de *Nereis diversicolor* (O. F. Müller). Comptes Rendus Hebdomadaire des Séances de l'Académie des Sciences, Série D, Sciences Naturelles 277:463–466.

Devauchelle, G., and C. Vago. 1971. Particules d'allure virale dans les cellules de l'estomac de la seiche *Sepia officinalis* L. (Mollusques, Céphalopodes). Comptes Rendus Hebdomadaire des Séances de l'Académie des Sciences, Série D, Sciences Naturelles 272:894–896.

Elston, R. 1979. Viruslike particles associated with lesions in larval Pacific oysters (*Crassostrea gigas*). Journal of Invertebrate Pathology 33:71–74.

Elston, R., and M. T. Wilkinson. 1985. Pathology, management and diagnosis of oyster velar virus disease (OVVD). Aquaculture 48:189–210.

Farley, C. A. 1976. Ultrastructural observations on epizootic neoplasia and lytic virus infection in bivalve mollusks. Progress in Experimental Tumor Research 20:283–294.

Farley, C. A. 1978. Viruses and viruslike lesions in marine mollusks. U.S. National Marine Fisheries Service Marine Fisheries Review 40(10):18–20.

Farley, C. A., and W. G. Banfield, G. Kasnic, and W. Forster. 1972. Oyster herpes-type virus. Science (Washington, D.C.) 178:759–760.

Fenner, F. 1976. Classification and nomenclature of viruses. Intervirology 7:1–115.

Ferreira, P. S., and A. A. Dias. 1973. Sur la répartition et l'évolution de l'altération des branchies de *Crassostrea angulata* dans le Tage, le Sado et l'Algarve. Conseil International pour l'Exploration de la Mer, C.M. 1973/K6, Copenhagen.

Franc, A., and L. Arvy. 1969. Sur *Thanatostrea polymorpha* n.g., n. sp., agent de destruction des branchies et des palpes de l'huître portugaise. Comptes Rendus Hebdomadaire des Séances de l'Académie des Sciences, Série D, Sciences Naturelles 268:3189–3190.

Gras, P. 1969. Recherches sur l'organisme responsable de la maladie des branchies. Revue des Travaux de l'Institut des Pêches Maritimes 32(2):161–164.

Hill, B. J. 1976. Properties of a virus isolated from the bivalve mollusc *Tellina tenuis* (da Costa). Pages 445–452 *in* L. A. Page, editor. Wildlife diseases. Plenum, New York.

His, E. 1969. Recherches d'un test permettant de comparer l'activitié respiratoire des huîtres au cours de l'évolution de la maladie des branchies. Revue des Travaux de l'Institut des Pêches Maritimes 33(2): 171–175.

Kelly, D. C., and T. W. Tinsley. 1972. The proteins of iridescent virus types 2 and 6. Journal of Invertebrate Pathology 19:273.

Leibovitz, L., R. Elston, V. P. Lipovsky, and J. Donaldson. 1978. A serious disease of larval Pacific oysters (*Crassostrea gigas*). Proceedings of the World Mariculture Society 8:603–615.

Marteil, L. 1968. La maladie des branchies. Conseil International pour l'Exploration de la Mer, C.M. 1968/K5, Copenhagen.

Marteil, L. 1969. Données générales sur la maladie des branchies. Revue des Travaux de l'Institut des Pêches Maritimes 33(2):145–150.

Matthews, R. E. F. 1979. Classification and nomenclature of viruses. Pages 1–296 *in* S. Karger, editor. Third report of international committee and taxonomy of viruses, Basel, Switzerland.

Menzel, R. W. 1974. Portuguese and Japanese oysters are the same species. Journal of the Fisheries Research Board of Canada 31:453–456.

Meyers, T. R. 1979. A reo-like virus isolated from juvenile American oysters (*Crassostrea virginica*). Journal of General Virology 46:203–212.

Meyers, T. R. 1980. Experimental pathogenicity of reovirus 13 p2 for juvenile American oysters *Crassostrea virginica* (Gmelin) and bluegill fingerlings *Lepomis macrochirus* (Rafinesque). Journal of Fish Diseases 3:187–201.

Reno, P. W., and B. L. Nicholson. 1981. Ultrastructure and prevalence of viral erythrocytic necrosis (VENO) virus in Atlantic cod, *Gadus morhua* L., from the northern Atlantic Ocean. Journal of Fish Diseases 4:361–370.

Reno, P. W., M. Phillipon–Fried, B. L. Nicholson, and S. W. Sherburne. 1978. Ultrastructural studies of piscine erythrocytic necrosis (PEN) in Atlantic herring (*Clupea harengus harengus*). Journal of the Fisheries Research Board of Canada 35: 138–154.

Rosenfield, A. 1969. Oyster diseases in North America and some methods for their control. Pages 67–78 *in* Proceedings of the conference on artificial propagation of commercially valuable shellfish. University of Delaware, College of Marine Studies, Newark.

Rungger, D., M. Rastelli, E. Braendle, and R. G.

Malsberger. 1971. A viruslike particle associated with lesions in the muscles of *Octopus vulgaris*. Journal of Invertebrate Pathology 17:72–80.

Stolz, D. B. 1973. The structure of icosahedral cytoplasmic desoxyriboviruses. Journal of Ultrastructure Research 43:58–74.

Walker, R., and S. W. Sherburne. 1977. Piscine erythrocytic necrosis virus in Atlantic cod, *Gadus morhua*, and other fish: ultrastructure and distribution. Journal of the Fisheries Research Board of Canada 34:1188–1195.

Walker, R., and R. Weissenberg. 1965. Conformity of light and electron microscopic studies on virus particle distribution in lymphocyctis tumor cells of fish. Annals of the New York Academy of Sciences 126:375–395.

American Fisheries Society Special Publication 18:38–46, 1988

Aber Disease of Edible Oysters Caused by *Marteilia refringens*

Antonio J. Figueras

Instituto de Investigaciones Marinas, Consejo Superior de Investigaciónes Cientificas
Muelle de Bouzas sin número, Vigo, Spain

Jaime Montes

Centro Experimental de Vilaxoan, Apartado 208, Villagarcia, Pontevedra, Spain

Abstract.—Aber disease of edible oysters *Ostrea edulis* is caused by *Marteilia refringens*, which has recently been placed into the protozoan orders Occlusosporida or Paramyxea. The life cycle of this pathogen is briefly reviewed; it involves a series of unique endogenous buddings that produce sporoplasms within sporoplasms. Aber disease causes severe pathology in edible oysters, grossly recognized by an emaciated brown digestive gland, a translucent mantle, and cessation of shell growth. Loss of edible oyster tissue weight adds to the impact of mortalities in commercial stocks. Aber disease appears in May, peaks in August, and continues into December. Subclinical infestations can persist through the winter, but new infestations occur in May through August. Prevalences may reach 100%, yet there is a puzzling inconsistency between mortalities and degree of infestation. An interesting situation in Galicia, Spain, indicates that spread of *M. refringens* can be restricted by 2 km of open water. Probably, *M. refringens* has alternate or reservoir hosts, since it has also been observed in *Mytilus edulis*, *Cardium edule*, and *Crassostrea gigas*. However, ultrastructural information is lacking for several species of *Marteilia*, which could confuse diagnosis.

Taxonomy

Recurrent and serious edible oyster mortalities have been recorded since 1968 in *Ostrea edulis* populations in France, particularly in the regions of Marennes and Bretagne. The disease syndrome soon became known as Aber disease, named after the estuaries in Bretagne where mortalities were first experienced, and was later named digestive gland disease after the main infection site in the edible oyster. The causative agent was first seen and superficially described by Comps (1970), who discussed its similarities to the haplosporidians. Herrbach (1971) suggested that it might be a member of the fungal order Chytridiales. Grizel and Tigé (1973) believed it to be a member of the genus *Labyrinthomyxa*. Grizel et al. (1974), who studied the organism in detail and named it *Marteila refringens*, related it to the Microspora. Perkins (1976), on the other hand, grouped it with the Haplosporea (now Balanosporida). Considering the arguments of Desportes and Ginsburger–Vogel (1977), Sprague (1979) eventually removed *M. refringens* from the latter group and assigned it to the newly created order Occlusosporida. A more recent and very acceptable classification is that of Desportes (1984), who places *M. refringens* in order Paramyxea.

Life Cycle

Marteilia refringens has an unusual life cycle previously not observed in protozoans. Its sporulation involves a series of endogenous buddings that produce sporoplasms within sporoplasms. Mature spores have a complete wall and do not possess an operculum as do the Balanosporida. The various stages of *M. refringens* have been studied intensively by means of electron microscopy (Grizel et al. 1974; Perkins 1976). Only a brief summary of the complex picture can be given here. However, a recent review has been completed by Lauckner (1983).

Presporulation vegetative stages of the parasite consist of ovoid plasmodia found in the lumina of digestive gland tubules and the intestine, between the epithelial cells of the tubules and intestine, and in the connective tissue surrounding both structures. These small (5–8-μm) plasmodia, which are delimited by a plasmalemma with no cell wall or fibrillar layer, have been termed primary cells by Grizel et al. (1974).

Sporulation commences with the delimitation (endogenous budding) of uninucleate segments within the plasmodial cytoplasm, and these become presporangia or sporangial primordia (secondary cells), about 8 μm along the long axis. At this stage, the plasmodium corresponds to a sporangiosorus. As sporulation progresses, the sporangiosorus enlarges from about 15 to 30 μm. About eight sporangial primordia are formed within each sporangiosorus. A single nucleus (sometimes two) remains in the plasmodial cyto-

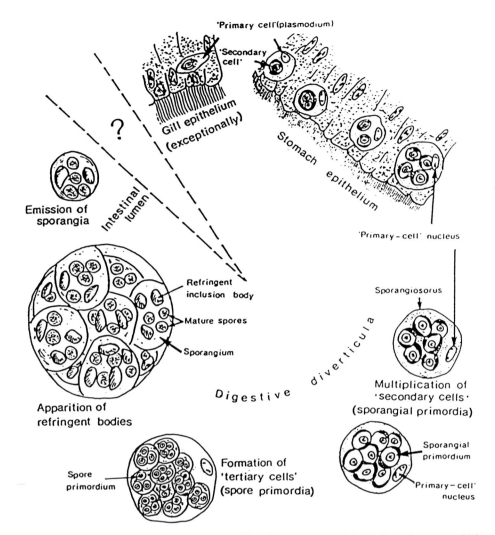

FIGURE 1.—Stages of *Marteilia refringens* observed in edible oysters, and the pathogen's presumed life cycle. (After Grizel et al. 1974.)

plasm (primary cell nucleus) and is not included in the sporangial primordia. A wall is then formed around each developing sporangium, and nuclear multiplication and cytoplasmic cleavage eventually result in the formation of three, or more often, four spore primordia (tertiary cells). Each spore primordium consists of three uninucleate sporoplasms of graded sizes. The smallest sporoplasm lies in an eccentrically located vacuole of the medium-sized sporoplasm, and the latter lies in an eccentrically situated vacuole of the largest sporoplasm. As spore maturity is attained, the sporangial cytoplasm not included within the spore walls degenerates, as does the primary cell nucleus. Polymorphic inclusion bodies, 2.3–4.0 μm along

the long axis, appear in the extraspore cytoplasm during spore maturation, eventually becoming a dominant feature of the sporangium. When viewed with light optics, these inclusion bodies are highly refringent; hence, the species name. Mature spores of *Marteilia refringens* are spherical to subspherical, with a diameter of 3.9 ± 0.08 μm (mean ± SE; $N = 22$ living cells; interference contrast optics). Nuclear diameters, as seen in electron micrographs, range from 1.02 to 1.52 μm (mean = 1.24, $N = 19$) (Perkins 1976).

From the occurrence of the various developmental stages in different tissues of edible oysters, Grizel et al. (1974) proposed a tentative life cycle for *M. refringens* (Figure 1). Primary infections presumably occur in epithelia of the gut, the gills, or both. Sporangia mature in the digestive diverticula and are discharged via the gut. How the edible oyster becomes infected and by what stage it becomes infected has not yet been determined. Experimental attempts to transmit the disease to healthy edible oysters in the laboratory have met with failure, although field experiments were successful.

Recent evidence suggests that the life cycle of *M. refringens* may be even more complex than outlined by Grizel et al. (1974). Franc (1980) maintained that the primary, secondary, and tertiary cells of Grizel et al. (1974) represent three independent categories of cells, which may be parts of several subcycles in the development of the organism. They observed schizogonic stages, as well as small amoebulae, 4–5.5 μm in size, which hatched from sporangia of a certain type. On the surface of the epithelium of mouth and esophagus, uninucleate cells, 9–13 μm in diameter, were seen in different stages of fusion. Paired cells were of different sizes. These and comparable stages, which are suggestive of a sexual phase, were also seen in the digestive gland. Furthermore, a new type of sporangium, believed to be derived from one of the three distinct cell categories, was seen in the visceral connective tissue of the digestive diverticula and the intestinal tract, that is, the sites harboring the hitherto known stages of the parasite. The presence of at least some of these stages appears to depend on the season. These very interesting observations have been made with light optics and should be confirmed by electron microscopy.

In samples examined bimonthly from Carantec in the Morlaix River during 1975, Balouet (1979) described an apparent sequence of parasite stages. During the winter period (December–February), less than 10% of specimens showed the presence of *M. refringens*. The stages present were mainly those with mature spores and refringent inclusion bodies and were located in small areas of the infected digestive gland. During the early spring, certain stages, referred to as plasmodial stages by Perkins (1976), became apparent in the stomach epithelia of animals with visible old infections and those without infections. Since the winter infections seem to consist of small, limited foci in one or two branches of the digestive gland, the problem of establishing levels of infection is great. Although the apparent level of infection may be only 10%, it is impractical to find every focus of infection because only limited numbers of sections can be made and examined from a single

animal. Therefore, many apparently new infections may simply represent further spread of a previously local infection.

The number of mature and maturing parasitic cells in the digestive epithelia also began to increase in the spring, and infection levels began to climb, eventually reaching 80% by July (Balouet 1979). Levels remained high until October, but there were relatively few stomach epithelial infections during the peak of the summer. An increased level of these infections was measured again in the autumn (September and October). After October, general levels of apparent infection began to fall, reaching 10% in January.

The occurrence of refringent inclusion bodies in the gut presumably indicates sporangial breakdown and spore release. Refringent bodies were present in the gut throughout most of the summer season. Occasionally, animals simultaneously released massive amounts of spores, and these were the only ones with really substantial tissue damage. Various extents of hemocytic reaction may be observed at different stages of the disease, but massive spore release is accompanied by breakdown of the tissues of edible oysters. Some physical initiator for such a mass spore release must exist. The tissue damage which accompanies this type of spore release contrasts sharply with the apparent lack of severe reaction to the presence of enormous numbers of mature parasites and may account for the catastrophic localized mortalities which characterize the progress of Aber disease.

Pathology

Aber disease is accompanied by severe pathological response. Infested edible oysters become progressively emaciated, and the digestive gland becomes brown to pale yellow in color. Upon depletion of its glycogen reserves, the mantle becomes translucent, and shell growth ceases. The visceral mass loses its pigmentation and, in heavily infected individuals, appears shrunken and slimy. First clinical signs become apparent in edible oysters in the autumn of the year of planting and intensify until death ensues in the autumn of the following year. These signs, however, are not specifically diagnostic, since they may also be exhibited by edible oysters not suffering from Aber disease. Moreover, *M. refringens* may also be found in apparently healthy individuals with good growth and normal gonads. The factors determining or influencing the virulence of *M. refringens* are not yet understood (Grizel et al. 1974; Grizel 1977; Cahour 1978).

Recordings of the valvular activity of edible oysters by means of an ostreograph revealed abnormal periods of shell closure by *M. refringens*-infected individuals (Grizel 1977). First signs of abnormal activity coincided with the appearance of parasite stages in the digestive gland. In the terminal stage of disease, the edible oysters were no longer capable of closing their valves (His et al. 1976; Grizel 1977).

The severe effect of *M. refringens* and its bearing on the edible oyster industry are best illustrated by comparison of the total wet weights of healthy and diseased individuals. Groups of 1,000 edible oysters each were weighed on March 26 and November 15, 1974. Healthy 18-month-old edible oysters weighed 16 and 31 kg, respectively, whereas diseased individuals of the same age weighed 11 and 23 kg, respectively. Two-year-old healthy oysters weighed 36 and 53 kg, respectively, whereas diseased ones weighed only 29 and 34 kg, respectively (Morel and Tigé 1974).

Initially, *M. refringens* was held responsible for causing gill disease in edible oysters, a syndrome similar to viral gill disease in *Crassostrea angulata*. With a single exception, *M. refringens* was never found again in the gills of numerous individuals of edible oysters examined over a period of several years. Gill disease of *O. edulis*, therefore, must have another etiology (Grizel et al. 1974), such as a virus (Comps 1978).

The related species of *Marteilia* described by Gingsburger-Vogel and Desportes (1979) in *Orchestia gammarella* causes a feminizing effect of its host. The apparent lack of a pathological effect explains why this form was not seen before.

Epizootiology

Mortalities from Aber disease usually commence in May, attain their maximum in June–August, and then gradually diminish, with decreasing losses persisting until December or January. In March and April, deaths are minimal. In edible oyster beds of southern Bretagne, *M. refringens* could not be found during this period, whereas in the colder waters of northern Bretagne, subclinical infestations persisted throughout the winter. From transplantation experiments it became evident that new infestations are acquired between May and August (Grizel and Tigé 1973, 1979). Long-term observations suggest that old mature sporangia containing refringent inclusions are present in May or June but are eliminated completely by the end of January. However, young plasmodia persist during the winter and reinitiate new clinical infections the

following May (Grizel 1976; Cahour 1978; Balouet et al. 1979).

Incidences of *M. refringens* infections ran as high as 100% in oysters from Aber Wrac'h and Aber Benoit, Bretagne. Similarly, high prevalences have been reported from other localities in France, whereas some areas have, thus far, remained unaffected (Bonami et al. 1971; Balouet et al. 1979). Initially, only edible oysters were found to be attacked; *C. angulata* and *C. gigas* were examined but were not infected. Therefore, the parasite was considered to be specific for *O. edulis* (Grizel and Tigé 1973; Marteil 1976). In May 1977, however, Balouet et al. (1979) detected young stages of *M. refringens* in individuals of *C. gigas* from Carantec near Roscoff (France). These findings were interpreted as a change in susceptibility of this oyster species.

Alderman (1979) described an interesting example of the epizootiological problems posed by Aber disease in Galicia, Spain (Figure 2A). As mentioned previously, edible oysters infected with *M. refringens* had been imported into Spain as early as 1969. There is no information as to exactly when the disease was introduced into Spanish waters, but, during the autumn of 1975, samples from the Ria de Muros showed high levels of *M. refringens* infection. Culture of *O. edulis* in Spain is limited to northwest Galicia, and bottom culture is practiced only at Ribadeo. Elsewhere, edible oysters are suspended either in trays or attached to ropes from rafts, such as those used in the same area for mussel culture. The two sites (Figure 2B, W and E) are operated by two separate companies. The company operating site W also has an edible oyster hatchery built to produce local spat, but both companies have imported large numbers of edible oysters from France. In autumn 1975, samples of edible oysters from five groups of animals, which had originally been imported from France, were examined from site W. These samples all showed levels of *M. refringens* infection between 80 and 100%.

A sixth sample of edible oysters, which consisted of hatchery spat produced in the winter of 1975 at the nearby hatchery, also showed 95% infection. Clearly, *M. refringens* was able to spread to local stocks at least within the limits of the site-W raft system. A seventh sample examined at the same time came from site E and showed no *M. refringens* infections. Samples taken from this site the following summer 1976, also showed no *M. refringens*.

There is, however, a complicating factor. Although site E showed no levels of Aber disease in

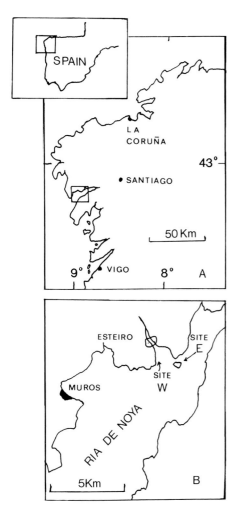

FIGURE 2.—Location map of study areas in northern Spain. (After Alderman 1979.)

the imported French edible oysters or in other stocks, mortalities still occurred, particularly in the French stocks. These edible oysters were originally imported into Ribadeo and remained for several months before being removed to the Muros rafts. After 4 months at Muros, mortalities were noted at around 10%/week and remained at this level throughout the summer. The edible oysters had been examined and found to be free of *M. refringens* before they left France, and they remained free of *M. refringens* during the period of examination at Muros. Although no indications of *M. refringens* were observed, some 30% of the animals examined showed evidence of a hematopoietic neoplasm (see Peters 1988, this volume).

The two Spanish sites at Muros are separated by 2 km of open water in a wholly marine estuary

with little input of fresh water and an adequate tidal exchange. Thus, it appears that *M. refringens* was able to become established in site W and infect local edible oysters, but it was unable to cross 2 km of open sea and infect edible oysters on site E rafts. Because the neoplasm, which has been implicated in substantial mortalities at site E, was also present at site W, the relative importance of *Marteilia* infection is uncertain in terms of site W mortalities. The inability of *Marteilia* to cross a 2 km gap in Spain is indicative of some of the problems that this disease presents.

The lack of published literature since 1982 on the epizootiology of this disease is surprising, if not striking. Could this lack be caused by a resistance developed by the host? Was *Marteilia* really the causative agent of the original mortalities first attributed to this parasite? Was *Bonamia* sp. present in these samples?

Alternate Hosts

During a search for possible alternate hosts expected to harbor unknown life-cycle stages of *M. refringens*, Comps et al. (1975) found different stages in 2% of blue mussels *Mytilus edulis* from Auray and Penzé, France, and in 10% of cockles *Cardium edule* from Auray.

Gutiérrez (1977a) observed a 3.3% infection in 92 *M. edulis* from Arosa on the northwest coast of Spain. Stages of the parasite were easily detected histologically by a modified staining technique (Gutiérrez 1977b). While the forms found in the blue mussels were indistinguishable from those in edible oysters, the cockles harbored, in addition to sporangia (tertiary cells), proteinaceous spherical masses, 0.3 μm in diameter. Corresponding forms found in edible oysters measured 3–4 μm. Although the latter may be identical with the amoebula described by Franc (1980), the importance of those found in the cockle remain obscure. Possibly, the parasite found in *C. edule* is a distinct species of *Marteilia*. Its effects on the cockle have not been studied, but there were many empty shells on the beds where the parasite occurred (Grizel 1977).

Viruslike particles, observed in the cockle parasite, appear to be identical with those previously reported by Bonami et al. (1971) and Grizel et al. (1974) in *M. refringens* in *O. edulis*. These particles probably represent haplosporosomes, unique cytoplasmic organelles of unknown importance, found exclusively in members of the Stellatosporea (Perkins 1976). Although Comps et al. (1975) and Gutiérrez (1977a) have shown that *M.*

refringens can exist in *M. edulis*, the observations of Tigé and Rabouin (1976) suggest that the blue mussel is merely an incidental host for the edible oyster pathogen. Apparently, the presence of *M. refringens* does not affect the health of blue mussels (Grizel 1977). Because *M. refringens* cannot be distinguished from *M. maurini* (another *Marteilia*-like parasite found in blue mussels) histologically or microscopically, it is difficult to interpret the occurrence of *M. refringens* in *Mytilus edulis*.

In *C. gigas*, the parasitic cells observed in the digestive system are similar to the primary cells of *M. refringens* observed in the early stages of infection of edible oysters (Perkins 1976). Localization in the apical part of the epithelial cells of the stomach is the same as in *M. refringens*. Cellular structure, cell size (5–15 μm), and the presence of two to four nuclei surrounded by a clear halo, are comparable. Cahour (1979) offered several suggestions concerning this *M. refringens*-like organism.

(1) The parasitic forms discovered in *C. gigas* may be considered as transient and nonpathogenic. There appears to be no ill effects at either the organismic or tissue level. Similar observations have been made on *M. refringens*-like parasites in *Cardium edule* and *Mytilus edulis* (Comps et al. 1975). In these cases, infection levels were low (4%); however, the parasites showed old or sporulating stages along with a wider distribution in the digestive tract.

(2) It may be assumed that the presence of these parasitic forms indicates infection of *C. gigas* by *M. refringens*. These infections may represent a de novo infection of *C. gigas,* because we have never observed advanced (sporulating) stages in the host. The cellular appearance of the *C. gigas* parasite is similar to that of the parasite of edible oysters, when the latter is placed in an infested area at the same time of year. In edible oysters, such cells indicate the onset of infection. The parasites are young or primary cells as described by Grizel et al. (1974) and are found in the cells of the stomach epithelium and its branches in as many as 45% of the edible oysters. At the beginning of an epizootic, only a low percentage of edible oysters contain *M. refringens*.

(3) The parasite of *C. gigas* may be similar to *M. sydneyi*, which is a parasite of the sidney rock oyster *C. commercialis*, because the latter parasite, described by Wolf (1972) and Perkins and Wolf (1976), is closely related to *M. refringens*. We are not, at present, in a position to critically evaluate such an association.

Four areas of study are indicated to determine the role and identity of the *C. gigas* parasite.

(1) Ultrastructural information is required concerning the substructure of the young cells, but this information is difficult to obtain because of the paucity of cells.

(2) We also need to determine if transmission of infections can occur from *C. gigas* to *O. edulis* or vice versa. Laboratory attempts were unsuccessful when infected oysters were held with uninfected ones in aquaria. Similarly, in natural waters, *C. gigas* was introduced into an *M. refringens*-endemic area, but after 3 months, no transmission of infection was observed.

(3) Further epizootiological data will be required concerning populations of *C. gigas* in which *M. refringens* has already been observed (Cahour 1979). It will be important to note whether the incidence of the parasite increases and whether pathogenicity is expressed. Whatever occurs will undoubtedly be linked to the resistance of *C. gigas*, the potential pathogenicity of the parasite, and the physical factors of the environment.

(4) The necessity for an intermediate host, which has been suspected for other haplosporidian parasites, has not been proved for *M. refringens*. The demonstration in *Orchestia gammarellus* (Ginsburger–Vogel and Desportes 1979) of cellular stages morphologically similar to *M. refringens* is an interesting lead, but transmission from this potential alternate host to edible oysters must be demonstrated.

Host–Parasite Interactions

The lack of consistent correlation between the degree of *M. refringens* infection and edible oyster mortality is puzzling. Edible oysters kept in areas of high prevalence for extended periods of time may show the characteristic signs of disease without yielding many parasites. Conversely, edible oysters heavily infected with young plasmodia and mature sporangia, may exhibit virtually no histological alterations. Therefore, the pathogenesis of *M. refringens* remains obscure.

The production of toxins by the parasite is one possible explanation. Also, the parasite may act synergistically with other, yet unidentified, pathogens. Environmental conditions that lead to stress may also play a prominent role (Cahour 1978; Ormiéres and Grizel 1979).

High mortalities have also been reported among *Ostrea edulis* imported into Spain from France. In 1974–1975, losses in some rías or estuaries on the northwest coast amounted to 50–70% and were most serious in areas where edible oysters and blue mussels were grown together. Microscopically, however, *M. refringens* was present in only a small fraction of the affected edible oysters. Massó (1978) concluded that poor adaptation of the seed oysters to the physical environment of the planting areas was the main reason for high mortalities. In the Netherlands, *M. refringens* introduced with seed oysters from Bretagne, failed to demonstrate any substantial virulence. Although the parasite remained alive and well in edible oysters of Zeeland, Netherlands, it did not cause abnormal mortalities among the imported lots (van Banning 1979a, 1979b). The differences in apparent virulence displayed by *M. refringens* in French and Dutch waters were unexplained. Much remains to be learned about the biology and epizootiology of this presumed pathogen of edible oysters.

Among physical factors, variations in temperature, salinity, and immersion depth seem of little importance in Bretagne. The overall consequences of these factors are only known when changes in them provoke disasters. In the future, we shall have to take them into account; perhaps Aber disease is not so much a microbial disease as one arising from unfavorable physicochemical factors (Balouet 1979).

Other *Marteilia* Species

Several species of *Marteilia* have been reported from several species of oysters. The first species, originally reported by Wolf (1972) as an unnamed haplosporidian from Sydney rock oysters *C. commercialis* (= *Saccostrea commercialis*) in Moreton Bay, Queensland (Australia), was later identified as a member of the genus *Marteilia* and named *M. sydneyi* (Perkins and Wolf 1976). Its developmental sequence, infection site, and size are similar to those of *M. refringens*, except that mature spores of *M. sydneyi* are surrounded by a heavy layer of concentric membranes, which are lacking in *M. refringens*. The pathogen has been found only in subtropical and tropical regions of the eastern Australian coast, not in the colder waters south of Richmond River. An organism, probably identical with *M. sydneyi*, occurs in *Crassostrea echinata* from the same area.

Marteilia sydneyi has been recovered alive from one moribund Sidney rock oyster and from four Sidney rock oysters that were in poor condition. The moribund specimen contained numerous spores, but fewer spores were found in the other specimens. High Sidney rock oyster mortalities

that occurred in Moreton Bay were believed to be caused by this parasite. Its incubation period was less than 60 d from early infestation to death of the host. *M. sydneyi* appears to prefer low salinities, at least for a part of its development. Higher salinity levels will prevent invasion or at least reduce the number of pathogens. As many as 80% of the Sidney rock oysters in southern Queensland and northern New South Wales were lost during severe epizootic outbreaks of this pathogen (Wolf 1972, 1979).

Marteilia lengehi parasitizes *Saccostrea cucullata* in the Persian Gulf. Plasmodia, 8–15 μm in diameter, and presporulation stages were found in the stomach epithelium and the digestive gland. Mature spores were not seen (Comps, 1976).

Marteilia maurini was found in *Mytilus galloprovincialis* imported into France from a lagoon in Venice, Italy. Its morphology and developmental sequence closely resemble those of *M. refringens* (Comps and Joly 1980; Comps et al. 1982).

Auffret and Poder (1985) conducted a histopathological study of wild bivalve populations in three culture areas of northern Bretagne. The parasite *M. maurini*, identified by electron microscopy, was observed for the first time in this geographical area and in a new host species, *Mytilus edulis*. It was occasionally present in samples from Aber Wrac'h, whereas high infestation rates (37–70%) were regularly noted in Paimpol. The parasitic cells, their distribution, and the absence of hemocytic response resemble those of the other parasites of the genus *Marteilia*. The ultrastructural observations made by these authors agree with Comps et al. (1975), but they could not find the elongated haplosporosomes, and the spherical haplosporosomes were larger than those described by Comps et al. (1975).

The last species of *Marteilia* is *M. christenseni* (Comps 1983), a parasite of the pelecypod *Scrobicularia piperata*. The morphological characteristics of the tricellular spore and the differentiation of the cytoplasmic contents of the sporants distinguish this parasite from other species of the genus *Marteilia*.

In relation to the structural similarities and parallel distributions of *M. refringens* and *Marteilia* sp. described by Ginsburger-Vogel and Desportes (1979) in *Orchestia gammarellus*, their association must be considered at different levels. Are they different species or are they two different forms of the same species? Do these species belong in the *Haplosporida*, as stated by Perkins (1976)? If not, to which group of protists do they belong? At present, answers to these questions are not definitive. Observed similarities between *M. refringens* and the species found in *O. gammarellus*, particularly as to the mode of budding by internal cleavage, suggest that these forms are very similar and may belong to the same genus. The differing cytological characters (i.e., presence of striated inclusions in the cytoplasm of primary cells of only *M. refringens*, and the occurrence of dense bodies in the spore primordia of only *Marteilia* sp.) seem insufficient to definitively conclude that they are different species. Different structures could be explained by different responses of the parasite in different hosts. The absence of centrioles in *M. refringens*, if confirmed, would present a strong argument favoring a separation into different species.

Marteilia spp. have been isolated from *Ostrea edulis* and *Mytilus edulis* and purified by density-gradient centrifugation (Mialhe et al. 1985). This accomplishment will allow the parasites to be characterized ultrastructurally, biochemically, and serologically. These steps are indispensable if we are to establish taxonomic affinities and to conduct enzyme-linked immunosorbent and immunofluorescent assays for these parasites. Such isolations also will facilitate experimental infections.

References

Alderman, D. J. 1979. Epizootiology of *Marteilia refringens* in Europe. U.S. National Marine Fisheries Service Marine Fisheries Review 47(1–2):67–69.

Auffret, M., and M. Poder. 1985. Recherches sur *Marteilia maurini*, parasite de *Mytilus edulis* sur les cotes de Bretagne Nord. Revue des Travaux de l'Institut des Pêches Maritimes 47(1–2):105–109.

Balouet, G. 1979. *Marteilia refringens*: Considerations of the life cycle and development of Aber disease in *Ostrea edulis*. U.S. National Marine Fisheries Service Marine Fisheries Review 47(1–2):64–66.

Balouet, G., A. Cahour, and C. Chastel. 1979. Épidémiologie de la maladie de la glande digestive de l'huître plate: hypothèse sur le cycle de *Marteilia refringens*. Haliotis 8:323–326.

Bonami, J. R., H. Grizel, C. Vago, and J. L. Duthoit. 1971. Recherches sur une maladie épizootique de l'huître plate, *Ostrea edulis* Linné. Revue des Travaux de l'Institut des Pêches Maritimes 35(4): 415–418.

Cahour, A. 1978. Épidémiologie de la maladie de l'huître plate, *Ostrea edulis* (L.). Thèse, 3eme cycle. Université de Bretagne Occidentale, Brest, France.

Cahour, A. 1979. *Marteilia refringens* and *Crassostrea gigas*. U.S. National Marine Fisheries Service Marine Fisheries Review 47(1–2):19–20.

Comps, M. 1970. Observations sur les causes d'une mortalité anormale des huîtres plates dans le bassin de Marennes. Revue des Travaux de l'Institut des Pêches Maritimes 34(3):317–326.

Comps, M. 1976. *Marteilia lengehi* n. sp., parasite de l'huître *Crassostrea cucullata* Born. Revue des Travaux de l'Institut des Pêches Maritimes 40(2): 347–349.

Comps, M. 1978. Evolution des recherches et études recentes en pathologie des huîtres. Oceanologica Acta 1:255–262.

Comps, M. 1983. Étude morphologique de *Marteilia christenseni* sp. n. parasite du lavignon *Scrobicularia piperata* P. (Mollusque Pellecypode). Revue des Travaux de l'Institut des Pêches Maritimes 47 (1–2):99–104.

Comps, M., H. Grizel, G. Tige, and J. L. Duthoit. 1975. Parasites nouveaux de la glande digestive des mollusques marins *Mytilus edulis* L. et *Cardium edule* L. Comptes Rendus Hebdomadaires des Séances de l'Academie des Sciences, Serie D, Sciences Naturelles 281:179–181.

Comps, M., and J. P. Joly. 1980. Contamination expérimentale de *Mytilus galloprovincialis* Lmk par *Marteilia refringens*. Science et Pêche 301:19–21.

Comps, M., Y. Pichot, and P. Papagianni. 1982. Recherche sur *Marteilia maurini* n. sp. parasite de la moule *Mytilus galloprovincialis* Lmk. Revue des Travaux de l'Institut des Pêches Maritimes 45:211–214.

Desportes, I. 1984. The Paramyxea Levine 1979: an original example of evolution towards multicellularity. Origins of Life 13:343–352.

Desportes, I. and T. Ginsburger–Vogel. 1977. Affinités du genre *Marteilia*, parasite d'huîtres (maladie des Abers) et du crustacé *Orchestia gammarellus* (Pallas), avec les myxosporidies actinomyxidies et paramyxidies. Comptes Rendus Hebdomadaire des séances de l'Academie des Sciences, Serie D, Sciences Naturelles 285:1111–1114.

Franc, A. 1980. Sur quelques aspects inedits du cycle de *Marteilia refringens* Grizel et coll. 1974, parasite de l'huître plate *Ostrea edulis* L. Cahiers de Biologie Marine 21:99–106.

Ginsburger-Vogel, T., and I. Desportes. 1979. Structure and biology of *Marteilia* sp. in the amphipod, *Orchestia gammarellus*. U.S. National Marine Fisheries Service Marine Fisheries Review 47(1–2):3–7.

Grizel, H. 1976. Développement et cycle du parasite responsable de l'épizootie de l'huître plate *Ostrea edulis* L. Haliotis 5:61–67.

Grizel, H. 1977. Parásitologie des mollusques. Oceanis 3:190–207.

Grizel, H., M. Comps, J. R. Bonami, F. Cousserans, J. L. Duthoit, and M. A. Le Pennec. 1974. Recherche sur l'agent de la maladie de la glande digestive de *Ostrea edulis* Linné. Bulletin de l'Institut des Pêches Maritimes du Maroc 240:7–30.

Grizel, H., and G. Tigé. 1973. La maladie de la glande digestive d'*Ostrea edulis* Linné. International Council for the Exploration of the Sea, C.M. 1973/ K:13, Copenhagen.

Grizel, H., and G. Tigé. 1979. Observations sur le cycle de *Marteilia refringens*. Haliotis 8:327–330.

Gutiérrez, M. 1977a. Nota sobre la marteiliasis en el mejillón *Mytilus edulis* (L.) de la costa noroeste de España. Investigación Pesquera 41:637–642.

Gutiérrez, M. 1977b. Técnica de colaboración del agente de la enfermedad de la glándula digestiva de la ostra plana. *Ostrea edulis* L. Investigación Pesquera 41:643–645.

Herrbach, B. 1971. Sur une affection parasitaire de la glande digestive de l'huître plate, *Ostrea edulis* Linné. Revue des Travaux de l'Institut des Pêches Maritimes 35:79–87.

His, E., G. Tigé, and M. A. Rabouin. 1976. Observations relatives á la maladie des huîtres plates dans le bassin d'Arcachon, vitesse d'infestation et réactions pathologiques. International Council for the Exploration of the Sea, C.M. 1976/K:17, Copenhagen.

Lauckner, G. 1983. Diseases of Mollusca: Bivalvia. Pages 553–558 *in* O. Kinne, editor. Diseases of marine animals, volume 2. Biologische Anstalt Helgoland, Hamburg, West Germany.

Marteil, L. 1976. La conchyliculture francaise. 2 partie. Biologie de l'huître et de la moule. Revue des Travaux de l'Institut des Pêches Maritimes 40: 149–345.

Massó, J. M. 1978. La enfermedad de la glándula digestiva de la ostra (*Ostrea edulis* L.) en las rias bajas. Boletín del Instituto Español de Oceanografía 4: 125–140.

Mialhe, E., E. Bachère, C. Le Bec, and H. Grizel. 1985. Isolement et purification de *Marteilia* (Protozoa: Ascetospora) parasites de bivalves marins. Comptes Rendus Hebdomadaires des Séances de l'Académie des Sciences, Serie III, Sciences de la Vie 301:137–142.

Morel, M., and G. Tigé. 1974. Maladie de la glande digestive de l'huître plate. Bulletin de l'Institut des Pêches Maritimes du Maroc 241:33–40.

Ormiéres, R., and H. Grizel. 1979. Les haplosporidies parasites de mollusques. Haliotis 8:49–56.

Perkins, F. O. 1976. Ultrastructure of sporulation in the European flat oyster pathogen, *Marteilia refringens*—taxonomic implications. Journal of Protozoology 23:64–74.

Perkins, F. O., and P. H. Wolf. 1976. Fine structure of *Marteilia sydneyi* sp. n.—haplosporidan pathogen of Australian oysters. Journal of Parasitology 62: 528–538.

Peters, E. C. 1988. Recent investigations on the disseminated sarcomas of marine bivalve molluscs. American Fisheries Society Special Publication 18:74–92.

Sprague, V. 1979. Classification of the Haplosporidia. U.S. National Marine Fisheries Service Marine Fisheries Review 41(1–2):40–44.

Tigé, G., and M. A. Raboiun. 1976. Étude d'un lot de moules transferées dans un centre touché par l'épizootie affectant l'huître plate. International Council for the Exploration of the Sea, C.M. 1976/K:21, Copenhagen.

van Banning, P. 1979a. Haplosporidian diseases of imported oysters, *Ostrea edulis*, in Dutch estuaries. U.S. National Marine Fisheries Service Marine Fisheries Review 41(1–2):8–18.

van Banning, P. 1979b. Protistan parasites observed in the European flat oyster (*Ostrea edulis*) and the cockle (*Cerastoderma edule*) from some coastal areas of the Netherlands. Haliotis 8:33–37.

Wolf, P. H. 1972. Occurrence of a haplosporidan in Sydney rock oysters (*Crassostrea commercialis*) from Moreton Bay, Queensland, Australia. Journal of Invertebrate Pathology 19:416–417.

Wolf, P. H. 1979. Diseases and parasites in Australian commercial shellfish. Haliotis 8:75–83.

American Fisheries Society Special Publication 18:47–63, 1988

Epizootiology of the Disease Caused by the Oyster Pathogen *Perkinsus marinus* and Its Effects on the Oyster Industry

J. D. ANDREWS

Virginia Institute of Marine Science, College of William and Mary
Gloucester Point, Virginia 23062, USA

Abstract.—*Perkinsus marinus* is a protozoan parasite that causes a major disease of eastern oysters *Crassostrea virginica* from Chesapeake Bay south along the Atlantic coast of the USA and throughout the Gulf of Mexico. It is a warm-season disease that kills eastern oysters at temperatures above 20°C. The pathogen requires salinities of at least 12–15‰ to be active, but it persists tenaciously when low temperatures and salinities occur during winter and spring. Prolonged droughts that increase salinities cause extensions of the range of disease. In the Chesapeake Bay, mortalities begin in June and end in October, and up to 50% of native susceptible eastern oysters are killed each year. Most infections are acquired by eastern oysters in proximity to disintegrating dead eastern oysters. Massive populations of prezoosporangia are released into marine waters and eventually produce thousands of zoospores, which are infective when ingested by eastern oysters. The disease is controlled by isolating new beds from infected eastern oysters and by early harvest before the pathogen becomes established. Seed areas in low-salinity waters usually provide disease-free eastern oysters, but beds must be monitored regularly to avoid transplanting infected eastern oysters. In the Gulf of Mexico where warm temperatures persist through much of the year, control is much more difficult. Eradication is difficult unless introductions of infected eastern oysters are avoided and summer warm seasons are short and relatively cool to prevent the pathogen from multiplying in eastern oysters. The presence of *P. marinus* in Chesapeake Bay has been monitored for 37 years; the disease is established in most eastern oyster-growing areas of Virginia and in many tributaries of Maryland. The disease thrives on densely planted private beds of eastern oysters but persists through wet periods of weather on sparsely populated public beds and artificial structures along shores where there is recruitment of eastern oysters. Unfavorable temperatures essentially eradicated it in Delaware Bay after importations of infected seed eastern oysters from Virginia were discontinued.

Extensive losses of eastern oysters *Crassostrea virginica* due to disease occur in the Gulf of Mexico and Chesapeake Bay, USA (Andrews and Hewatt 1957; Mackin 1962). Extensive mortality and prevalence data from eastern oyster beds and intensively monitored trays of eastern oysters substantiate these losses. Management methods for control and avoidance of disease caused by the protozoan *Perkinsus marinus* were developed long ago (Andrews 1966, 1979), but oystermen and regulatory agencies are not applying these practices. Harvesting and transplanting policies are restated in this paper together with recommendations for laborious, but necessary, seasonal testing of seed and market eastern oysters of individual beds. Accurate histories of beds are required for proper analysis of management techniques. A summary of the epizootiology of the disease was given by Andrews (1976a).

Perkinsus marinus was the dominant pathogen of eastern oysters in Chesapeake Bay during the 1950s. In 1959, a highly pathogenic parasite, *Haplosporidium nelsoni* (MSX) (Haskin et al. 1965), was apparently introduced into Chesapeake Bay (Andrews and Wood 1967). It spread rapidly in 1960 and caused high mortalities that stopped all eastern oyster culture where salinities exceeded 20‰ in late summer (Andrews and Frierman 1974). The rapid decline of eastern oyster populations throughout lower Chesapeake Bay caused *P. marinus* to decline in activity in high-salinity areas where eastern oysters were no longer planted. To show the interactions between the two diseases, new prevalence and mortality data are presented for periods before and after the new disease became enzootic. The contrasts in parasitic activities and methods of control of these two eastern oyster pathogens provided many interesting epizootiological studies (Mackin 1962; Andrews 1965, 1967). The consequences for the eastern oyster industries are discussed; 27 years after the invasion of MSX in Chesapeake Bay, beds in high-salinity areas have not been replanted, and private culture continues to decline, partly from fear of these two diseases. In years when MSX abates, *P. marinus* increases in importance.

The stages of *P. marinus* are well known (Mackin and Boswell 1956; Perkins 1976, 1988, this volume), but their functions in the life cycle are unclear. Zoospores develop in sporangia outside of eastern oysters, yet they have not been seen in stained tissues of the hosts. They may be the major stage for infection in nature particularly from distant sources. Confusion persists in the names of stages seen in eastern oysters and during thioglycollate tests of tissues for diagnostic purposes. Now that the taxonomic affinity of *P. marinus* is firmly established in the phylum Apicomplexa (Perkins 1988), clarification of the names and functions of the life cycle stages may be expected soon.

History of Disease Research

The disease caused by *P. marinus* was discovered in the Gulf of Mexico in 1948 (Mackin et al. 1950). For 30 years it was called *Dermocystidium marinum* in the belief that it had fungal affinities. The signet-ring stage seen most commonly in eastern oysters, called trophozoites or prezoosporangia (aplanospores of Mackin 1962 and Lauckner 1983), resemble funguslike protistans parasitic on freshwater vertebrates. These vegetative cells have a large vacuole with a dense inclusion and a peripheral nucleus, and they multiply in eastern oysters by successive division into multinucleated clusters of trophozoites. The discovery by electron microscopy of the production of biflagellated zoospores with organelles called apical complexes (Perkins and Menzel 1966; Perkins 1976) led to the recognition of the true affinity of *P. marinus* to Apicomplexa (Levine 1978). A recent review of the epizootiology of this warm-season pathogen and its new classification was given by Lauckner (1983).

The search for the causes of eastern oyster mortalities in the Gulf of Mexico began in 1948 after oystermen filed damage suits for millions of dollars against oil companies (Mackin et al. 1950; Mackin and Sparks 1962). Four biologists, J. G. Mackin, S. H. Hopkins, R. W. Menzel, and H. M. Owen, who worked at the Virginia Fisheries Laboratory, Gloucester Point, during World War II, organized extensive literature and field researches in Louisiana. The Texas A&M Research Foundation, College Station, which was supported financially by some oil companies, published many reports on purported disease agents and on literature searches (see Mackin 1962 for list). Mackin and his associates came to Virginia during the summer of 1949 and found *P. marinus* in eastern oysters growing in environments away from oil fields. Consequently, another associate, W. G. Hewatt of Texas Christian University, Fort Worth, came to Virginia during the summer of 1950 to initiate eastern oyster mortality studies by the tray method (Hewatt and Andrews 1954). The first trays were established in June 1950 at the Virginia Institute of Marine Science (VIMS) pier, and *P. marinus* has been monitored continuously since 1950 (Andrews 1980). Many thousands of eastern oyster tissues were sectioned or tested for diseases both in the Gulf of Mexico (Mackin and Sparks 1962) and in Virginia, and these regions became centers for disease studies. Stained tissue sections of gapers (dying eastern oysters) and live animals amounted to 170,000 at VIMS by 1983 when I retired.

The researchers in the Gulf of Mexico encountered two problems in their studies of *P. marinus* (Mackin 1953). First, the pathogen was so widespread geographically in the Gulf of Mexico that disease-free eastern oysters had to be obtained from New England for their experiments. In Virginia, large, annual freshwater discharges in rivers with large drainage areas, such as James River, provided disease-free oysters from low-salinity areas around the year. Also, the diagnosis of disease was difficult, and few eastern oysters had been sectioned anywhere in the U.S. because the process was expensive and tedious. Furthermore, the usefulness of stained tissues was limited to advanced infections because only they could be diagnosed accurately for incidence and intensity.

In 1952, S. M. Ray attempted to culture *P. marinus* at Rice Institute, Houston, Texas, and discovered the thioglycollate test, which is a sensitive method for diagnosis and for rating the intensity of infection (Ray 1952, 1954a, 1966). Use of this growth medium causes prezoosporangia to enlarge and develop walls that stain blue-black when Lugol's iodine solution is added. Mackin (1962) devised a rating system that combines prevalence and intensity of infections into a single expression, weighted incidence (WI). Values of one, three, and five are assigned to light, moderate, and heavy infections, respectively (Ray 1954a), and the total divided by the number of eastern oysters gives an average. A value of 0.5 signifies mostly light infections with little or no mortality; 1.0 means generally light infections, but some cases will be severe, and some mortality may occur; a value of 1.5 or higher for live hosts indicates that most eastern oysters are infected, and those with severe cases die. Deaths tend to keep WI from rising much higher than 1.5 in live eastern oysters, but gapers typically exhibit values around 4.0 when 80–90% of eastern oyster deaths are caused by *P. marinus*. These values are useful for predicting the extent of mortalities, when they are combined with the date of sampling and the prevailing water temperatures. The larger the number sampled, the more accurate WI becomes; samples of 25 live eastern oysters were routinely taken at VIMS. Gapers are difficult to collect on eastern oyster beds in summer because crabs, worms, and small fishes regularly remove meats in about 1 d after eastern oysters gape. A much higher proportion of gapers can be obtained during winter, when scavengers are dormant, or by searching trays daily as I did during warm seasons of the 1950s at VIMS (Andrews 1980).

Infection experiments began in the Gulf of Mexico and in Virginia as soon as Ray's (1952) thioglycollate test was available. The maceration of infected gapers in a blender provided infective inoculant that was injected into the mantle cavity through holes bored in the shells of eastern oysters or that was fed to eastern oysters in closed aquaria (Ray 1954a; Andrews and Hewatt 1957). Numerous infection experiments conducted during the 1950s were summarized by Ray (1954a), Ray and Chandler (1955), and Mackin (1962).

Distribution of *Perkinsus marinus*

P. marinus ranges from Tampico Bay, Mexico, along the southeastern coast of the U.S. to Delaware Bay (Mackin 1962; Andrews 1976b, 1979). Delaware Bay is now believed to be free of the disease it causes owing to an embargo placed on imports of eastern oysters from more southern areas. In the Gulf of Mexico, the distribution of the disease includes all coastal areas (Quick and Mackin 1971; Andrews and Ray 1988, this volume). The gulf coast is subject to wide annual and seasonal variations in rainfall with corresponding fluctuations in salinities, but few areas retain salinities low enough to preclude *P. marinus* from persisting during wet periods and thriving during dry ones (Mackin and Hopkins 1962; Andrews and Ray 1988).

Northward along the Atlantic coast, most estuaries or bays where eastern oysters are grown have high salinities until Chesapeake Bay is reached. *Perkinsus marinus* is prevalent in most areas including intertidal beds. South Carolina eastern oysters exhibited considerable resistance to the disease in Chesapeake Bay (Andrews and McHugh 1956). However, the seaside bays of the eastern shore of Virginia are free of *P. marinus* for unexplained reasons. This absence of *P. marinus* is usually true of Chincoteague Bay, which is mostly in Maryland (Andrews 1980).

In Chesapeake Bay (Figure 1), the pathogen is firmly established in the lower bay, all rivers in Virginia, Pocomoke Sound, and northward into Maryland as far as the St. Marys River, Choptank River, Patuxent River, and Eastern Bay (Andrews and Hewatt 1957; Andrews 1980; S. Otto, Maryland Department of Fisheries, unpublished data). The upper parts of major rivers, including the James, Rappahannock, and Potomac rivers, are free of the disease because large discharges of fresh water during winter and spring reduce the summer period of favorable salinities for *P. marinus*. In contrast, the smaller rivers with only coastal plain drainage areas are dominated by bay salinities which are often favorable to the pathogen. These small estuaries include the Great Wicomico, St. Marys River on the Potomac River, Choptank River, and all small tributaries in lower Maryland draining from eastern shore land. *Perkinsus marinus* activity fluctuates up and down with dry and wet years in these waters that have marginal salinities for the maintenance of the disease (Andrews 1980). There was a major extension of the disease throughout Chesapeake Bay during the prolonged drought of 1985–1987, including all seed areas and much of the James River seed area. Flushing and dilution of infective particles tend to limit infections to local beds where recruitment of eastern oysters is regular; therefore, large flushing-type rivers are not as favorable to the pathogen as small, shallow coastal plains tributaries. However, this normal slow pattern of spread changed during the recent droughts to one of rapid spread from bed to bed and area to area as the disease invaded more densely populated seed areas, probably because of an increase in abundance of infective stages. In Delaware Bay, the disease disappeared from commercial beds a few years after importations of infected eastern oysters from Virginia ceased.

During the 1980s, three winter–spring droughts (1981, 1985, and 1986) in the Chesapeake Bay drainage area were followed by dry, warm summers that caused exceptionally high mortalities attributed mostly to *P. marinus*. The successive dry years of 1985–1987 caused severe losses mostly attributable to *P. marinus* throughout Virginia, and few surviving eastern oysters were left on public or private beds for harvest or broodstock. Since the advent in 1959 of the disease caused by MSX in Chesapeake Bay (Andrews and

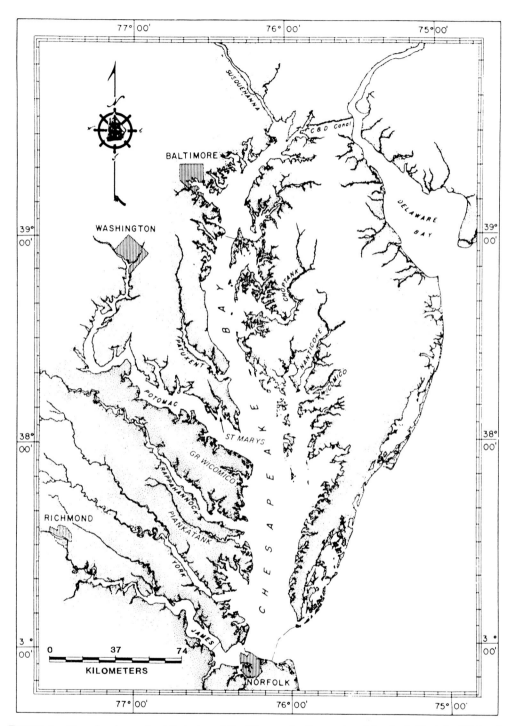

FIGURE 1.—Map showing major tributaries of the Chesapeake Bay region, Maryland–Virginia, USA, where *Perkinsus marinus* is enzootic. The upper ends of the large rivers of the bay are still free of the disease despite a great increase in abundance and a wider distribution of the parasite during the two drought periods of the 1980s.

Wood 1967), eastern oysters are grown only in areas where summer salinities do not exceed 20‰. However, even these low-salinity areas (<10‰ in spring, which allows eastern oysters to expel MSX) are not safe in Chesapeake Bay from either disease during drought periods. Mortalities have been so severe in Virginia that during the 1986–1987 and 1987–1988 eastern oyster harvests, Virginia tongers worked in the James River seed area catching small eastern oysters (sold as seed in 1985–1986 for $3/bushel) for market shucking at $12–$22/bushel (1 bushel = 35.2 L). Unfortunately, *Perkinsus*-infected eastern oysters from the lower section of the seed area were transplanted to private beds in the Machodoc and the Yeocomico rivers (Potomac tributaries usually of low salinity). This transplantation resulted in heavy mortalities from *Perkinsus* disease during summer and fall of 1986 and 1987 (E. M. Burreson, VIMS, personal communication). Several wet or normal years of fallowing these beds may eradicate the pathogen. Transplanting infected seed eastern oysters is the most rapid and frequent method by which *Perkinsus* disease is spread.

Host Species of *Perkinsus marinus*

Besides *Crassostrea virginica*, *P. marinus* infects *Dendostrea frons* and *Ostreola equestris*, which live offshore in the Gulf of Mexico (Ray and Chandler 1955). Uzmann (U.S. Fish and Wildlife Service) at Milford, Connecticut, transmitted it to *Ostreola conchaphila* (Ray 1954a). Probably, *Perkinsus* will infect almost any species of oyster because dosage is so intensive, but the degree of pathogenicity remains to be determined. The persistent effort to import *Crassostrea gigas* to New England and Canada suggests that testing of susceptibility of that species and of *Ostrea edulis*, grown in Maine, should be conducted.

Perkinsus marinus, as known in southeastern USA, provides no threat to eastern oyster culture on temperate coasts with oceanic-type climates, such as western Europe and western North America. It requires temperatures of 20°C or higher to multiply in eastern oysters; below this temperature eastern oysters expel the pathogen. Eastern oysters infected with *P. marinus* were introduced into Hawaiian waters and caused mortalities (Kern et al. 1973). *Perkinsus marinus* was also reported from Adriatic waters in the Mediterranean area (Da Ros and Canzonier 1985).

The occurrence of *Perkinsus*-like cells in a large variety of other bivalve molluscs, some scavengers such as mud crabs (xanthids), and nereid worms has caused much speculation about new species or alternate hosts (Ray 1954b). Andrews (1955) found these cells in 12 of 16 bivalves tested from the York River. Often, all specimens in samples of 25 bivalves had a few iodine-stained cells, and these cells persisted in some species for long periods without evidence of pathogenicity or seasonality. Most other bivalves were tested in 1954, a year of intensive culture of eastern oysters in the York River and of severe *Perkinsus* mortalities. The large quantity of prezoosporangia released by eastern oyster gapers plus the estimated 1,000–2,000 zoospores produced per large sporangium (Perkins 1976) make it likely that scavengers and filter-feeding organisms would encounter *P. marinus* stages. Yet these organisms do not appear to be suitable hosts or reservoirs (Ray 1954a). Because the *Perkinsus*-like parasite in *Macoma balthica* was abundant and multiplying (Valiulis and Mackin 1969), it was described as a new species (Mackin and Ray 1966).

The series of labyrinthulid species described by Mackin and Ray (1966) were probably contaminants and not *P. marinus*. There may be races of *P. marinus* as well as races of eastern oysters along the Atlantic coast, but they have not been satisfactorily defined. Growth patterns and resistance to diseases are measurable genetic traits of eastern oysters that have been observed to vary from one region to another (Andrews 1968; Haskin and Ford 1979). Other interactions between hosts and pathogens can be expected to vary accordingly.

Epizootiology of the Disease Caused by *Perkinsus marinus*

Life Cycle and Regulatory Factors

The two most important environmental factors regulating the life cycle of *P. marinus* are temperature and salinity (Mackin and Boswell 1956). These factors have been extensively studied in the Gulf of Mexico and in Virginia (Hewatt and Andrews 1956; Mackin 1956; Andrews and Hewatt 1957; Mackin and Hopkins 1962). This pathogen causes a warm-season disease, although obscure infections (few cases or of low intensity) persist through winter at water temperatures of 0–5°C. It also survives winter salinities less than 5‰ although 12–15‰ are required for multiplication in eastern oysters (Andrews and Hewatt 1957). Seasonal and annual variations of temperature and salinity between the Gulf of Mexico and Chesapeake Bay cause important differences in the respective seasonal cycles of the disease (Mackin 1956; Andrews 1980).

In Virginia, *P. marinus* begins its annual cycle when overwintering infections, not easily disclosed by thioglycollate tests, begin multiplying actively in June at temperatures above 20°C. These overwintering infections are difficult to monitor, and much more study of them is needed. From January through May, thioglycollate tests show only rare cases of infection in eastern oysters which had up to 100% infection the previous October. An exception occurred in 1987 when 80–90% of eastern oysters carried infections through the winter and began development in June (Burreson, personal communication). If such eastern oysters are placed in heated aquaria (30°C) in spring, a few infections develop, and these eastern oysters begin dying in about 1 month. Because live eastern oysters may discharge infective stages and gapers may deteriorate before their removal, satisfactory testing for overwintering cases requires isolation of each eastern oyster in warm water. Most eastern oysters with severe infections do not survive winter conditions, and most other infections are rare or nonclinical by late December when winter temperatures occur. There is no reason to expect undetected stages, but how the pathogen overwinters and how abundant it is are unknown. Advanced infections typically persist during early winter when most light cases have disappeared. Eastern oysters with advanced cases fail to recover from the disease in late fall, and most appear to die during the winter stress period. The rare survivors of advanced cases may be carriers of overwintering infections. Overwintering of the disease requires more study. The few eastern oysters with overwintering cases develop severe infections by late July or early August, and their deaths initiate a second generation of disease which is fatal in late August or early September. Often, all acclimated eastern oysters (having spent 1 or 2 years in an endemic area) in crowded trays have *Perkinsus* infections by 1 November of the second or third year. When the critical water temperature decreases to 20°C, deaths cease. The physiological balance between host and pathogen is altered in favor of the host, and most light infections are eliminated by mid-December. The persistence of infections through winter and the extent of survival are not clear.

The weather in September and October is a critical factor that determines the mortality level during late summer and fall. If water temperatures remain high, more infections develop into lethal ones. The pathogen multiplies fastest at temperatures between 25 and 30°C which persist for 3 months or longer during Virginia summers and much longer in the Gulf of Mexico.

In Virginia, eastern oyster planters typically obtain disease-free oysters from low-salinity seed areas which are transplanted in fall and winter. If growing beds are cleaned of old eastern oysters or fallowed 1–2 years to allow them to die, local infection sources are eliminated, and few if any infections are acquired from distant sources the first summer of culture. Under these conditions, mortality is low. If infections are acquired from local foci during the first summer in endemic areas, they cause deaths in late August or September; therefore, second-generation infections are too late to become serious before temperatures decline. If overwintering infections occur from the first summer of exposure, they become clinical the following June, and first deaths occur in July or early August during the second year. Prevalences and intensities increase rapidly with each additional generation; therefore, mortalities are highest in September and October.

In the Gulf of Mexico, winter temperatures often remain high enough for most infections to persist as clinical cases throughout the winter, although intensities usually decline (Mackin 1953). Infection cycles begin earlier and continue longer with higher mortalities there than in Chesapeake Bay. Planters there have learned to obtain large seed eastern oysters which are planted after summer temperatures decline and harvested before the next summer when pathogen activity exacts a heavy toll.

Salinity is almost as critical a factor as temperature in Chesapeake Bay; *P. marinus* tolerates quite low salinities during the cold season, although in normal years, most eastern oysters overcome patent infections in late fall. A large sector of the lower bay and adjacent tributaries provide adequate salinities (above 12–15‰) throughout warm seasons during all years. Nearly all areas where eastern oysters are grown in Virginia and many tributaries in lower Maryland provide adequate salinities during late summer; if they remain below 15‰ during early summer, few deaths occur, and disintegrating gapers do not materialize to provide high infective dosages. The cycles of infection and mortality are delayed, and late infections do not achieve lethal levels because fall temperatures stop multiplication of the pathogen. In the Gulf of Mexico, even wider fluctuations of salinity that occur seasonally and annually cause critical conditions for control and management of the disease.

Transmission of Perkinsus marinus

In the Gulf of Mexico, many infection experiments were conducted in aquaria and in trays in open waters, as were reviewed by Mackin (1962). For the most part, experiments in Virginia confirmed results obtained in the Gulf of Mexico (Andrews and Hewatt 1957). The proximity studies of Ray (1954a) and Andrews (1965, 1967) were attempts to assess dosage and timing of infections. Mackin (1962) found that a dose of 500 prezoosporangia was required to produce substantial mortality from *P. marinus* infections. Increasing the dosage to 5×10^5 decreased the lag time before infected eastern oysters died. Pathogen cells were cultured in thioglycollate medium for 24 h or more before injection to enlarge them for staining and easy counting and to stop reproduction. M. H. Roberts (VIMS, personal communication) found that 1×10^5 zoospores were required to induce infections. These dosages may be comparable if each sporangium produces 1,000–2,000 zoospores (Perkins 1976). This conclusion does not imply that continuous smaller dosages may not induce infections over a period of time. With high dosages, serious infections are produced in 3 or 4 weeks (Roberts, personal communication). In vitro production of zoospores from prezoosporangia enlarged in thioglycollate culture occurs regularly in seawater outside of eastern oysters. After 1 or 2 d in thioglycollate culture, zoosporangia take 4–5 d to produce zoospores at warm temperatures; the process is temperature dependent (F.-L. E. Chu, VIMS, personal communication). What happens inside live eastern oysters when they are injected with macerated tissues from infected gapers is unclear. The 3- to 4-μm zoospores have not been seen in stained tissues; however, they may be easily overlooked because of their small size.

Infection occurs typically through the digestive tract as indicated by the location of foci of infection in sectioned live eastern oysters (Mackin 1951). Epithelial cells of the digestive tubules and hemocytes phagocytize the pathogens and probably facilitate transport to connective tissues and to blood sinuses through which the disease is spread to all tissues of the body. Multiplication in tissues is rapid by successive fissions of trophozooites (aplanospores of Mackin 1962 and Lauckner 1983) in clusters or clumps which separate to become typical pathogen cells engulfed by hemocytes. At temperatures of 25–30°C, the parasite develops rapidly; occlusion of blood sinuses and lysis of tissues cause death in about 1 month. The density of pathogen cells is great upon death of eastern oysters. Deaths are hastened among eastern oysters that are near disintegrating infected gapers; this phenomenon was observed in tray studies where positions of eastern oysters were fixed. Infections from remote foci may occur but probably develop more slowly in normal years (Mackin 1962). Studies to determine the relationship between infective dosage, distance from source, and rate of expulsion by eastern oysters at various temperatures are urgently needed to determine effective isolation distances for commercial plantings. The massive spread of *P. marinus* throughout Chesapeake Bay during the 1980–1982 and 1985–1987 droughts provided excellent opportunities to study the rapid, long-range transmission of the disease in nature. Experiments with trays and aquaria, which are necessarily point sources of infective particles, and fixed times of exposure probably do not simulate all conditions in natural waters.

In nature, dosage is difficult to estimate. Long experience with tray eastern oysters that are separated various distances to avoid *P. marinus* infection, provides some conclusions on the effectiveness of isolation. For 20 years, about 60 trays of eastern oysters were monitored annually for diseases at Gloucester Point, Virginia, on an old public eastern oyster bed in the York River. The legged trays were set by stakes about 15 m apart in three rows. New lots of disease-free eastern oysters were imported each year from low-salinity areas of the James River. All ages of eastern oysters from spat to 10-year-old survivors of diseases were monitored. After the second or third year at this enzootic site, most tray lots acquired *P. marinus* infections, and deaths rapidly depleted the stocks.

Eastern oysters were routinely grown to market size during 1 or 2 years after importation without serious losses due to the pathogen. Thousands of gapers and frequent live eastern oyster samples were tested by thioglycollate culture to confirm causes of deaths. Some native eastern oysters on the bottom harbored *P. marinus*, and older infected tray lots were interspersed with new importations. Yet, distances of 15 m between trays prevented or delayed infection for 2 years while data on MSX were being collected.

During the summer of 1957, a proximity (or isolation) experiment was conducted in open waters near Gloucester Point on bottom that was free of eastern oysters. One control and one experi-

TABLE 1.—Proximity experiment in nature near Gloucester Point, Virginia, USA, with disease-free eastern oysters held on sandy bottom in trays 15 m apart to observe transmission of *Perkinsus marinus*. Eastern oysters infected in the laboratory were added to the experimental trays only. Monthly observations of mortality were made 8 July–4 November 1957; then samples of live eastern oysters were taken to determine prevalence and intensity of the disease. Intensity ratings: H = heavy, M = moderate, L = light, and N = no infection.

Group	Number tested in group	Frequency (%) of infections by intensity				Weighted incidence[a]	Mortality 8 Jul–4 Nov (%)
		H	M	L	N		
Control							
1	25	0	4	16	80	0.28	5.6
2	25	0	4	20	76	0.32	
Experimental							
1	25	4	12	52	32	1.08	26.9
2	25	8	12	64	16	1.40	
Laboratory-infected[b]							
1	14	0	64	36	0	2.29	32.7
2	19	0	37	63	0	1.74	

[a]Weighted incidence combines intensity and incidence ratings by assigning values of 5, 3, and 1 to heavy, moderate, and light infections, respectively, and dividing the total by the number of eastern oysters rated.

[b]Laboratory-infected eastern oysters were marked with paint to distinguish them from originally disease-free experimental animals in the same trays.

mental tray of disease-free eastern oysters were spaced 15 m apart. Each tray had a divider in the middle to segregate two groups of animals. Forty-nine eastern oysters exposed to *P. marinus* infection in aquaria were added to the experimental tray, some to each group, to induce infections by proximity. The trays were examined monthly to record mortalities (Table 1). The control tray had low prevalences and low mortality when live eastern oysters were sampled in November. The experimental groups exhibited moderate values for prevalences, mortalities, and weighted incidences. The survivors among the laboratory-infected eastern oysters had higher values for all three measures of disease activity. The timing of the experiment was a little late for the pathogen to cause first-generation mortalities in nature, but the effects of the pathogen's proximity to, and isolation (by 15 m) from, potential hosts were demonstrated clearly.

An infection experiment in open waters was conducted at Gloucester Point in 1965 at the peak of a 3-year drought that favored *P. marinus* even in low-salinity areas where most eastern oysters were being grown (Andrews 1967). Four trays were spaced 5 m apart on sandy bottom with 400 disease-free eastern oysters in each tray and separated into two lots by a divider board. Eastern oysters marked with paint were fed tissue from infected gapers that was macerated in a blender; then 2, 10, or 61 of these eastern oysters were added on 18 June to one end of three trays, respectively. The central tray was kept as a control. Eastern oysters infected with both MSX and *P. marinus* were found in all compartments,

and mortality was high. Incidence and intensity of *P. marinus* infection increased in proportion to the number of paint-marked eastern oysters introduced, as did total mortality. Most early deaths were caused by MSX, but by late August, *P. marinus* began killing about 0.5 of the dying eastern oysters, as gapers collected by scuba divers showed. Two additional trays of disease-free eastern oysters, from the same James River collection, placed 100 m offshore on a barren, sandy bottom, had only 1 of 50 eastern oysters infected with *P. marinus*. Remote outside sources of infection probably did not contribute to this locally induced epizootic. This experiment provided evidence that transmission of *P. marinus* tends to be localized and that isolation of beds is a useful strategy for controlling the disease during normal years.

The distances required between beds of eastern oysters to prevent epizootic mortalities by *P. marinus* are not known. Extensive sampling for 37 years has revealed that most public beds, where some recruitment occurs, are major reservoirs of infected eastern oysters. Overharvesting and reduction of setting rates have greatly depleted and thinned stocks on public beds in Virginia in recent decades; yet the disease persists. Piers and pilings also serve as foci of infection if any eastern oysters are attached. In small rivers where eastern oyster beds are crowded near the shore, these structures can be a continuous source of infection. *Perkinsus marinus* will never be eradicated from estuaries with marginal salinities because the disease can persist with few or no deaths for several years until the next dry period. Low temperatures do not extirpate the disease either.

In Virginia, young eastern oysters (2 years old) usually do not acquire *P. marinus* infections until their second summer in an endemic area (Ray 1954b; Andrews and McHugh 1956). If dosage is high, eastern oysters of any size or age can become infected (Andrews and Hewatt 1957). It is not clear whether small eastern oysters do not collect enough infective stages or their rapid metabolism permits expulsion of the pathogen more rapidly. Most eastern oysters can overcome patent infections when temperatures are too low for the pathogen to multiply. Why overwintering infections are not expelled too is a mystery. Mackin (1962) believed that sporangia overwintered on the bottom and released infective stages in early summer. If this were true, eastern oysters imported from low-salinity areas in spring should develop infections in June and July rather than in late July or early August when eastern oysters with overwintering infections die. Because transmission is direct from one eastern oyster to another, proximity usually is required to ensure the large dosage required to establish infection.

A host of scavengers live on eastern oyster beds to feed on oysters killed by predators or disease. Blue crabs and mud crabs (xanthids) kill small eastern oysters, and nereid worms, spider crabs, and several small fishes, such as blennies, gobies, and clingfishes, are also quick to snatch bits of flesh out of gaping oysters. Recently, *Boonea impressa* has been added to the list of organisms found to have *Perkinsus*-type cells in their tissues (White et al. 1987). This finding is not unexpected, because these ectoparasitic snails puncture the edge of eastern oyster mantles to suck juices. The role that any of these scavengers has as a source of infection or as a reservoir for *Perkinsus* disease remains to be shown. Probably, scavengers cannot make major contributions to the high dosage necessary to produce infections.

Prevalence and Mortality Data

The clearest data on prevalences and deaths caused by *P. marinus* in Virginia were collected during the 1950s before MSX caused epizootic mortalities in Chesapeake Bay. Many private and public beds were sampled during the 1950s for thioglycollate tests, but box counts provided unsatisfactory mortality data because it was difficult to determine the period during which eastern oysters died. In contrast, tray studies provided detailed and accurate data on prevalences and

TABLE 2.—Average annual mortalities in 3–10 trays of disease-free eastern oysters imported from nonenzootic areas to Virginia Institute of Marine Science pier, Gloucester Point, each spring to monitor mortality and prevalence of *Perkinsus marinus* infections in gapers during the 1950s.

Year	Mortality (%)	Number dead	Number of gapers tested	Infection in gapers (%)
1952	25	56	33	94
1953	35	275	251	87
1954	56	675	625	92
1955	37	455	383	87
1956	24	191	143	71
1957	43	103	49	88
1958	29	155	96	78
1959	59	241	129	84
Average	38.5			85.1

deaths. Furthermore, it was possible to select disease-free stocks and to transplant them to enzootic areas at optimal times for determining infection periods and duration of morbidity.

The pier at VIMS was selected as the site for *P. marinus* disease studies for convenience in examining trays and collecting dying eastern oysters or gapers. Sea-Rac trays (Chesapeake Corporation, West Point, Virginia) were suspended from three catwalks to hold about 100 market-size eastern oysters. Water depth was about 1.5 m at low tide. Many trays were examined daily throughout the warm seasons for 8 years during the 1950s to recover gapers. Trays of eastern oysters exposed in an enzootic area for one or more summer seasons and newly transplanted disease-free eastern oysters were suspended adjacent to each other. This arrangement provided an optimal density and proximity of eastern oyster groups for *P. marinus* to flourish. This situation was believed to approximate regularly planted eastern oyster beds with some old survivors and plenty of new susceptible eastern oysters. During these years, *P. marinus* caused 80–90% of all deaths in the trays (Andrews and Hewatt 1957). The ratio was always lower on commercially planted beds where smothering and dredge injuries caused some deaths.

Total annual mortality varied from 24 to 59% during 8 years of tray monitoring at the VIMS pier. Average annual mortalities in groups of 3–10 trays of eastern oysters are given in Table 2; this summary excludes lots in their first year of exposure in an enzootic area. Remote sources of infection were not necessary to keep *Perkinsus* epizootics active at the VIMS pier because 85% of the gapers had serious infections (mostly heavy, but some moderate). The two worst mor-

talities (1954 and 1959) were caused by early
warm water temperatures in late spring and high
temperatures that continued into September and
October (Andrews 1980).

Seasonal prevalence and intensity data for live
eastern oysters in 1954 are shown in Figure 2.
Eastern oysters from three sources in two rivers
were sampled monthly for thioglycollate tests. All
lots were exposed to *P. marinus* infection in enzootic areas for at least 1 year before the 1954
sampling, and winter samples in 1953–54 showed
that the parasite was established. Mortality in the
VIMS pier lot (James River seed eastern oysters in
trays) for the warm season was 51%, and 94% of
the gapers were infected and had a WI of 4.85, an
indication of nearly all heavy infections. At peak
prevalence on 5 October, 96% of live eastern
oysters had infections with a WI of 2.40 or an
average intensity of nearly moderate infection. If
warm waters had prevailed for another month,
nearly all eastern oysters would have died. The
apparent absence or low prevalence of the pathogen in late winter and early spring is typical for
thioglycollate tests, because these tests do not
detect overwintering infections; yet, the method is
highly sensitive. Samples were taken near the first
of every month shown, yet three winter–spring
months showed no infections (Figure 2). Patent
infections appeared early in 1954 with 16% infection on 1 June; this event allowed three generations
of infections and deaths to occur before declining
water temperatures stopped multiplication about 1
November. Heavy infections always remained in
low numbers because deaths removed infected
individuals continuously during late summer. Remission of infections was slower in late fall and
winter than development which occurred during
early summer, and often a few severely infected
individuals persisted into winter before dying.

Native eastern oysters were sampled from
Ferry pier pilings at Gloucester Point, Virginia,
for 11 months through 1954 (Figure 2). These were
2-year-olds at a site where older eastern oysters
were scarce, yet it is clear that they acquired
infections during the summer of 1953 despite their
small size as yearlings. Density of eastern oysters
was much lower on the Ferry Pier where no trays
of older, infected eastern oysters were located,
yet overwintering of patent infections was more
frequent than usually occurs in enzootic areas.
Hoghouse Bar in the Rappahannock River is a
public eastern oyster bed that usually has a sparse
population of large, old eastern oysters and limited recruitment. Spring salinities at this site are

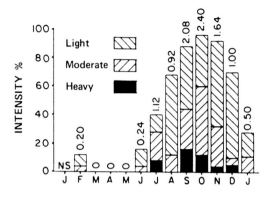

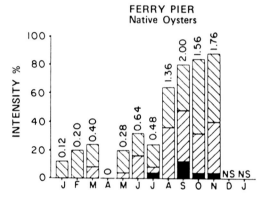

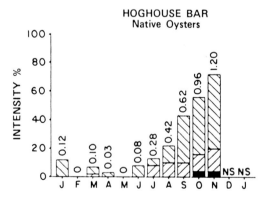

FIGURE 2.—Monthly intensities and prevalences of
Perkinsus marinus infections in live Virginia eastern
oysters during 1954 at three sites where the disease is
enzootic. Note the frequent absence of the disease in
samples of 25 eastern oysters during late winter and
early spring. VIMS = Virginia Institute of Marine
Science, York River, Gloucester Point. Ferry pier, York
River, is at Gloucester Point. Hoghouse Bar, Rappahannock River, is near Towles Point. Categories of
heavy, moderate, and light intensities as defined by Ray
(1954a). NS = no samples.

TABLE 3.—Average monthly mortality in tray eastern oysters at Virginia Institute of Marine Science, Gloucester Point, and average monthly prevalence of serious infections (moderate and heavy) of *Perkinsus marinus* in gapers over eight pre-MSX years (1952–1959). Three to 10 trays were monitored each year. For histories of tray lots, see Hewatt and Andrews (1954) and Andrews and Hewatt (1957).

Month	Mortality (%)	Number dead	Number of gapers tested	Prevalence of serious infections (%)
Jan	0.4	22	19	26
Feb	0.4	20	15	13
Mar	0.6	35	24	4
Apr	0.4	27	11	0
May	1.4	78	35	14
Jun	2.1	86	75	39
Jul	6.8	351	262	87
Aug	13.6	723	631	94
Sep	13.1	539	443	97
Oct	6.4	311	171	96
Nov	1.1	29	22	96
Dec	0.9	17	9	89
Total	47.2	2,238	1,717	

often near the lower limit (12–15‰) for *P. marinus,* and development was delayed in 1954, which became a favorable year (high salinities and an extensive warm season) for the disease. The disease did not reach lethal intensities (average WI > 0.50 in a sample of 25 eastern oysters) until about 1 September; therefore, mortality was low for the summer (few boxes). *Perkinsus marinus* persisted for 37 years in the Rappahannock River despite the marginal spring salinities and several hurricane floods that reduced salinities to less than 10‰ for weeks or months (e.g., 1955 and 1972, Andrews 1973).

Disease in the Rappahannock River is typical of many small estuaries with coastal plain drainage basins which exhibited *P. marinus* mortalities mostly when dry years occurred over the Chesapeake Bay watershed. These estuaries include such systems as the Great Wicomico River, the lower Potomac River and its tributaries including St. Marys River (a former eastern oyster seed area now enzootic for *Perkinsus*), the Patuxent and Choptank rivers in Maryland, and Pocomoke Sound. Nether low temperatures nor low salinities will ever eradicate *Perkinsus* from these estuaries if experience during the past 37 years is indicative.

The monthly progression of *P. marinus* was monitored for eight years in 3–10 trays of eastern oysters at VIMS pier (Table 3). The average mortality was low in 8 months of the year. Over eight successive years of daily monitoring, 77% of all dead animals were recovered as gapers. The annual cycle of activity for the disease is shown in Table 3. From July through December, a high proportion of gapers had advanced infections, which indicates that *P. marinus* was the overwhelming cause of deaths in eastern oysters protected in trays from other adverse agents and events. During winter and early spring months when death rates were low (Table 3), the percentage of *Perkinsus*-infected eastern oysters declined (Table 4), but by late May and June, when temperatures were favorable for the pathogen, the level of intensity of the disease increased to begin a new warm season of epizootic mortality. During

TABLE 4.—Prevalence of *Perkinsus marinus* in winter gapers at Virginia Institute of Marine Science pier, Gloucester Point, for 3 years, 1952–1954. Infections were diagnosed by the thioglycollate test as heavy, moderate, light, and negative. Gapers were collected from 18 trays containing about 5,000 live eastern oysters.

Month	Number of gapers collected	Gapers infected (%)	Number of infected gapers by intensity				Infected gapers with heavy or moderate infection (%)
			Heavy	Moderate	Light	None	
Jan	15	53	1	4	3	7	62
Feb	13	31	1	1	2	9	50
Mar	9	44	0	1	3	5	25
Apr	9	33	0	1	2	6	33
May	23	44	3	1	6	13	40
Jun	52	63	17	8	8	19	76

TABLE 5.—First-year prevalence of *Perkinsus marinus* in gapers from trays of disease-free James River (Virginia, USA) eastern oysters transplanted to old eastern oyster beds to monitor MSX. Most dead eastern oysters were killed by MSX. All sites were enzootic for *Perkinsus* disease, but few eastern oysters were left on the beds.

Year	Location	Number of gapers tested	Perkinsus marinus infection (%)	Weighted incidence[a]
1966	James River[b]	54	24	0.80
	York River	381	3	0.13
	Mobjack Bay	156	1	0.06
	Rappahannock River	11	0	0.00
1967	James River[c]	85	3	0.08
	York River	529	4	0.08
	Mobjack Bay	114	0	0.00
	Rappahannock River	3	0	0.00
1968	James River[c]	70	0	0.00
	York River	477	7	0.19
	Mobjack Bay	92	0	0.00
	Rappahannock River	5	0	0.00

[a]Weighted incidence combines intensity and incidence ratings by assigning values of 5, 3, and 1 to heavy, moderate, and light infections, respectively, and dividing the total by the number of eastern oysters rated.
[b]Brown shoal public bed where recruitment continued and *Perkinsus*-infected eastern oysters persisted.
[c]Some trays were at Wreck Shoal, a nonenzootic area where *Perkinsus* is rare.

the pre-MSX period of the 1950s, gapers seldom exhibited light infections (Table 3), but this condition changed after the introduction of MSX disease in 1959–1960 when concurrent *Perkinsus* and MSX infections occurred. These eastern oysters were all acclimated to the enzootic area at least 1 year before the records were collected. About 5,000 eastern oysters were monitored each year to obtain these records.

When MSX disease invaded Chesapeake Bay in 1959, tray operations were moved away from VIMS pier to avoid *Perkinsus* interference with studies of its epizootiology. Legged trays were constructed and were set beside stakes on eastern oyster beds all over Virginia. MSX infects eastern oysters everywhere in lower Chesapeake Bay whereas *P. marinus* tends to be localized and can be avoided by isolation for 1–3 years. Furthermore, *Perkinsus* requires dense populations to cause epizootic mortalities, but MSX is not affected by eastern oyster population size or density. It was not feasible to inspect trays daily at these scattered sites. After MSX had reduced eastern oyster populations in areas enzootic for *P. marinus*, trays of disease-free eastern oysters on outlying public eastern oyster beds showed little activity by *Perkinsus* during the first 2 years of exposure to the two diseases (Table 5). The low number of gapers collected in the Rappahannock River reflected the low mortality caused by both diseases during most years of normal rainfall and the once-per-month examinations for gapers. Mobjack Bay trays were examined monthly too, but they had high mortality from MSX, and many

gapers were collected. More gapers are recovered usually during winter when decay is slow and scavengers are inactive than in warmer season when the opposite is true. The disparity is especially great when offshore trays are visited only at 2- or 3-week intervals during the year. In winter, MSX occurs in about 50% of gapers in trays. After MSX killed more than 90% of all eastern oysters on beds in about 3 years, *P. marinus* declined in abundance in the lower bay. Table 6 shows the proportion of eastern oysters killed by the two diseases in 1979 after 20 years of MSX activity. MSX killed most eastern oysters during the first 2 years of exposure in enzootic areas, but, once

TABLE 6.—Causes of death in gapers collected from trays of susceptible James River eastern oysters at Tillages Bed in open waters of York River, Virginia, USA, 1979.[a]

Date imported	Number of gapers collected	Number of gapers dead, by cause	
		MSX	*Perkinsus marinus*
4 Mar 1976	3	0	3
1 Sep 1976	5	0	2
17 Aug 1977	25	7	10
11 Oct 1977	10	5	0
24 Mar 1978	11	9	0
13 Mar 1978	16	12	1
13 Mar 1978	17	7	0
13 Mar 1978	7	6	0
28 Aug 1978	28	12	0
Total	122	58	16

[a]These tray eastern oysters had been exposed in an enzootic area for both diseases 1 or 2 years before these 1979 data were collected.

TABLE 7.—Number and intensity of infection by *Perkinsus marinus* and *Haplosporidium nelsoni* (MSX) in mostly native eastern oysters from public eastern oyster beds in Virginia, 1986. Samples of live eastern oysters contained 25 specimens. Data were provided by G. Burreson, Virginia Institute of Marine Science.

Location	Site	Month	Prevalence[a] MSX Number H	M	L	%	*Perkinsus marinus* Number H	M	L	%
James River	Wreck Shoal	Aug	0	0	2	8	0	0	0	0
	Horsehead	Aug	0	0	1	4	0	0	0	0
	Miles beds	Sep	3	0	4	28	1	3	20	96
York River	Poropotank River	Aug	0	0	0	0				67[b]
	Poropotank River	Aug	0	0	0	0				72
	Tucker Ground	Sep	0	1	0	4	16	3	6	100
	Skimino Creek	Sep	1	0	3	16	2	1	5	32
Mobjack Bay[c]										
1985		Dec	2	2	7	44	0	1	8	36
1984		Dec	4	6	7	68	0	1	1	8
Chesapeake Bay	Deep Rock	Aug	8	3	4	60	0	1	0	4
Rappahannock River	Drummond Creek	Oct	0	1	3	16	2	4	15	84
	Morattico Bar	Oct	0	0	1	4	0	0	1	4
	Sharps Wharf	Oct	0	1	0	4	0	0	0	0
Potomac River area	Machodoc Creek	Sep	0	0	0	0	11	8	6	100[c]
	Yeocomico River	Sep	0	0	0	0	13	5	6	96[c]
	Coan River	Sep	0	0	0	0	7	8	8	92[d]
Great Wicomico River	Fleeton Point	Sep	0	1	1	8	6	5	9	80
	Haynie Bar	Sep	1	0	3	16	2	2	16	80
	Cranes Creek	Sep	0	0	4	16	0	1	7	32

[a]Prevalence of disease is the level of infection at a given time: H = heavy, M = moderate, L = light. Prevalences are based on samples of 25 live eastern oysters.
[b]Sample of 21 eastern oysters.
[c]James River transplants in year noted.
[d]*Perkinsus marinus*-infected seed eastern oysters were imported from lower James River in 1984 and 1985.

Perkinsus established residence in older lots, it became the primary cause of deaths (Table 7).

Severe drought in the Chesapeake Bay area from 1985 to 1987 caused serious eastern oyster mortalities by *P. marinus* in low-salinity fringe areas for MSX (Burreson, personal communication). Samples of live eastern oysters were taken from the James River seed area and several low-salinity estuaries where *Perkinsus* disease was prevalent. Because MSX did not extend its range upbay during the first 2 years of drought, *P. marinus* was able to cause enzootic mortalities (Table 7). Prevalences are expressed as cases per 25 live eastern oysters. Only the lower part of the James River seed area had *Perkinsus* infections in 1986, but the disease spread widely in 1987. Private beds in the lower sector were the source of seed eastern oysters transplanted to Potomac River tributaries, where serious mortalities occurred as indicated by the prevalences. The pathogen was severely active in the Great Wicomico River, too. Prevalences in the York River were no surprise because salinities through-out this river are usually favorable for both diseases, but MSX was not as active in 1986 as in typical years. MSX was a serious problem in 1986 only in Mobjack Bay, which is usually highly saline. The data for 1986 prove that *P. marinus* is more persistent in low-salinity estuaries than MSX and more regular in producing mortalities in such areas (Burreson, personal communication). During 1987, the 3rd year of drought, MSX and *P. marinus* both became active in all areas listed in Table 7 and in most of Maryland.

Effect of *Perkinsus* on Eastern Oyster Culture in Chesapeake Bay

During the 1950s, Virginia produced annually 3–4 million bushels of market eastern oysters. A large proportion of these eastern oysters were grown on private beds with seed eastern oysters from the James River. The three largest producers planted eastern oysters in high-salinity waters in Hampton Roads, Chesapeake Bay (deep beds off Egg Island, New Point Comfort, and Wolf Trap), and Mobjack Bay. Other important producing areas were York

River, Rappahannock River, and smaller estuaries along the western shore of Chesapeake Bay from Old Point Comfort up bay and along the Potomac River. Eastern oysters are no longer planted in many of these areas, particularly below the mouth of the Rappahannock River. Maryland produced almost as many eastern oysters but mostly on public grounds; however, losses to the two major diseases were substantial during dry years.

During the period from 1946 to 1960, *P. marinus* was the major cause of mortality among eastern oysters 2 years or more of age. During the early part of this period, oystermen typically held 2- and 3-year-old seed eastern oysters on growing beds for three additional years. Yields were strongly reduced during the 3rd year on beds in high-salinity areas, which comprised about two-thirds of the total acreage of private eastern oyster beds. Yields in areas enzootic for *Perkinsus* were often 0.5 bushel for each bushel of seed eastern oysters planted. In bushel counts, seed with 1,500–2,000 small eastern oysters yielded 300–400 market eastern oysters, which represented an 80% mortality from all causes over 3 years. Most spat and many yearlings were smothered or killed by predators (flatworms, crabs, and oyster drills). During the 1950s, oystermen began to harvest eastern oysters after 2 years on growing beds, which provided better yields. Early harvesting is an important method of disease control.

During eastern oyster company surveys in the 1950s to determine which private beds were ready to be harvested, samples were taken for thioglycollate tests for *P. marinus,* and box counts were made of dead eastern oysters with valves still attached at the hinge. Box counts were typically lower on beds planted only 1 year than those with 2 years of growth. Typical box counts on acclimated eastern oysters (i.e., grown 1 year in an area enzootic for *Perkinsus*) reached as high as 50% in 1954 and 1959.

Yields were often much higher (1 or 2 bushels harvested per bushel planted) from eastern oysters planted in low-salinity areas of the Rappahannock and Potomac rivers. Usually, these areas were free of disease-produced mortalities, except in periods of drought. Oystermen have never considered it practical first to transplant eastern oysters into low-salinity areas to avoid diseases and predators and then to transplant into high salinities for final growth and fattening. Once seed eastern oysters are planted, they are nearly always left on the same bed until ready for harvest. The cost of transplanting is substantial, and often nearly one-third of the eastern oysters planted are not recovered by the dredges used.

When MSX disease, caused by *H. nelsoni,* invaded and spread in Chesapeake Bay in 1959–1960, it quickly became the dominant disease killing eastern oysters (Andrews 1967, 1976b, 1984). In the lower bay, private eastern oyster beds were last planted in the spring of 1960. In 2 or 3 years, MSX had killed over 90% of 2 million bushels of eastern oysters in Mobjack Bay, in the Chesapeake Bay down to Old Point Comfort, and in Hampton Roads. During 26 years of waiting for the disease to subside, oystermen made only small trial plantings in these areas. Eastern oysters resistant to MSX have not developed naturally except in the lower York River and Mobjack Bay. Most beds are too barren of shells and eastern oysters to catch spat; therefore, *Perkinsus* is absent too, but it persists in nearby creeks and on man-made structures along the shores where setting occurs.

Production has declined drastically on both public and private beds (Andrews, in press). With brood stocks in low abundance, setting has declined too. Yet, tongers working on public beds have been allowed to overfish and further deplete the brood stocks. The climax in Virginia was reached in 1986–1987 when mismanagement (i.e., no cull or size limit) and disease induced by drought combined to deplete all public beds, including the seed area in James River. In 1987–1988, this seed area was opened again to remaining tongers. Consequently few seed eastern oysters were available for planters during the 1986–1987 and 1987–1988 seasons. The extension of *Perkinsus* upriver into the lower James River seed area in 1985–1987 produced disastrous results for those planters in low-salinity areas who transplanted infected eastern oysters.

Meanwhile, few private beds are being planted because economics, politics, and high risks of eastern oyster culture have changed management objectives (Andrews, in press). Disease is only one of the reasons why so few private beds are planted in Virginia, even in low-salinity areas where risks are lower in most years. The state should never allow *Perkinsus*-infected eastern oysters in seed areas to be transplanted in such a way as to spread disease and discourage private planting. The seed area should be sampled for prevalence of MSX and *Perkinsus* to determine where and when eastern oysters can be safely transplanted. Monitoring private growing beds for *Perkinsus* is a more difficult program, but must be pursued. Because *Perkinsus* cannot be adequately assessed from December through June or July, tests must be made during the peak-incidence period in September and October each year. Unfortunately, unlike MSX, *Perkinsus* activity must be tested on each planted bed because it is localized and variable with age, exposure, weather, and location of site.

Private investment on a planted bed of eastern oysters may be as high as $3,000/acre (1 acre = 0.4 hectare), which justifies more attention to diseases and other biological problems than is usually given. The market price of eastern oysters is favorable, but more attention must be given to seed supply and methods of culture and management to revive the industry. Large areas of public and private eastern oyster beds suitable for culture are unproductive in Chesapeake Bay. The states control most seed areas and must insure availability of seed supply. Off-bottom culture, practiced in other regions of the world, would be more expensive in labor and supplies without avoiding problems of disease and fouling. Harvesting of poor-condition seed eastern oysters from James River for marketing has caused a decline in quality of marketed meats. A large proportion of shucked eastern oysters distributed by Chesapeake Bay packers is imported from the Gulf of Mexico and the west coast of the U.S. Public eastern oyster beds are severely overfished, and their recovery from natural setting appears questionable. The states continue to subsidize public oystering by transplanting seed eastern oysters and planting shells for substrate. The huge acreage of barren public grounds cannot be made productive by limited state financing. Eastern oyster planters with leased private grounds seem to be discouraged by high costs and risks of eastern oyster culture. Extreme fluctuations of weather seem to compound these problems by encouraging diseases. Now, seed eastern oysters are scarce, and no strategy for survival of private culture has been found. Rapid human population growth along the shore of Chesapeake Bay has multiple effects on the waters in addition to pollution. The eastern oyster industry is at a crossroad in terms of survival, and diseases are an important facet of efforts to revive it.

Strategies to control the diseases caused by MSX and *Perkinsus* are quite different. Control of MSX depends on movement to low-salinity areas, and, ultimately, on developing and breeding resistant eastern oysters (Ford and Haskin 1988, this volume). Resistance has occurred naturally in

Delaware Bay where nearly all stocks of eastern oysters were exposed to the disease and intensive selection has occurred over many years. In Chesapeake Bay, seed eastern oyster stocks are located in relatively low-salinity areas, where little selection occurred; therefore, selection and breeding for resistance must be done at high cost in hatcheries.

Control of *Perkinsus* disease depends on isolation and manipulation of seed stocks before and after they are transplanted to growing areas. The following management procedures are detailed in Andrews and Ray (1988): (1) Avoid use of infected seed eastern oysters. This precaution can set back infection by a full year. (2) Isolate newly planted beds from those with infected eastern oysters. (3) Harvest and fallow beds to allow all infected eastern oysters to die before replanting. (4) Harvest early if beds become infected. (5) Diagnose for disease on public and privately planted beds in September and October each year in enzootic areas, or earlier if summer mortality occurs. If seed eastern oysters are infected, the area should be closed to all transplanting until the beds recover from the disease.

Acknowledgments

I thank M. H. Roberts, Jr., Virginia Institute of Marine Sciences (VIMS), for critical review of the manuscript, computer draft of Figure 2, and unpublished data on dosage of zoospores required for infection. F.-L. E. Chu, VIMS, provided information on the culture of zoospores. My gratitude goes to E. M. Burreson, VIMS, for advice and for the data of Table 7. This is contribution 1489 of the Virginia Institute of Marine Science, College of William and Mary.

References

Andrews, J. D. 1955. Notes on fungus parasites of bivalve mollusks in Chesapeake Bay. Proceedings National Shellfisheries Association 45:157–163.

Andrews, J. D. 1965. Infection experiments in nature with *Dermocystidium marinum* in Chesapeake Bay. Chesapeake Science 6:60–67.

Andrews, J. D. 1966. Oyster mortality studies in Virginia. V. Epizootiology of MSX, a protistan pathogen of oysters. Ecology 47:19–31.

Andrews, J. D. 1967. Interaction of two diseases of oysters in natural waters. Proceedings National Shellfisheries Association 57:38–49.

Andrews, J. D. 1968. Oyster mortality studies in Virginia. VII. Review of epizootiology and origin of *Minchinia nelsoni*. Proceedings National Shellfisheries Association 58:23–36.

Andrews, J. D. 1973. Effects of tropical storm Agnes on epifaunal invertebrates in Virginia estuaries. Chesapeake Science 14:223–234.

Andrews, J. D. 1976a. Epizootiology of *Dermocystidium marinum (Labyrinthomyxa marina)* in oysters. Proceedings International Colloquium on Invertebrate Pathology 1:172–174.

Andrews, J. D. 1976b. Epizootiology of oyster pathogens *Minchinia nelsoni* and *M. costalis*. Proceedings International Colloquium on Invertebrate Pathology 1:169–171.

Andrews, J. D. 1979. Oyster diseases in Chesapeake Bay. U.S. National Fisheries Service Marine Fisheries Review 41(1–2):45–53.

Andrews, J. D. 1980. *Perkinsus marinus* (= *Dermocystidium marinum,* "Dermo") in Virginia, 1950–1980. Virginia Institute of Marine Science Data Report 16, Gloucester Point.

Andrews, J. D. 1984. Epizootiology of diseases of oysters (*Crassostrea virginica*) and parasites of associated organisms in eastern North America. Helgolaender Meeresuntersuchungen 37:149–166.

Andrews, J. D. In press. The oyster fishery of eastern North America based on *Crassostrea virginica.* CRC Press, Boca Raton, Florida.

Andrews, J. D., and M. Frierman. 1974. Epizootiology of *Minchinia nelsoni* in susceptible wild oysters in Virginia, 1959–1970. Journal of Invertebrate Pathology 24:127–140.

Andrews, J. D., and W. G. Hewatt. 1957. Oyster mortality studies in Virginia. II. The fungus disease caused by *Dermocystidium marinum* in oysters of Chesapeake Bay. Ecological Monographs 27:1–25.

Andrews, J. D., and J. L. McHugh. 1956. The survival and growth of South Carolina seed oysters in Virginia waters. Proceedings National Shellfisheries Association 47:3–17.

Andrews, J. D., and S. M. Ray. 1988. Management strategies to control the disease caused by *Perkinsus marinus.* American Fisheries Society Special Publication 18:257–264.

Andrews, J. D., and J. L. Wood. 1967. Oyster mortality studies in Virginia. VI. History and distribution of *Minchinia nelsoni,* a pathogen of oysters in Virginia. Chesapeake Science 8:1–13.

Da Ros, L., and W. J. Canzonier. 1985. *Perkinsus,* a protistan threat to bivalve culture in the Mediterranean basin. Bulletin of European Fish Pathology 5:23–27.

Ford, S. E., and H. H. Haskin. 1988. Management strategies for MSX (*Haplosporidium nelsoni*) disease in oysters. American Fisheries Society Special Publication 18:249–256.

Haskin, H. H., W. J. Canzonier, and J. L. Myhre. 1965. The history of "MSX" on Delaware Bay oyster grounds, 1957–1965. Bulletin of the American Malacological Union Incorporated 32:20–21.

Haskin, H. H., and S. E. Ford. 1979. Development of resistance to *Minchinia nelsoni* (MSX) mortality in laboratory-reared and native oyster stocks in Delaware Bay. U.S. National Fisheries Service Marine Fisheries Review 41(1–2):54–63.

Hewatt, W. G., and J. D. Andrews. 1954. Oyster mortality studies in Virginia. I. Mortalities of oysters in trays at Gloucester Point, York River. Texas Journal of Science 6:121–133.

Hewatt, W. G., and J. D. Andrews. 1956. Temperature control experiments on the fungus disease *Dermocystidium marinum* of oysters. Proceedings National Shellfisheries Association 46:129–133.

Kern, F. G., L. C. Sullivan, and M. Takata. 1973. *Labyrinthomyxa*-like organisms associated with mass mortalities of oysters, *Crassostrea virginica* from Hawaii. Proceedings National Shellfisheries Association 63:43–46.

Lauckner, G. 1983. Diseases of Mollusca: Bivalvia. Pages 457–1038 *in* O. Kinne, editor. Diseases of marine animals, volume 2. Biologische Anstalt Helgoland, Hamburg, West Germany.

Levine, N. D. 1978. *Perkinsus* gen. n. and other new taxa in the protozoan phylum Apicomplexa. Journal of Parasitology 64:549.

Mackin, J. G. 1951. Histopathology of infection of *Crassostrea virginica* (Gmelin) by *Dermocystidium marinum* Mackin, Owen and Collier. Bulletin of Marine Science of the Gulf and Caribbean 1:72–87.

Mackin, J. G. 1953. Incidence of infection in oysters by *Dermocystidium* in the Barataria Bay area of Louisiana. Proceedings of the National Shellfisheries Association 42:22–35. (Convention addresses, 1951.)

Mackin, J. G. 1956. *Dermocystidium marinum* and salinity. Proceedings National Shellfisheries Association 46:116–128.

Mackin, J. G. 1962. Oyster diseases caused by *Dermocystidium marinum* and other microorganisms in Louisiana. Publications of the Institute of Marine Science, University of Texas 7:132–229.

Mackin, J. G., and J. L. Boswell. 1956. The life cycle and relationships of *Dermocystidium marinum.* Proceedings National Shellfisheries Association 46:112–115.

Mackin, J. G., and S. H. Hopkins. 1962. Studies on oyster mortality in relation to natural environments and to oil fields in Louisiana. Publications of the Institute of Marine Science, University of Texas 7:1–131.

Mackin, J. G., H. M. Owen, and A. Collier. 1950. Preliminary note on the occurrence of a new protistan parasite, *Dermocystidium marinum* n. sp. in *Crassostrea virginica* (Gmelin). Science (Washington, D.C.) 111:328–329.

Mackin, J. G., and S. M. Ray. 1966. The taxonomic relationships of *Dermocystidium marinum* Mackin, Owen and Collier. Journal of Invertebrate Pathology 8:544–545.

Mackin, J. G., and A. K. Sparks. 1962. A study of the effects on oysters of crude oil loss from a wild well. Publications of the Institute of Marine Science, University of Texas 7:230–319.

Perkins, F. O. 1976. Zoospores of the oyster pathogen, *Dermocystidium marinum.* I. Fine structure of the conoid and other sporozoan-like organelles. Journal of Parasitology 62:959–974.

Perkins, F. O. 1988. Structure of protistan parasites found in bivalve molluscs. American Fisheries Society Special Publication 18:93–111.

Perkins, F. O., and R. W. Menzel. 1966. Morphological and cultural studies of a motile stage in the life cycle of *Dermocystidium marinum*. Proceedings National Shellfisheries Association 56:23–30.

Quick, J. A., and J. G. Mackin. 1971. Oyster parasitism by *Labyrinthomyxa marina* in Florida. Florida Department of Natural Resources, Marine Research Laboratory, Professional Papers Series 13:1–55.

Ray, S. M. 1952. A culture technique for the diagnosis of infections with *Dermocystidium marinum*, Mackin, Owen and Collier, in oysters. Science (Washington, D.C.) 166:360–361.

Ray, S. M. 1954a. Biological studies of *Dermocystidium marinum*. Rice Institute Pamphlet, Special Issue. (The Rice Institute, Houston, Texas.)

Ray, S. M. 1954b. Studies on the occurrence of *Dermocystidium marinum* in young oysters. Proceedings of the National Shellfisheries Association 44:80–92. (Convention addresses, 1953.)

Ray, S. M. 1966. Effects of various antibiotics on the fungus *Dermocystidium marinum* in thioglycollate cultures of oyster tissues. Journal of Invertebrate Pathology 8:433–438.

Ray, S. M., and A. C. Chandler 1955. *Dermocystidium marinum*, a parasite of oysters. Experimental Parasitology 4:172–200.

Valiulis, G. A., and J. G. Mackin. 1969. Formation of sporangia and zoospores by *Labyrinthula* sp. parasitic in the clam *Macoma balthica*. Journal of Invertebrate Pathology 14:268–270.

White, M. E., E. M. Powell, S. M. Ray, and E. A. Wilson. 1987. Host-to-host transfer of *Perkinsus marinus* in oysters (*Crassostrea virginica*) populations by the ectoparasitic snail *Bonnea impressa* (Pyramidellidae). Journal of Shellfish Research 6:1–6.

American Fisheries Society Special Publication 18:64–73, 1988

Mytilicola intestinalis, a Copepod Parasite of Blue Mussels

J. T. Davey and J. M. Gee

Natural Environment Research Council, Plymouth Marine Laboratory
Prospect Place, The Hoe, Plymouth PL1 3DH, England

Abstract.—*Mytilicola intestinalis* is a copepod inhabiting the digestive tract of blue mussels *Mytilus edulis* and, for nearly 40 years, has been regarded as an important pest in relation to European blue mussel cultivation. The biological basis for this view has been progressively undermined during the last 20 years as researchers have examined in greater detail the ecology, physiology, and pathogenesis of the association between copepod and blue mussel. The authors review these developments and discuss the evidence for a commensal rather than a parasitic relationship. This evidence includes the relatively simple adaptation of cyclopoid morphology and gut histochemistry, indicating a life of herbivorous grazing in the host's gut, the life history with its succession of temperature-controlled developmental stages, producing a predictable annual population cycle, and the swamping of any detectable effect of the copepod on its host's physiological performance by every other measurable variable affecting host individuals and populations. However, further work is required in the area of pathogenicity and biological and anthropogenic synergisms, particularly in relation to other pathogens and to pollutants, before the pest status of *M. intestinalis* can be finally decided.

Since its earliest appearance in north European waters, *Mytilicola intestinalis* Steuer has been perceived as a threat to the mariculture of blue mussels *Mytilus edulis* and possibly other shellfish. It acquired a reputation as a pest that is hard to dispel even after at least two decades of mounting evidence that it may have more of the characteristics of a commensal than of a parasite. In a critical review of the research on *M. intestinalis* to produce a defensible and informed position on its current disease status, we begin by asking a few simple questions, to which we return at our conclusion, in order to assess both our present position and possible directions of future research. Does *M. intestinalis* exhibit epizootic patterns of occurrence through its host populations? Does it measurably affect its host's physiological performance? Is it a cause for concern from an economic point of view?

These three questions also usefully channel our thinking into three aspects of any host–parasite system: the ecology, physiology, and pathology of the relationship, corresponding to the population, individual, and cellular levels of the system's functioning.

Background

The copepod was first described by Steuer (1902) from *Mytilus galloprovincialis* in the Mediterranean Sea. Its subsequent occurrence in populations of the blue mussel in northwestern Europe has been documented many times and most

recently reviewed by Lauckner (1983). Between the mid-1930s and the 1950s, "every renowned shellfish biologist from each of the marine biological laboratories along the north European coasts commented on this parasite" (Lauckner 1983) and, to varying degrees, helped build the reputation of *M. intestinalis* as a serious pest of commercial blue mussel culture. The concensus was always that *M. intestinalis* had been spread from the Mediterranean Sea, chiefly by Mediterranean mussels fouling ships and, although slow, this spread was inexorable and probably impossible to check. Meyer and Mann (1950) alone proposed that *M. intestinalis* had been present enzootically but undetected until epizootic outbreaks had occurred due to some combination of environmental conditions peculiarly favorable to the parasite.

Whatever the origin and mode of spread of the copepod, initial recordings of infestation in blue mussels in Holland (Korringa 1950), Britain (Cole and Savage 1951), Germany (Meyer 1951), and Denmark (Thiesen 1966) were all described in esssentially epizootic terms. These descriptions noted the sudden appearance of high numbers of parasites in stocks of blue mussels previously uninfested, or not known to be infested, and ascribed deleterious effects to varying numbers of parasites per host. Many of these studies lacked the experimental protocols and statistical rigor that would be demanded today, and thus doubt is cast upon their credibility. In the 1950s, these researchers did not have the benefit of the immense advances in understanding of blue mussel physiology, which have been gained over the last 20 years. Lauckner (1983) commented that all the authors of the 1950s paid scant attention to the possible role of other factors, whether environment or disease agents (microbial or other pathogens), that might have had a bearing on the correlations they presented between the parasite and the condition of the hosts. For example, Thiesen (1966) described sharp breaks in the growth of blue mussels in the Limfjord, Denmark, which he took as indicating the first exposure of the animals to heavy infestation with *M. intestinalis.* The year was 1963, and yet no mention was made of the exceptionally cold weather and reduced sea temperatures of the period, December 1962–March 1963, which could have equally produced such an effect. Dollfus (1951) alone recognized the possibility that other pathogens were being overlooked, but his warning went unheeded.

Dynamics of the Host–Parasite System

Life Cycle

To properly assess the impact of a parasite on its host population, a detailed knowledge of the parasite and host life cycles and population dynamics is a necessary starting point. The life cycle of *M. intestinalis* was not fully elucidated until after the establishment of its reputation as an economic pest. Pesta (1907) described the free-living stages, and Caspers (1939) and Ahrens (1937, 1939a, 1939b) gave descriptions of some of the parasitic copepodite stages. Grainger (1951), Hockley (1951), Bolster (1954), Costanzo (1959), and Williams (1969) all provided further information.

Gee and Davey (1986a) carried out experimental studies on the life cycle involving the examination of many thousands of copepods from both naturally and artificially infested blue mussels. They showed that the life cycle includes a free-living nauplius, a metanauplius, and a first copepodite stage. This first copepodite then infests the blue mussel and passes through four more copepodite stages before becoming the adult (Figure 1). Copepodite V molts to the adult when it reaches approximately 2.5 mm; males become sexually mature at about 2.9 mm, and females at 4.5 mm. Males reach a maximum length of 4.5 mm, and females 9.0 mm, indicating considerable growth of the adults, a feature not usually found in free-living copepods.

From the time that *M. intestinalis* was implicated in the crash of the Dutch blue mussel fisheries, the tendency to refer to mean numbers of parasites per blue mussel without further qualification as to the stage of development of these parasites has clouded this whole subject. Copepodites I–III are less than 1 mm long. We have found that blue mussels in the Lynher River population in Cornwall, England, frequently contain over 30, and sometimes as many as 90, parasites. In such cases, up to two-thirds of the parasites are copepodites I–III, and all of these are found either in the stomach or the direct intestine of the blue mussel. Stages larger than copepodite V are invariably found further back in the host's intestine than the style sac (junction of direct and recurrent intestines).

Population Dynamics

The most detailed description of the population dynamics of the copepod is that of Davey et al. (1978) for a heavily infested population of blue mussels in the Lynher River. When all developmental stages were taken into account, they reported that copepod numbers in the blue mussels fluctuated greatly throughout the year but with a population peak in autumn and a trough in spring. Subsequent monitoring of this population has shown that these fluctuations are consistent from year to year (Figure 2).

In the Lyhner population, female *M. intestinalis* breed twice, and two generations of parasites coexist for most of the year; recruitment takes place in summer and autumn. One generation contributes its first brood to the autumn recruits before it overwinters and contributes its second brood to the following summer's recruits. The other generation overwinters as juvenile and immature stages and contributes its two broods successively to the summer and autumn recruits (Figure 3). This cycle of overlapping generations occurs because temperature is the controlling environmental variable for the developmental rates of all stages in the cycle from egg to adult. The mild climate of southwest England never entirely halts breeding; it merely slows the maturation of the latest autumn recruits so that they overwinter as juveniles.

This Cornish population does not typify what happens elsewhere. However, the temperature-dependent developmental rates throughout the parasite's life cycle could be expected to account for the differences observed over its geographical range. Thus, more northerly populations are limited to a single generation per year (Bolster 1954), with an even sharper seasonal variation in numbers of copepods per host. Continuous breeding of

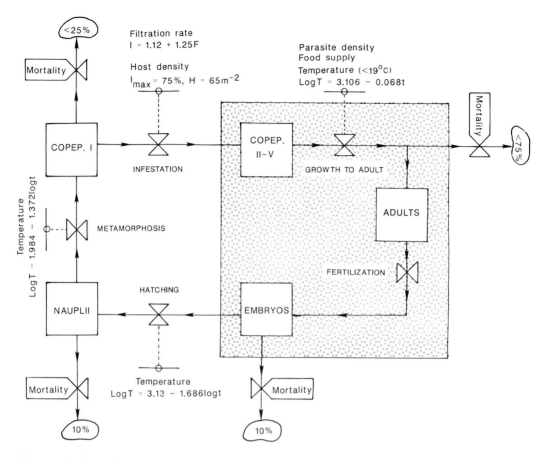

FIGURE 1.—Flow diagram of the life cycle of *Mytilicola intestinalis* with experimentally derived regression equations (or Michaelis–Menton equation in the case of host density) for critical processes in the life cycle. Shaded area represents in vivo development. Copep. = copepodite, F = filtration rate, H = host density, I = infestation success, T = time in days, t = temperature in °C.

the parasite occurs in more southerly populations of Spain and the Mediterranean Sea (Figueras and Figueras 1981; Paul 1983), with less seasonal variation in parasite numbers because generations overlap continuously through the year.

Evidence for the controlling role of temperature emerged from the experimental studies of Gee and Davey (1986a, 1986b), in which the free-living and parasitic phases of the life cycle were studied over the temperature range 7–22°C in the laboratory (Figure 1). Growth to maturation of the parasites in these experimental infestations was achieved at the higher temperatures of 18 and 22°C in 70–80 d; below 18°C, maturation times increased exponentially.

Infestation

Gee and Davey (1986b) recognized the critical importance of the infestation process, but they were

unable to find evidence for chemical attraction. They concluded that infestation depended on a chance encounter between the copepodite and the blue mussel's field of filtration. The tendency for the infestive copepodite stage to swim downward and away from light (Bolster 1954), increases the chances of such encounters under natural conditions. This helps to explain why blue mussels growing off the bottom—on pier pilings, for example, or in raft-and-rope culture—have lower infestation rates than bottom dwellers (Hockley 1951). The strength of a blue mussel's inhalant current will further enhance the chance of infestation; possibly, this phenomenon explains why larger blue mussels have more parasites (Davey and Gee 1976). The essentially passive nature of the infestation process, coupled with the short duration and poor swimming ability of the infestive stage, also explains why levels of infestation are lower in turbulent regions of

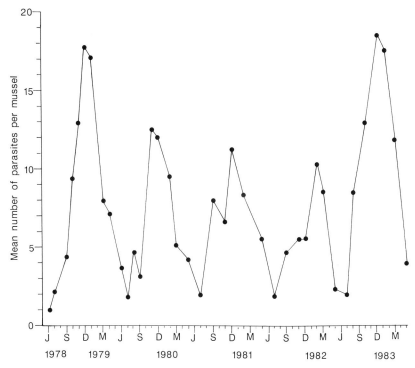

FIGURE 2.—Mean numbers of *Mytilicola intestinalis* (all stages) per blue mussel in size-representative subsamples of 50 blue mussels from 0.1-m² samples from Beggars Island, Lynher River, Cornwall, England, over the 5-year period June 1978–May 1983. M = March, J = June, S = September, D = December.

the open coast and higher in calm regions such as estuaries and harbors. In the latter habitats, the size and density of hosts are the main factors determining overall infestation success. Mature *M. intestinalis* have not been found in blue mussels under 10 mm in shell length and are rare in those under 20 mm. This relationship may be due partly to the relative strength of the inhalant current and also to the operation of a density-dependent mortality of parasites during their growth.

These findings from experimental studies on the life cycle by Gee and Davey (1986b) contradict the earlier work that demonstrated the high infestivity of *M. intestinalis*. Meyer and Mann (1950) obtained infestations exhibiting all stages of parasite development within 8 d of exposing healthy blue mussels in aquaria to suspensions of nauplii or after simply placing healthy blue mussels in the tanks with infested mussels. However, Gee and Davey (1986b) demonstrated that nauplii would only have developed to copepodite I–III in this time period, even under the most ideal conditions of temperature and salinity. Perhaps these "healthy" mussels were already infested with *M. intestinalis*. We do not know of any recorded

instance of adult copepods leaving the host and invading another; adult *M. intestinalis* dissected out of blue mussels and placed in water with new hosts did not enter them (Davey and Gee, unpublished observations).

Epizootiology

Adult *M. intestinalis* are only found in the hind gut, and early copepodite stages are found only in the stomach and digestive tubules. Increased body size and the requirement of the embryo-bearing females to release their hatching nauplii into the sea may require adult copepods to move toward the posterior intestine of the mussel. The importance to the infestive copepodites of remaining in the stomach or direct intestine has implications which have not been fully investigated.

Couteaux–Bargeton (1953) and Moore et al. (1978) recorded the encapsulation of *M. intestinalis* larvae in the tissues of the blue mussel digestive gland, which suggests that the smallest copepodites may be feeding on the epithelial cells of the digestive tubules. Some larvae evidently cause sufficient irritation that the encapsulation response is elicited, and they are killed. This outcome is not

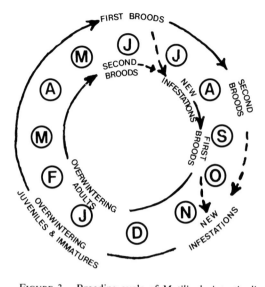

FIGURE 3.—Breeding cycle of *Mytilicola intestinalis* in blue mussels from the Lynher River, Cornwall, England. Solid lines and arrows denote growth and development of a parasitic generation; broken lines and arrows show free-living phases linking one generation to the next; circled letters are months of the year. (From Davey et al. 1978.)

in the parasite's interest. The questions that arise are whether these stages of *M. intestinalis* are obligate host tissue feeders and whether feeding is necessary as a trigger to their further development as they make the transition from a free-living existence to the endozoic environment of the blue mussel gut. This situation is analogous to the exsheathing stimulus provided by mammalian gut environments for invading nematodes (Rogers and Sommerville 1963). Certainly, development beyond the first copepodite is never seen outside the host in rearing experiments.

This speculation raises the question of the extent to which *M. intestinalis* may be considered a parasite in the sense of physiological dependency on the host (Davey et al. 1977; Dare 1985). *Mytilicola intestinalis* is not confined to the two species, *Mytilus edulis* and *M. galloprovincialis;* infestations have been recorded in *Ostrea edulis, Cerastoderma edule,* and *Tapes decussatus.* However, these hosts have only been found to harbor low levels of infestation. Experimental attempts to infest *Scrobicularia plana, Pecten (Chlamys) varius, P. maximus,* and *Macoma balthica* were unsuccessful (Baird et al. 1951; Hepper 1953, 1955, 1956). A single record exists of the infestation of *Crassostrea gigas* in France (His 1979). Experimental efforts to infest this

species achieved such low levels of infestation that the author concluded that *C. gigas* seed presented no risk to the transmission of *M. intestinalis,* because one or two overwintering juvenile stages in relatively larger *C. gigas* were unlikely to mature to breeding condition (Dare 1982).

The body of *M. intestinalis* is modified morphologically, but when the bizarre morphological modifications of most copepods parasitic on fishes are considered (see Kabata 1970), it might be conceded that these are relative matters. The body shape of *M. intestinalis* is cylindrical, that is, it is adapted for occupation of the cylindrical gut space of its host. There is an overall reduction in the complexity of the appendages as compared with free-living cyclopoids. This modification is particularly evident in the mouthparts; the mandibles are entirely lacking, and the maxillulae, maxillae, and maxillipedes are extremely simplified. The antennae are modified as a pair of stout hooks that are used as anchors for resisting peristaltic expulsion from the host.

Giusti (1967) claimed that *M. intestinalis* removes the microvillar border of the intestinal epithelium of *Mytilus galloprovincialis.* However, Moore et al. (1978) found that gut cells of *M. intestinalis* were equipped for intracellular digestion of carbohydrates, not proteins, and they concluded that the copepods were more likely to be herbivores that feed on host gut contents than on host tissue. Dethlefsen (1970) dissected out adult specimens of *M. intestinalis* from blue mussels and kept them alive in seawater at various salinities and temperatures. Some specimens lived for 76 days at 4°C and 20‰ salinity, and this in vitro longevity does not suggest a highly evolved physiological dependence on the host's internal environment.

The picture that emerges is not that of a parasite with the capacity to produce epizootic infestations. Like the copepods that parasitize fish (Kabata 1981), *M. intestinalis* seems incapable of endogenous regulation of its life cycle, which is subject to environmental factors, especially temperature. Whereas Williams (1969) and Bolster (1954) thought specific temperature thresholds were necessary to initiate breeding, work of Gee and Davey (1986a, 1986b) shows that reproduction is more a matter of developmental rates. Breeding probably only ceases at temperatures of 4°C and below when, due to exponential increases in development times, maturation would only be achieved in the order of hundreds of days. Over the range of 10–20°C breeding, infestation and growth to maturity can take place continuously within the normal time spans of the passage of the seasons. The concept of a lethal level of infestation with *M. intestinalis* no longer appears tenable. The sharp drop in the mean number of parasites per blue mussel that occurs each spring in the Lynher river population (Figure 2) clearly results from parasite deaths and not host deaths (Davey et al. 1978). An environmentally controlled relationship between host and parasite populations of this sort contrasts with the host-regulatory (inhibitory) mechanisms, which are usually in evidence in well-integrated, single host–parasite systems. In these systems, the parasite has evolved to a sophisticated level of physiological dependence on the host, and acute pathological effects may result from overinfection (Ractliffe et al. 1969; Lester and Adams 1974).

Physiology of Host–Parasite System

Korringa's papers published 1950–1968 (see Korringa 1968) did much to establish the pest status of *M. intestinalis*. Korringa (1968) reviewed almost every aspect of the parasite. He was convinced that one parasite could adversely affect the host and that more than 10 could kill, but he conceded that the mechanism of the effect was still not understood.

Meyer and Mann (1950) were the first to determine the effects of the copepod's presence on blue mussel physiology. These effects included reduced total flesh weight; reduced relative proportions of ash, lipid, and protein in the flesh; reduced relative weight of the digestive gland compared to total flesh weight; increased protein digestion; and increased filtration and respiration rates. Only the measured strength of the byssus threads was found to be unaffected by infestation with the copepods. These authors gave no details of the numbers of host animals examined, no sites of origin or time of year of the experiments, and no statistical tests of the data, but all the effects that they listed were found in the presence of eight, or fewer, copepods per blue mussel.

Cole and Savage (1951) examined blue mussels from Blyth in Northumberland, England, in January 1948 and found them to be 100% infested. All the parasites were found in the region of the gut behind the style sac, which would be consistent with an overwintering, nonrecruiting population of parasites (Davey et al. 1978), notwithstanding that some females carried egg sacs. The flesh weight of these mussels was deemed to show the deleterious effect of the copepods, but Cole and Savage (1951) compared them (nonstatistically) with similar-sized, uninfested blue mussels taken from Conway, north Wales, in June.

Bolster (1954) obtained samples from Llanelli (Wales), that failed to show the negative correlation between condition and number of parasites. He suggested that the real relationship was masked by particular factors, such as the increase of infestations when the host's gonads were developing or had already developed. This factor would enable blue mussels to hold their condition in spite of the infestation.

As the years went by, the failure of the infestation rate of many stocks to decline (i.e., for the "epidemic" to wane) and the manifest ability of many blue mussel stocks to sustain growth in the face of continued infestation began to raise doubts regarding the virulent pest status of *M. intestinalis*. Williams (1969) failed to demonstrate substantial effects at the biochemical level between infested and uninfested mussels at Lee-on-Solent (England). Dethlefsen (1975) could not detect statistically significant differences between meat content and infestation rate for Waddensea (Netherlands) blue mussels. Several other reports showed blue mussel populations thriving in the face of the infestation (Andreu 1963; Brienne 1964; Campbell 1970). Even where the infestation has begun to affect the raft-cultivated blue mussels of the Rias of Galicia, Spain, that had previously been less vulnerable, the Dutch experience (Korringa 1950) was not repeated. Paul (1983) could find no ill effects on condition index for up to nine parasites per host (the maximum found in that study).

Researchers in marine invertebrate physiology established, during the 1970s, a rigorous set of protocols for measuring the physiological performance of *Mytilus edulis* in relation to environmental stress. In particular, they measured the ratio of oxygen consumed to nitrogen (ammonia) excreted (O:N ratio) and the ratio of carbohydrate to protein being metabolized, and they developed methods for assessing an individual blue mussel's "scope for growth" as a measure of its energy balance (Bayne et al. 1977; also see Newell and Barber 1988, this volume). These studies showed that most physiological processes "have a marked seasonal variability, are size dependent, and are responsive both in the short and long term to changes in temperature, salinity, tidal periodicity and even the degree of satiation in the digestive system. Even when variability within a population can be accounted for, variability between populations can be considerable" (Bayne 1980).

Bayne et al. (1978) used their physiological protocols to determine the effects of *M. intestinalis* on host performance within each of two populations of blue mussels from southwest England, the Lyhner estuary (infested) and the Erme estuary (uninfested). For the Lyhner population, they compared heavy and light natural infestations of blue mussels from different levels on the beach; for the Erme population they compared artificially infested blue mussels with uninfested blue mussels that had been subjected to the same handling procedures. They detected no effect of the copepod on rates of oxygen consumption, and a slight positive effect on nitrogen excretion was of doubtful statistical significance because of the interaction with both time and temperature. Variation in the O:N ratio did not correlate with infestation; the only positive effect that was statistically significant was a slight depression of feeding rates in those animals with the very highest infestations. This depression was observed in blue mussels with 10–26 copepods each and was only evident at high temperatures (22 or 23°C) and at a maintenance or submaintenance level of food ration. A combination of such conditions in the field would be unusual; summer temperatures and winter food levels would have to coincide with autumn peaks in parasite density.

Use of the condition index, as a measure to assess the impact of *M. intestinalis,* has been misleading. The index is the ratio of the dry flesh weight to the shell cavity volume. Gee et al. (1977) studied the contribution of parasitic infestation in relation to environmental factors and host gametogenic stage and how these factors determined the variability of the condition index. From the two populations of blue mussels studied by Bayne et al. (1978), they examined nearly 2,500 blue mussels over 2 years. In every case, the magnitude of effects on the condition index due to variations in shell length, stage of gonad development, seasonal cycles, and environmental factors, including thermal and starvation stress, was greater than that due to parasitism. This observation is in accord with the view of Bayne et al. (1985) that the condition index is insensitive, has a relatively low signal-to-noise ratio, and can only reflect significant changes in measurements made over the long term (months or years). Further, Mallet et al. (1987) showed that changes in shell length and tissue weight (parameters of the condition index) are influenced by different factors; shell length is not limited by food availability, but tissue growth is.

Pathogenesis of Host–Parasite System

Our knowledge of the pathogenicity of infestation by *M. intestinalis* is rudimentary compared with that of the physiology of the host–parasite relationship. Couteaux–Bargeton (1953) described in detail the effects of *M. intestinalis* on the digestive tubule epithelium of *Mytilus.* The replacement of the normal grouping of tall columnar cells and small attendant cells by small uniform cells increased the size of the tubule lumen. No major difference could be found histochemically between the effects of parasitism and starvation. In both conditions, the cells contained a large number of bulky granules at the free end of the cell, and these were probably glycoproteins. They were also the site of alkaline phosphatase activity. Couteaux–Bargeton (1953) claimed that the presence of the parasite caused a greatly increased utilization of the glycogen reserves of the connective tissue and a change in the appearance of the parenchyma, such as is found in postspawning oysters. Infestation was also associated with increased activity in the renal epithelium and a greater production by the renal cells of granules that were stained by alkaline phosphatase.

Moore et al. (1978) also examined the histopathology of infested blue mussels. They described local metaplastic changes in the epithelium of the intestine and rectum associated with the presence of *M. intestinalis;* normal ciliated columnar cells were reduced to cuboidal nonciliated types. Only in a few instances of very high infestation were there indications of a more sustained metaplastic response. A large parasite or several smaller ones apparently produced some erosion of gut epithelial cells, but this erosion and the metaplastic changes were so localized that Moore et al. (1978) concluded that repair would be rapid. Although evidence was found of encapsulated larval parasites in connective tissue of the digestive gland, there was no fibrosis of the tissues. However, fibrosis was reported in similar circumstances by Sparks (1962) for the related copepod species *M. orientalis* in Pacific oysters *Crassostrea gigas* in North America. The observed encapsulation reaction was limited in this instance to hemocytes in the immediate vicinity of the dead larva.

Biological and Anthropogenic Synergisms

Mawdesley-Thomas (1972), addressing the diseases of fish, said that "disease, *per se,* is not an entity or an end in itself. Disease is the end result of an interaction between a noxious stimulus and a biological system and to understand disease is to understand all aspects of the biology of the species." Snieszko (1974) illustrated this statement in a simple diagram of three converging circles representing an organism, a pathogen, and the environment. In the region where all three circles overlap lies the domain of disease. In such a figure, *M. intestinalis* has been viewed as the pathogen, and the debate has concerned itself with the degree of virulence of the pathogenicity and, secondarily, with the possible role of environmental factors in interacting with the parasite to the detriment of the host. We cannot find in favor of any view that *M. intestinalis* is, in itself, a virulent pathogen. The true cause of the blue mussel mortalities in Holland in 1949 and 1950 remains unknown, but we can concede that some environmental stress (the summer of 1949 was particularly hot) enhanced the impact of a first-time infestation with *M. intestinalis.*

Such synergistic effects between environmental stresses and metazoan parasites are not unknown. Vernberg and Vernberg (1963) found that eastern mudsnails *Ilyanassa obsoleta* naturally infected with cercariae were less able to withstand thermal stress than uninfected specimens. In general, however, infectious disease agents, like viruses, bacteria, and other microorganisms, are more commonly associated with such synergisms (Wedemeyer 1970). Laird (1961) proposed that oyster epizootics associated with saprobic microorganisms were related to the development of microenvironments, in particular those initiating anaerobic conditions. Such microenvironments led to a general deterioration in the condition of the shellfish, which then succumbed to disease organisms to which they would otherwise be resistant. The possibility that stressful environmental conditions acted synergistically with *M. intestinalis* infestation in the Dutch example cannot be disregarded, but neither can the possibility that another infection was involved, because there are no data to consult. Section IV of this volume addresses some aspects of environmental influence on molluscan host defenses.

The interaction of stress and parasitism is a neglected area. Noble (1970) noted that the word stress was used in many branches of scientific literature but not in the field of animal parasitism. Esch et al. (1975) proposed a definition of stress to consider, theoretically and experimentally, the impact of stress on the dynamics of host–parasite systems. Studies that have followed this lead suggest that it is chiefly when the stressor is a pollutant that effects are found. Sindermann (1979) recorded, however, that the bulk of this literature relates to freshwater fishes and that, for marine species, good evidence relating pollutants with changes in parasite abundance is scarce. Moller (1977) was unable to find clear-cut relationships between parasitism and pollution for fish in the North Sea, but Yevich and Barszcz (1983) claimed that "a great need exists for information about synergistic effects of low-level, long term multiple pollutants on marine organisms. It has been shown that exposure of an organism to multiple pollutants can lower its resistance to a variety of parasites. While the parasites do not kill the host, they do reduce its ability to reproduce, which could have a detrimental effect on the population over a period of time." Their statement was based on blue mussels from the USA that were examined as part of the 1976 mussel watch program; the parasites included larval trematodes, nematodes, ciliates, microsporidia, and viruslike organisms, but not copepods. Nevertheless, copepods could be the vectors of more virulent pathogens, such as protozoa, fungi, bacteria or viruses.

Conclusions

Almost every answer we can give to the first two of our opening three questions is negative: *M. intestinalis* is neither epizootic nor a physiological burden in mussels. Our contention is that *M. intestinalis* and blue mussels form more of a commensal association in which the copepod uses the bivalve to concentrate algal food and then feeds on that fraction which is not utilized by the blue mussel. Additional work should be done to understand the copepod's feeding mechanism in more detail, and thereby to shed more light on the physiology of the transition from the free-living to the endozoic life-style.

Our reservations in answering the third question (economic concern) in the negative lies in the possibility that *M. intestinalis* could be responsible for introducing more virulent pathogens or that synergistic effects operate to enhance the response of infested blue mussels to such pathogens or to pollutants. The area of stress and pollution, even without regard to *M. intestinalis* infestation, warrants further research to maximize commercial blue mussel production.

References

Ahrens, W. 1937. Welche Tatsachen konnen zur Beurteilung der Meiosis als gesichert vorausgesetzt werden? Zoologischer Anzeiger 120:241–267.

Ahrens, W. 1939a. Die Entwicklung des primaren Spaltes der Copepod-Tetraden nach Untersuchungen über der Meiosis (Oögenese) von *Mytilicola intestinalis*. Zeitschrift für mikroskopisch-anatomische Forschung (Leipzig) 46:68–120.

Ahrens, W. 1939b. Über des Auftreten von Nukleolenchromosomen mit enstandigem Nucleolus in der Oögenese von *Mytilicola intestinalis*. Zeitschrift für wissenschaftliche Zoologie 152:185–220.

Andreu, B. 1963. Propagacion del copepodo parasito *Mytilicola intestinalis* en el mejillon cultivado de las rias gallegas (NW de España). Investigación Pesquera 24:3–20.

Baird, R. H., G. C. Bolster, and H. A. Cole. 1951. *Mytilicola intestinalis* Steuer in the European flat oyster (*Ostrea edulis*). Nature (London) 168:560.

Bayne, B. L. 1980. Physiological measurements of stress. Rapports et Procès-Verbaux des Réunions, Conseil International pour l'Exploration de la Mer 179:56–61.

Bayne, B. L., J. M. Gee, J. T. Davey, and C. Scullard. 1978. Physiological responses of *Mytilus edulis* L. to parasitic infestation by *Mytilicola intestinalis*. Journal du Conseil, Conseil International pour l'Exploration de la Mer 38:12–17.

Bayne, B. L., and nine coauthors. 1985. The effects of stress and pollution on marine animals. Praeger, New York.

Bayne, B. L., J. Widdows, and R. I. E. Newell. 1977. Physiological measurements on estuarine bivalve molluscs in the field. Pages 57–68 *in* B. F. Keegan, P. O. Ceidigh, and P. J. S. Boaden, editors. Biology of benthic organisms. Pergamon Press, Elmsford, New York.

Bolster, G. C. 1954. The biology and dispersal of *Mytilicola intestinalis* Steuer, a copepod parasite of mussels. Fishery Investigations, Series II, Marine Fisheries, Great Britain Ministry of Agriculture, Fisheries and Food 18:1–30.

Brienne, H. 1964. Observations sur l'infestation des moules du pertuis breton par *Mytilicola intestinalis* Steuer. Revue des Travaux de l'Institut des Pêches Maritimes 28(3):205–230.

Campbell, S. A. 1970. The occurrence and effects of *Mytilicola intestinalis* in *Mytilus edulis*. Marine Biology (Berlin) 5:89–95.

Caspers, H. 1939. Über Vorkommen und Metamorphose von *Mytilicola intestinalis* Steuer (Copepoda paras.) in der sudlichen Nordsee. Zoologischer Anzeiger 126:161–171.

Cole, H. A., and R. E. Savage. 1951. The effect of the parasitic copepod, *Mytilicola intestinalis* (Steuer), upon the condition of mussels. Parasitology 41:156–161.

Costanzo, G. 1959. Sullo sviluppo di *Mytilicola intestinalis* Steuer (Crost., Cop.). Archivio Zoologico Italiano 44:151–163.

Couteaux-Bargeton, M. 1953. Contribution a l'étude de *Mytilus edulis* L. parasite par *Mytilicola intestinalis* Steuer. Journal du Conseil, Conseil International pour l'Exploration de la Mer 19:80–84.

Dare, P. J. 1982. The susceptibility of seed oysters of *Ostrea edulis* L. and *Crassostrea gigas* Thunberg to natural infestation by the copepod *Mytilicola intestinalis* Steuer. Aquaculture 26:201–211.

Dare, P. J. 1985. The status of *Mytilicola intestinalis* Steuer as a pest of mussel fisheries: harmful parasite or benign commensal? International Council for the Exploration of the Sea, C.M. 1985/K:39, Copenhagen.

Davey, J. T., and J. M. Gee. 1976. The occurrence of *Mytilicola intestinalis* Steuer, an intestinal copepod parasite of *Mytilus*, in the south-west of England. Journal of the Marine Biological Association of the United Kingdom 56:85–94.

Davey, J. T., J. M. Gee, B. L. Bayne, and M. N. Moore. 1977. *Mytilicola intestinalis*—serious pest or harmless commensal of mussels? Parasitology 75:xxxv.

Davey, J. T., J. M. Gee, and S. L. Moore. 1978. Population dynamics of *Mytilicola intestinalis* in *Mytilus edulis* in south-west England. Marine Biology (Berlin) 45:319–327.

Dethlefsen, V. 1970. On the parasitology of *Mytilus edulis* (L. 1758). International Council for the Exploration of the Sea, C.M. 1970/K:16, Copenhagen.

Dethlefsen, V. 1975. The influence of *Mytilicola intestinalis* Steuer on the meat content of the mussel *Mytilus edulis* L. Aquaculture 6:83–97.

Dollfus, R. P. 1951. Le copepode, *Mytilicola intestinalis* Steuer, peut-il etre la cause d'une maladie épidémique des moules? Revue des Travaux de l'Institut des Pêches Maritimes 17(2):81–84.

Esch, G. W., J. W. Gibbons, and J. E. Bourke. 1975. An analysis of the relationship between stress and parasitism. American Midland Naturalist 93:339–353.

Figueras, A., and A. J. Figueras. 1981. *Mytilicola intestinalis* Steuer en el mejillon de la ria de Vigo (NO de España). Investigación Pesquera 45:263–278.

Gee, J. M., and J. T. Davey. 1986a. Stages in the life history of *Mytilicola intestinalis* Steuer, a copepod parasite of *Mytilus edulis* (L.) and the effect of temperature on their rates of development. Journal du Conseil, Conseil International pour l'Exploration de la Mer 42:254–264.

Gee, J. M., and J. T. Davey. 1986b. Experimental studies on the infestation of *Mytilus edulis* (L.) by *Mytilicola intestinalis* Steuer (Copepoda, Cyclopoida). Journal du Conseil, Conseil International pour l'Exploration de la Mer 42:265–271.

Gee, J. M., L. Maddock, and J. T. Davey. 1977. The relationship between infestation by *Mytilicola intestinalis* Steuer (Copepoda, Cyclopoida) and the condition index of *Mytilus edulis* in southwest England. Journal du Conseil, Conseil International pour l'Exploration de la Mer 37:300–308.

Giusti, F. 1967. The action of *Mytilicola intestinalis* Steuer on *Mytilus galloprovincialis* Lam. of the Tuscan Coast. Parasitology 28:17–26.

Grainger, J. N. R. 1951. Notes on the biology of the copepod *Mytilicola intestinalis* Steuer. Parasitology 41:135–142.

Hepper, B. T. 1953. Artificial infection of various molluscs with *Mytilicola intestinalis* Steuer. Nature (London) 172:250.

Hepper, B. T. 1955. Environmental factors governing the infection of mussels, *Mytilus edulis,* by *Mytilicola intestinalis.* Fishery Investigations, Series II, Marine Fisheries, Great Britain Ministry of Agriculture, Fisheries and Food 20:1–21.

Hepper, B. T. 1956. The European flat oyster, *Ostrea edulis* L., as a host for *Mytilicola intestinalis* Steuer. Journal of Animal Ecology 25:144–147.

His, E. 1979. Mytilicolides et myicolides parasites des lamellibranches d'intérêt commercial du bassin d'Arcachon. Haliotis 8:99–102.

Hockley, A. R. 1951. On the biology of *Mytilicola intestinalis* (Steuer). Journal of the Marine Biological Association of the United Kingdom 30:223–232.

Kabata, Z. 1970. Crustacea as enemies of fishes. Pages 1–171 *in* S. F. Snieszko, and H. R. Axelrod, editors. Diseases of fish, book 1. T.F.H. Publications, Neptune City, New Jersey.

Kabata, Z. 1981. Copepoda (Crustacea) parasitic on fishes: problems and perspectives. Advances in Parasitology 19:2–71.

Korringa, P. 1950. De aanval van de parasite *Mytilicola intestinalis* op de Zeeuwse mosselcultuur. Visserijnieuws 7 (supplement):1–7.

Korringa, P. 1968. On the ecology and distribution of the parasitic copepod *Mytilicola intestinalis* Steuer. Bijdragen tot de Dierkunde 38:47–57.

Laird, M. 1961. Microecological factors in oyster epizootics. Canadian Journal of Zoology 39:449–485.

Lauckner, G. 1983. Diseases of Mollusca: Bivalvia. Pages 477–961 *in* O. Kinne, editor. Diseases of marine animals, volume 2. Biologische Anstalt Helgoland, Hamburg, West Germany.

Lester, R. J. G., and J. R. Adams. 1974. A simple model of a *Gyrodactylus* population. International Journal for Parasitology 4:497–506.

Mallet, A. L., C. E. A. Carver, S. S. Coffen, and K. R. Freeman. 1987. Winter growth of the blue mussel *Mytilus edulis* L.: importance of stock and site. Journal of Experimental Marine Biology and Ecology 108:217–228.

Mawdesley-Thomas, L. E. 1972. Diseases of fish. Academic Press, London.

Meyer, P. F. 1951. Epidemische Erkrankungen der Meismuschelbestande in Nordwesteuropa. Fischereiwelt 3:82–83.

Meyer, P. F., and H. Mann. 1950. Beitrage zur Epidemiologie and Physiologie des parasitischen Copepoden *Mytilicola intestinalis*. Archiv für Fischereiwissenschaft 2:120–134.

Moller, H. 1977. Distribution of some parasites and diseases of fishes from the North Sea in February, 1977. International Council for the Exploration of the Sea, C.M. 1977/E20, Copenhagen.

Moore, M. N., D. M. Lowe, and J. M. Gee. 1978. Histopathological effects induced in *Mytilus edulis* by *Mytilicola intestinalis* and the histochemistry of the copepod intestinal cells. Journal du Conseil, Conseil International pour l'Exploration de la Mer 38:6–11.

Newell, R. I. E., and B. J. Barber. 1988. A physiological approach to the study of bivalve molluscan diseases. American Fisheries Society Special Publication 18:269–280.

Noble, G. A. 1970. Stress and parasitism. Journal of Parasitology 56:250.

Paul, J. D. 1983. The incidence and effects of *Mytilicola intestinalis* in *Mytilus edulis* from the rias of Galicia, northwest Spain. Aquaculture 31:1–10.

Pesta, O. 1907. Die Metamorphose von *Mytilicola intestinalis* Steuer. Zeitschrift für wissenschaftliche Zoologie 88:78–98.

Ractliffe, L. H., H. M. Taylor, J. H. Whitlock, and W. R. Lynn. 1969. Systems analysis of a host–parasite interaction. Parasitology 59:649–662.

Rogers, W. P., and R. I. Sommerville. 1963. The infective stage of nematode parasites and its significance in parasitism. Advances in Parasitology 1:109–177.

Sindermann, C. J. 1979. Pollution-associated diseases and abnormalities of fish and shellfish: a review. U.S. National Marine Fisheries Service Fishery Bulletin 76:717–749.

Snieszko, S. F. 1974. The effects of stress on outbreaks of infectious diseases of fishes. Journal of Fish Biology 6:197–208.

Sparks, A. K. 1962. Metaplasia of the gut of the oyster *Crassostrea gigas* (Thunberg) caused by infection with the copepod *Mytilicola orientalis* Mori. Journal of Invertebrate Pathology 4:57–62.

Steuer, A. 1902. *Mytilicola intestinalis* n.gen. n. sp. aus den Darme von *Mytilus galloprovincialis* Lam. Zoologischer Anzeiger 25:635–637.

Thiesen, B. F. 1966. *Mytilicola intestinalis* Steuer in Danish waters 1964–65. Meddelelser fra Danmarks Fiskeri- og Havundersogelser 4:327–337.

Vernberg, W. B., and F. J. Vernberg. 1963. Influence of parasitism on thermal resistance of the mud-flat snail, *Nassarius obsoleta* Say. Experimental Parasitology 39:445–451.

Wedemeyer, G. 1970. The role of stress in the disease resistance of fishes. American Fisheries Society Special Publication 5:30–35.

Williams, C. S. 1969. The life history of *Mytilicola intestinalis*. Journal of Natural History 1:299–301.

Yevich, P. P., and C. A. Barszcz. 1983. Histopathology as a monitor for marine pollution. Results of the histopathologic examination of the animals collected for the U.S. 1976 mussel watch program. Rapports et Procès-Verbaux des Réunions, Conseil International pour l'Exploration de la Mer 182:96–102.

American Fisheries Society Special Publication 18:74–92, 1988

Recent Investigations on the Disseminated Sarcomas of Marine Bivalve Molluscs

Esther C. Peters

Registry of Tumors in Lower Animals, National Museum of Natural History
Smithsonian Institution, Washington, D.C. 20560, USA

Abstract.—Almost 20 years have passed since the discovery of possible hematopoietic neoplasms in estuarine and marine bivalve molluscs. This enigmatic group of diseases, recognized as disseminated sarcomas, is presently the focus of research efforts in the USA and in Europe. Although these neoplasms vary in their development and cellular morphology, they are characterized by the appearance of hypertrophied cells with hyperchromatic nuclei in the connective tissue and in blood vessels and sinuses of the muscle and mantle tissue. These diseases can be diagnosed by histopathologic examinations or histocytologic techniques; however, the methods need to be standardized for interlaboratory data comparisons. The origin of the atypical cells has been variously attributed to hemic or connective tissues, but little is known about hemopoiesis in the bivalves. Several studies are attempting to address this question in at least two different species. Mammalian leukemias may be caused by viruses or other factors, such as chemical or radiation exposures. Similarly, several studies indicate that the disseminated sarcomas of bivalves may arise from contact with viruses, from exposure to adverse environmental conditions, or from the synergistic effects of several etiologic agents, but more research is needed to test these hypotheses. This paper reviews the literature and current research and suggests further investigations to meet the challenge of the disseminated sarcomas.

Beginning in the late 1960s, disseminated sarcomas of possible hemic origin have been reported in field populations of 15 species of marine and estuarine bivalve molluscs (Table 1) from around the world (Figure 1). In the last decade, there have been a number of investigations into the biology, ecology, and etiology of these diseases, particularly because of reported associations with shellfish mortalities on the east and west coasts of the USA and in Europe (Alderman et al. 1977; Emmett 1984; Farley et al. 1986b).

The classification of these conditions, as well as the results of many of the associated studies, have been quite controversial (see reviews by Dawe 1969, 1980; Farley and Sparks 1970; Harshbarger and Dawe 1973; Mix 1975b, 1976b, 1986; Rosenfield 1976; Farley 1977, 1988; Mix et al. 1977, 1979a; Harshbarger et al. 1980; Harshbarger 1982; Lauckner 1983; Cosson-Mannevy et al. 1984; Balouet and Poder 1985a, 1985b; Sparks 1985; Rasmussen 1986b). Reviewers' objections have referred to poorly designed field surveys, confusing experimental results, few standardized techniques, and the lack of basic information on molluscan biology. Despite these criticisms, there have been definite advances toward understanding these diseases. The sarcoma of softshell *Mya arenaria* has received much attention from two groups in New England, one from the University of Rhode Island, Kingston (Brown et al. 1976, 1977, 1979; Brown 1980; Oprandy et al. 1981; Cooper and Chang 1982; Cooper et al. 1982a, 1982b; Oprandy and Chang 1983; summarized in Appeldoorn et al. 1984), and one group now working at Tufts University, Boston, and the Woods Hole Oceanographic Institution, Woods Hole, Massachusetts, (represented by C. L. Reinisch, R. M. Smolowitz, and colleagues). Another group, at the National Marine Fisheries Service's Oxford Laboratory, Oxford, Maryland (C. A. Farley and colleagues), has been investigating the recent Chesapeake Bay epizootic. On the northwest coast of the USA, M. C. Mix and colleagues at Oregon State University, Corvallis, have investigated the sarcoma in blue mussel *Mytilus edulis* and Olympic oyster *Ostreola conchaphila* (formerly *Ostrea lurida*). R. A. Elston and colleagues at Battelle Marine Research Laboratory, Sequim, Washington, are continuing the studies on blue mussels in Puget Sound, Washington. European scientists are directing their efforts toward understanding similar diseases in blue mussel, edible oyster *Ostrea edulis,* and *Cerastoderma edule.* They include D. J. Alderman, M. Auffret, A. Cahour, D. M. Lowe, M. N. Moore, M. Poder, L. P. D. Rasmussen, E. Twomey, and the late Georges Balouet, whose contributions greatly enhanced the field of comparative tumor pathology. The importance and challenge of the sarcoma problem has been underscored in recent years by a series of Shellfish Mortality Conferences/Soft Clam Neoplasia workshops. Researchers from the eastern USA have come together to report current research, compare techniques, and discuss new approaches. Presentations at international conferences on comparative aspects of leukemia have also brought the attention of human and veterinary pathologists to these proliferative disorders of suspected hemic origin in bivalve molluscs.

There have been many published reviews of these diseases, but most reviews have discussed research done on only one or a few species, and there have not been any recent reviews which have discussed new directions for research. This paper will synthesize our present knowledge of these pathological conditions, based on studies with all species, and will address the lingering concerns and questions of invertebrate pathologists in three specific areas of needed research. In the fall of 1987, as this manuscript was being prepared, additional studies were underway in which modern cellular and molecular techniques were used.

Diagnosis and Techniques

These diseases were first discovered by histopathological examinations of fixed, embedded, and sectioned tissues. This method is still the primary means of diagnosis, although a number of nonlethal clinical techniques have been explored to allow long-term monitoring of the progress of the disease (see below). The Registry of Tumors in Lower Animals (RTLA), Smithsonian Institution, Washington, D.C., now has documented over 50 cases of these diseases and similar conditions (Table 2). Qualified investigators may request microslides for study.

The disseminated sarcomas have been known as hematopoietic neoplasms: diffuse neoplasm of hyaline hemocyte origin, diffuse sarcoma of possible amoebocyte stem cell origin, hemocytoblastic sarcoma, hemocytosarcoma, or disseminated hemic sarcoma. These conditions have also been termed undifferentiated sarcomas: poorly differentiated sarcoma of undetermined origin, diffuse mesenchymal sarcoma. The cytology and histopathology have been reviewed by Farley (1976a), Frierman (1976), Brown et al. (1977), Mix et al. (1977, 1979a), Mix and Breese (1980), Balouet et al. (1983), Green and Alderman (1983), Appeldoorn et al. (1984), and Sparks (1985). These sarcomas are characterized by the appearance of unusual cells in the connective tissue, blood vessels, and sinuses of the visceral mass, muscle, and mantle tissue (Figure 2A–E). The cells can be focal (Farley and Sparks 1970; Green and Alderman 1983) or widely disseminated (Farley 1969a; Farley and Sparks 1970; Brown et al. 1976; Alderman et al. 1977; Yevich and Barszcz 1977; Lowe and Moore 1978; Mix and Breese 1980). They are hypertrophied, usually enlarged to two to four times the diameters of normal hemocytes, and have a hyperchromatic nucleus. The nucleus can be pleomorphic or spheroid, depending on the bivalve species. There is a high nucleus:cytoplasm ratio. Although two separate morphological types of these atypical cells were reported in softshell (Brown et al. 1976) and in blue mussel (Mix et al. 1977; Lowe and Moore 1978), it is now believed that, at least in blue mussel, these types are actually endpoints of a continuum representing one type of abnormal cell (Mix et al. 1979a; Elston et al. 1988). Two or more nuclei and multiple necleoli may be found (Farley and Sparks 1970). Mitotic figures may be present in some of the cells. Unusual tri- or tetrapolar mitoses, suggestively of polyploidy, have also been reported. These cells are not phagocytic.

Three other abnormal cellular proliferative conditions of marine bivalves are histologically similar to disseminated sarcomas and this similarity has contributed to some of the confusion in the literature (e.g., Lauckner 1983). The first is a mesenchymal neoplasm which appears to arise from vesicular connective tissue (Leydig) cells. This condition has been reported in Olympic oyster, blue mussel, and eastern oyster *Crassostrea virginica* (Farley 1969a, 1969b; Farley and Sparks 1970; Newman 1972). The cells have hypertrophied nuclei and infiltrate the mantle and other tissues to form numerous focal lesions, but the cells remain attached and are not present in the hemolymph. The second neoplasm is a rare condition that has been identified in eastern oyster: 3 cases out of 25,000 animals collected from the Chesapeake Bay (Couch 1969, 1970; Harshbarger et al. 1979; but see also RTLA 1435, Table 2). In this condition, blastoid cells of the vesicular tissues, bearing hypertrophied nuclei with one or more prominent nucleoli and with numerous mitoses evident (some tripolar or atypical), form focal masses of basophilic attached spindle-shaped cells, infiltrating, replacing, and causing the degeneration of normal reticular fibers (Figure 2F). These cells do not appear to circulate in the

TABLE 1.—Published reports of disseminated sarcomas in various species of bivalve molluscs (see also Rosenfield 1976; Lauckner 1983; Mix 1986).[a]

Species	Location	Frequency[b]	Reference
Adula californiensis[c] California datemussel	West coast USA (?)	Unknown	Goner (personal communication, cited in Mix 1975a, 1975b)
Cerastoderma edule	Ireland	41/103	Twomey and Mulcahy (1984)
	18 sites	0–60%	Twomey and Mulcahy (in press)
	France		
	3 sites	31/752	Poder et al. (1983), Poder and Auffret (1986)
	1 site	46%	
Crassostrea commercialis	Australia	3/?	Wolf (1979)
Crassostrea gigas Pacific oyster	Matsushima Bay, Japan	1/?	Farley (1969a), Farley and Sparks (1970)
	Breton, France	0/8,000	Balouet et al. (1986)
Crassostrea rhizophorae	Todos os Santos Bay, Bahia, Brazil	1/986	Nascimento et al. (1986)
Crassostrea virginica Eastern oyster	Chesapeake Bay, Maryland	5/30,000	Farley (1969a), Farley and Sparks (1970)
		12/20,000	Harshbarger et al. (1979)
		70/?	Frierman (1976), Frierman and Andrews (1976)
	Appalachicola Bay, Florida	1/373	Couch and Winstead (1979)
	Pensacola Bay, Florida	20/4,486	Couch (1985)
	Mobile Bay, Mississippi	3/2,336	Couch (1985)
	Pascagoula Harbor, Mississippi	1/2,461	Couch (1985)
Macoma calcarea Chalky macoma	Baffin Island, Canada	1/519	Neff et al. (1987)
Macoma nasuta Bent-nose macoma and *M. inquinata*[d] Stained macoma	Yaquina Bay, Oregon	"Small numbers"	Goner (personal communication, cited in Mix et al. 1977a), Farley (1976a, 1977)
Mya arenaria Softshell	Rhode Island	49/147	Brown et al. (1976)
	Maine (oil spill)	48/440	Yevich and Barszcz (1976)
	Maryland–Nova Scotia	0–40%	Yevich and Barszcz (1977)
	Maine, Massachusetts,	159/1,609 (0–39%)	Brown et al. (1977)
	Rhode Island (5/10 sites)	152/1,325 (0–64%)	Brown et al. (1979)
	Allen Harbor, Rhode Island	20–40%	Cooper et al. (1982a, 1982b)
	Annisquam River, Massachusetts	12%	Farley (1976a)
	New Bedford Harbor, Massachusetts	10–90%	Reinisch et al. (1984)
	Chesapeake Bay, Maryland (5 sites)		
	1979–1983	4/3,584	Farley et al. (1986a)
	1983–1984	42–65%	Farley et al. (1986b)
	Long Island Sound, Connecticut (3 sites)	99/1,621 (peak 45%)	Brousseau (1987)
		164/1,262 (peak 59%)	
		136/1,075 (peak 60%)	
Mya truncata Truncate softshell	Baffin Island, Canada	3/856	Neff et al. (1987)
Mytilus edulis Blue mussel	Yaquina Bay, Oregon	7 and 12%	Farley (1969b)
	1972–1973	0/80	Mix et al. (1977, 1981)
		0–20%	Mix (1982, 1983)
	River Lynher, Plymouth, England	16/994	Lowe and Moore (1978)
	England (21 sites)	0–4.3%	Green and Alderman (1983)
	Saanich Inlet, British Columbia (3 sites)	0–29%	Cosson-Mannevy et al. (1984)
	Departure Bay and Bamfield, Vancouver Island, British Columbia	0–45%	Emmett (1984)
	Denmark	2/320	Rasmussen et al. (1985)
		44/8,400	Rasmussen (1986b)
	Puget Sound, Washington	0–40%	Elston et al. (1988)
	Furuskar, Tvärminne, Finland	1/205	Sunila (1987)

Table 1.—Continued.

Species	Location	Frequency[b]	Reference
Ostrea chilensis	Chiloe, Chile	1/48 1/100 (4/78)	Mix and Breese (1980)
Ostrea edulis Edible oyster	France Spain and Yugoslavia Mediterranean and Atlantic coasts	20–45%	Franc (1975) Alderman et al. (1977)
	7-year study	<1%	Balouet and Poder (1978), Balouet et al. (1983)
	Breton coast, France	184/40,500	Balouet et al. (1986)
Ostreola conchaphila[c] Olympia oyster	Yaquina Bay, Oregon	Unknown	Farley and Sparks (1970), Mix (1975a, 1975b, 1976), Farley (1976a)
	Imports	0/1,426, 0/399	Mix et al. (1977)
	Indigenous	2/112, 3/150	

[a] Unpublished reports indicate that the number of species affected and the range of locations where the condition exists are greater than those cited here.

[b] Fractions indicate the number of affected animals per total examined; percentages indicate either the number of animals examined is unknown or a range of peak prevalence was reported.

[c] Formerly *A. californica*.

[d] Formerly *M. irus*.

[e] Formerly *Ostrea lurida*.

hemolymph, although some of the abnormal cells were found on the walls of the blood vessels and occasionally in the lumen of the vessels. In-creased reticular fiber deposition was found in sections of two animals archived as RTLA 1763–1764 (Table 2). Couch (1969) thought that these

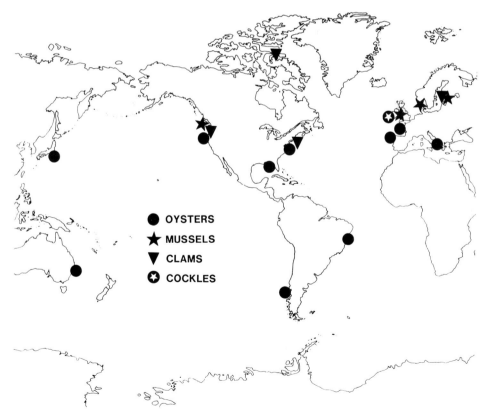

Figure 1.—Worldwide locations of reported cases of the disseminated sarcomas in various species of marine bivalve molluscs.

TABLE 2.—Representative cases of the disseminated sarcomas and similar conditions of marine bivalve molluscs accessioned in the Registry of Tumors in Lower Animals, Smithsonian Institution, Washington, D.C.

Species	RTLA number and collection site[a]	Contributor
Arctica islandica Ocean quahog	1392 Rhode Island Sound	P. P. Yevich
Cerastoderma edule	2798, 2801 Ireland 2891, 2892 Brittany, France	E. Twomey M. Poder and M. Auffret
Crassostrea commercialis	767, 983, 1117, 1794 Sydney, Australia	P. Wolf
Crassostrea gigas Pacific oyster	225 Yaquina Bay, Oregon (?)	C. A. Farley
Crassostrea virginica Eastern oyster	150[b] Chesapeake Bay 224 Chesapeake Bay (?) 502[c] Chesapeake Bay 844–887 Chesapeake Bay, Virginia 1435[b] ? 1599, Appalachicola Bay, Florida 1749–1750, 1752–1754, 1757–1761, 1763[b], 1764[b], 1765 Chesapeake Bay 1818.5 Laboratory study 2132 Pascagoula, Mississippi 4040[c] New Haven harbor, Connecticut 4042 Chesapeake Bay	C. A. Couch C. A. Farley M. S. Newman E. M. Frierman P. P. Yevich J. A. Couch S. V. Otto P. P. Yevich J. A. Couch M. W. Newman M. Frierman
Macoma balthica Baltic macoma	1004[d] Chesapeake Bay 3284 Finland 4032[d] Fox Hole Bar, Tred Avon River, Chesapeake Bay	F. G. Kern M. Pekkarinen D. Christensen, C. A. Farley, and F. G. Kern
Macoma inquinata[e] Stained macoma	4034 Idaho Point, Yaquina Bay, Oregon	F. G. Kern
Macoma nasuta Bent-nose macoma	4033 Yaquina Bay, Oregon	F. G. Kern
Mya arenaria Softshell	1169 ? 1413 ? 1698 Searsport, Maine 2641 Rhode Island 3523 Boston harbor, Massachusetts 4035 Jones Creek, Annisquam River, Massachusetts 4036 Umpqua Bay, Oregon 4044 Harpswell, Maine	P. P. Yevich R. S. Brown P. P. Yevich P. W. Chang M. A. Stockmayer F. G. Kern F. G. Kern P. P. Yevich and C. A. Barszcz

TABLE 2.—Continued.

Species	RTLA number and collection site[a]	Contributor
Mytilus edulis Blue mussel	235 Yaquina Bay, Oregon	C. A. Farley
	1833 Menai Straits, North Wales	D. J. Alderman
	4030[c], 4031[c] Lower Yaquina Bay, Oregon	C. A. Farley
Ostrea edulis Edible oyster	1779 Spain	D. J. Alderman
	1807–1809 Yugoslavia	P. Van Banning
	1855 ?	H. Grizel
	1940–1946 France	G. Balouet
Ostreola conchaphila[f] Olympia oyster	930, 981, 1029 Yaquina Bay, Oregon (?)	M. C. Mix
	4038[c], 4039[c] Oysterville Flats, Yaquina Bay, Oregon	C. A. Farley
Scallop (species unknown)	1834 ?	R. S. Brown

[a] Accession number of specimens in Registry of Tumors in Lower Animals (RTLA); ? = doubtful or unknown.
[b] Identified as a possible reticulum cell sarcoma, hemangioendothelioma (hemangiosarcoma).
[c] Identified as a sarcoma of possible vesicular connective tissue origin.
[d] Identified as a highly invasive gill carcinoma.
[e] Formerly *M. irus*.
[f] Formerly *Ostrea lurida*.

lesions resembled a reticulum cell sarcoma. Farley and Sparks (1970) remarked on the apparent formation of new hemolymph vessels and the reticulin borders of the neoplastic cells. They referred to the condition as a hemangioendothelioma (hemangiosarcoma) that originated from the reticuloendothelial cells of the hemolymph vessels. Harshbarger and Dawe (1973) acknowledged that these cases were of doubtful hemic origin. Other examples have not been found for more thorough study. The third similar condition is the "highly invasive gill carcinoma" described by Christensen et al. (1974) and Farley (1976a, 1977) in specimens of *Macoma balthica* from the Chesapeake Bay, Maryland. These abnormal cells closely resemble those of the disseminated sarcoma, because they have hypertrophied nuclei and multiple nucleoli suggestive of polyploidy and are found circulating in the hemolymph. However, Farley (1976b) reported that these cells originated from the gill epithelium, crossed the basement membrane, and caused a highly invasive systemic disease that was lethal to the animals. Similar conditions have been noted in *Macoma* spp. from Yaquina Bay, Oregon (J. J. Goner, Oregon State University, personal communications, cited in Mix 1975b; Mix et al. 1977), and in a specimen of Baltic macoma *M. balthica* from Finland (RTLA 3284, Table 2). The gill epithelial cell origin has not been confirmed in these cases, which may represent other diseases, possibly disseminated sarcomas. Other cells with abnormal morphologies are also seen in neoplasms originating from the germinal epithelium and filling the gonadal follicles in northern quahog *Mercenaria mercenaria* (Barry and Yevich 1972, 1975) softshell (Yevich and Barszcz 1976, 1977) and blue mussel (Cosson-Mannevy et al. 1984).

In the disseminated sarcomas, as the number of atypical cells increase, the number of normal granular and agranular amoebocytic hemocytes (leukocytes) decrease, and other pathological changes in the tissues become evident. Specimens with numerous atypical cells are in poor condition: disruptions of normal tissue architecture include gonad atrophy and collagenous fibrosis in Pacific oyster *Crassostrea gigas* (Farley 1969a); displacement, compression, and necrosis of normal gill, gonad, and connective tissues (Brown et al. 1977; Cooper et al. 1982b; Farley et al. 1986b); arrest of gametogenesis in blue mussel, eastern oyster, and softshell (Farley 1969b; Farley and Sparks 1970; Cosson-Mannevy et al. 1984; Emmett 1984; Brousseau 1987); and widespread degeneration and necrosis of tissues with death of the organism (Farley and Sparks 1970; Alderman et al. 1977; Farley et al. 1986b; others). Pycnosis and karyolysis of the atypical cells may occur in advanced cases (Farley

1969a, 1969b; Mix et al. 1977; Appeldoorn et al. 1984). Frierman (1976) and Appeldoorn et al. (1984) noted that shell growth and body weight increases were inhibited in eastern oysters and softshells with these diseases.

In the late 1970s, a clinical technique was developed for repeated sampling of living bivalves to diagnose the disease in studies of softshells and allowed large numbers of the animals to be screened quickly (Farley et al. 1986b). Hemolymph is removed directly from the pericardial cavity or adductor muscle sinus with a hypodermic needle and syringe. The cells are placed directly on glass slides and are examined by phase-contrast microscopy with a hemocytometer, or they are fixed, stained with Giemsa or Wright's stain, and examined by bright-field microscopy (Brown et al. 1976; Cooper et al. 1982b). Farley et al. (1986) termed this technique "histocytology" (later designated "hemocytology," see Elston et al. 1988) and modified it by mixing one part of the blood, as it is collected, into nine parts of sterile saline (ambient 15‰) and placing this solution in a small plastic "Farley chamber" affixed to a microscope slide coated with 0.1% poly-L-lysine. After the blood cells settled, the diluent was carefully washed off and the slides were fixed, stained (usually with Feulgen picromethyl blue, FPB), and examined by light microscopy. Smolowitz and Reinisch (1986) used murine monoclonal antibodies prepared against the atypical cells (Reinisch et al. 1983) to develop an immunoperoxidase (IP) staining technique to apply to cells prepared in this manner.

The numbers of atypical cells and hemocytes were compared, and five, six, or seven stages of the disease were recognized (Table 3). Appeldoorn et al. (1984) reported that the disease was diagnosed correctly 94% of the time, and they obtained slightly better results with the Giemsa-stained preparations than with the histological techniques.

However, the ability to correctly diagnose this condition was strongly correlated with the severity of the disease (Table 3). Unless the slides are first coated with poly-L-lysine, the abnormal cells will not stick to the slides due to the loss of membrane attachment sites (Brown et al. 1977; Mix et al. 1979; Appeldoorn et al. 1984; others); therefore, studies with noncoated slides are invalid. Also, fresh preparations will not keep for later comparisons and confirmation of experimental results. Histopathological examinations of the tissues were not always done (Cooper et al. 1982b), and false positive diagnoses have occurred (Appeldoorn et al. 1984). This observation was confirmed by Smolowitz and Reinisch (1986), who found that the hemocytometer method of examining fresh cells was only 53% as accurate as, and more time-consuming than, the IP and FPB staining methods. The high specificity of the IP staining technique also removes any doubt that similar-appearing normal cells rarely will be misidentified as abnormal in the FPB method.

There are potential problems in the frequency of repetitive bleedings and the need to use proper controls in experiments. For example, Cooper et al. (1982b) bled *Mya arenaria* at monthly intervals and found no difference in mortality from that of unbled controls. But they did not report if any tissue changes occurred in the animals which were regularly bled. In other cases, sham bleedings would be desirable controls for the stress of handling and injection (e.g., introduction of bacteria into the puncture wounds). The histocytology technique may also be limited in value because of the restricted availability of the monoclonal antibodies and the possibility of cross-reaction with the atypical cells of other species. However, these antibodies should prove useful in distinguishing cell types, such as hemic from gonadal or connective tissue cells, when more

FIGURE 2.—Photomicrographs of the disseminated sarcomas and the reticulum cell sarcoma in marine bivalve molluscs. All sections were stained with hematoxylin and eosin. Bar = 100 μm for A–F and 25 μm for inset of D. **A.** Neoplastic cells filling hemolymph vessels in the gills of ocean quahog *Arctica islandica*, Registry of Tumors in Lower Animals (RTLA) 1392. Note pycnotic nuclei at arrow. **B.** Gills of softshell *Mya arenaria* RTLA 3523, showing sarcoma cells with pleomorphic nuclei (large arrow) and normal hemocytes (small arrow). **C.** Disseminated sarcoma of blue mussel *Mytilus edulis*, RTLA 1833, filling connective tissue of mantle. Granular cells (small arrow) are nutritive cells of the vesicular connective tissue. **D.** Neoplastic cells in disrupted connective tissue of *Cerastoderma edule*, RTLA 2798. Note mitotic figures (small arrows), unusual tripolar mitosis at point of large arrow, and another cell in mitosis at tail of large arrow. Inset shows enlargement of these last two mitotic figures with tripolar mitosis at point of large arrow. **E.** Disseminated sarcoma cells filling the connective tissue and blood vessel (arrow) in the digestive gland of eastern oyster *Crassostrea virginica*, RTLA 1818.5. **F.** Reticulum cell sarcoma (large arrow) in vesicular connective tissue (Leydig cells) of eastern oyster, RTLA 1764. Note absence of neoplastic cells in blood vessel (small arrow).

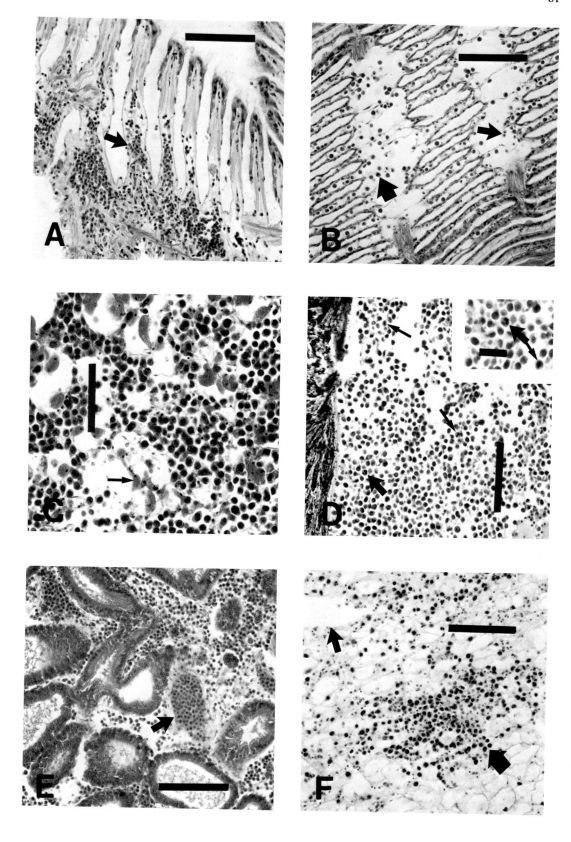

TABLE 3.—Comparison of different systems of classifying the clinical (histocytology) diagnostic stages of the disseminated sarcomas in softshells *Mya arenaria*.

Diagnostic stage	Classification system			
	Number abnormal cells/mL hemolymph (Cooper et al. 1982a, 1982b)[a]	Number abnormal cells/ 100 normal cells		Number abnormal cells/number normal cells (Farley 1988)[a,b,c]
		(Farley et al. 1986a, 1986b)	(Smolowitz and Reinisch 1986)	
0		0	0	
		Early stages		
1	0–1 × 10⁴ (66–71%)	0.01–0.09	0.01–0.09	0–9/100,000 [1] (15%)
2	2–10 × 10⁴ (76–93%)	0.1–0.9	0.1–0.9	10–99/100,000 [1] (30%)
		Intermediate stage		
3	11–100 × 10⁴ (100%)	1–49	1–50	1–9/1,000 [2] (71%)
		Advanced stages		
4	101–1,000 × 10⁴ (100%)	50–89	50.1–90.0	1–9/100 [3] (95%)
5	>1,000 × 10⁴ (100%)	90–100	90.1–100	10–49/100 [3] (100%)
6				50–89/100 [4] (100%)
7				90–100/100 [5] (100%)

[a] Percent accuracy in parentheses determined by comparison with histological examinations (Cooper et al.) or live monolayer preparations (Farley).

[b] Number in brackets refers to earlier scale of Farley et al. (1986a).

[c] The seven stages were proposed by Farley (1986a) to diagnose other hemolymph disorders (e.g., MSX in oysters) histologically. Stages 1 and 2 = number abnormal cells/100,000 normal cells per slide; stage 3 = 1–9 abnormal cells per 200× field.

than one type of neoplasm is suspected in a particular species (e.g., softshell).

Farley et al. (1986b) considered that the histocytological technique was superior to histopathological techniques for diagnosis of early stages of the disease. However, Green and Alderman (1983) and many other investigators believed that careful examination of properly prepared sections taken from the whole animal can detect mild cases and also demonstrate other lesions which may be the cause of mortality. Mix (1983) proposed a qualitative four-part staging scale for histopathological diagnosis that was based on relative infiltration of the cells through the connective tissue and into the hemolymph sinuses. Appeldoorn et al. (1984) reported a quantitative five-step staging scale based on the number of atypical cells seen in a 100× magnification field of connective tissue. The need to standardize these techniques and apply them uniformly in all investigations is urgent. In particular, suitable fixatives and staining procedures should be used to produce the best quality sections; a consensus among investigators must be reached on an appropriate, easily applied cell and tissue staging scheme so that data comparisons can be made among laboratories. Standard histopathological examinations should still be used to confirm cases identified by histocytological (hemocytological) methods, and files of fixed and stained preparations must be maintained for reference.

Cell of Origin

Despite numerous observations, much doubt remains that these abnormal cells, which resemble the anaplastic cells of mammalian leukemias, are actually altered circulating hemocytes. Moreover, the cell of origin may differ in the various species. Farley (1969a) noted that the cells in *Crassostrea* spp. appeared to be atypical immature hyaline hemocytes; however, he suggested that they might also be altered gonadal germ cells which were invading other body tissues (see also Cosson-Mannevy et al. 1984; Rasmussen 1986b). Other investigators have proposed that the cells are dedifferentiated vesicular connective tissue cells (Farley and Sparks 1970) that later become disseminated (Harshbarger and Dawe 1973), transformed mesenchymal cells of loose connective tissue (Mix et al. 1979a), or mesenchymal cells whose origins and functions are not known

(Sparks 1985). The cells were also suspected of being protozoan parasites (Mackin and Schlict 1976), but this suggestion has been ruled out (Farley 1969a, 1976a; Farley and Sparks 1970; Couch and Winstead 1979; Dawe 1980). Lowe and Moore (1978), Balouet and Poder (1980), Balouet et al. (1983), Appeldoorn et al. (1984), and other recent authors have agreed with the anaplastic hyaline hemocyte interpretation of cytogenesis. Furthermore, the tumor cells vary, within and between species, from large undifferentiated blastlike (stem cell) forms to smaller well-differentiated hyaline hemocyte forms, yet they are still distinguishable from normal hemocytes (Harshbarger et al. 1980).

Little is known about hemopoiesis in bivalves. Feng (1967) suggested that the vesicular connective tissue, mesenchymal tissues, and kidney epithelium were possible sites for stem cell differentiation and hemocyte renewal. Farley (1969a) believed that other possible sites of hemopoiesis were the reticuloendothelial lining of hemolymph vessels and germinal epithelium; Mix and Tomasovic (1973) added the pericardial membrane to this list. Cheng (1981) reviewed the question of bivalve hemopoiesis and noted that current accounts have been based on interpretive evaluations and not on direct evidence. In particular, he argued that Mix's (1976a) design was too complex and was developed from reviews of incorrect data. Auffret (1988, this volume) provides a review and synthesis of much of this material. Although most investigators have reported finding granular (granulocytes) and agranular (hyalinocytes) hemocytes (also called leucocytes), they have reported one to seven different types according to species. Most of these differences can be attributed to the types of fixatives and stains that were used. Although some of these cell types may be ontogenetic stages, this has not been demonstrated (Cheng 1981). Thus, the terminology related to observations of normal cellular structure and function is perplexing.

Several indications suggest that hemocytoblasts and fibroblasts (connective tissue cells) are closely related and that the blood cells might arise from fibroblasts or vice versa (Mix 1976a; Appeldoorn et al. 1984). Lowe and Moore (1978) and Green and Alderman (1983) observed, in early cases, a few altered cells in the basal connective tissue of the gut and digestive diverticula, which are also sites of increased accumulations of granular hemocytes in chronic inflammatory responses to parasites (Farley 1976a) or pollutants

(Rasmussen 1982; Rasmussen et al. 1985; Neff et al. 1987). Some authors believe that the cells originate by a replication of normal DNA that produces polyploidy (Lowe and Moore 1978) or aneuploidy (R. A. Elston, personal communication). These observations were made by flow cytometry and transmission electron microscopy of the cells. The latter studies also revealed a variety of anomalous and abnormal cytoplasmic organelles that do not clearly link them with normal hemocytes (Brown et al. 1977; Mix et al. 1977; Cahour and Balouet 1984; Auffret and Poder 1986). In cytochemistry studies, Lowe and Moore (1978) found that the blue mussel sarcoma cells were periodic acid–Schiff (PAS)-negative and usually, but not always, negative for the lysosomal hydrolases found in normal hemocytes. Reinisch et al. (1983) used monoclonal antibodies developed against the atypical softshell cells and reported that only 1 of 10 antigens on the surface of these cells matched antigens on the normal hemocytes, an indication perhaps of different tissue origins. Elston and colleagues (personal communication) have found qualitative and quantitative differences of hemolymph and cellular proteins between normal and affected blue mussels.

Most invertebrate pathologists believe that the abnormal cells are derived from one or more lines of bivalve stem cells and that some of the variation in cell structure is directly related to the process of neoplastic transformation (Cahour and Balouet 1984). Because these cells mimic the morphology and pathogenic behavior of cells found in mammalian leukemias, most investigators have agreed that the atypical disseminated cells of bivalves represent a neoplastic disease. These atypical cells exhibit (1) anaplasia (similar to vertebrate neoplastic cells), (2) rapid proliferation (Mix and Tomasovic 1973; Mix 1975a; Mix et al. 1977), (3) invasiveness (Brown et al. 1977), (4) clonal alteration of genetic material (probably polyploidy), (5) transplantability (Farley 1988; Twomey and Mulcahy 1988; Elston et al., in press), and (6) progression and malignancy (Frierman 1976; Brown et al. 1977; Sparks 1985; Farley et al. 1986b).

However, Mix et al. (1977, 1979a) argued that many of the properties exhibited by the atypical cells were characteristic of normal cells undergoing rapid proliferation (hyperplasia), for example, in response to bacterial pathogens. Thus, the atypical cells may be a non-neoplastic cellular response to chronic chemical exposure or other types of environmental stress with functional capabilities associated with the detoxification of organic body burdens. The cells could also be initially nonneoplastic with subsequent neoplastic transformation, or they could be entirely neoplastic. Furthermore, Cheng (1976) and Mix (1976b) cautioned that the condition cannot be defined as a sarcoma until the nature of the cells is determined. In assessing the arguments for the occurrence of neoplasia in molluscs, Stewart (1976) remarked that there are "no fixed and immutable criteria for the diagnosis of cancer in all species," and there are many examples of misdiagnoses in all groups of animals that have become evident only after many years of research. Still, he believed that the current evolutionary thinking was important to establish principles in comparative pathology. The behavior and morphology of the abnormal cells are characteristic of a neoplastic disease and consistent with the interpretation of a disseminated sarcoma. The diffuse appearance of the cells suggests a hemic origin, but there has not been any definite evidence to support this hypothesis.

The Massachusettes group is investigating cell line precursors in softshell with antibody probes, sodium dodecylsulfate–polyacrylamide gel electrophoresis, and western blot techniques to determine similarities and differences in proteins between the hemocytes and sarcoma cells (which they are calling leukemia, see Mioski et al. 1989). Smolowitz et al. (1989) reported using IP staining techniques with a monoclonal antibody that is cross-reactive (i.e., shares an antigen) with both normal hemocytes and the neoplastic cells. In those softshells in which no neoplasia was detected by the histocytological techniques, the antibody reacted with the surface of a normal but slightly different hemocyte (perhaps a granulocyte at an early stage of maturation or a precursor cell?). In the tissue sections there was also a reaction with connective tissue cells lining hemolymph sinuses. This latter reaction was seen only occasionally in "disease-free" clams, increasing with the intensity of the neoplasia. Also, fewer reactive normal hemocytes were seen as the number of neoplastic cells increased. These observations suggest the possible site of origin of the hemocytes and neoplastic cells, but more work needs to be done. The Batelle group is initiating similar studies of the sarcoma of blue mussel and quantitatively comparing changes in cell function using flow cytometry to search for anomalous DNA. They plan to examine the DNA content of the cells and to determine the cell lineage by applying monoclonal antibodies or other antibodies, specific molecular probes, and hybridization techniques to investigate gene sequences and protooncogene expression (Elston, personal communication).

Mix (1975a) used ^{3}H-thymidine to label two affected Olympia oysters and prepared autoradiographs of histological sections to demonstrate more rapid proliferation of the abnormal cells; he suggested that such techniques might be useful to search for the originating cell type. D. Leavett (Coastal Research Laboratory, Woods Hole Oceanographic Institution, personal communication) labeled softshell leukemia cell membranes with radioisotopes to study the monoclonal antibody epitopes. However, no other reports have appeared in which radioactive labeling has been applied to bivalve tissues to trace the development of normal or neoplastic cells (as proposed by Balouet et al. 1986) or to identify cell populations during transplantation and transmission experiments. Although Brown et al. (1977) expected that the isolation and growth of normal and neoplastic cells in cell culture should be easy, this procedure has not yet met with success. The advanced biomedical techniques used to study mammalian leukemias and other cellular proliferative disorders should be applied to the exploration of cell origins in the disseminated sarcomas of bivalves.

There is still much work to be done on the basic cellular biology and ontogeny of bivalves. Besides discovering the site(s) of hemopoiesis, we also need to learn what mechanisms control the process of blood cell formation. Also, we need to know how contact with foreign organisms (parasites and pathogens), requirements for wound and shell repair, chemical insults, and abiotic factors such as temperature or oxygen concentrations may alter the process and trigger increased hemocyte production or release of hemocytes from tissues (hemocytosis or leucocytosis, an inflammatory response). Furthermore what factors can decrease normal hemocyte production or alter the blast cells to produce neoplastic disease? What are the cellular identities and relationships of the various types of neoplastic diseases that have been discovered? Some exciting and perhaps controversial revelations on these topics may be forthcoming in the next few years.

Case for an Etiologic Agent

Dameshek (1970) remarked that the etiology of leukoproliferative disorders must be differentiated from their pathogenesis. In mammalian leukemias, there may be chronic or acute self-limiting conditions (such as mononucleosis). Although a variety of viruses have been associated with many leukemias, Dameshek warned that the viral etiology was not clear-cut and that somatic mutations (perhaps produced by exposure to radiation or chemicals) and the development of abnormal clones of cells could be enhanced by immunodeficiencies. Today, medical researchers are linking viral and cellular oncogenes to neoplastic diseases, and understanding the process of transformation of a normal cell to a neoplastic cell now seems within reach. Recent investigations of the bivalve disseminated sarcomas also point to complex and perhaps varied etiologies.

Field studies indicated that there were substantial differences in effects of neoplasia on populations in different localities and probably within and between different species (Brown et al. 1977; Mix et al. 1977; Mix 1983; Rasmussen 1986b). Limited samplings demonstrated that the severity of the disease increased with age in softshells (Appeldoorn et al. 1984) and edible oysters (Balouet et al. 1983, 1986). Emmett (1984) reported the occurrence of the sarcoma in young (1-year-old) seed blue mussels at one site off Vancouver Island, British Columbia, in the same frequency as in 2-year-old wild blue mussels sampled in July. However, there were no deaths in the younger specimens, whereas smaller individuals of the adult populations suffered up to 90% mortality. Emmett (1984) believed that adverse environmental conditions (but not pollution) and handling stress probably contributed most to the mortality in the older blue mussels. Farley (1988) found the disease in Chesapeake Bay softshells of less than 50 mm shell length (1-year-old), and attributed mortality to the disease. Thus, factors of resistance present in bivalves may mediate the progression of the disease. No differences in prevalence have been reported in relation to sex (Emmett 1984; but see Twomey and Mulcahy, in press) or gametogenesis and frequency of spawning (Mix et al. 1981; Cooper et al. 1982b).

The data have been derived from various field surveys that have often lacked appropriate methodology for statistical analysis and interpretation of the quantitative effects of the disease. To examine these aspects more thoroughly, B. J. Barber (Rutgers Shellfish Research Laboratory, unpublished data) examined individually marked softshells for the presence of the sarcoma by histocytology, placed them in Goldberg cages in the Shrewsbury River, New Jersey, and reexamined them at 2-week intervals for 3 months. Preliminary results indicated significantly higher mortalities in those clams originally diagnosed with the disease. Fifteen percent of disease-free animals developed the disease, and 22% of those with light cases of the disease apparently lost the neoplastic condition over the course of the summer. Average time to death decreased as the severity of the disease increased. The actual cause of death was not determined.

Similar results have been obtained in laboratory experiments (Appeldoorn et al. 1984; Farley et al. 1986b; Elston et al. 1988; Farley 1988). For example, Cooper et al. (1982b) reported significantly higher mortality in neoplastic softshells; stage 5 animals survived less than 3 months, and all died; stage 3 and 4 animals survived 6.5 months. Sixty percent of stage 1 individuals survived, and only 50% of low stage bivalves were alive at the end of the experiment. Forty percent of the animals remained at the same neoplastic stage at which they were diagnosed; 10% of the nonneoplastic clams developed the disease, and 10% of the neoplastic clams became nonneoplastic.

The above reports suggest that organismal defense mechanisms may operate to cause remission of the disease in certain individuals within a population of bivalves (Cooper 1976). Some investigators have suggested that early stages of neoplasia may be mistaken for hemopoiesis occurring in response to infections; therefore reversals are noted when the animal successfully fights the pathogen. Elston et al. (1988) found granuloma-like formations of normal hemocytes surrounding the abnormal cells in blue mussels. Farley and Sparks (1970), Couch and Winstead (1979), and others have reported pycnosis of the neoplastic cells, so true reversals may occur. Similar stages may occur in mammalian leukemias, which are known to be affected by the nutritional status of the organisms and by external stressors, including environmental conditions and pathogens. This area should be a fertile one for research into bivalve sarcomas. Elston (personal communication) is studying remission and the physiological basis of mortality in blue mussels.

The disseminated sarcomas occur sporadically at any particular site, such as background levels of 0.01% in Chesapeake Bay and less than 0.45% in gulf coast oysters (Mix et al. 1977; Couch 1985; Mix 1986). However, epizootics are not uncommon and have been associated with mortalities, particularly in softshell (Reinisch et al. 1984; Farley et al. 1986b). Certain environmental conditions, such as seasonal cycles, can mediate the appearance and course of the disease. For example, Farley and Sparks (1970) attributed winter mortality of Olympic oysters to the sarcoma; however, few cases were actually diagnosed, and most of the deaths were probably caused by reduced salinities in winter and by parasite infections (Mix 1975a; 1976b; Mix et al. 1977). Whereas Mix et al. (1977) found no conclusive evidence for sarcoma seasonality in blue mussel in Yaquina Bay, Oregon, Cosson-Mannevy et al. (1984), Emmett (1984), and Rasmussen (1986b) noted that there were seasonal cycles in prevalence that differed according to site. On the east coast, the sarcoma of softshell exhibited a biphasic cycle in New England animals that peaked in May and October and decreased in summer and winter (Cooper and Chang 1982; Cooper et al. 1982b), but Brousseau (1987) found peak prevalences in the late fall and winter of Long Island Sound specimens. She further noted that the prevalence varied among the three sites (Table 1) and also that the peak prevalence varied among sampling years (i.e., the Stonington site prevalence was higher in 1983–84 than in 1984–85 or the fall of 1985, and the Old Mill site peak occurred in winter of 1984–85). These data support the earlier contention of Farley (1976a) that all epizootics develop during the fall and continue through the spring, but the reasons for this occurrence are obscure. Additional long-term monitoring of populations in the field could reveal cyclic variations according to species and environment at each site (e.g., Twomey and Mulcahy, in press).

Most collection sites of affected bivalves were relatively unpolluted (Farley 1969a, 1969b; Mix and Breese 1980; Emmett 1984; Poder and Auffret 1986). Scientists have found correlations between the presence of high levels of polycyclic aromatic hydrocarbons or other potentially carcinogenic hydrocarbons in the substrata and tissues of blue mussels and the high incidence of sarcomas at two sites, one in Yaquina Bay, Oregon, and one in England (Lowe and Moore 1978; Mix 1979; Mix et al. 1979b, 1981; Mix 1980, 1982). However, Mix (1986) noted no sarcomas in Coos Bay, Oregon, softshells that had similar high levels of polycyclic aromatic hydrocarbons (the samples were not analyzed for other xenobiotics), whereas other animals from pristine areas exhibited the disease (Mix 1979, 1986; Elston et al. 1988). Although 64% of softshells at an oil spill site in Bourne, Massachusetts, had sarcoma (Brown et al. 1977), there was no correlation of the disease with the levels

of petroleum-derived hydrocarbons in tissues or sediments (Brown 1980). Reinisch et al. (1984) reported epizootic levels of the sarcoma in softshells from the polluted New Bedford harbor in Massachusetts. Other workers have noted similar contradictions (Mix 1986; Rasmussen 1986a).

After many years of debate and conflicting data, Parry et al. (1981) observed that mussels do possess limited cytochrome P_{450} enzymes and, therefore, could activate certain mutagens to carcinogens. Additional work on these metabolic systems in bivalves has been reported by Anderson (1978, 1985), Anderson and Doos (1983), and Kurelec and Britvic (1985). Laboratory exposures of marine bivalves to carcinogens have produced only toxic effects (Rasmussen 1982; Couch and Harshbarger 1985; Rasmussen et al. 1985). Khudoley and Syrenko (1978) induced hemic neoplasia in the digestive gland, kidneys, gills, gonads, and lacunar tissue of freshwater mussels, *Unio pictorum*, by exposure to carcinogenic nitrosamines. Their results require confirmation.

The relationship of the sarcoma to hydrocarbon pollution was tested in a University of Rhode Island laboratory (see Appeldoorn et al. 1984). Softshells that were exposed to the effluent from a tank of infected animals developed varying degrees of the sarcoma under conditions of no sediment (a stressful situation for these molluscs) and in 50:50 hydrocarbon-contaminated sediment:clean sediment, but not in 100% clean sediment or 100% polluted sediment. Softshells exposed to effluent from head tanks that contained no clams, or only clams deemed free of the disease, did not develop the sarcoma after 6 months of exposure. Brown et al. (1977) concluded that the type and degree of hydrocarbon pollution may be related to the development of sarcoma, but other causative factors must also be operative. A field experiment that simulated oil spills and treatments off the northwest coast of Baffin Island, Canada (Neff et al. 1987), yielded 4% of truncate softshell *Mya truncata* and 1% of chalky macomas *Macoma calcarea* with hemopoietic neoplasms from an undispersed spill site 1 year later after the oil was released. There were few unexposed animals for comparison with those from the oil spill sites, and no clear-cut effects of oil exposure were noted in the histopathology studies.

Other observations have indicated that an infectious agent may be involved. Farley et al. (1986a, 1986b) reported that 42–65% of softshells from sampled sites in the Chesapeake Bay were diseased during the epizootic of December 1983 to May 1984. The disease had not been detected in

Chesapeake Bay animals prior to 1978, and only a few cases were reported before 1983. He hypothesized that softshells from New England had been transplanted to the bay after hurricane Agnes destroyed native softshells in 1972. He also noted that there was antigenic similarity of the bay softshells to the New England ones. Maryland softshells exposed to diseased Rhode Island animals developed the disease (Appeldoorn et al. 1984), although K. R. Cooper (Rutgers University, personal communication) had recommended transplanting Chesapeake Bay animals (thought to be resistant) to New England to protect commercial softshell interests there. Alderman et al. (1977) and Green and Alderman (1983) were also able to induce the disease in healthy edible oysters by laying them near infected animals in the field. In a flow-through laboratory experiment, cohabitation of neoplastic and disease-free blue mussels resulted in 40% neoplasia in the control group (Elston et al., in press).

Because of these observations, some investigators suspected that the disseminated sarcomas might have a viral etiology. Several viruses have been detected in numerous marine bivalves, sometimes infecting hemocytes (Farley et al. 1972; Farley 1976a, 1977, 1978, 1981; Harshbarger et al. 1979; Johnson 1984; Rasmussen 1986c [but not Rasmussen 1986b]). A synchronous transformation of hemocytes has been reported in blue mussels (Elston et al. 1988), which also suggests an infectious etiology. However, no viruses were seen by transmission electron microscopy of the atypical cells of Olympic oysters, blue mussels, or *Cerastoderma edule* (Farley 1977; Sonstegard 1980; Balouet et al. 1983; Mix 1983; Twomey and Mulcahy 1984; Auffret and Poder 1986). Oprandy et al. (1981) and Oprandy and Chang (1983) reported the isolation of a B-type retrovirus from homogenates of affected softshell tissues. Cooper and Chang (1982) were able to infect small numbers of normal softshells with the sarcoma by injections of this isolate, and they reisolated the virus from these clams. They also noted that replication of the virus and development of the sarcoma were induced by exposure to 5-bromodeoxyuridine. However, the authors could not demonstrate the virus in thin sections of the neoplastic tissue, nor did they observe capsular material in negative-stained preparations of viral particles. Also, the softshells were diagnosed for sarcoma by fresh hemocytometer preparations, and this technique has since been questioned as to its accuracy in detecting the disease correctly; no

permanent hemolymph or histological preparations were made. Mix (1982) could not isolate a virus from similarly afflicted blue mussels; therefore, the viral etiology has not been confirmed.

Appeldoorn et al. (1984) also noted that the atypical cells could be transplanted into healthy softshells and produce the sarcoma. Others repeated this experiment successfully (Farley 1988; Twomey and Mulcahy 1988; Elston et al., in press), but Reinisch et al. (1983) failed. Farley (1988) obtained as much as 90% infection in softshell within 82 d of injection. Because of the short period of observation, he believed that he had transplanted cells that multiplied in the host and that he did not transmit a virus. In all these cases, it appeared that the molluscs failed to reject allografts, and this failure suggested a lack of operating immunologic factors (Cooper 1976). None of these studies used radioactive-labeled cells to examine whether the cells that were in the diseased animals were replicated from those introduced in the injection or whether an introduced virus may have activated the healthy animal's own cells.

Farley (personal communication) and Elston (personal communication) are continuing studies to isolate and characterize a virus from infected bivalves and to retest for disease in healthy bivalves exposed to hemolymph, cell filtrates, and effluents from affected individuals. Elston et al. (in press) have now reported the transmission of the disease by cell-free homogenates of blue mussel. A number of interesting questions remain. Is the virus activated only at certain temperatures; is it destroyed at high temperatures (Steward 1976)? Mix (1975a, 1976b) proposed that the atypical cells in Olympic oysters were normal stem cells that did not differentiate because of low water temperatures but continued to divide and produce the sarcomatous condition. Are there any immunological factors that may mediate the presence of the disease? Could there be any genetic factors that affect the development of the sarcoma? Frierman (1976) and Frierman and Andrews (1976) noted that two inbred lots of eastern oyster had a high frequency of the sarcoma (31/ 369 versus 39/51,000 eastern oysters from other lots). The distribution and appearance of the disease suggest that a genetic susceptibility may operate at some sites (Mix et al. 1977; Poder and Auffret 1986; Twomey and Mulcahy, in press). Cellular protooncogenes have been detected in a variety of marine invertebrates (Barnekow and Schartl 1984) and implicated in the development of neoplastic diseases in many vertebrate species.

Such genes could also be present in bivalves and be activated by chromosome alterations following exposure to mutagens, carcinogens, or viral oncogenes that induce neoplasia by amplification of host cell gene products (Gilden et al. 1984). The long latency period between viral infection and neoplasia development may also help explain the absence of detectable disease in the Chesapeake Bay (Farley et al. 1986b) and possibly in the west coast bivalves (Mix et al. 1977). Development of cell cultures will be necessary to study viruses (Johnson 1984). More work is needed on the relationship of environmental stress to metabolic changes and to depression of defense mechanisms of bivalves (Stainken 1978; Mix 1986; Neff et al. 1987). The genotoxic effects of organic pollutants and heavy metals should also be explored.

Prospectives

Mix (1986) thought that the evidence to support a viral etiology was strong for some cases of disseminated sarcoma but perhaps not all. No doubt the etiology could differ among the various species of bivalves, as has been observed in neoplastic diseases of mammals. Brown (1980) hypothesized that a multifactoral complex of chemical carcinogens may enhance the induction of neoplasms by viruses in polluted areas. Couch and Harshbarger (1985) also suggested that an interaction of viruses, chemicals, and genetics may produce the sarcomas, depending on the species and the site.

Resistant bivalves may occur naturally and could be protected from commercial exploitation to repopulate areas. Our lack of knowledge of the effects of neoplastic and other diseases on commercial species has led to embargoes against the importing and planting of Yaquina Bay Olympia oysters (Mix et al. 1977) and other species. Similarly, no information is available on the public health significance of the neoplasms found in shellfish populations (Mix 1986). Whereas these diseases have received more attention than other neoplastic diseases of the marine bivalves molluscs, the etiology and ecology of these conditions are still not understood. However, the research of the past 20 years has laid a framework for further investigations with techniques developed to investigate mammalian leukemias. With intense effort by several laboratories, the results of such research should increase our knowledge and understanding in the comparative pathology of marine bivalves.

Multidisciplinary team studies will be most important. The etiology and ecology of these diseases should be compared among all species affected to distinguish the contributions of genetics, pathogens, and pollutants in the development and course of the disease. Histopathological, histocytological, physiological, and biochemical techniques must be standardized. Scientists must be aware of the latest applications of field survey methods and statistics and adhere to optimum aquacultural methods during laboratory studies. Major contributions are waiting in the studies of cellular ontogeny in bivalves, hemopoiesis, and neoplastic transformations. Clearly, much work remains to be done on these poorly known disseminated sarcomas.

Acknowledgments

B. J. Barber, R. A. Elston, C. A. Farley, and C. L. Reinisch shared their unpublished research for this report. C. A. Farley, J. C. Harshbarger, H. B. McCarty, M. C. Mix, and P. P. Yevich provided helpful comments and critical reviews of the manuscript. B. Emmett, R. E. Hillman, S. Sherburne, and the staff of the Registry of Tumors in Lower Animals contributed assistance and additional information. The author was supported by a National Institutes of Health, National Research Service Award Training Fellowship in the Registry of Tumors in Lower Animals (award 1 F32 CA08427-01 from the National Cancer Institute) during the preparation of this paper.

References

Alderman, D. J., P. van Banning, and A. Perez-Colomer. 1977. Two European oysters (*Ostrea edulis*) mortalities associated with an abnormal haemocytic condition. Aquaculture 10:355–340.

Anderson, R. S. 1978. Benzo[a]pyrene metabolism in the American oyster *Crassostrea virginica*. National Technical Information Service, EPA-600/3-78-004, Springfield, Virginia.

Anderson, R. S. 1985. Metabolism of a model environmental carcinogen by bivalve mollusks. Marine Environmental Research 17:137–140.

Anderson, R. S., and J. E. Doos. 1983. Activation of mammalian carcinogens to bacterial mutagens by microsomal enzymes from a pelecypod mollusk *Mercenaria mercenaria*. Mutation Research 116: 247–256.

Appeldoorn, R. S., and eight coauthors. 1984. Field and laboratory studies to define the occurrence of neoplasia in the soft shell clam, *Mya arenaria*. American Petroleum Institute Publication 4345, Washington, D.C.

Auffret, M. 1988. Bivalve hemocyte morphology. American Fisheries Society Special Publication 18:169–177.

Auffret, M., and M. Poder. 1986. Sarcomatous lesion in the cockle *Cerastoderma edule*. II. Electron microscopical study. Aquaculture 58:9–15.

Balouet, G., A. Cahour, and M. Auffret. 1983. Haemocytosarcomas in marine bivalve molluscs: hypothesis on histogenesis in European flat oyster *Ostrea edulis* L. Pages 59–60 *in* M. A. Rich, editor. Leukemia reviews international. Marcel Dekker, New York.

Balouet, G., and M. Poder. 1978. Hyperplasie hémocytaire chez *Ostrea edulis* L. Haliotis 9:99–102.

Balouet, G., and M. Poder. 1980. A proposal for classification of normal and neoplastic types of blood cells in molluscs. Pages 205–208 *in* D. S. Yohn, B. A. Lapin, and J. R. Blakeslee, editors. Advances in comparative leukemia research 1979. Elsevier, Amsterdam.

Balouet, G., and M. Poder. 1985a. A consideration of the cellular reactions in bivalve molluscs with emphasis on haemocytic diseases. Pages 371–378 *in* A. W. Ellis, editor. Fish and shellfish pathology. Academic Press, London.

Balouet, G., and M. Poder. 1985b. Current status of parasitic and neoplastic diseases of shellfish: a review. Pages 381–386 *in* A. W. Ellis, editor. Fish and shellfish pathology. Academic Press, London.

Balouet, G., M. Poder, A. Cahour, and M. Auffret. 1986. Proliferative hemocytic condition in European flat oysters: a 6-year survey. Journal of Invertebrate Pathology 48:208–215.

Barnekow, A., and M. Schartl. 1984. Cellular *src* gene product detected in freshwater sponge *Spongilla lacustris*. Molecular and Cellular Biology 4:1179–1181.

Barry, M. M., and P. P. Yevich. 1972. Incidence of gonadal cancer in the quahog *Mercenaria mercenaria*. Oncology (Basel) 26:87–96.

Barry, M., and P. P. Yevich. 1975. The ecological, chemical and histopathological evaluation of an oil spill site. III. Histopathological studies. Marine Pollution Bulletin 6(11):171–173.

Brousseau, D. J. 1987. Seasonal aspects of sarcomatous neoplasia in *Mya arenaria* (soft-shell clam) from Long Island Sound. Journal of Invertebrate Pathology 50:269–276.

Brown R. S. 1980. The value of the multidisciplinary approach to research on marine pollution effects as evidenced in a three-year study to determine the etiology and pathogenesis of neoplasia in the softshell clam, *Mya arenaria*. Rapports et Procès-Verbaux des Réunions Conseil International pour l'Exploration de la Mer 179:125–128.

Brown, R. S., R. E. Wolke, C. W. Brown, and S. B. Saila. 1979. Hydrocarbon pollution and the prevalence of neoplasia in feral New England soft-shell clams (*Mya arenaria*). Pages 41–51 *in* Animals as monitors of environmental pollutants. National Academy of Sciences, Washington, D.C.

Brown, R. S., R. E. Wolke, and S. B. Saila. 1976. A preliminary report on neoplasia in feral populations of the soft-shell clam, *Mya arenaria*, prevalence, histopathology and diagnosis. Proceedings of the International Colloquium on Invertebrate Pathology 1:151–158.

Brown, R. S., R. E. Wolke, S. B. Saila, and C. Brown. 1977. Prevalence of neoplasia in 10 New England populations of the soft-shelled clam (*Mya arenaria*).

Annals of the New York Academy of Sciences 298:522–534.

Cahour, A., and G. Balouet. 1984. Hémocytosarcomes de l'huître plate *Ostrea edulis* L. Observations ultrastructurales. Revue Internationale d'Oceanographie Médicale 75/76:65–76.

Cheng, T. C. 1976. Identification of proliferative lesions in mollusks. U.S. National Marine Fisheries Service Marine Fisheries Review 38(10):5–6.

Cheng, T. C. 1981. Bivalves. Pages 233–300 *in* N.A. Ratcliffe and A. F. Rowley, editors. Invertebrate blood cells, volume 1. Academic Press, London.

Christensen, D. J., C. A. Farley, and F. G. Kern. 1974. Epizootic neoplasms in the clam *Macoma balthica* (L.) from the Chesapeake Bay. Journal of the National Cancer Institute 52:1739–1749.

Cooper, E. L. 1976. Immunity and neoplasia in mollusks. Israel Journal of Medical Sciences 12:479–494.

Cooper, K. R., R. S. Brown, and P. W. Chang. 1982a. Accuracy of blood cytological screening techniques for the diagnosis of a possible hematopoietic neoplasm in the bivalve mollusc, *Mya arenaria*. Journal of Invertebrate Pathology 39:281–289.

Cooper, K. R., R. S. Brown, and P. W. Chang. 1982b. The course and mortality of a hematopoietic neoplasm in the soft-shell clam, *Mya arenaria*, Journal of Invertebrate Pathology 39:149–157.

Cooper, K. R., and P. W. Chang. 1982. A review of the evidence supporting a viral agent causing a hematopoietic neoplasm in the soft-shelled clam, *Mya arenaria*. Proceedings of the International Colloquium on Invertebrate Pathology 3:271–272.

Cosson-Mannevy, M. A., C. S. Wong, and W. J. Cretney. 1984. Putative neoplastic disorders in mussels (*Mytilus edulis*) from southern Vancouver Island waters, British Columbia. Journal of Invertebrate Pathology 44:151–160.

Couch, J. A. 1969. An unusual lesion in the mantle of the American oyster (*Crassostrea virginica*). National Cancer Institute Monograph 31:557–562.

Couch, J. A. 1970. Sarcoma-like disease in a single specimen of the American oyster. Bibliotheca Haematologica 36:647.

Couch, J. A. 1985. A prospective study of infectious and noninfectious diseases in oysters and fishes in the Gulf of Mexico estuaries. Diseases of Aquatic Organisms 1:59–82.

Couch, J. A., and J.C. Harshbarger. 1985. Effects of carcinogenic agents on aquatic animals: an environmental and experimental overview. Environmental Carcinogenesis Reviews 3(1):63–105.

Couch, J. A., and J.T. Winstead. 1979. Concurrent neoplastic protistan disorders in the American oyster (*Crassostrea virginica*). Haliotis 8:249–253.

Dameshek, W. 1970. Leukemia—definition and characterization from the comparative viewpoint. Bibliotheca Haematologica 36:1–10.

Dawe, C. J. 1969. Neoplasms of blood cell origin in poikilothermic animals: a review. National Cancer Institute Monograph 32:7–28.

Dawe, C. J. 1980. Rapporteur's resume and comments on session on leukemias and related diseases in poikilotherms. Pages 193–204 *in* D. S. Yohn, B. A. Lapin, and J. R. Blakeslee, editors. Advances in comparative leukemia research 1979. Elsevier, Amsterdam.

Elston, R. A., M. L. Kent, and A. S. Drum. 1988. Progression, lethality and remission of hemic neoplasia in the bay mussel, *Mytilus edulis*. Diseases of Aquatic Organisms 4:135–142.

Elston, R. A., M. L. Kent, and A. S. Drum. In press. Transmission of hemic neoplasia in the bay mussel, *Mytilus edulis*, using whole cells and cell homogenate. Developmental and Comparative Immunology.

Emmett B. 1984 The analysis of summer mortality in cultured mussels, *Mytilus edulis*. Report to the British Columbia Science Council, Victoria, Canada.

Farley, C. A. 1969a. Probable neoplastic disease of the hemopoietic systems in oysters (*Crassostrea virginica* and *Crassostrea gigas*). National Cancer Institute Monograph 31:541–555.

Farley, C. A. 1969b. Sarcomatid proliferative disease in a wild population of blue mussels (*Mytilus edulis*). Journal of the National Cancer Institute 43:509–516.

Farley, C. A. 1976a. Proliferative disorders in bivalve mollusks. U.S. National Marine Fisheries Service Marine Fisheries Review 38(10):30–33.

Farley, C. A. 1976b. Ultrastructural observations on epizootic neoplasia and lytic virus infection in bivalve mollusks. Progress in Experimental Tumor Research 20:283–294.

Farley, C. A. 1977. Neoplasms in estuarine molluscs and approaches to ascertain causes. Annals of the New York Academy of Sciences 298:225–232.

Farley, C. A. 1978. Viruses and viruslike lesions in marine mollusks. U.S. National Marine Fisheries Service Marine Fisheries Review 40(10):18–20.

Farley, C. A. 1981. Phylogenetic relationships between viruses, marine invertebrates and neoplasia. Pages 75–87 *in* C. J. Dawe, J. C. Harshbarger, S. Kondo, T. Sugimura, and S. Takayama, editors. Phyletic approaches to cancer. Japan Scientific Societies Press, Tokyo.

Farley, C. A. 1989. Selected aspects of neoplastic progression in mollusks. Pages 24–31 *in* H. E. Kaiser, editor. Cancer growth and progression, volume 5. Martinez Nijhoff, New York

Farley, C. A., W. G. Banfield, G. Kasnic, Jr., and W. S. Foster. 1972. Oyster herpes-type virus. Science (Washington, D.C.) 178:759–760.

Farley, C. A., S. V. Otto, and C. L. Reinisch. 1986a. Epizootic sarcoma in Chesapeake Bay soft-shell clams—a virus? Proceedings of the International Colloquium on Invertebrate Pathology 4:436–440.

Farley, C. A., S. V. Otto, and C. L. Reinisch. 1986b. New occurrence of epizootic sarcoma in Chesapeake Bay soft shell clams *Mya arenaria*. U.S. National Marine Fisheries Service Fishery Bulletin 84:851–857.

Farley, C. A., and A. K. Sparks. 1970. Proliferative diseases of hemocytes, endothelial cells and connective tissue cells in mollusks. Bibliographica Haematologica 36:610–617.

Feng, S. Y. 1967. Responses of molluscs to foreign

bodies with special reference to the oyster. Federation Proceedings 26:1685–1692.

Franc, A. 1975. Hyperplasie hémocytaire et lesions chex l'huître plate, *Ostrea edulis* L. Compte Rendus Hebdomadaires de Séances de l'Academie de Sciences, Série D, Sciences Naturelles 280:495–498.

Frierman, E. M. 1976. Occurrence of hematopoietic neoplasms in Virginia oysters (*Crassostrea virginica*). U.S. National Marine Fisheries Service Marine Fisheries Review 38(10):34–36.

Frierman, E. M., and J. C. Andrews. 1976. Occurrence of hematopoietic neoplasms in Virginia oysters. Journal of the National Cancer Institute 56:319–322.

Gilden, R. V., N. R. Rice, and R. M. McAllister. 1984. Oncogenes. Gene Analysis Techniques 1:23–33.

Green, M., and D. J. Alderman. 1983. Neoplasia in *Mytilus edulis* L. from United Kingdom waters. Aquaculture 30:1–10.

Harshbarger, J. C. 1982. Epizootiology of leukemia and lymphoma in poikilotherms. Pages 39–46 *in* D. S. Yohn and J. R. Blakeslee, editors. Advances in comparative leukemia research 1981. Elsevier, Amsterdam.

Harshbarger, J. C., and C. J. Dawe. 1973. Hematopoietic neoplasms in invertebrate and poikilothermic vertebrate animals. Bibliotheca Haemotologica 39:1–25.

Harshbarger, J. C., E. R. Jacobson, C. E. Smith, and J. A. Couch. 1980. Hematopoietic neoplasms in invertebrates and cold-blooded vertebrates. Pages 223–225 *in* D. S. Yohn, B. A. Lapin, and J. R. Blakeslee, editors. Advances in comparative leukemia research 1979. Elsevier, Amsterdam.

Harshbarger, J. C., S. V. Otto, and S. C. Chang. 1979. Proliferative disorders in *Crassostrea virginica* and *Mya arenaria* from the Chesapeake Bay and intranuclear virus-like inclusions in *Mya arenaria* with germinomas from a Maine oil spill site. Haliotis 8: 243–248.

Johnson, P. T. 1984. Viral diseases of marine invertebrates. Helgoländer Meeresuntersuchungen 37: 65–98.

Khudoley, V. V., and O. A. Syrenko. 1978. Tumor induction by *N*-nitroso compounds in bivalve mollusks *Unio pictorum*. Cancer Letters 4:349–354.

Kurelec, B., and S. Britvic. 1985. The activation of aromatic amines in some marine invertebrates. Marine Environmental Research 17:141–144.

Lauckner, G. 1983. Neoplasia. Pages 863–879 *in* O. Kinne, editor. Diseases of marine animals, volume 2. Biologische Anstalt Helgoland, Hamburg, West Germany.

Lowe, D. M., and M. N. Moore. 1978. Cytology and quantitative cytochemistry of a proliferative atypical hemocytic condition in *Mytilus edulis* (Bivalvia, Mollusca). Journal of the National Cancer Institute 60:1455–1459.

Mackin, J. G., and F. G. Schlicht. 1976. A proteomyxan amoeba stage in the development of *Labyrinthomyxa patuxent* (Hogue) Mackin and Schlicht, with remarks on the relation of proteomyxids to the neoplastic disease of oysters and clams. U.S. National Marine Fisheries Service Marine Fisheries Review 38(10):16–18.

Mioski, D., R. Smolowitz, and C. L. Reinisch. 1989. Leukemia cell specific protein of the bivalve mollusc *Mya arenaria*. Journal of Invertebrate Pathology 53:32–40.

Mix, M. C. 1975a. Proliferative characteristics of atypical cells in native oysters (*Ostrea lurida*) from Yaquina Bay, Oregon. Journal of Invertebrate Pathology 26:289–298.

Mix, M. C. 1975b. The neoplastic disease of Yaquina Bay bivalve mollusks. Pages 369–384 *in* J. C. Hampton, editor. The cell cycle in malignancy and immunity. National Technical Information Service, CONF-731005, Springfield, Virginia.

Mix, M. C. 1976a. A general model for leucocyte cell renewal in bivalve mollusks. U.S. National Marine Fisheries Service Marine Fisheries Review 38(10): 37–41.

Mix, M. C. 1976b. A review of the cellular proliferative disorders of oysters (*Ostrea lurida*) from Yaquina Bay, Oregon. Progress in Experimental Tumor Research 20:275–282.

Mix, M. C. 1979. Chemical carcinogens in bivalve mollusks from Oregon estuaries. U.S. Environmental Protection Agency, EPA-600/3-79-034, Gulf Breeze, Florida.

Mix, M. C. 1980. Utilisation of bivalve molluscs for monitoring carcinogenic polynuclear aromatic hydrocarbons in estuarine environments. Pages 33–44 *in* C. J. Dawe, J. C. Harshbarger, S. Kondo, T. Sugimura, and S. Tokayama, editors. Phyletic approaches to cancer. Japan Scientific Societies Press, Tokyo.

Mix, M. C. 1982. Cellular proliferative disorders in bay mussels (*Mytilus edulis*) from Oregon estuaries. Proceedings of the International Colloquium on Invertebrate Pathology 3:266–267.

Mix, M. C. 1983. Haemic neoplasms of bay mussels, *Mytilus edulis* L., from Oregon: occurrence, prevalence, seasonality and histopathological progression. Journal of Fish Diseases 6:239–248.

Mix, M. C. 1986. Cancerous diseases in aquatic animals and their association with environmental pollutants: a critical literature review. Marine Environmental Research 20:1–141.

Mix, M. C., and W. P. Breese. 1980. A cellular proliferative disorder in oysters (*Ostrea chilensis*) from Chiloe, Chile, South America. Journal of Invertebrate Pathology 36:123–124.

Mix, M. C., J. W. Hawkes, and A. K. Sparks. 1979a. Observations on the ultrastructure of large cells associated with putative neoplastic disorders of mussels, *Mytilus edulis,* from Yaquina Bay, Oregon. Journal of Invertebrate Pathology 34:41–56.

Mix, M. C., H. J. Pribble, R. T. Riley, and S. P. Tomasovic. 1977. Neoplastic disease in bivalve mollusks from Oregon estuaries with emphasis on research on proliferative disorders in Yaquina Bay oysters. Annals of the New York Academy of Sciences 298: 356–373.

Mix, M. C., R. L. Schaffer, and S. J. Hemingway. 1981.

Polynuclear aromatic hydrocarbons in bay mussels (*Mytilus edulis*) from Oregon. Pages 167–177 *in* C. J. Dawe, J. C. Harshbarger, S. Kondo, T. Sugimura, and S. Takayama, editors. Phyletic approaches to cancer. Japan Scientific Societies Press, Tokyo.

Mix, M. C., and S. P. Tomasovic. 1973. The use of high specific activity tritiated thymidine and autoradiography for studying molluscan cells. Journal of Invertebrate Pathology 21:318–320.

Mix, M. C., S. R. Trenholm, and K. I. King. 1979b. Benzo[a]pyrene body burdens and the prevalence of proliferative disorders in mussels (*Mytilus edulis*) in Oregon. Pages 52–64 *in* Animals as monitors of environmental pollutants. National Academy of Sciences, Washington, D.C.

Nascimento, I. A., D. H. Smith, F. Kern II, and S. A. Pereira. 1986. Pathological findings in *Crassostrea rhizophorae* from Todos os Santos Bay, Bahia, Brazil. Journal of Invertebrate Pathology 47:340–349.

Neff, J. M., R. E. Hillman, R. S. Carr, R. L. Buhl, and J. I. Lahey. 1987. Histopathologic and biochemical responses in Arctic marine bivalve molluscs exposed to experimentally spilled oil. Arctic 40(supplement 1):220–229.

Newman, M. W. 1972. An oyster neoplasm of apparent mesenchymal origin. Journal of the National Cancer Institute 48:237–243.

Oprandy, J. J., and P. W. Chang. 1983. 5-bromodeoxyuridine induction of hemopoietic neoplasia and retrovirus activation in the soft-shell clam, *Mya arenaria*. Journal of Invertebrate Pathology 42: 196–206.

Oprandy. J. J., P. W. Chang, A. D. Pronovost, K. R. Cooper, R. S. Brown, and V. J. Yates. 1981. Isolation of a viral agent causing hemopoietic neoplasia in the soft-shell clam, *Mya arenaria*. Journal of Invertebrate Pathology 38:45–51.

Parry, J. M., M. Kadhim, W. Barnes, and N. Danford. 1981. Assays of marine organisms for the presence of mutagenic and/or carcinogenic chemicals. Pages 141–166 *in* C. J. Dawe, J. C. Harshbarger, S. Kondo, T. Sugimura, and S. Takayama, editors. Phyletic approaches to cancer. Japan Scientific Societies Press, Tokyo.

Poder, M., and M. Auffret. 1986. Sarcomatous lesion in the cockle *Cerastoderma edule*. I. Morphology and population survey in Brittany, France. Aquaculture 58:1–8.

Poder, M., M. Auffret, and G. Balouet. 1983. Etudes pathologiques et épidémiologiques de lésions parasitaires chez *Ostrea edulis* L. Premiers résultats d'une recherche prospective comparative chez principales espèces de Mollusques des zones ostréicoles de Bretagne Nord. IFREMER (Institut Francais de Recherche pour l'Exploitation de la Mer) Actes de Colloques 1:125–138.

Rasmussen, L. 1982. Light microscopical studies of the acute toxic effects of *N*-nitrosodimethylamine on the marine mussel, *Mytilus edulis*. Journal of Invertebrate Pathology 39:66–80.

Rasmussen, L. P. D. 1986a. Cellular reactions in molluscs with special reference to chemical carcinogens

and tumors in natural populations of bivalve molluscs. Proceedings of the International Colloquium
on Invertebrate Pathology 4:441–443.

Rasmussen, L. P. D. 1986b. Occurrence, prevalence
and seasonality of neoplasia in the marine mussel
Mytilus edulis from three sites in Denmark. Marine
Biology (Berlin) 92:59–64.

Rasmussen, L. P. D. 1986c. Virus-associated granulo-
cytomas in the marine mussel, *Mytilus edulis* from
three sites in Denmark. Journal of Invertebrate
Pathology 48:117–123.

Rasmussen, L. P. D., E. Hage, and O. Karlog. 1985.
Light and electron-microscope studies of the acute
and long-term toxic effects of *N*-nitrosodipropyl-
amine and *N*-methylnitrosurea on the marine mussel *Mytilus edulis*. Marine Biology (Berlin) 85:
55–65.

Reinisch, C. L., A. M. Charles, and A. M. Stone. 1984.
Epizootic neoplasia in softshell clams collected from
New Bedford Harbor. Hazardous Waste 1:73–81.

Reinisch, C. L., A. M. Charles, and J. Troutner. 1983.
Unique antigens on neoplastic cells of the softshell
clam *Mya arenaria*. Developmental and Comparative Immunology 7:33–39.

Rosenfield, A. 1976. Recent environmental studies of
neoplasms in marine shellfish. Progress in Experimental Tumor Research 20:263–274.

Smolowitz, R., D. Mioski, and C. L. Reinisch. 1989.
Ontogeny of leukemic cells of the soft shell clam.
Journal of Invertebrate Pathology 53:41–51.

Smolowitz, R. M., and C. L. Reinisch. 1986. Indirect
peroxidase staining using monoclonal antibodies
specific for *Mya arenaria* neoplastic cells. Journal
of Invertebrate Pathology 48:139–145.

Sonstegard, R. A. 1980. Virus associated hematopoietic
neoplasia in shellfish and fish. Page 227 *in* D. S.
Yohn, B. A. Lapin, and J. R. Blakeslee, editors.

Advances in comparative leukemia research 1979.
Elsevier, Amsterdam.

Sparks, A. K. 1985. Synopsis of invertebrate pathology,
exclusive of insects. Elsevier, Amsterdam.

Stainken, D. 1978. Effects of uptake and discharge of
petroleum hydrocarbons on the respiration of the
soft-shell clam *Mya arenaria*. Journal of the Fisheries Research Board of Canada 35:637–642.

Stewart, H. L. 1976. Some observations on comparative vertebrate and invertebrate pathology: a summary discussion of the workshop. U.S. National
Marine Fisheries Service Marine Fisheries Review
38(10):46–48.

Sunila, I. 1987. Histopathology of mussels (*Mytilus
edulis* L.) from the Tvärminne area, Gulf of Finland (Baltic Sea). Annales Zoologici Fennici 24:
55–69.

Twomey, E., and M. F. Mulcahy. 1984. A proliferative
disorder of possible hemic origin in the common
cockle, *Cerastoderma edule*. Journal of Invertebrate Pathology 44:109–111.

Twomey, E., and M. F. Mulcahy. 1988. Transmission of
a sarcoma in the cockle *Cerastoderma edule* (Bivalvia: Mollusca) using cell transplants. Developmental and Comparative Immunology 12:195–200.

Twomey, E., and M. F. Mulcahy. In press. Epizootiological aspects of a sarcoma in *Cerastoderma edule*.
Diseases of Aquatic Organisms.

Yevich, P. P., and C. A. Barszcz. 1976. Gonadal and
hemopoietic neoplasms in *Mya arenaria*. U.S. National Marine Fisheries Service Marine Fisheries
Review 38(10):42–43.

Yevich, P. P., and C. A. Barszcz. 1977. Neoplasia in
soft-shell clams (*Mya arenaria*) collected from oil-
impacted areas. Annals of the New York Academy
of Sciences 298:409–426.

Wolf, P. H. 1979. Diseases and parasites in Australian
commercial shellfish. Haliotis 8:75–83.

American Fisheries Society Special Publication 18:93–111, 1988
© Copyright by the American Fisheries Society 1988

PARASITE MORPHOLOGY, STRATEGY, AND EVOLUTION

Structure of Protistan Parasites Found in Bivalve Molluscs[1]

FRANK O. PERKINS

Virginia Institute of Marine Science, School of Marine Science, College of William and Mary
Gloucester Point, Virginia 23062, USA

Abstract.—The literature on the structure of protists parasitizing bivalve molluscs is reviewed, and previously unpublished observations of species of class Perkinsea, phylum Haplosporidia, and class Paramyxea are presented. Descriptions are given of the flagellar apparatus of *Perkinsus marinus* zoospores, the ultrastructure of *Perkinsus* sp. from the Baltic macoma *Macoma balthica,* and the development of haplosporosome-like bodies in *Haplosporidium nelsoni.* The possible origin of stem cells of *Marteilia sydneyi* from the inner two sporoplasms is discussed. New research efforts are suggested which could help elucidate the phylogenetic interrelationships and taxonomic positions of the various taxa and help in efforts to better understand life cycles of selected species.

Studies of the structure of protistan parasites found in bivalve molluscs have been fruitful to the morphologist interested in comparative morphology, evolution, and taxonomy. A diversity of species and higher taxa contain a great range of structures and developmental stages of considerable complexity. The most intensive studies have centered on pathogens of commercially important hosts, and the results have contributed to the management of these valuable hosts, primarily oysters, by identifying the agent causing mortality. Thus, the investigator can identify the causative agent and conduct epizootiological studies to determine (1) geographical range, (2) frequency and periodicity of disease expression, and (3) controlling physical factors such as salinity and temperature. Then managers and users of the hosts can be advised how to avoid the disease and how to minimize the impact of the disease. Unless protistan pathogens can be distinguished from one another, the ecological relationships of parasite to host may be misinterpreted.

Morphologists have not always been successful in their attempts at species characterization, because the protists within genera have relatively few distinguishing structures. The problem is compounded by the inability to establish pure cultures of all but a very few parasitic protista or of host cells that could serve as the culture medium. The lack of cultures is the greatest barrier to more complete morphological characterization of the parasite species, to elucidation of many parasite life cycles, and to knowledge of parasite metabolism. The latter, especially, is important for determining the effects of prophylaxis and therapy.

In most cases, the parasitic state, its pathogenicity, or both cannot be directly transmitted experimentally. In cases of successful direct transmission, for example, of *Perkinsus marinus* or (perhaps) *Bonamia ostreae,* host specificity of the parasite is often not known because morphologically distinguishing structures are lacking and transmission experiments with a single population of parasites in several host species have not been attempted. Because infections of bivalves with most parasites cannot be transmitted, investigations should be conducted on host defense mechanisms, identification of intermediate hosts, and whether the organisms are truly parasitic or merely opportunistic.

The availability of transmission electron microscopes beginning in the early 1960s was a major advance for protistologists. Most cell stages are 2–10 μm in size, and distinguishing structures were often impossible to find by light microscopy. Electron microscopes have provided the ability to distinguish major taxa, for example, coccoid unicells of the thraustochytriacean protists from the species of class Perkinsea.

In this presentation, I shall provide an overview of the major groups of protistan parasites with emphasis on their structure, phylogeny, and geographical distribution. Literature citations will direct readers to published descriptions and micrographs of morphological structures.

---

[1]Contribution 1414 VIMS/SMS of the College of William and Mary.

Class Perkinsea

Perkinsus marinus

Levine (1978) established the class Perkinsea for the phylum Apicomplexa to accommodate the observations of Perkins (1976c), who found that the flagellated cells of *Dermocystidium marinum*, the name then given to a pathogen of the eastern oyster *Crassostrea virginica*, have organelles very similar to those of other species in the Apicomplexa. The phylum essentially includes protistan species formerly classified in the Sporozoa. Perkins (1974) noted that flagellated cells are not formed in *Dermocystidium* sp. from salmon, nor have other cyst-forming species of *Dermocystidium* in fish and amphibia been observed to form flagellated cells. Therefore, Levine (1978) created a new genus for the pathogen of the eastern oyster and renamed the species *Perkinsus marinus*.

This systematic revision has stimulated debate on the validity of the Apicomplexa and Perkinsea by several workers, notably Vivier (1982) and myself as related by Canning (1986). Until more is known about Perkinsea and other Protista with apicomplexan characteristics, the validity of the class will remain unsettled. However, it appears that the eastern oyster pathogen is not properly placed in the genus *Dermocystidium*.

The structures found in the flagellated cells (zoospores) of *Perkinsus* spp., which are similar to those found in the Apicomplexa, particularly the Coccidia, are the apical complex, micropores, rhoptries, micronemes, and subpellicular complex described in detail in *Perkinsus marinus* by Perkins (1976c). They have also been observed in *Perkinsus* sp. found in *Macoma balthica* (my unpublished data) and in *Tridacna* sp. (R. J. G. Lester, University of Queensland, unpublished data). Unlike those of other known Protista, the anterior flagellum of *Perkinsus* spp. has a row of long, filamentous mastigonemes and a row of short, spurlike structures along the full length of the flagellum. The repeating units are one spur and five clustered mastigonemes grouped together (Perkins 1987). The posterior flagellum lacks mastigonemes.

The cell body (2–3 × 4–6 µm) is rounded at the anterior end and tapers to a pointed posterior end (Plate 1, Figure 1). The anterior flagellum emerges from a groove located about one-third of the way from the anterior end toward the posterior end. The posterior flagellum emerges from a cul-de-sac in the cell surface at right angles to the anterior flagellum.

The kinetosome and base of the flagellum are complex. There are no cartwheels in the kinetosome lumen; however, a large electron-dense inclusion body is found in the lumen with cross bridging to the A and B microtubules of the triplet blades (Plate 1, Figure 2 [d]). Each triplet blade is linked to the adjacent blade by a single bridge. The proximal end of the kinetosome is blocked by an electron-dense plate; however, there is no basal plate in the interfacial zone between the kinetosome and flagellum (Plate 1, Figure 2). An electron-dense, hollow cylinder is present at the base of the flagellum; it is closed at the distal end but may be partially open or closed at the proximal end. A secondary, thin-walled cylinder is found outside the primary one (Plate 1, Figures 2, 2 [b]).

The central doublet of flagellar microtubules terminates on a plate which is free in the flagellar lumen and located close to the cylinders mentioned above. At right angles to the plate and extending distally is a structure that is either a stack of rings or a helically wound filament (Plate 1, Figures 2, 2 [a]).

Two or more structures, perhaps microtubules, coated with electron-dense material, extend at right angles to the long axis of the cell body (Plate 1, Figure 2). They originate from a wedge-shaped, electron-dense structure located at the base of the two kinetosomes and between them. If this filamentous extension were shown to be homologous to the presumptive vestigal flagella of the microgametes of *Toxoplasma gondii* and *Eimeria magna* (Pelster and Piekarski 1971; Speer and Danforth 1976), the evidence for including *P. marinus* in the Apicomplexa would be strengthened and questions would be raised as to whether or not the zoospores of *P. marinus* are isogametes.

The complement of the kinetosomal–flagellar structures in *Perkinsus* spp. has also been observed in the *Perkinsus* sp. found in *Macoma balthica* (Perkins 1968; unpublished data). Fur-

PLATE 1.—Flagellar apparatus of *Perkinsus marinus* zoospores. FIGURE 1.—Phase-contrast micrograph of ▶ *Perkinsus marinus* zoospore. A = anterior flagellum; P = posterior flagellum. Bar = 3 µm. FIGURE 2.— Kinetosome and base of anterior flagellum of *P. marinus*. Bar = 0.5 µm. Lines a–d indicate the sites of cross sections of flagella illustrated in inserts a–d; insert bar = 0.1 µm. E = electron-dense core in lumen of kinetosome; C = hollow cylinder in flagellar base; P = plate at base of central pair of flagellar axonemes; Fi = helical filament (?); F = bundle of filaments or microtubules.

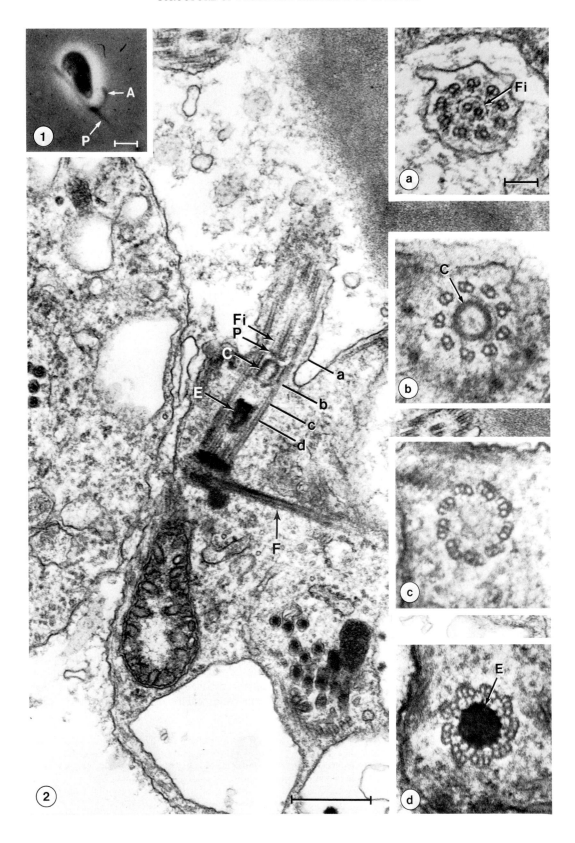

thermore, the electron-dense cylinder in the proximal end of the kinetosome lumen has been observed in the thraustochytriaceous–labyrinthulid protists (Perkins 1974). The phylogenetic affinity between *Perkinsus* spp. and the latter group and the importance of the other kinetosomal–flagellar structures remain to be determined.

Mastigonemes that are filamentous rather than tubular are found scattered throughout the protistan phyla (Moestrup 1982); therefore, they appear to have no phylogenetic importance, at least at the phylum level. If they were shown to be important, the alliance of the Perkinsea species with the Apicomplexa would be placed in doubt. The only flagellated cells in the Apicomplexa are microgametes which appear to lack mastigonemes; however, no high-resolution micrographs of negatively stained or shadowed flagella have been published.

Zoospores probably establish infections in the gills or mantle by penetrating the gill, labial palp, or mantle epithelium after losing the flagella and encysting within or between the cells. The details have not been determined. I have induced infections in uninfected eastern oysters using zoospores (my unpublished data) and have observed the earliest foci of infections in those tissues; however, the cellular details were not observed. Infections were detected using the Ray culture method involving fluid thioglycollate medium (Ray 1952).

The earliest cell type observed in eastern oysters (Perkins 1969, 1976a) consists of a 2- to 4-μm uninucleate coccoid trophozoite, which is presumed to be immature (Plate 2, Figures 3, 5). It is often found within the phagosome of a hemocyte (Plate 2, Figure 5). The trophozoite is delineated by a fibrogranular wall in and around which are found what appear to be phagosome membranes of the host. The cytoplasm contains two centrioles with electron-dense cylinders as in the kinetosomes, lomasome, tubulovesicular mitochondria, smooth endoplasmic reticulum, and a few lipoid droplets.

The immature trophozoite enlarges to about 10–20 μm and, in the process, acquires an eccentrically situated vacuole (Plate 2, Figure 6), often with a prominent, refringent vacuoplast which floats freely in the vacuole fluid. This inclusion is useful in locating the cells in fresh squash preparations of eastern oyster tissue observed in bright field microscopy because it has characteristic Brownian movement and is refringent. It has cytochemical characteristics similar to volutin as found in yeast cells. The nucleus has a centrally located endosome and is located near the cell wall, giving the characteristic signet configuration which led Mackin et al. (1950) to name the pathogen *Dermocystidium marinum*.

The life cycle within the eastern oyster consists of successive bipartitioning of the protoplast of the mature trophozoite (i.e., repeating cycles of karyokinesis followed by cytokinesis) to yield 8- to 32-cell (rarely 64-cell) sporangia within a mother cell wall which then ruptures to yield coccoid or cuneiform immature trophozoites (Plate 2, Figures 4, 6–9; Plate 3, Figures 10, 11). The vacuoplast and large vacuole presumably subdivide during the formation of multicellular sporangia, or sporangium formation occurs from stem cells lacking a large vacuole. Progressive cleavage (i.e., repeated karyokinesis followed by synchronous cytokinesis) may also be involved.

When the cells are placed in fluid thioglycollate medium prepared in seawater with antibiotics to retard bacterial growth, or on rarely observed occasions in moribund eastern oyster tissue, any stage may enlarge markedly to form 15- to 100-μm-diameter cells (Plate 3, Figures 12, 13). Such cells have an extremely large vacuole which compresses the cytoplasm into a thin layer against the cell wall. No vacuoplast remains after the enlargement is completed. The nucleus assumes a sausage-shaped configuration in the thin layer of cytoplasm. Numerous small lipoid droplets are scattered throughout the cytoplasm.

These large cells, called hypnospores after the phycologists' term for resting cells (Mackin 1962), are actually prezoosporangia and are labile in fluid thioglycollate medium, where they survive only about 10 days. Upon release into seawater, the prezoosporangia initiate zoosporulation. The lipoid droplets dissolve, and a discharge pore (blocked by a plug of secondary wall material) and a discharge tube are formed (Plate 4, Figure 14). The protoplast condenses within the cell wall, and the large vacuole subdivides and loses its identity. Presumably,

PLATE 2.—Schizogony in *Perkinsus marinus*. FIGURE 3.—Immature trophozoite. N = nucleus; M = mitochondrion. Bar = 1 μm. FIGURE 4.—Eight-cell division stage. V = forming vacuoplast; Ce = centriole. Bar = 1 μm. FIGURE 5.—Immature trophozoite (arrow) in oyster hemocyte. N = nucleus of hemocyte. Bar = 5 μm. FIGURE 6.—Mature trophozoite. Arrow indicates large eccentric vacuole without a vacuoplast. N = nucleus. Bar = 5 μm. FIGURES 7, 8, 9.—Two-, four-, and eight-cell stages, respectively, resulting from successive bipartitioning of the protoplast. Arrows indicate cleavage furrows. Bar = 5 μm.

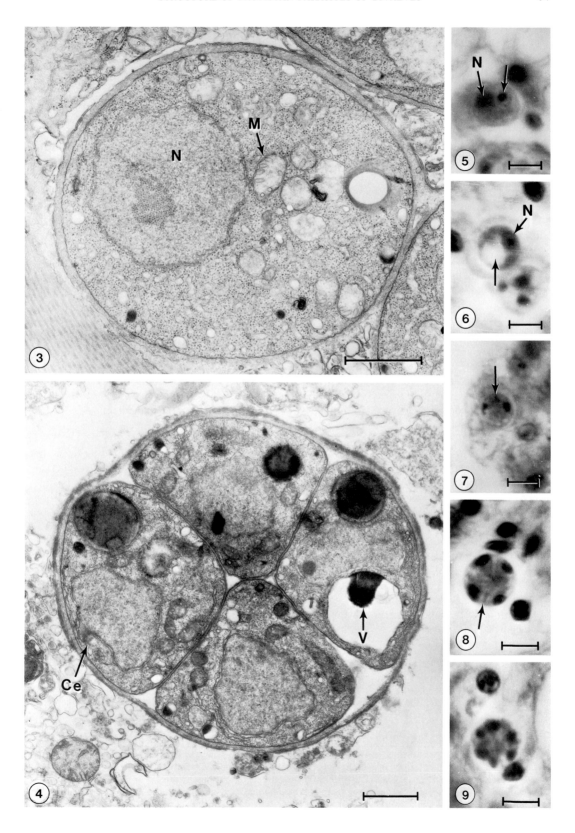

much of the vacuolar fluid is forced into the space between the cell wall and protoplast. Successive bipartitioning follows and yields a mass of zoospores, which swim around within the mother cell wall. Then they escape through the discharge pore and tube upon dissolution of the plug (Perkins and Menzel 1966, 1967) (Plate 4, Figures 14, 15).

Zoospores have been shown to infect eastern oysters and eastern oyster organ explants (Perkins and Menzel 1966); however, their importance in the life cycle is not clear. They have only been observed experimentally, never in nature. The presumed prezoosporangia are so rarely observed that they have not been isolated and induced to zoosporulate, in contrast to *Perkinsus* sp. in *Macoma balthica*. The trophozoites are probably the normal agents of disease transmission, as suggested by the ease with which infections can be transmitted with infected eastern oyster tissue in which there are no zoospores (Mackin 1962). I have attempted to determine whether the zoospores are isogametes; however, there is no reason to believe they are anything but vegetative stages except for the presence of a possible third (vestigal) flagellum (see above and Plate 1, Figure 2). The few quadriflagellated cells observed appeared to result from incomplete cleavage. It is unlikely that isogametic copulation had to occur before the flagellated cells infected eastern oysters under experimental conditions. However, one wonders why so few zoospores successfully establish foci of infections. When I exposed eastern oysters or eastern oyster tissue explants to millions of zoospores (my unpublished data), only very light infections resulted. At present, zoospores cannot be assumed to be a primary infective cell type in the life cycle of the pathogen.

Perkinsus sp. in Macoma balthica

The species of *Perkinsus* found in *M. balthica* can cause mortality in the Baltic macoma, but it is not usually as virulent as *P. marinus*. The host often pseudoencapsulates the parasite with connective tissue fibers and cellular elements, resulting in small isolated groups of parasitic cells in the connective tissue (Plate 4, Figure 18). This is in contrast to *P. marinus*, the cells of which are distributed without encapsulation throughout almost all organs and in the hemolymph.

The cells of *Perkinsus* sp. may be coccoid, pyriform, or egg-shaped and tend to have larger lipoid droplets in the cytoplasm than those of *P. marinus* (Plate 4, Figures 16, 17). In addition, the mature trophozoites are larger (9–48 μm) than those of *P. marinus*. Otherwise, their ultrastructure is the same as described for *P. marinus*. They resemble trophozoites of *P. marinus* seen in the first 24 h of incubation in fluid thioglycollate medium.

When *Perkinsus* sp. is isolated from the host and placed in seawater, zoosporulation occurs in mature trophozoites as in *P. marinus* (Valiulis and Mackin 1969) and yields zoospores with the same surface structure (Perkins 1968). An identical response occurs if the cells are induced to enlarge in fluid thioglycollate medium and then isolated in seawater to induce zoosporulation.

Cells of *P. marinus* lose their mitochondria after 24 h in thioglycollate medium and re-form them during zoosporulation. Mature trophozoites and prezoosporangia of *Perkinsus* sp. retain their mitochondria (Plate 3, Figure 12; Plate 4, 16) if they are not treated with the medium. It is not known if the mitochondria disappear when *Perkinsus* sp. cells are placed in thioglycollate medium.

Perkinsus sp. in Other Bivalve Molluscs

There are only two described species in the genus *Perkinsus*, *P. marinus* and *P. olseni* (Lester and Davis 1981); the latter is found in the blacklip abalone *Haliotis ruber* in South Australia. However, after treatment in fluid thioglycollate medium (FTM), cells presumed to be prezoosporangia of other *Perkinsus* species have been observed in 34 species of bivalve molluscs by myself (unpublished data) and other workers (Andrews 1954; Ray 1954; Da Ros and Canzonier 1985). The hosts occur in temperate, subtropical, and tropical waters of the Atlantic and Pacific oceans and Mediterranean Sea. It is unlikely that protists other than *Perkinsus* spp. were observed because (1) the cell structure of prezoosporangia is easy to recognize and is unlike that of any other known protist cells, (2) the cells stain blue or blue-black in Lugol's iodine solution without acid hydrolysis, and (3) cells from selected

---

PLATE 3.—Unicellular development in *Perkinsus* spp. FIGURES **10, 11.**—Enlarging trophozoites of *P. marinus* (Figure 10; bar = 1 μm) after rupture of sporangium (Figure 11; bar = 5 μm). V = forming vacuoplast material. FIGURES **12, 13.**—Prezoosporangia, each cell with an enlarged vacuole, in host tissue not treated with fluid thioglycollate medium. Figure 12 = *Perkinsus* sp. in *Macoma balthica*; Figure 13 = *P. marinus*. N = nucleus; M = mitochondrion; L = lipoid droplet. Bar of Figure 12 = 1 μm; bar of Figure 13 = 5 μm.

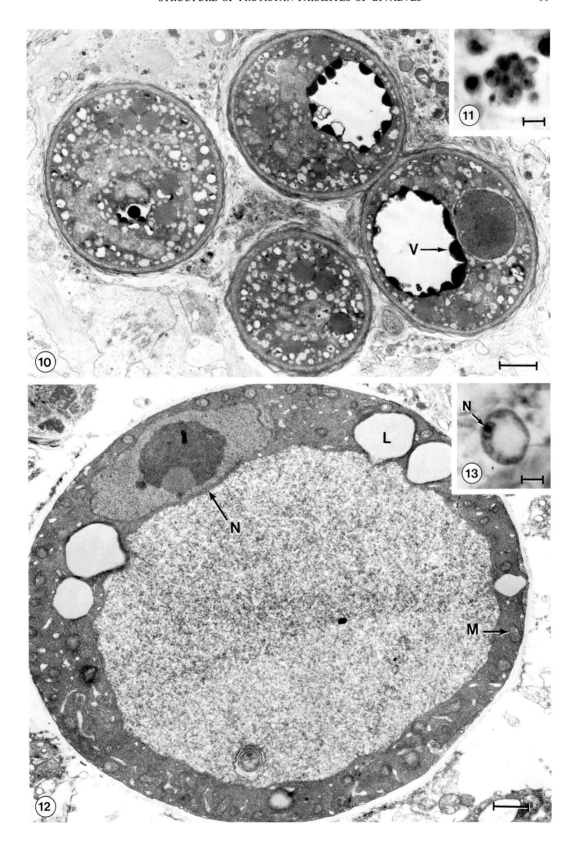

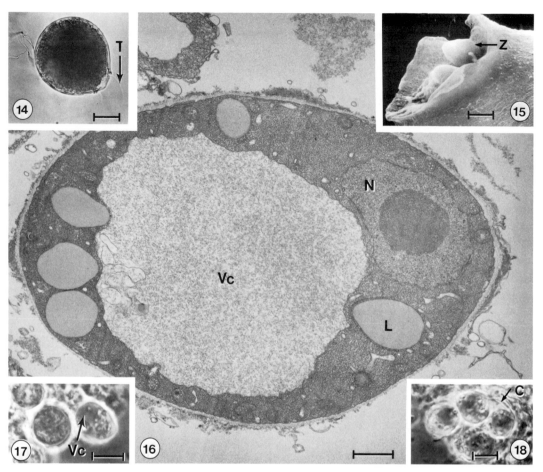

PLATE 4.—*Perkinsus* sp. from *Macoma balthica*. **FIGURE 14.**—Zoosporangium with mature zoospores about to be released. T = discharge tube. Bar = 10 μm. **FIGURE 15.**—Scanning electron micrograph of zoosporangium discharge pore. Discharge tube has been removed during specimen preparation. Z = zoospore. Bar = 0.2 μm. **FIGURE 16.**—Mature trophozoite. N = nucleus; L = lipoid droplet; Vc = vacuole. Bar = 1 μm. **FIGURE 17.**—Phase-contrast micrograph of mature trophozoite. Vc = large vacuole. Bar = 10 μm. **FIGURE 18.**—Phase-contrast micrograph of emerging mature trophozoites. C = capsule wall. Bar = 10 μm.

FTM cultures that have been isolated and incubated in seawater yielded zoospores typical of *P. marinus* (i.e., biflagellated with filamentous mastigonemes on the anterior flagellum) as a result of progressive cleavage. I have observed such zoospore formation in *Perkinsus* spp. from *Arca umbonata*, *Pseudochama radians*, and *Tridacna maxima* (my unpublished data) in addition to *M. balthica* and *C. virginica*. Obviously, prezoosporangia from each host species should be treated as in step (3) to confirm the identity of the parasite.

Significant structural differences have not been recognized among the *Perkinsus* spp. parasitizing the 34 species of bivalve molluscs, probably due to the lack of intensive research. It is difficult to

believe that only one species is involved because of the considerable geographical and host isolation that characterizes the group. In addition, numerous attempts to transmit infections from host to host have failed, even when the hosts lived in the same estuary (Ray 1954). *Perkinsus* species probably will have to be described and distinguished on the basis of fine structural differences.

After examining prezoosporangia of *Perkinsus* spp. isolated from 14 different hosts, I believe that there are at least two basic groups. One group resembles *P. marinus*, and the other group resembles *Perkinsus* sp. as seen in *M. balthica*. The mature trophozoites of *P. marinus* are smaller and have a smaller vacuole:cytoplasm ratio than those of *Perkinsus* sp.; they are coccoid rather than pyriform or egg-shaped. *Perkinsus marinus* cells are scattered throughout the host tissues and are not confined to pseudocapsules as in *Perkinsus* sp. These are weak distinctions, and further study is required; however, there appears to be a basis for dividing the species along those lines. It will be interesting to see whether the larger cells of the *Perkinsus* sp. group behave as prezoosporangia when released into seawater (known to occur in *Perkinsus* sp. of *Macoma balthica*) and if cells from the *P. marinus* group do not, unless treated with fluid thioglycollate medium.

Phylum Haplosporidia

Since their discovery in the late 1800s, the Haplosporidia have been a troublesome group for taxonomists and phylogenists. Numerous taxonomic schemes have been proposed with little agreement as to their taxonomic affinities. The Haplosporidia, Paramyxea, and Myxozoa may be more closely related to each other than to other Protista if one accepts the hypothesis that haplosporosomes (described below) are unique organelles found only in these taxa. Perkins (in press) reviewed the pertinent literature and proposed the following classification scheme based on that of Corliss (1984).

Phylum Haplosporidia
  Class Haplosporea
    Order Haplosporida
      Family Haplosporidiidae

I included 33 species and three genera (*Haplosporidium, Minchinia* and *Urosporidium*) within the family. The first two genera are relevant to this presentation. Larsson (1987) described a fourth genus (*Claustrosporidium*) from an amphipod after I reviewed the literature. It is unique because the sporoplasm is completely enclosed by a wall without an orifice.

Haplosporidium spp. and Minchinia spp.

Spores.—Six species of *Haplosporidium* and *Minchinia* have been described from bivalve molluscs (Azevedo 1984; La Haye et al. 1984). Nine other reports deal with parasites that may or may not be already described (Katkansky and Warner 1970; Armstrong and Armstrong 1974; Mix and Sprague 1974; Kern 1976; Pichot et al. 1979; Vivarès et al. 1982; Bachere and Grizel 1983; Bonami et al. 1985; Bachere et al. 1987). The two genera should be distinguished on the basis of the presence and visibility of prominent extensions (tails) of the spore wall under the light microscope. It is

not clear whether V. Sprague (Chesapeake Biology Laboratory, unpublished observations; 1978) meant to include in *Minchinia* only those species with one anteriorly and one posteriorly directed extension or those species with any extensions visible under the light microscope. I suggest that the latter characteristic be accepted. This point may be relevant in future generic designations of haplosporidian parasites of bivalve molluscs. Currently it is of importance only with respect to the limpet parasite *Haplosporidium lusitanicum* which should be placed in the genus *Minchinia* (Azevedo 1984). The only other species of *Minchinia* parasitic in bivalve molluscs is *M. armoricana,* which is already properly named.

The spores of *Haplosporidium* spp. and *Minchinia* spp. are basically alike in structure. The cell wall is cup-shaped and has a flange at the anterior end on which rests a lid of wall material continuous with the flange along a limited extent of the flange, thus forming a hinge. The sporoplasm inside the spore wall is uninucleate and has an unusual organelle, the spherule of unknown function at the anterior end. Membrane-bound, electron-dense organelles, the haplosporosomes, with bipartite substructure are also found scattered in the cytoplasm.

Spore structure has been described in detail (Perkins 1979; in press; Perkins and van Banning 1981). The chief structural feature for distinguishing species within a genus has been the size of the spores. Because the size ranges of species overlap considerably, host, geographical range, and epizootiological differences are additional characters used to distinguish species. The structural difference in fibers (ornaments) found attached to and around the spore wall is another distinguishing characteristic (Perkins, in press). When specimen preparation and resolution are adequate, periodicities are often found which I believe will be of value in distinguishing species (Perkins and van Banning 1981). All species not yet adequately examined need to be observed in thin sections of well-fixed, mature spores as well as negatively stained whole mounts. Inadequate attention has been given to the ornamentation, and, generally, only micrographs of thin sections are available.

Preplasmodial Stages.—The earliest stage of newly established infections consists of a uni- or binucleated naked cell about 5 μm in diameter usually found in the gills or labial palps beneath the epithelium or in the connective tissue near the basal membrane of the upper gut or midgut epithelium. Nuclear structure is the same as in plasmodia. None of the life cycles of the Haplosporidia have been elucidated; therefore, the precursor cell stage is unknown. Possibly, the early stages arise from excystment of spores entangled in the gills or labial palps or from ingested spores. However, direct transmission of infections by means of spores has not been documented.

Plasmodia.—Nuclear divisions of the preplasmodial stages, coupled with increase in cellular mass, result in formation of plasmodia (i.e., naked cells with more than two nuclei). Nuclear division (Plate 5, Figure 19) involves a mitotic spindle with spindle pole bodies free of the nuclear envelope located in the nucleoplasm and a nuclear membrane which is persistent throughout division (Perkins 1975b; Bonami et al. 1985). Cytokinesis apparently occurs by multiple fission to yield daughter cells with variable numbers of nuclei. However, this occurrence has been deduced after observations of cell clusters; no cytoplasmic divisions have been observed in progress.

The cytoplasm contains numerous electron-dense organelles with a bipartite substructure, termed haplosporosomes. They are generally spheroidal but may have other forms (Perkins 1979). The substructure is unique enough to lead me to believe that they are useful phylogenetic indicators of interrelationships. Haplosporosomes have been found in the Haplosporidia, Paramyxea (see below), and, possibly, the Myxozoa (Current and Janovy 1977; Kent and Hedrick 1985).

Another inclusion of unknown function resembles haplosporosomes because it has a delimiting membrane with a separate internal membrane. It differs from haplosporosomes because it has a granular, electron-dense core free of the internal membrane and separated from it by an electron-light zone (Plate 5, Figure 25). These haplosporosome-like bodies are formed from flattened cisternae of the Golgi bodies which, in turn, are derived from the nuclear envelope (Plate 5, Figures 20–24). During the synthesis, a cisterna arches (Plate 5, Figure 20) and forms electron-dense material on the surface of one membrane. The future core of the body is synthesized free in the cytoplasm. The arch becomes more pronounced until

the cisterna encloses the forming core by fusing to create a sphere (Plate 5, Figures 20–25). The membranes of these bodies are difficult to resolve and thus appear different from those of the haplosporosomes, in which the membrane's tripartite substructure is easily resolved. They do not appear to differentiate into haplosporosomes.

Sporulation.—Spore formation occurs generally after the parasite has spread throughout the connective tissue of the host. The plasmodia enlarge, acquire numerous nuclei (>20), and form a thin, electron-dense wall about the thickness of the plasmalemma. The haplosporosomes disappear, and the protoplast completely subdivides into uninucleate sporoblasts within the wall. Meiosis may occur during this process because nuclear pairing, large nuclei, and presumptive synaptonemal complexes and polycomplexes have been observed (Perkins 1975a). However, the events are not well understood, and more detail in more species is required before generalizations can be made.

How spores are then formed from the uninucleate sporoblasts is uncertain. I have suggested that invagination of the sporoblast plasmalemma or fusion of cytoplasmic vesicles occurs to delimit the sporoplasm primordium within a cup-shaped unit of anucleate epispore cytoplasm (Perkins 1975a). Desportes and Nashed (1983) suggested that the sporoblast elongates and the posterior, anucleate half envelops the anterior, nucleated half and becomes separate from it after envelopment. The suggestion of Desportes and Nashed is most convincing; however, I have not yet seen the pertinent stages in the species of haplosporidians I have studied.

After the spore primordium is differentiated into the two halves, first the spore wall and lid, then the spore wall ornaments, are synthesized in the anucleated epispore cytoplasm. Haplosporosomes and the spherule are formed in the sporoplasm early in the process of spore wall formation. Progressive thickening of the cell wall and degeneration of the epispore cytoplasm yield a spore which is capable of survival for at least a few weeks in seawater. I have observed that

PLATE 5.—Organelles of *Haplosporidium nelsoni* in *Crassostrea virginica*. FIGURE **19.**—Anaphase nucleus. ▶
Sp = spindle pole bodies; Mi = microtubules of spindle; G = golgi body; H = haplosporosome; HB = haplosporosome-like body. Bar = 0.5 μm. FIGURE **20.**—Haplosporosome-like bodies (arrows) forming from cisternae of Golgi body. Bar= 0.1 μm. FIGURES **21–24.**—Serial sections through two developing haplosporosome-like bodies (arrows). In Figure 24 cisternae arising from nuclear envelope are visible near the top of the micrograph. The cisternae form the Golgi bodies and haplosporosome-like bodies. Bar = 0.2 μm. FIGURE **25.**—Mature haplosporosome-like body. Co = granular core. Arrows indicate two membranes. Bar = 0.5 μm.

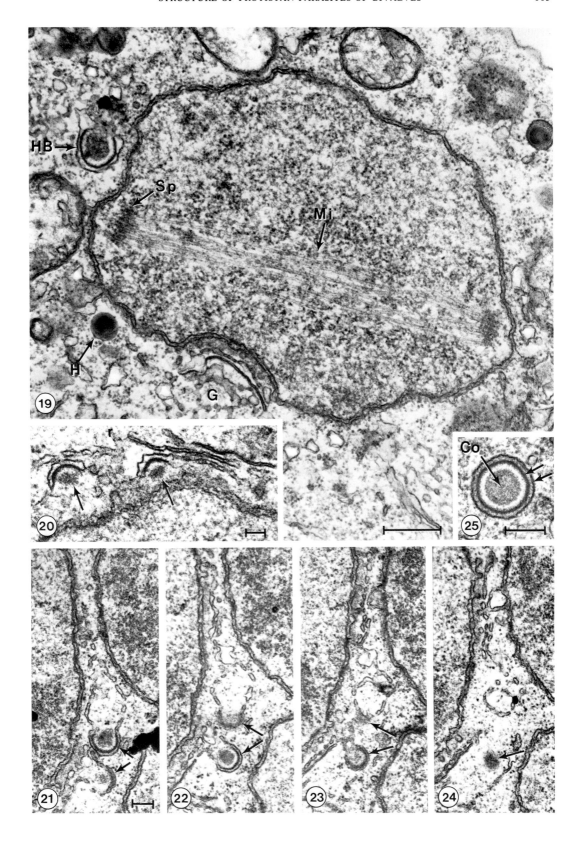

mature spores of *H. costale*, isolated from host tissue, have maintained their structural integrity in seawater with antibiotics (0.1 mg/mL each of streptomycin sulphate and penicillin G) at 4°C for up to 1 month. Because the spores were not used to induce infections in organisms or tissue, I do not know whether the spores were viable. Death can be detected in unfixed preparations because the sporoplasm pulls away from the wall and condenses. Observations of the fine structure confirm that the protoplasm is dead, based on the loss of structural integrity.

In attempts to elucidate the life cycles of haplosporidians, it is important to know if spores can exist for extended periods of time in sediments or in seawater. Dye-exclusion tests facilitate detection of spore death before the spore loses structural integrity (Perkins et al. 1975). *Haplosporidium nelsoni* continues to infect imported oysters at a more or less steady level even where resident populations of oysters decline to low levels. Thus, there may be a reservoir of infective elements independent of the oysters. There are no data to indicate whether a similar state may exist with other species of haplosporidians.

Given the structure of spores, it is unlikely that long residence outside the hosts is possible. The sporoplasm is not encased in a continuous wall (except perhaps in *Claustrosporidium gammari*). The cap rests on the spore cup flange and is attached to it by ligaments or fused by secondary wall material, an arrangement which appears to lack great strength because slight mechanical pressure induces the cap to swing open and provide an opening for the sporoplasm to exit from the spore casing. Haplosporidia appear to be obligate parasites without a saprozoic phase in the life cycle, and it is not likely that a resistant, host-free state is present as is found in fungi.

Class Paramyxea

The taxonomy of the class Paramyxea is in a state of flux. Desportes (1984) has considered its phylogenetic affinities and whether it should be a class or phylum. There are three genera, *Para-*

myxea, Marteilia, and *Paramarteilia,* in the class and six species. The only species thus far described from bivalve molluscs are all members of the genus *Marteilia: M. refringens* (Grizel et al. 1974a) in *Ostrea edulis, M. sydneyi* (Perkins and Wolf 1976) in *Saccostrea* (= *Crassostrea*) *commercialis* and possibly *Crassostrea echinata* (Wolf 1979), *M. maurini* (Comps et al. 1982) in *Mytilus galloprovincialis,* and *M. lengehi* (Comps 1977) in *Crassostrea cucullata.* An unidentified species of *Marteilia* has been found in the cockle *Cardium edule* (Comps et al. 1975).

The life cycles of *Marteilia* spp. are not known; however, the development of spores within the hosts thus far identified has been described in detail for *M. refringens* (Grizel et al. 1974a, 1974b; Perkins 1976b) and *M. sydneyi* (Perkins and Wolf 1976). Cell multiplication other than in spore formation has not been noted in any of the five species.

Desportes (1984) believed that the simplest cell is a stem cell which is uninucleate and amoeboid. It then supposedly undergoes one nuclear division followed by internal cleavage to yield a secondary cell. It has been suggested that a plasmodium is the earliest stage (Perkins 1976b; Perkins and Wolf 1976). I now question whether the simplest or stem cell is as represented by Desportes (1984) and Perkins and Wolf (1976). Grizel et al. (1974a, 1974b) may have been correct when they stated that the earliest stage is a uninucleate cell containing a uninucleate, secondary cell in a vacuole within its cytoplasm (i.e., primary and secondary cells). The evidence for a stem cell or plasmodium preceding the primary–secondary cell combination is not convincing. Due to its size, thin sections of the primary cell could easily not include the secondary cell. The plasmodium presented by Perkins and Wolf (1976, their Figure 1) could have been a later stage in sporulation in which the internal cleavage membranes were not visible. Further study is required to identify the earliest stage in infections.

The primary cell with its secondary cell forms spores by a series of secondary cell divisions followed by internal cleavages (Plate 6, Figures

---

PLATE 6.—Sporulation in *Marteilia sydneyi*. FIGURE **26.**—Primary cell engaged in formation of secondary cells (S). N_1 = primary cell nucleus; N_2 = secondary cell nuclei; H_v = vermiform haplosporosomes. Bar = 1 μm. FIGURE **27.**—Vermiform haplosporosomes of primary cell. Bar = 0.1 μm. FIGURE **28.**—Mature spore. N_m = nucleus of middle sporoplasm; N_1 = nucleus of inner sporoplasm; H = haplosporosomes in cytoplasm of outer sporoplasm; H_v = vermiform haplosporosome of middle sporoplasm. Bar = 1 μm. FIGURE **29.**—Vermiform haplosporosomes of Figure 28. Note two delimiting membranes and incompletely synthesized electron dense matrix. Bar = 0.1 μm. FIGURE **30.**—Haplosporosome of outer sporoplasm. Arrows indicate membranes. Bar = 0.1 μm.

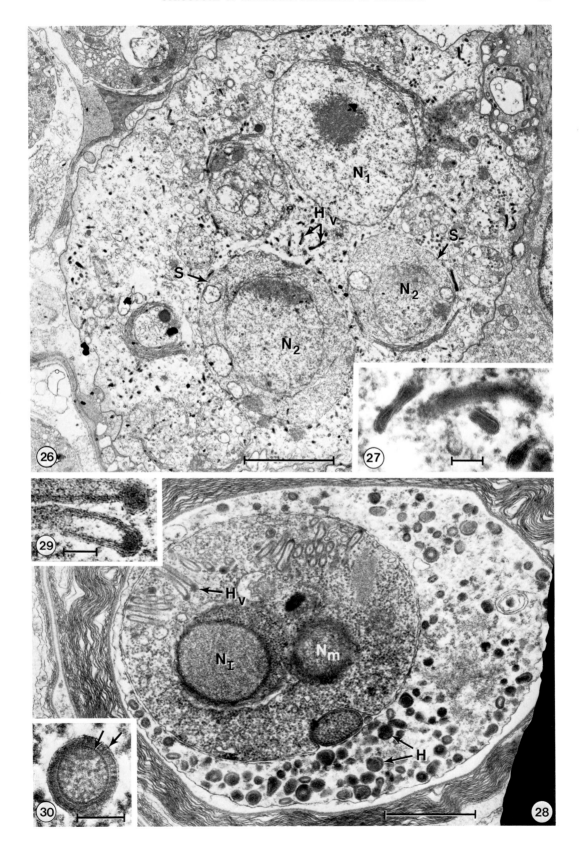

26, 28). The process has been described by Grizel et al. (1974a), Perkins (1976b), and Perkins and Wolf (1976). The essence of the process is the multiplication of the secondary cell to form 8–16 secondary, uninucleate cells which then form three to four spores within each secondary cell. Each spore consists of three uninucleate sporoplasms (cells), one within the other and contained within a cell wall (Plate 6, Figure 28). Prior to rupture and release of the tripartite spores, the full cellular complement includes the uninucleate, primary cell with its internal complement of uninucleate secondary cells each enclosed by a wall and each containing a complement of spores. Presumably, upon release, the spores disperse into the seawater and infect another host. The host tissue is in such a degraded state, and mortality among the hosts is so high at the time of sporulation, that one must assume that dispersion occurs when crabs or other predators or scavengers eat the gaping bivalves. The next host is not known. Attempts have been made by myself (unpublished data) and Grizel (1985) to infect edible oysters *Ostrea edulis* by injecting and feeding spore suspensions of *M. refringens* contained in minced, infective edible oyster tissue. As with the haplosporidians, transmissions of infections did not occur. As suggested by Grizel (1985), an alternate host may be necessary in the life cycle or a period of maturation may be required in the seawater or sediments before the spores are infective.

If edible oysters are infected by the spores, it is possible that the excystment from the spore wall could result in formation of the earliest primary–secondary cell stage. The outermost sporoplasm could degenerate upon entry into the host tissues, leaving the middle and innermost sporoplasms, which then become the primary and secondary cell complement. This suggestion has some credibility when one notes that (1) the middle sporoplasms of *M. refringens* and *M. sydneyi* have vermiform, haplosporosome-like organelles as does the primary cell (Plate 6, Figures 26–29), (2) the innermost sporoplasm has no haplosporosomes as is the case with the secondary cell, and (3) the outermost sporoplasm is often degenerate in mature spores (Plate 6, Figure 28). Possibly, the spheroidal haplosporosomes (Plate 6, Figures 28, 30) of the outermost sporoplasm are involved in host invasion by inducing either phagocytosis by host cells or lysis of host cells as a result of chemical release from the organelles.

Desportes (1981) observed that the multicellular (cells within cells) condition of the Paramyxea is similar to that of the Myxozoa, the differences being that the former lacks polar capsules and filaments as well as a spore membrane with valves. Previously, I believed that the presence of haplosporosomes in *Marteilia* spp. warranted inclusion of the species in the Haplosporidia; however, I now recognize that the multicellular condition is a more important characteristic which warrants inclusion of the species in the Paramyxea. In addition, it is now clear that haplosporosomes are formed in the Paramyxea and Haplosporidia and probably in the sporoplasms of Myxozoa (Current and Janovy 1977; Desser and Paterson 1978; Lom et al. 1983), although I have not seen micrographs of high enough resolution to make a judgment. PKX cells, the causative agents of proliferative kidney disease (PKD), appear to be stages in the life cycle of a myxozoan and, as expected, structures which appear to be haplosporosomes are found in the primary cell and in the sporoplasm (Seagrave et al. 1980; Kent and Hedrick 1985).

Because the function of haplosporosomes is unknown, the organelles cannot be precisely defined; however, I believe that their unique substructure can serve as a phylogenetic marker. They are not found in very dissimilar groups of protists. An important step toward their definition will be to isolate them and characterize them biochemically. It would also be helpful to find the hosts that are infected by the spores and, by means of the electron microscope, examine their involvement in the establishment of infections when the spores penetrate the host epithelium or are carried into the host by hemocytes. If host cell cultures could be used to culture the stem or primary–secondary cells, the role of haplosporosomes could be established. I suspect that they could be involved in lysis of host cells in the case of *H. nelsoni*, because I have observed them in oyster cells adjacent to plasmodia.

Ginsburger-Vogel and Desportes (1979) suggested that haplosporosomes of *Paramarteilia orchestiae* from the amphipod *Orchestia gammerellus* participate in formation of a wall or loosely consolidated layer of material around the spore by emptying the contents onto the cell surface. The contents appear to include mucopolysaccharides as determined by cytochemistry. Azevedo and Corral (1985) determined that haplosporosomes of *H. lusitanicum* contain glycoproteins. Whether the electron-dense inclusions that Ginsburger-Vogel and Desportes observed are haplosporosomes is questionable, since the typical substructure of haplosporosomes was not clearly demonstrated. Much

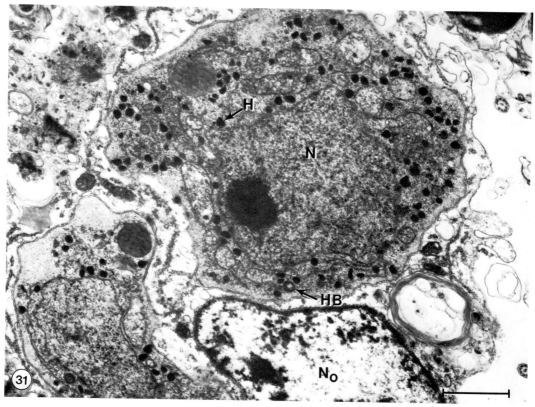

PLATE 7.—FIGURE 31.—*Bonamia* sp. from New Zealand oyster. H = haplosporosome; N = parasite nucleus; N_o = oyster hemocyte nucleus; HB = haplosporosome-like body. Bar = 1 μm.

smaller organelles that I would characterize as haplosporosomes (Ginsburger-Vogel and Desportes 1979, their Figure 24) were present in the cytoplasm. I have not found evidence of haplosporosome participation in wall formation in *Marteilia* spp. or in haplosporidians, nor have other workers noted such an involvement.

Bonamia ostreae

Bonamia ostreae causes serious mortalities of *Ostrea edulis* in France, England, The Netherlands, Spain, and Denmark. It appears to have been transported in oysters from California (USA) to Europe, where it was first observed in 1979 in oysters from the port of L'Ile-Tudy in Brittany (Comps et al. 1980; Elston et al. 1986). In addition, a species of *Bonamia* (Plate 7, Figure 31) has been reported to be associated with heavy mortality of the New Zealand oyster *Tiostrea lutaria* from Foveaux Strait along the south coast of New Zealand (M. Hine, Fisheries Research Center, personal communication).

The taxonomic affinities of *B. ostreae* are not known, but some workers relate it to the Haplosporidia due to the presence of haplosporosomes (Pichot et al. 1980) and multinucleate plasmodia (Brehélin et al. 1982). I believe that the species is probably one of the Haplosporidia. Because cells within cells are not observed, affinities with the Paramyxea or Myxozoa appear unlikely.

The dominant cell type is a uninucleate, naked cell with haplosporosomes, lipoid droplets, tubulovesicular mitochondria, and a variable amount of smooth endoplasmic reticulum (Figure 31). It is usually found in granular hemocytes where it multiplies by binary fission. The nuclear membrane remains intact during mitosis, which involves an intranuclear mitotic apparatus consisting of two spindle pole bodies and microtubules. Published micrographs of nuclear division (Pichot et al. 1980; Balouet et al. 1983) resemble those of *H. nelsoni* (Perkins 1975b). Cytokinesis may also occur by multiple fission since multinucleate plasmodia have been observed; however, there is a possibility that the plasmodia observed represented the earliest stage leading to sporulation. No spore formation has been observed, but spores are rarely observed in *H. nelsoni* (Andrews 1982). The haplosporosome-like inclusions (see phylum Haplosporidia, above) are also formed in the New Zealand *Bonamia* sp. (Figure 31, HB). Whether they are formed by Golgi bodies or by the smooth endoplasmic reticulum has not been determined.

In contrast to species of the Haplosporidia and Paramyxea, the life cycle of *B. ostreae* can be completed experimentally. Direct transmission of infection can be induced by placing infective oysters in aquaria with uninfected ones or by injecting infective oyster cells into healthy oysters (Poder et al. 1982; Grizel 1985). Whether spores or another host are part of the life cycle in nature remains to be determined.

Other Parasites

Sparks (1985) reviewed the literature concerning parasites of noninsect invertebrates, and Lauckner (1983) reviewed the diseases of bivalve molluscs. The two reviews cover the structure of diverse bivalve mollusc protists including ciliates, sarcodinids, coccidians, zooflagellates, gregarines, microsporidians, and unidentified protists. However, because the literature lacks detailed structural analyses and contains little ultrastructural information, it is not being considered here. The reader is referred to the two reviews for literature citations and summaries of what is known of species morphology. In addition, Papayanni and Vivarès (1987) described the fine structure of the parasitic zooflagellate *Hexamita nelsoni*, found in the Mediterranean oysters *Ostrea edulis* and *Crassostrea gigas*. They established the parasite in axenic culture.

One group of protists was considered by Lauckner (1983) but not Sparks (1985): the Labyrinthomorpha, a group of mostly saprophytic protists which are ubiquitous in marine and estuarine waters all over the world. Olive (1975) reviewed their structure in detail. The cells form ectoplasmic networks from distinctive cell cortex organelles termed sagenogenetosomes. The networks are used in gliding motility and are believed to serve in the digestion of food substrates and assimilation of dissolved nutrients. The cell covering consists of circular plates synthesized in the Golgi apparatus. Motile heterokont zoospores or isogametes are formed.

The cells of most of the species of *Thraustochytrium*, *Labyrinthuloides*, *Schizochytrium*, and other genera are coccoid, mostly in the size

range of 3–10 μm (except they are larger when cultured). I have found that at least one of the many species is always associated with the surface of the mantle and gill tissues of the oyster, *Crassostrea virginica*. Therefore, they have been confused with *Perkinsus marinus* (Mackin and Ray 1966) after oyster tissue infected with *P. marinus* was placed in nutrient medium and proliferation of labyrinthomorphid protists occurred. Based on these observations, the authors suggested that the name *P. marinus* be changed to *Labyrinthomyxa marina*. This study was followed by others involving descriptions of labyrinthomorphids derived from bivalve molluscs and suspected of being parasites (see Lauckner 1983 for literature review). Because the protists are easily cultured in a diversity of nutrient media and substrates (Perkins 1973) and are difficult to eliminate from the animal tissue even when the visceral mass tissues (as opposed to gill and mantle explants) are used, I believe that their designation as parasites must be regarded with doubt, and the relationships should be reevaluated. However, species of presumptive *Thraustochytrium* and *Labyrinthuloides* have been observed to parasitize an octopus, squid, nudibranch, and abalone (Polglase 1980; Jones and O'Dor 1983; Bower 1987; McLean and Porter 1987). All hosts except the nudibranch were being held in aquaculture conditions. Thus, it appears reasonable to consider the protists as facultative parasites. Nevertheless, more studies clearly demonstrating whether parasitism occurs in the bivalve molluscs are needed. The isolation and cultivation of labyrinthomorphids from sick individuals cannot be regarded as prima facie evidence that they are parasites or agents of disease. Oyster tissue explants serve as excellent substrates for growth of the organisms (Perkins 1973); therefore, a sick oyster with lysis of tissues would be expected to support proliferation of the ubiquitous protists.

Acknowledgments

The author thanks P. M. Hine for providing useful information and the micrograph concerning *Bonamia* sp. and acknowledges Susan Arnold and Patrice Mason for their able technical assistance.

References

Andrews, J. D. 1954. Notes on fungus parasites of bivalve molluscs in Chesapeake Bay. Proceedings National Shellfisheries Association 45:157–163.

Andrews, J. D. 1982. Epizootiology of late summer and fall infections of oysters by *Haplosporidium nel-*

soni, and comparison to annual life cycle of *Haplosporidium costalis*, a typical haplosporidan. Journal of Shellfish Research 2:15–23.

Armstrong, D. A., and J. L. Armstrong. 1974. A haplosporidian infection in gaper clams, *Tresus capax* (Gould), from Yaguina Bay, Oregon. Proceedings National Shellfisheries Association 64:68–72.

Azevedo, C. 1984. Ultrastructure of the spore of *Haplosporidium lusitanicum* sp. n. (Haplosporida, Haplosporidiidae), parasite of a marine mollusc. Journal of Parasitology 70:358–371.

Azevedo, C., and L. Corral. 1985. Cytochemical analysis of the haplosporosomes and vesicle-like droplets of *Haplosporidium lusitanicum* (Haplosporida, Haplosporidiidae), parasite of *Helicon pellucidus* (Prosobranchia). Journal of Invertebrate Pathology 46:281–288.

Bachere, E., D. Chagot, G. Tige, and H. Grizel. 1987. Study of a haplosporidian (Ascetospora) parasitizing the Australian flat oyster *Ostrea angasi*. Aquaculture 67:266–268.

Bachere, E., and H. Grizel. 1983. Mise en evidence d'*Haplosporidium* sp. (Haplosporida–Haplosporidiidae) parasite de l'huître plate *Ostrea edulis* L. Revue des Travaux de l'Institut Pêches Maritimes 46:226–232.

Balouet G., M. Poder, and A. Cahour. 1983. Haemocytic parasitosis: morphology and pathology of lesions in the French flat oyster, *Ostrea edulis* L. Aquaculture 34:1–14.

Bonami, J.-R., C. P. Vivarès, and M. Brehélin. 1985. Etude d'une nouvelle haplosporidie parasite de l'huître plate *Ostrea edulis* L.: morphologie et cytologie de différents stades. Protistologica 21:161–173.

Bower, S. M. 1987. The life cycle and ultrastructure of a new species of thraustochytrid (Protozoa: Labyrinthomorpha) pathogenic to small abalone. Aquaculture 67:269–272.

Brehélin, M., J.-R. Bonami, F. Cousserans, and C. P. Vivarès. 1982. Existence de formes plasmodiales vraies chez *Bonamia ostreae* parasite de l'huître plate *Ostrea edulis*. Comptes Rendus de l'Academie des Sciences, Série III, Sciences de la Vie 295:45–48.

Canning, E. U. 1986. Terminology, taxonomy, and life cycles of Apicomplexa. Insect Science and its Application 7:319–325.

Comps, M. 1977. *Marteilia lengehi* n. sp., parasite de l'huître *Crassostrea cucullata* Born. Revue des Travaux de l'Institut Pêches Maritimes 40:347–349.

Comps, M., H. Grizel, G. Tige, and J.-L. Duthoit. 1975. Parasites nouveaux de la glande digestive des mollusques marins *Mytilus edulis* L. et *Cardium edule* L. Comptes Rendus Hebdomadaires des Séances de l'Academie des Sciences, Série D, Sciences Naturelles 281:179–181.

Comps, M., Y. Pichot, and P. Papagianni. 1982. Recherche sur *Marteilia maurini* n. sp. parasite de la moule *Mytilus galloprovincialis* Lmk. Revue des Travaux de l'Institut Pêches Maritimes 45:211–214.

Comps, M., G. Tige, and H. Grizel. 1980. Etude ultrastructurale d'un protiste parasite de l'huître plate

Ostrea edulis L. Comptes Rendus Hebdomadaire des Séances de l'Academie des Sciences, Série D, Sciences Naturelles 290:383–384.

Corliss, J. O. 1984. The kingdom Protista and its 45 phyla. Biosystems 17:87–126.

Current, W. L., and J. Janovy. 1977. Sporogenesis in *Henneguya exilis* infecting the channel catfish: an ultrastructural study. Protistologica 13:157–167.

Da Ros, L., and W. J. Canzonier. 1985. *Perkinsus*, a protistan threat to bivalve culture in the Mediterranean basin. Bulletin of the European Association of Fish Pathologists 5:23–27.

Desportes, I. 1981. Etude ultrastructural de la sporulation de *Paramyxa paradoxa* chatton (Paramyxida) parasite de l'annelide polychete *Poecilochaetus serpens*. Protistologica 17:365–386.

Desportes, I. 1984. The Paramyxea Levine 1979: an original example of evolution towards multicellularity. Origins of Life 13:343–352.

Desportes, I., and N. N. Nashed. 1983. Ultrastructure of sporulation in *Minchinia dentali* (Arvy), an haplosporean parasite of *Dentalium entale* (Scaphopoda, Mollusca); taxonomic implications. Protistologica 19:435–460.

Desser, S. S., and W. B. Paterson. 1978. Ultrastructural and cytochemical observations on sporogenesis of *Myxobolus* sp. (Myxosporida: Myxobolidae) from the common shiner *Notropis cornutus*. Journal of Protozoology 25:314–325.

Elston, R. A., C. A. Farley, and M. L. Kent. 1986. Occurrence and significance of bonamiasis in European flat oysters *Ostrea edulis* in North America. Diseases of Aquatic Organisms 2:49–54.

Ginsburger-Vogel, T., and I. Desportes. 1979. Etude ultrastructurale de la sporulation de *Paramarteilia orchestiae* gen. n., sp. n., parasite de l'amphipode *Orchestia gammarellus* (Pallas). Journal of Protozoology 26:390–403.

Grizel, H. 1985. Etude des recentes épizooties de l'huître plate *Ostrea edulis* Linné et de leur impact sur l'ostreiculture Bretonne. Thèse. Académie de Montpellier, Université des Sciences et Techniques du Languedoc, France.

Grizel, H., M. Comps, J. R. Bonami, F. Cousserans, J. L. Duthoit, and M. A. Le Pennec. 1974a. Recherche sur l'agent de la maladie de la glande digestive de *Ostrea edulis* Linné. Science et Pêche 240:7–30.

Grizel, H., M. Comps, F. Cousserans, J. R. Bonami, and C. Vago. 1974b. Etude d'un parasite de la glande digestive observé au cours de l'épizootie actuelle de l'huître plate. Comptes Rendus Hebdomadaires des Séances de l'Academie des Sciences, Série D, Sciences Naturelles 279:783–784.

Jones, G. M., and R. K. O'Dor. 1983. Ultrastructural observations on a thraustochytrid fungus parasite in the gills of squid (*Illex illecebrosus* Lesueur). Journal of Parasitology 69:903–911.

Katkansky, S. C., and R. W. Warner. 1970. Sporulation of a haplosporidian in a Pacific oyster (*Crassostrea gigas*) in Humboldt Bay, California. Journal of the Fisheries Research Board of Canada 27:1320–1321.

Kent, M. C., and R. P. Hedrick. 1985. PKX, the causative agent of proliferative kidney disease (PKD) in Pacific salmonid fishes and its affinities with the Myxozoa. Journal of Protozoology 32:254–260.

Kern, F. G. 1976. Sporulation of *Minchinia* sp. (Haplosporida, Haplosporidiidae) in the Pacific oyster *Crassostrea gigas* (Thunberg) from the republic of Korea. Journal of Protozoology 23:498–500.

La Haye, C. A., N. D. Holland, and N. McLean. 1984. Electron microscope study of *Haplosporidium comatulae* n. sp. (phylum Ascetospora: class Stellatosporea), a haplosporidian endoparasite of an Australian crinoid, *Oligometra serripinna* (phylum Echinodermata). Protistologica 20:507–515.

Larsson, J. I. R. 1987. On *Haplosporidium gammari*, a parasite of the amphipod *Rivulogammarus pulex,* and its relationships with the phylum Acetospora. Journal of Invertebrate Pathology 49:159–169.

Lauckner, G. 1983. Diseases of Mollusca: Bivalvia. Pages 520–615 *in* O. Kinne, editor. Diseases of marine animals, volume 2. Biologische Anstalt Helgoland, Hamburg, West Germany.

Lester, R. J. G., and G. H. G. Davis. 1981. A new *Perkinsus* species (Apicomplexa, Perkinsea) from the abalone *Haliotis ruber*. Journal of Invertebrate Pathology 37:181–187.

Levine, N. D. 1978. *Perkinsus* gen. n. and other new taxa in the protozoan phylum Apicomplexa. Journal of Parasitology 64:549.

Lom, J., I. Dyková, and S. Lhotáková. 1983. Fine structure of *Sphaerospora renicola* Dykova and Lom, 1982 a myxosporean from carp kidney and comments on the origin of pansporoblasts. Protistologica 18:489–502.

Mackin, J. G. 1962. Oyster disease caused by *Dermocystidium marinum* and other microorganisms in Louisiana. Publications of the Institute of Marine Science, University of Texas 7:132–229.

Mackin, J. G., H. M. Owen, and A. Collier. 1950. Preliminary note on the occurrence of a new protistan parasite, *Dermocystidium marinum* n. sp., in *Crassostrea virginica* (Gmelin). Science (Washington, D.C.) 111:328–329.

Mackin, J. G., and S. M. Ray. 1966. The taxonomic relationships of *Dermocystidium marinum* Mackin, Owen, and Collier. Journal of Invertebrate Pathology 8:544–545.

McLean, N., and D. Porter. 1987. Lesions produced by a thraustochytrid in *Tritonia diomedea* (Mollusca: Gastropoda: Nudibranchia). Journal of Invertebrate Pathology 49:223–225.

Mix, M. C., and V. Sprague. 1974. Occurrence of a haplosporidian in native oysters (*Ostrea lurida*) from Yaquina Bay and Alsea Bay, Oregon. Journal of Invertebrate Pathology 23:252–254.

Moestrup, O. 1982. Flagellar structure in algae: a review, with new observations particularly on the Chrysophyceae, Phaeophyceae (Fucophyceae), Euglenophyceae, and *Reckertia*. Phycologia 21:427–528.

Olive, L. S. 1975. The mycetozoans. Academic Press, New York.

Papayanni, P., and C. P. Vivares. 1987. Etude cytologique de *Hexamita nelsoni* Schlicht et Mackin, 1968 (Flagellata, Diplomonadida) parasite des huîtres méditerranéennes. Aquaculture 67:171–177.

Pelster, B., and G. Piekarski. 1971. Elektronenmikroskopische Analyse der Mikrogametenentwicklung von *Toxoplasma gondii*. Zeitschrift für Parasitenkunde 37:267–277.

Perkins, F. O. 1968. Fine structure of zoospores from *Labyrinthomyxa* sp. parasitizing the clam *Macoma balthica*. Chesapeake Science 9:198–208.

Perkins, F. O. 1969. Ultrastructure of vegetative stages in *Labyrinthomyxa marina* (= *Dermocystidium marinum*), a commercially significant oyster pathogen. Journal of Invertebrate Pathology 13:199–222.

Perkins, F. O. 1973. Observations of thraustochytriaceous (Phycomycetes) and labyrinthulid (Rhizopodea) ectoplasmic nets on natural and artificial substrates—an electron microscope study. Canadian Journal of Botany 51:485–491.

Perkins, F. O. 1974. Phylogenetic considerations of the problematic thraustochytriaceous–labyrinthulid–*Dermocystidium* complex based on observations of fine structure. Veröffentlichungen des Instituts für Meeresforschung in Bremerhaven, supplement 5:45–63.

Perkins, F. O. 1975a. Fine structure of *Minchinia* sp. (Haplosporidia) sporulation in the mud crab, *Panopeus herbstii*. U.S. National Marine Fisheries Service Marine Fisheries Review 37(5–6):46–60.

Perkins, F. O. 1975b. Fine structure of the haplosporidian *Kernstab*, a persistent, intranuclear mitotic apparatus. Journal of Cell Science 18:327–346.

Perkins, F. O. 1976a. *Dermocystidium marinum* infection in oysters. U.S. National Marine Fisheries Service Marine Fisheries Review 38(10):19–21.

Perkins, F. O. 1976b. Ultrastructure of sporulation in the European flat oyster pathogen, *Marteilia refringens*—taxonomic implications. Journal of Protozoology 23:64–74.

Perkins, F. O. 1976c. Zoospores of the oyster pathogen, *Dermocystidium marinum*. I. Fine structure of the conoid and other sporozoan-like organelles. Journal of Parasitology 62:959–974.

Perkins, F. O. 1979. Cell structure of shellfish pathogens and hyperparasites in the genera *Minchinia, Urosporidium, Haplosporidium,* and *Marteilia*—taxonomic implications. U.S. National Marine Fisheries Service Marine Fisheries Review 41(1–2): 25–37.

Perkins, F. O. 1987. Protistan parasites of commercially significant marine bivalve molluscs—life cycles, ultrastructure, and phylogeny. Aquaculture 67:240–243.

Perkins, F. O. In press. The Haplosporidia. *In* L. Margulis, J. D. Corliss, M. Melkonian, and D. Chapman, editors. Handbook of Protoctista. Jones and Bartlett, Boston.

Perkins, F. O., and R. W. Menzel. 1966. Morphological and cultural studies of a motile stage in the life cycle of *Dermocystidium marinum*. Proceedings National Shellfisheries Association 56:23–30.

Perkins, F. O., and R. W. Menzel. 1967. Ultrastructure

of sporulation in the oyster pathogen *Dermocystidium marinum.* Journal of Invertebrate Pathology 9: 205–229.

Perkins, F. O., and P. van Banning. 1981. Surface ultrastructure of spores in three genera of Balanosporida, particularly in *Minchinia armoricana* van Banning, 1977—the taxonomic significance of spore wall ornamentation in the Balanosporida. Journal of Parasitology 67:866–874.

Perkins, F. O., and P. H. Wolf. 1976. Fine structure of *Marteilia sydneyi* sp. n.—haplosporidan pathogen of Australian oysters. Journal of Parasitology 62: 528–538.

Perkins, F. O., D. E. Zwerner, and R. K. Dias. 1975. The hyperparasite, *Urosporidium spisuli* sp. n. (Haplosporea), and its effects on the surf clam industry. Journal of Parasitology 61:944–949.

Pichot, Y., M. Comps, and J.-P. Deltreil. 1979. Recherches sur *Haplosporidium* sp. (Haplosporida—Haplosporidiidae) parasite de l'huître plate *Ostrea edulis* L. Revue des Travaux de l'Institut Pêches Maritimes 43:405–408.

Pichot, Y., M. Comps, G. Tige, H. Grizel, and M.-A. Rabouin. 1980. Recherches sur *Bonamia ostreae* gen. n., sp. n., parasite nouveau de l'huître plate *Ostrea edulis* L. Revue des Travaux de l'Institut Pêches Maritimes 43:131–140.

Poder, M., A. Cahour, and G. Balouet. 1982. Hemocytic parasitosis in European oyster *Ostrea edulis* L.: pathology and contamination. Proceedings of the International Colloquium on Invertebrate Pathology 3:254–257.

Polglase, J. L. 1980. A preliminary report on the thraustochytrid(s) and labyrinthulid(s) associated with a pathological condition on the lesser octopus *Eledone cirrhosa.* Botanica Marina 23:699–706.

Ray, S. M. 1952. A culture technique for the diagnosis

of infections with *Dermocystidium marinum* Mackin, Owen, and Collier in oysters. Science (Washington, D.C.) 116:360–361.

Ray, S. M. 1954. Biological studies of *Dermocystidium marinus,* a fungus of oysters. Rice Institute Pamphlet, special issue. (The Rice Institute, Houston, Texas.)

Seagrave, C. P., D. Bucke, and D. J. Alderman. 1980. Ultrastructure of a haplosporean-like organism: the possible causative agent of proliferative kidney disease in rainbow trout. Journal of Fish Biology 16:453–459.

Sparks, A. K. 1985. Synopsis of invertebrate pathology—exclusive of insects. Elsevier, Amsterdam.

Speer, C. A., and H. D. Danforth. 1976. Fine structural aspects of microgametogenesis of *Eimeria magna* in rabbits and in kidney cell cultures. Journal of Protozoology 23:109–115.

Sprague, V. 1978. Comments on trends in research on parasitic diseases of shellfish and fish. U.S. National Marine Fisheries Service Marine Fisheries Review 40(10):26–30.

Valiulis, G. A., and J. G. Mackin. 1969. Formation of sporangia and zoospores by *Labyrinthomyxa* sp. parasitic in the clam *Macoma balthica.* Journal of Invertebrate Pathology 14:268–270.

Vivarès, C. P., M. Brehélin, F. Cousserans, and J.-R. Bonami. 1982. Mise en evidence d'une nouvelle Haplosporidie parasite de l'huître plate *Ostrea edulis* L. Comptes Rendus de l'Academie des Sciences, Série III, Sciences de la Vie 295:127–130.

Vivier, E. 1982. Réflexions et suggestions à propos de la systématique des sporozoaires: création d'une classe des Hematozoa. Protistologica 18:449–457.

Wolf, P. H. 1979. Life cycle and ecology of *Marteilia sydneyi* in the Australian oyster, *Crassostrea commercialis.* U.S. National Marine Fisheries Service Marine Fisheries Review 41(1–2):70–72.

American Fisheries Society Special Publication 18:112–129, 1988

Strategies Employed by Parasites of Marine Bivalves to Effect Successful Establishment in Hosts[1]

THOMAS C. CHENG

Marine Biomedical Research Program and Department of Anatomy and Cell Biology
Medical University of South Carolina
Post Office Box 12559 (Fort Johnson), Charleston, South Carolina 29412, USA

Abstract.—The infective stages of protozoan and metazoan parasites of marine bivalves are discussed to stress the specific adaptive features that enhance parasitization. Also, the phases of parasitism are reviewed to emphasize the genetic endowment of parasites with a variety of phenotypic expressions that enhance successful parasitism. Such adaptive features include structural, physiological, ethological, and biochemical elements, and those that enable the parasites to overcome the immunological resistance of their hosts. Research to elucidate the invasive strategies of parasites of marine bivalves has not been extensively conducted, and little is known about these adaptive mechanisms. Therefore, examples of known successful strategies of parasites of other hosts, especially mammals, should guide us in identifying the survival mechanisms of parasites of marine bivalves. Survival strategies of parasites are interesting in themselves and also are important practically because, by interrupting them, the parasitic diseases of economically important hosts can be reduced.

Considerable information is available on the invasiveness and survival ability of parasites of humans (Bloom 1979; Trager 1986; Cheng et al. 1988), but not of marine bivalves. This aspect of parasitology is an integral part of the disease-producing process, and elucidation of the mechanisms involved will contribute to our understanding of pathogenesis and provide clues to disease control.

Parasites, as living organisms, are genetically endowed for survival. Generally, the biotic potential of parasites is greater than that of free-living organisms, i.e., parasites produce more zygotes which may attain the reproductive stage and hence are *r*-strategists (Cheng 1986a). This characteristic enhances survival of the species because most parasites have complex life cycles commonly involving two or more hosts which they must contact and then become attached to, enter, or both. Furthermore, many parasitic species include one or more free-living stages in their life cycles, and these are subjected to hazardous environmental factors: fluctuations in salinity, temperature, pH, partial pressures of gases, etc. This combination of host alternation (often involving habitat shifts) and environmental hazards commonly leads to high mortality of the parasites. Successful species, however, have become adapted through a variety of structural, ethological, physiological, and biochemical survival strategies, including enhanced biotic potentials.

Survival mechanisms can be recognized in many parasitic species but have not been investigated intensively for marine bivalve parasites. To stimulate additional research in this area, the known strategies of certain parasites of marine bivalves are reviewed here. In addition, known strategies employed by closely related parasites of other hosts are discussed to suggest related mechanisms that might be used by bivalve parasites.

Types of Parasites and Their Infective Stages

Parasites of marine bivalves belong to several phyla and subtaxa (Cheng 1967). The infective stages in the diverse life cycles of these parasites are reviewed briefly below.

Protozoans

Six groups (phyla) of protozoans are known to parasitize marine pelecypods: amoebae, flagellates, ciliates, apicomplexans, ascetosporans, and microsporans. Although additional species have been added since the publication by Cheng (1967), that monograph still provides the most complete compendium available.

Amoebae.—Although most parasitic amoebae encyst, and cysts usually represent the infective stage, essentially nothing is known about the life cycles of amoebae from marine bivalves. Hogue (1921) reported *Vahlkampfia patuxent* from the alimentary tract of eastern oysters *Crass-*

[1]The original information included herein reflects research supported by contract DE-A509-83ER60132, U.S. Department of Energy.

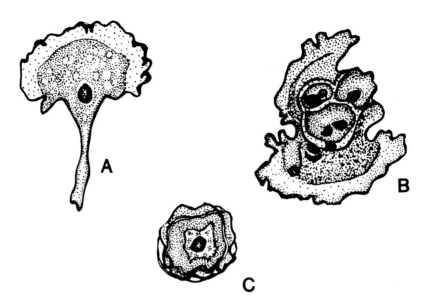

FIGURE 1.—*Flabellula calkensi*. **A.** Trophozoite showing typical broad forward–moving ectoplasmic pseudopodium and elongated trailing portion. Note rounded karyosome surrounded by clear zone in nucleus. **B.** Formation of endogenous buds in multinucleated form. **C.** Typical cyst showing three irregular cyst wall layers. (Redrawn from Hogue 1915.)

ostrea virginica in Chesapeake Bay. This species, together with *V. calkensi*, described by Hogue (1915) from the liquid content of the digestive tract of eastern oysters found on a tramp steamer at Woods Hole, Massachusetts, were transferred to the genus *Flabellula* by Schaeffer (1926). Both species are generally considered to be facultative invaders. How these apparently nonpathogenic amoebae infect oysters remains undetermined; however, both species are capable of encystation (Figure 1). Because of the feeding mechanism of oysters (Nelson 1938, 1960), amoeboid trophozoites must overcome the extraintestinal physical deterrents prior to reaching their permanent habitats. Probably, the encysted forms of *F. patuxent* and *F. calkensi* represent the passively infective stage. Thus, the ability of these facultative parasites to encyst is an adaptive feature that enhances successful parasitization.

Unlike *F. patuxent* and *F. calkensi*, *Hartmannella tahitiensis*, a facultative amoebic parasite of *Crassostrea commercialis* in Tahiti (Cheng 1970), is not an intestinal parasite; rather, its trophozoites invade essentially all other tissues of the oyster host. It has been observed to enter the host by invading surface epithelia, especially that on the ctenidial surface. Thus, it appears that *H. tahitiensis* may be adapted to invading molluscan hosts by synthesizing and secreting enzymes that

facilitate host penetration in an aquatic environment. This adaptation is considered a biochemical strategy that permits parasitization. This interpretation is supported by other species of closely related, free-living amoebae, e.g., *Naegleria gruberi* and *Acanthamoeba culbertsoni*, that are facultative parasites of mammals and include a cystic stage (Warhurst 1985). Hydrolases have been reported in numerous species of protozoans, including amoebae (Müller 1967), but these function largely in intracellular digestion. Secreted enzymes do play a role in invasion by the human pathogen *Entamoeba histolytica* (Cheng 1986a).

Richards (1968) reported that two other species of *Hartmannella*, *H. biparia* and *H. quadriparia*, can parasitize freshwater gastropods. However, unlike *H. tahitiensis*, these species do encyst, and it is the encysted stage that is infective.

The ability of *Hartmannella* spp. to engage in facultative parasitism also represents an adaptive strategy, i.e., they are preadapted to parasitism (Baer 1952), but their preadaptive features have yet to be characterized. The same reasoning apparently holds true for the amoeboid organisms reported by Sawyer (1966) from eastern oysters.

Flagellates.—Although certain parasitic flagellates are capable of encysting (e.g., *Giardia* spp.), the common invasion mechanisms are (1) introduction by a vector, often an arthropod (e.g., trypano-

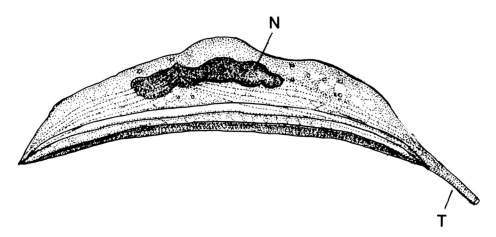

FIGURE 2.—*Sphenophyra dosiniae*, lateral view, showing flattened body with concave ventral surface. N = lobed nucleus, T = tentacle. Note absence of cilia.

somes) or (2) introduction through a body opening (mouth, urogenital orifice) of the host in the flagellated form (e.g., *Trichomonas* spp.) (Cheng 1986a). Neither the introduction of parasitic flagellates into a molluscan host in the encysted form nor the involvement of a vector has yet been reported.

Probably the best known parasitic flagellate of marine bivalves is *Hexamita nelsoni*, an intestinal parasite that causes the so-called pit disease (Mackin et al. 1952). This protozoan is normally a free-living saprobic member of the community of organisms living at the seawater–substrate interphase in estuaries. However, it can become a facultative parasite of oysters (eastern oyster, edible oyster *Ostrea edulis*, and Olympia oyster *Ostreola conchaphila*) when the hosts are under environmental stress, and it can even become pathogenic (Laird 1961; Stein et al. 1962; Scheltema 1962). *Hexamita nelsoni* enters the body of oysters with food and multiplies rapidly if the internal defense mechanisms of the host, primarily phagocytosis by granulocytes, are compromised as a result of environmental stress (e.g., reduced temperature, altered salinity, etc.) (Scheltema 1962; Feng and Stauber 1968). Because *H. nelsoni* can become a facultative parasite, it is probably genetically preadapted to parasitism (Baer 1952), but nothing is presently known about its preadapted phenotypic expressions.

Ciliates.—Some thigmotrichous ciliates were reported attached to the ctenidia and mantle of estuarine bivalves (Cheng 1967; Corliss 1979). The nature of the relationships between these protozoans and their hosts is unknown, but their use of positive thigmotropism to attach to their hosts indicates that they are genetically endowed for symbiosis. Certain species of these ciliates, e.g., *Crebricoma hozloffi* associated with blue mussel *Mytilus edulis* and *Ancistrocoma pelseneeri* with Baltic macoma *Macoma balthica*, were reported to feed on host cells (Kozloff 1946a, 1946b; Mackinnon and Hawes 1961).

Although Fenchel (1965) considered the initiation of contact between thigmotrichous ciliates and marine bivalves to be a passive process, the thigmotropic behavior may be considered an adaptive feature. Furthermore, Fenchel (1966) reported that among the 42 species of ciliates that he identified in the mantle of marine bivalves, 21 exhibited strict host specificity. This observation suggests the occurrence of some adaptive mechanism(s) which regulates the apparent specificity.

One or more species of the ciliate genus *Sphenophyra* was reported parasitic in eastern oysters (Sindermann 1970). Infected oysters present large cysts on their gills. *Sphenophyra* spp. have structural modifications that facilitate attachment to the gills of their hosts, and these modifications may be interpreted as adaptations to parasitism. Specifically, their bodies lack cilia and are flattened and concave ventrally, permitting attachment by suction to the gill surfaces of the hosts. In addition, a tentacle is present which also functions in attachment (Figure 2).

Apicomplexans.—Several representatives of the Apicomplexa are parasitic in marine bivalves. The most pathogenic is *Perkinsus marinus*, the causative agent of warm-weather mortalities of eastern oysters and perhaps other bivalves from the coast of Virginia southward and along the Gulf coast of the

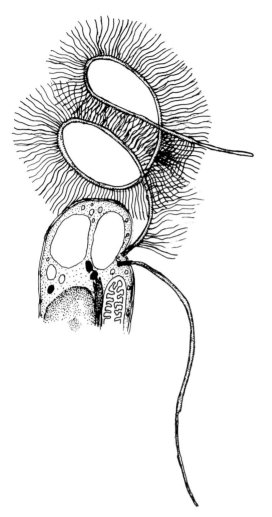

FIGURE 3.—*Perkinsus marinus*. Portion of zoospore (or planont) showing two protruding flagella, one of which is tinseled. (Redrawn from Perkins and Menzel 1967.)

USA. The infective stage of this parasite is a flagellated zoospore (or planont). After escaping from the spores in which they are formed, zoospores can infect adjacent tissues or other oysters. White et al. (1987) reported that, in some instances, the ectoparasitic gastropod *Boonea impressa* can serve as a vector, transmitting *P. marinus* from oyster to oyster.

The pair of flagella of *P. marinus*, one of which is tinseled (Figure 3), possibly represents adaptive organelles for infecting bivalves by providing locomotion for the infective zoospore. The ultrastructure of these organelles has been reported by Perkins and Menzel (1967). No homologous or analogous structure occurs in any other group

belonging to the Apicomplexa. It is a major distinguishing characteristic of the class Perkinsea to which *P. marinus* belongs.

Several species of *Nematopsis* and of the related *Porospora*, *Pseudoklossia*, and *Hyaloklossia*, all members of the class Sporozoa, are heteroxenous. These apicomplexans alternate between a crustacean and a bivalve host. The stage infective for the mollusc is known as a gymnospore; it is located in the hindgut of the crustacean host and is voided in the feces. When gymnospores, arranged as rosettes, come in contact with the molluscan host, they are engulfed by phagocytes in the gills, mantle, or digestive tract of the host. The active penetration of phagocytes by gymnospores has also been reported (Prytherch 1940); however, this observation requires confirmation.

The biotic potential of *Nematopsis ostrearum*, which alternates between a mud crab and an eastern oyster host, is great. Prytherch (1940) reported that as many as 8 million gymnospores may be released from a single crab between two molts. This example of *r*-strategy must be considered an adaptive feature because a high percentage of the gymnospores probably never enter oyster phagocytes. Although high levels of infection in oysters with *N. ostrearum* occur along the Atlantic and Gulf coasts of the USA, because the gregarious habit of the mud crab within oyster beds provides for recurring contacts, there is little evidence that this apicomplexan is pathogenic under normal conditions. This apathogenicity is probably the result of a dynamic equilibrium in the oyster between elimination of the parasite and reinfection by it (Feng 1958).

What has been stated about *N. ostrearum* is also true for *N. prytherchi*, which alternates between crab and mussel or oyster hosts, and for *Porospora gigantea*, which alternates between lobster and mussel hosts.

There is another adaptive feature of *Nematopsis* spp. that enhances successful parasitism of the bivalve host. The phagocytosis of gymnospores upon contact with a mollusc implies that the gymnospores are recognized as nonself by molluscan hemocytes (probably granulocytes), i.e., that surface molecules on gymnospores are recognized by specific sites on the surface of phagocytes. This reflection of molecular specificity (Cheng 1986b) must be considered a genetically endowed strategy which permits successful entry into the molluscan host. Even if gymnospores actively penetrate host phagocytes (Prytherch 1940), based on what is known about the entry of

Plasmodium spp. merozoites into mammalian erythrocytes (Trager 1986), molecular specificity is involved at the penetration site.

Ascetosporans.—Representatives of the ascetosporan genus *Haplosporidium* are well-known parasites of American oysters. Specifically, *H. nelsoni* and *H. costalis* are extensively studied pathogens of eastern oysters along the Atlantic coast of the USA. Although the life cycles of these parasites have been discussed by Farley (1968), the mode of infection of oysters remains unknown. Consequently, it is not known if special adaptive features are associated with the infective stages. However, based on the life cycles of other ascetosporans (Cheng 1986a), the infective stage is probably a spore which is ingested by the host. An amoebula escapes from each spore and invades the host. Thus, a resistant spore represents an adaptation favoring infection. Although sporulation by *H. nelsoni* was reported by Couch et al. (1966), these spores are noninfective to other oysters. Histological evidence suggests that *H. nelsoni* invades oysters through the surface epithelium.

Microsporans.—*Steinhausia ovicola* and *S. mytilovum* are probably the best known microsporans that parasitize marine bivalves. *Steinhausia ovicola* is a parasite of eggs of the edible oyster (Léger and Hollande 1917); *S. mytilovum* is a common parasite of eggs of blue mussel along the Atlantic coast of the USA. Similar microsporans have been reported in Pacific oyster *Crassostrea gigas* from Humboldt Bay, California (Becker and Pauley 1968), Pacific oyster from Korea (Sparks 1985), and blacklipped oyster *Crassostrea echinata* from Northern Territory, Australia (Wolf 1977).

Steinhausia spp. have an unusual habitat: they occur in the cytoplasm of eggs of marine bivalves (Figure 4). Although the route of infection is unknown, sporulation and mature spores have often been found in bivalve eggs. This finding implies that these parasites may be transmitted transovarially. Sprague (1963, 1965) reported the occurrence of *S. mytilovum* within egg nuclei and in undifferentiated gonadal cells. If this method of transmission is verified, then this phase in the life cycle of *Steinhausia* spp. must be considered an adaptive feature facilitating parasitization.

Helminths

Representatives of several taxa of helminths parasitize marine bivalves. These include members of the Platyhelminthes and Nematoda.

Platyhelminths.—Three classes of platyhelminths are known to parasitize marine pelecypods:

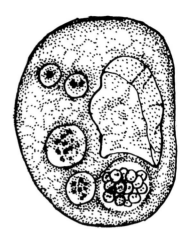

FIGURE 4.—*Steinhausia mytilorum* cysts in a *Mytilus* egg showing five different stages of development. (Redrawn from Sprague 1965.)

Turbellaria, Trematoda, and Cestoidea. Although several species of turbellarians, especially members of the orders Rhabdocoela and Alloeocoela, are parasitic in the mantle cavities, digestive tracts, and other internal organs of marine bivalves (Lauckner 1983), little is known about the invasive stage of these flatworms to permit comment on adaptive features.

Among the Trematoda, only certain members of the order Digenea have been reported from marine bivalves, although species of the order Aspidobothrea, such as *Aspidogaster conchicola*, parasitize freshwater clams (Pauley and Becker 1968; Huehner and Etges 1977, 1981). Because digeneans have complex life cycles involving two or more hosts, with sexually mature adults occurring in vertebrate definitive hosts, the role of marine molluscs is that of either the first or second intermediate host.

If the bivalve serves as the first intermediate host, it becomes infected either by ingesting a trematode egg, from which a ciliated miracidium hatches in the digestive tract, or by penetration of its surface epithelium by a free-swimming miracidium. In either case, adaptive features are involved and are discussed below. If the bivalve serves as the second intermediate host, it harbors the metacercarial stage of the parasite. Usually the molluscan second intermediate host becomes parasitized when the free-living cercarial stage released from the first intermediate host successfully penetrates it and becomes encysted as a metacercaria. The adaptive strategies involved in effecting successful penetration by cercariae are also discussed below.

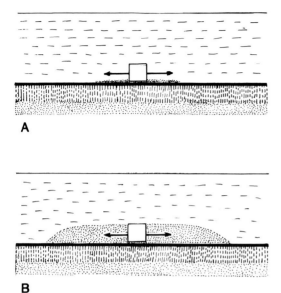

A

B

FIGURE 5.—Migratory distribution of cercariae after they have left the first intermediate host. **A.** *Holorchis pynoporus*. **B.** *Lepocreadium pegorchis*. (Redrawn from Bartoli and Combes 1986.)

A study by Bartoli and Combes (1986) illustrates adaptive behavior of platyhelminth parasites in the marine environment. Bartoli and Combes documented the migratory behavior of 10 species of marine cercariae, of which two, *Holorchis pycnoporus* and *Lepocreadium pegorchis*, employ marine bivalves as the second intermediate host. The cercariae of *H. pycnoporus* develop in the prosobranch *Barleeia rubra*. After they leave this host, the cercariae migrate horizontally abutting the substrate (Figure 5), where the second intermediate host, *Parvicardium papillosum*, occurs. *Lepocreadium pegorchis* cercariae also restrict their migration to the lower stratum of the lagoon (Figure 5); however, they are not limited to the water–substrate interface. Thus, these cercariae can readily make contact with suitable second intermediate hosts, which include various species of lamellibranchs, especially *Venerupis aurea* and *Cerastoderma glaucum*. Thus, the innate, restricted migratory patterns of *H. pycnoporus* and *L. pegorchis* cercariae must be considered as adaptive strategies that enhance their ability to contact the bivalve hosts.

Among members of the Cestoidea known to parasitize marine bivalves, it is the metacestode stage that occurs in such hosts. Thus, the molluscs act as intermediate hosts in the life cycles of these tapeworms. Metacestodes metamorphose

from a free-living postegg stage, usually a coracidium, which enters the mollusc.

Nematodes.—Cheng (1978) briefly reviewed the limited available information about nematode parasites of marine bivalves. In all known cases, these nematodes are larvae, the oldest being third-stage larvae. Although still uncertain, it appears that the egg or first-stage larva invades the molluscan host. Thus, the role of marine bivalves in the life cycles of nematodes is, again, that of intermediate hosts. The route of entry of the infective stage has not been studied.

Molluscs

Several species of gastropods have been reported as ectoparasites of marine bivalves, particularly mussels and oysters. Most of these ectoparasites are members of the Pyramidellidae. They are usually situated at the lips of the mantle, where they pierce their hosts with a stylet and suck the fluids and tissues (Fretter and Graham 1949). It is unknown whether pyramidellids attach to their bivalve hosts as veligers or as young adults. Nonetheless, their possession of a stylet must be considered as an adaptation to ectoparasitism.

Boss and Merrill (1965) reported that two pyramidellid gastropods, *Odostomia seminuda* and *O. bisuturalis*, exhibit host preference rather than host specificity. Earlier, Hopkins (1956) reported that *Odostomia impressa* parasitizes mainly old oysters along the southern Atlantic coast of the USA and that *O. bisuturalis* occurs mainly on young oysters along the northern coast. It is not known if the age of the host is of survival value to these pyramidellids. Nevertheless, the demonstration of preferences for host species (Boss and Merrill 1965) and host age (Hopkins 1956) may represent manifestations of genetically endowed strategies that perpetuate the parasitic associations.

Arthropods

A number of species of arthropods, particularly copepods, have been reported to parasitize marine bivalves. The molluscs serve as the final host because the parasites attain sexual maturity within the molluscs. How the hosts become infected, i.e., whether the arthropods invade as preadults or as adults, is uncertain. Davey and Gee (1976) reported that the parasite burden of the copepod *Mytilicola intestinalis* in *Mytilus* sp. off the coast of southwest England increases with age of the host. Although the survival value of parasitizing older mussels is not known, the preference for older hosts may represent a strategy for optimal survival. On the other hand, the larger number of copepods in older hosts may merely reflect the longer period of time during which the hosts can become infected.

Mytilicola intestinalis not only occurs in larger numbers in older bivalves, it also prefers certain species of molluscan hosts (Hepper 1956). Although the mechanism governing host selection remains unexplored, the occurrence of the phenomenon probably reflects an endowed strategy that at least favors the copepod.

Another well-known group of arthropodan parasites of oysters and other marine bivalves are the pinnotherid crabs. They live in the mantle cavities of their hosts and are considered by some investigators to be commensals. However, at least the crab stages I–V of Christensen and McDermott (1958) are metabolically dependent on their molluscan hosts, i.e., they feed on gill tissues of their hosts and retard growth, and hence qualify as being true parasites (Cheng 1986a).

During the life cycle of *Pinnotheres* spp., the preparasitic larval stages are planktonic. These include four zoeae followed by one megalops. After molting, the megalops develops into the first crab stage which is parasitic (Stauber 1945; Christensen and McDermott 1958).

Phases of Parasitism and Adaptive Features

To achieve a critical analysis of parasitism, a host–parasite association must be dissected into its constituent parts (Cheng 1968), commencing with the mechanisms responsible for attraction between the two associates. The phases of parasitism (after Cheng 1986) will be discussed in the following order.

Host–parasite contact.
Entry into host.
Establishment of the parasite.
  Habitat selection and attachment.
  Overcoming the host's defense mechanisms.
  Acquisition of nutrients.
  Developmental and growth stimuli of host origin.
Escape of parasite.

Host–Parasite Contact

In spite of the enhanced biotic potential of almost all parasitic species, parasite survival still depends on additional, commonly subtle, genetically endowed adaptive features. This is especially true of the infective stages which often are subjected to fluctuating environmental factors that mediate against chance contact with potential hosts. During the course of evolution, parasites have developed a variety of mechanisms that enhance host–parasite contact.

Protozoans.—An encysted stage occurs in the life cycles of most species of amoebae. The ability of species that parasitize marine bivalves to encyst (e.g., *Flabellula* spp., certain *Hartmannella* spp.) represents a genetically endowed advantage to effect parasitism. The exception is *Hartmannella tahitiensis*, which apparently is cable of secreting lytic enzymes that facilitate entry into stressed oysters (Cheng 1970).

The flagellate *Hexamita nelsoni* is not known to encyst and hence must enter its host as trophozoites. It may be assumed that the amoebae, flagellates, and ciliates that associate with aquatic bivalves have been drawn to the vicinity of their hosts by the water currents created by the hosts. However, the behavioral pattern of thigmotrichous ciliates assures an intimate association with their hosts. Thus, their positive thigmotropism must be considered an adaptive strategy favoring host–parasite contact.

Platyhelminths.—Host-finding by trematode miracidia represents the most studied model in this aspect of parasitology (Cheng 1967; Ulmer 1971; MacInnis 1976). Such studies, however, have been limited primarily to the location of freshwater gastropods by the miracidia of human schistosome parasites. Generally, positive kineses are involved; however, even before such kineses become operative, innate taxes of the miracidia bring them to the immediate proximity of their snail hosts. The chemical signals emitted from snails that attract schistosome miracidia have not been isolated and identified; however, the term "miraxone" has been proposed by Chernin (1970) to designate this complex of molecules.

Host-finding by trematode miracidia that utilize bivalves as first intermediate hosts has yet to be explored. Study of this phenomenon should prove interesting in view of the reports by Sponholtz and Short (1976) and Stibbs et al. (1976) that the ratio of calcium to magnesium and the concentration of magnesium chloride, respectively, affect miracidial host-finding behavior.

Nematodes.—As yet, no studies have been conducted to ascertain how nematodes parasitic in marine bivalves find their hosts.

Molluscs.—It remains unknown whether host preference of the parasitic gastropods *Odostomia seminuda* and *O. bisuturalis* (Boss and Merrill 1965) reflects host selection by chemical signals or abandonment of less-suitable hosts.

Arthropods.—A chemical signal that attracts *Pinnotheres maculatus* to its host is suspected as a result of behavioral studies (Sastry and Menzel 1962), but the nature of the signal and how the parasite recognizes it remain undetermined. It also is unknown if chemical attraction occurs between other arthropodan parasites and their bivalve hosts.

The characterization of chemotactic and possibly other types of attraction between the infective stage of a parasite and its bivalve host could be of practical value. Alteration of the chemical or physical characteristics of the immediate environment might deter host–parasite contact.

Entry into Host

In some instances, morphogenetic alterations of the parasite must occur prior to its entry into the host. Although such prerequisites to successful entry into a host remain uninvestigated among parasites of marine bivalves, examples are known among other host–parasite associations. For example, prior to successful penetration of the molluscan host, the miracidium of *Fasciola hepatica* must be stimulated to shed its ciliated coat (Dawes 1960). Furthermore, the apical papilla must become invaginated to form a suction cup into which histolytic enzymes are secreted from gland cells in the anterior region of the miracidium. Cheng (1968) demonstrated that such preparative changes of the parasite are stimulated by a host factor.

Endoparasites enter their hosts passively via ingestion or another vector or they actively penetrate host. Either process can reflect adaptive strategies that enhance establishment of the parasite.

Except in the case whereby *Perkinsus marinus* is transmitted to eastern oysters by *Boonea impressa* (White et al. 1987), vectorial transmission is not known in parasitism of marine bivalves. The ability of amoebae such as *Flabellula patuxent* and *F. calkensi* to encyst suggests that encystment is an adaptation for passive entry.

Active entry is exemplified by the apparent ability of *Hartmannella tahitiensis* to secrete penetration enzymes (Cheng 1970). Also, the occurrence of an apical complex in the infective stages suggests that protozoan species of *Perkinsus*, *Porospora*, and *Nematopsis* are equipped for active entry.

The cephalic gland of marine trematode miracidia and the penetration glands of marine cercariae have not been studied, but homologous structures in freshwater species (Erasmus 1972) are functional structures that permit active entry. Similarly, the production by some helminths of microscopic eggs that are infective when ingested by the host and the ability of trematode miracidia to penetrate the integument of their molluscan host by aid of lytic secretions must be considered genetically endowed strategies that enhance parasitism.

Nothing is known at this time as to how endoparasitic nematodes and arthropods effect entry into marine bivalves. Understanding the mechanisms employed by endoparasites to enter their bivalve hosts could lead to the development of methods to interfere with successful entry.

Establishment of the Parasite

After successful entry, the parasite still has a number of hurdles to overcome before it can become established. Although these aspects of parasitism have been little studied for parasites of marine bivalves, the following information from investigations of terrestrial and freshwater parasites will suggest areas of research.

Habitat selection and attachment.—The migration of parasites within the host en route to their normal microhabitats is well known; however, little is known about the mechanisms governing these migrations. Probably, chemotaxis, as well as other mechanisms, is involved. Invertebrate hosts should provide ideal models for research in this area.

Overcoming the host's defense mechanisms.—Much effort has gone into understanding the strategies that the parasites of mammals, especially of humans, employ in avoiding destruction by the immune systems and related mechanisms of their hosts (Bloom 1979; Cheng et al. 1988). Cheng (1987) and Cheng et al. (1988) distinguished between "camouflage mechanisms" and "avoidance mechanisms" employed by parasites within their hosts. Camouflage mechanisms are those by which the parasite avoids the direct or indirect deleterious effects of host immunoglobulins. Avoidance mechanisms are those by which the parasite avoids being destroyed by the internal defense mechanisms of the host that do not involve immunoglobulins. Invertebrates do not synthesize immunoglobulins; therefore, only avoidance mechanisms are employed by parasites of marine bivalves.

Protozoa.—Several avoidance mechanisms are known to operate in four protozoan parasites of marine bivalves, *Bonamia ostreae*, *Marteilia refringens*, *Haplosporidium nelsoni*, and *Perkinsus marinus*. *Bonamia ostreae* is an intracellular parasite of *Ostrea edulis*, usually within the phagocytes. The microhabitat of *B. ostreae* is similar to that of three human parasites, *Leishmania donovani*, *Toxoplasma gondii*, and *Trypanosoma cruzi*; therefore, some clues to how *B. ostreae* may escape destruction by its molluscan host can be gleaned by examining how the three human pathogens accomplish this feat.

After the promastigote of *Leishmania donovani* enters the host cell, it transforms into an amastigote which is situated in a parasitophorous vacuole. Subsequently, lysosomes fuse with the parasitophorous vacuolar membrane, and lysosomal hydrolases are released into the phagolysosome. Because the amastigote survives and divides (Chang and Dwyer 1978), the compatible substrates for the lysosomal hydrolases do not occur on the surface of the amastigote. The operation of such an avoidance mechanism in *B. ostreae* is not known, and its occurrence has not yet been studied.

After the sporozoite (or merozoite) of *Toxoplasma gondii* enters the host cell, it also becomes situated within a parasitophorous vacuole. Endodyogeny occurs at this site, and 8–16 tachyzoites are formed. By some unexplained mechanism, the occurrence of tachyzoites within the parasitophorous vacuole prevents the fusion of lysosomes with the vacuolar membrane (Jones et al. 1972), and the tachyzoites are not destroyed. In addition, Sibley et al. (1986) reported that the intracellular form of *T. gondii* secretes membranous vesicles shortly after entering the host cell. This vesicular network also contributes to the prevention of intracellular digestion of the parasite. Thus, *T. gondii* avoids destruction by inducing alterations in the parasitophorous vacuolar membrane of the host and by the secretion of protective vesicles. It is not known whether *B. ostreae* employs a similar strategy.

Most likely, *B. ostreae* employs a strategy similar to that of *Trypanosoma cruzi*. Amastigotes of this species, as they transform from invading trypomastigotes, initially occur within parasitophorous vacuoles where they divide; however, they rupture out of the vacuole and remain in the host cell cytoplasm until the host cell ruptures. The amastigotes are not destroyed in the cytoplasm of the host cell, possibly due to the low level of lysosomal enzymes (Bloom 1979). When the host cell ruptures, the newly metamorphosed trypomastigotes that are released must rapidly enter another host cell or be destroyed by circulating antibodies.

The intracellular phase in the life cycle of *B. ostreae* closely parallels that of *T. cruzi*. The invading microcell initially occurs within a parasitophorous vacuole, where it undergoes division.

Subsequently, the daughter cells escape into the cytoplasm of the host cell and are not destroyed. Apparently, the strategy employed by *B. ostreae* to avoid destruction within host cells resembles that of *T. cruzi* amastigotes; at least there are morphological parallels.

Marteilia refringens, a pathogen of edible oysters, *Haplosporidium nelsoni*, a pathogen of eastern oysters, and *Perkinsus marinus*, a parasite which can be highly pathogenic to eastern oysters, are all usually extracellular parasites. These three protists, especially *P. marinus*, may occur within the phagocytes of their bivalve hosts. These parasites may become phagocytosed only when their surfaces become altered stereochemically or molecularly so that they are recognized by the host as being nonself (Cheng 1986a). Endocytosed *M. refringens*, *H. nelsoni*, and *P. marinus* are usually degraded intracellularly. Cheng (1987) presented a theoretical explanation of how some surface-altering factor(s), such as a lectin, could cause the alteration of self to nonself. Cheng and Howland (1979) proposed that microorganisms pathogenic to molluscs usually are recognized as self by phagocytes and hence not resisted.

Although flagellates, e.g., *Hexamita* spp., are known to parasitize marine bivalves, the mechanism they use to overcome the immunologic responses of their hosts and survive is unknown. In the human-infecting flagellates of Africa—*Trypanosoma brucei brucei*, *T. b. rhodesiense*, *T. b. gambiense*, and probably most trypanosomes that develop in the salivary glands of their vectors—a phenomenon known as antigenic variation occurs which protects a small number of mutants in the bloodstream of the host. Vickerman (1978) and Cross (1978) showed that certain trypomastigotes undergo antigenic variation, i.e., the antigens on their surfaces become altered, and the parasites are not killed by the binding antibodies. The unharmed variants divide, and their progeny represent almost all of the individuals at the next peak of the cyclic parasitemia (Borst and Cross 1982; Cheng et al. 1988).

Because molluscs do not synthesize immunoglobulins, it is doubtful that antigenic variation occurs in their flagellate parasites. However, other types of humoral factors occur in molluscs (Cheng and Combes, in press), and determining how flagellates avoid destruction by such factors is a worthwhile endeavor.

Helminths.—Our knowledge of the strategies employed by helminth parasites to resist the immune mechanisms of their hosts is based largely on the schistosome–mammal relationship. However, these strategies are not the only ones employed by helminths. Cheng et al. (1988) critically reviewed all known mechanisms. Studies designed to investigate the strategies employed by parasites (especially helminths) of marine molluscs to avoid destruction by immune mechanisms of their hosts have yet to be conducted. The strategies employed by helminth parasites of vertebrates, especially mammals, will now be briefly reviewed.

Intestinal nematodes of vertebrates represent the largest group of parasites that possess a body surface that is partially or totally resistant to both enzymatic degradation and immunologic reactions of the host. Adult *Ascaris lumbricoides* have antienzymes in the cuticle that prevent their being digested. Green (1957) demonstrated the presence of a trypsin and a chymotrypsin inhibitor, and Peanasky and Laskowski (1960) isolated and purified the chymotrypsin inhibitor.

Three other mechanisms also protect parasites from being digested in their hosts: (1) the chemical incompatibility of the body surface, (2) the protective role of the glycocalyx, and (3) the selective impermeability of the body surface. Evidence for these mechanisms were provided by Cheng (1986a) and Cheng et al. (1988).

Damian (1964) proposed that schistosomes have evolved antigenic determinants similar or identical to those of the host. As a consequence, the host recognizes the schistosome surfaces as self and does not attack the parasite. This proposal is referred to as the "natural selection" hypothesis.

Alternatively, Smithers and Terry (1969) proposed that the similarity between the surface antigens of schistosomes and host antigens is the result of adsorption or absorption of host antigens. This proposal is referred to as the "antigen-masking" hypothesis. A third theory, known as the "host induction" hypothesis, proposes that the host is capable of inducing the parasite to produce hostlike antigens (Damian 1979).

Evidence available up to this time favors the antigen-masking and natural selection hypotheses. These three hypotheses are collectively grouped as host or molecular mimicry. Cheng et al. (1988) reviewed the evidence for and against all three hypotheses.

From the standpoint of adaptation of parasites of their mollusc hosts, Yoshino and Cheng (1978) reported *Biomphalaria glabrata*-like antigens associated with the surface membranes of *Schistosoma mansoni* miracidia. Subsequently, Yoshino and Bayne (1983) reported that both miracidia and

sporocysts of *S. mansoni*, cultured in vitro and never exposed to *B. glabrata*, shared antigens with the gastropod host. Probably, these common antigens are of parasite origin rather than adsorbed from the host, thus agreeing with Damian's (1964) natural selection hypothesis. The importance of these findings is that the major histocompatibility complex has not been demonstrated in invertebrate hosts, including molluscs, although the hosts can discriminate between self (isografts) and nonself (allografts and xenografts) as demonstrated by tissue transplantation studies (Jourdane and Cheng 1987; Cheng and Jourdane 1987).

Van der Knaap et al. (1985) searched for shared antigens between mother and daughter sporocysts of *Trichobilharzia ocellata* and swamp lymnaea *Lymnaea stagnalis*, the gastropod host, by employing immunocytochemistry. They could not demonstrate any shared antigens; however, the immunocytochemical reaction product occurred in the postacetabular glands and on the surface of emerged cercariae. Earlier, Roder et al. (1977) reported the presence of hostlike antigens on *T. ocellata* cercaria. It is not known whether these antigens on the cercarial surface serve a protective role, and it is puzzling that van der Knaap et al. (1985) did not find such surface antigens on *T. ocellata* sporocysts.

Immune mechanisms, nutrition, respiration, and other physiologic processes are similar in all bivalves and gastropods that have been studied (Bishop et al. 1983; De Zwaan 1983; Livingstone and De Zwaan 1983; Voogt 1983; Ratcliffe et al. 1985). Therefore, the basic mechanisms responsible for the compatibility between the hundreds of species of larval digeneans and marine molluscs (Lauckner 1983) are probably similar to those between larval schistosomes and their molluscan hosts. This aspect of parasitology has yet to be studied for marine bivalve hosts.

In *Trichinella spiralis* infections of mammals, a variety of pronounced immunologic reactions, both humoral and cell-mediated, develop in the host (Love et al. 1976; Despommier 1981; Campbell 1983; Ortega-Pierres et al. 1984). Severe host immunologic reactions are directed at both adults and larvae, but some of the larval worms survive by penetrating into the cytoplasm of striated myofibers, where they are protected from the host's immune mechanisms by nurse cells.

This strategy by which a metazoan parasite avoids being destroyed by immune mechanisms of its host has yet to be demonstrated in an invertebrate host–parasite combination. Because invertebrates do not synthesize immunoglobulins, it is doubtful that this type of camouflage mechanism will be found in them. However, because cell-mediated and nonimmunoglobulin humoral factors do occur in molluscs (and other invertebrates), an analogous strategy is possible. Jourdane and Mounkassa (1986) reported that if the snail *Biomphalaria glabrata* (bloodfluke planorb) is infected with both *Schistosoma mansoni* and *Echinostoma caproni*, most of the *S. mansoni* primary sporocysts avoid the deleterious effects of competition by shifting from the host's head-foot, the usual habitat, to deeper tissues of the host's cephalopedal zone (cerebral ganglia, sinus, and kidney) where they continue to develop normally. A comparable example has yet to be reported in a marine bivalve.

Arrested development of the larvae of certain species of nematodes is a well-known phenomenon (Schad 1977). This mechanism is an adaptive feature for survival because the worms within a host that temporarily cease to develop do not exit from the host and are protected from external environmental hazards. When such deleterious factors ameliorate (e.g., when temperature increases or rainfall resumes), the arrested parasites resume their development. Soulsby (1982) considered arrested development as a mechanism by which certain parasites avoid the deleterious effects of their hosts' immune mechanisms, because both focal and systemic host responses to arrested (or dormant) parasites usually do not occur. The underlying mechanism(s) remains unexplored. Arrested development has not been conclusively demonstrated in an invertebrate host, with one possible exception. The encystment and encapsulation of *Himasthla quissetensis* metacercariae in various marine bivalves (Cheng et al. 1966) may be interpreted to represent arrested development.

The ability of certain parasites to escape destruction by their hosts need not always be a reflection of their ability to camouflage themselves from the immunologic attacks of the hosts, their ability to reduce the level of attack, or their ability to avoid being destroyed by other categories of antiparasitic substances (e.g., lysosomal or digestive enzymes). Certain parasites may survive because the host does not respond. Such nonresponsiveness does not mean that the host recognizes the parasite as self sensu strictu; rather, it is a temporary state during which the host responds to immunologic challenge slightly or not at all. This mechanism differs from host mimicry.

Soulsby (1982) emphasized that vertebrate hosts do not consistently react to parasites throughout their lives. Neonatal mammals are often unresponsive to parasitic invasions, e.g., young ruminants parasitized by gastrointestinal helminths. The reasons for this unresponsiveness may include feedback inhibition of passively acquired colostral antibodies, transfer of tolerogenic substances via colostrum, presence of suppressor factors in the serum, and induction of suppressor cell formation following precocious exposure to parasitism.

Among parasites of marine molluscs, the nonresponsiveness of hosts probably represents a major method by which parasites avoid being attacked. A frequently cited example of host nonresponsiveness is that of eastern oyster parasitized by *Haplosporidium nelsoni*. Usually, this protozoan is not phagocytosed by host cells, but, when it becomes moribund, it is phagocytosed (Cheng 1986a). However, in both intermediate and advanced stages of haplosporidiasis, hyalinocytes infiltrate the diseased tissues. Thus, cellular reaction, without phagocytosis, occurs in the form of hyalinocyte infiltration. Consequently, the eastern oyster–*H. nelsoni* association may not be a true example of nonresponsiveness of the host to the parasite. Furthermore, Feng and Canzonier (1970) found that a serum protein fraction increased in *H. nelsoni*-infected eastern oysters. On the other hand, experiments with young oysters have not been performed; hence, hyalinocyte infiltration and quantitative alteration of a serum protein fraction may reflect the response only of older (mature) oysters.

Richards (1973) reported that a strain of bloodfluke planorb juveniles susceptible to infection by *Schistosoma mansoni* miracidia become refractory when they reach sexual maturity. In crossing experiments, he demonstrated that insusceptibility develops in adults by the simple Mendelian inheritance of a single dominant gene. Neither the phenotypic manifestation responsible for the change nor comparable examples involving marine bivalve hosts have been studied.

During the developing phase of certain metacestodes in mammals, specific immunoglobulin(s) is adsorbed onto the surface of the parasite; it is nondamaging but it prevents attack by granulocytes. Because this mechanism is probably associated with anticomplement factors that inhibit complement-mediated cytotoxicity (Soulsby 1982), it cannot exist in molluscs; invertebrates do not possess complement. However, the absorption of molecules onto the body surface to inhibit attack by host cells may occur and should be sought among parasites of marine bivalves.

Escape by lysis of the host's cells is an inadequately understood mechanism that apparently prevents certain parasites from being destroyed. It does not involve the retardation of the effects of immunoglobulins. This parasite strategy has only been reported in insect hosts of larval cestodes. Ubelaker et al. (1970) reported that *Hymenolepis diminuta* cysticercoids in *Tribolium confusum*, the beetle intermediate host, possibly protect themselves by secreting a lytic substance from their tegumental microvilli which lyses the hemocytes of the host. The same phenomenon probably occurs with *Hymenolepis citelli* cysticercoids in *T. confusum* (Collin 1970).

In summary, various parasites employ different strategies to avoid being deleteriously affected by the internal defense mechanisms of their hosts. The examples are primarily parasites of mammals, and, although the strategies used are principally camouflage mechanisms, avoidance mechanisms also exist. Probes into parasites of marine bivalves should prove to be intellectually exciting and may lead to practical applications. Thus, by tilting a demonstrated avoidance mechanism in favor of the host, the parasite could be killed.

Acquisition of nutrients.—The acquisition and metabolism of nutrients by endoparasites of vertebrates has been explored extensively, but this subject is essentially unexplored for parasites of marine bivalves. Gutteridge and Coombs (1977) and Barrett (1981) authoritatively synthesized the biochemical information on nutrition and metabolism of parasitic protozoa and helminths, respectively.

Carbohydrates.—The pioneering work of Axmann (1947) demonstrated that glycogen abounds in larval trematodes (miracidia and cercariae). Subsequently, the occurrence of somatic glycogen was reported in several species of larval trematodes (Ginecinskij 1960; Cheng and Snyder 1962a; Ginecinskij and Dobrovalskij 1962; Cheng 1963a, 1963b, 1963c). What is the source of the glucose from which the stored glycogen is polymerized? Cheng and Snyder (1963) demonstrated that, in the presence of trematode larvae (sporocysts and rediae), glycogen stored in the digestive gland of molluscan hosts is degraded to glucose. The glucose molecule is small enough to permeate not only the surface membrane of the acinar cells of the digestive gland but also that of the larval trematodes. Electron microscope studies revealed that the tegumental surface of these parasites consists of an outer layer of microvilli (Lee 1966,

1972). The adsorptive surface, thus increased, may enhance the uptake of glucose, possibly by phosphorylative transport (Cheng and Snyder 1963; Cheng 1964). Glucose taken into sporocysts and rediae become polymerized into glycogen (Cheng 1963a). Thus, the specialized tegumental surface of sporocysts and rediae, coupled with the presence of phosphatases, may be considered to be adaptive strategies that permit the uptake of glucose and probably other molecules (Cheng 1964).

Some information is available for parasites of marine bivalves. Cheng and Burton (1966) reported a gradual reduction of glycogen in the cells of the digestive gland of eastern oysters parasitized by the dendritic sporocysts of *Bucephalus* sp. This reduction results from the breakdown of glycogen to glucose which permeates the host digestive gland cell membrane and becomes incorporated in the developing cercariae, where it serves as a source for glycogen synthesis.

Lipids.—Ginecinskij (1961) apparently was the first to demonstrate the occurrence of lipids in intramolluscan larval trematodes. This demonstration was followed by the report of Cheng and Snyder (1962b), who demonstrated histochemically that free fatty acids are transported from the acinar cells of the molluscan digestive gland through the tegumental surface of sporocysts and are deposited in the parenchyma of sporocysts and developing cercariae. Because the environment in the digestive gland of molluscs is essentially anaerobic, Cheng and Snyder (1962b) suggested that lipids are probably not metabolized in energy production in intramolluscan larval trematodes. Nonetheless, the ability of intramolluscan larval trematodes to take up and store fatty acids (which are utilized for energy production in free-living cercariae) must be considered a biochemical adaptive strategy.

Cheng (1965) demonstrated the presence of fatty acids in the sporocyst wall of *Bucephalus* sp. parasitizing eastern oysters. The pattern of uptake of fatty acids of host origin by *Bucephalus* sporocysts parallels that by sporocysts in freshwater molluscs (Cheng 1967). This one-way transport of nutrient molecules from host cells to parasite cells must be considered another physiological adaptive mechanism that enhances successful parasitism.

Proteins.—Cheng (1963a) documented the utilization of host proteins and amino acids by intramolluscan larval trematodes. Although these nitrogen compounds are not employed as primary energy sources in the essentially anaerobic conditions of the digestive glands of their hosts, they are undoubtedly utilized for the formation of new soma (e.g., in developing cercariae) and other categories of functional molecules (e.g., enzymes, hormones, etc.). The ability to derive amino acids from their molluscan hosts must be considered a survival strategy for these endoparasites.

An example of structural adaptation favoring nutrient uptake was given by Kent et al. (1987), who reported on an *Isonema*-like flagellate that is pathogenic to larvae of the geoduck clam *Panopea abrupta*. This apparently saprobic free-living flagellate is a facultative parasite. Its preadaptive ability as an endoparasite includes the ingestion of host cells by means of an ingestion apparatus consisting of a pharynx surrounded by a microtubular complex.

Developmental and growth stimuli of host origin.—Most helminths undergo true metamorphosis or give rise to the subsequent generation by asexual reproduction while in their molluscan hosts. The cestode procercoids and plerocercoids from marine bivalves are examples of true metamorphosis. The procercoid of *Tetragonocephalum* sp. (= *Tylocephalum*) from the pearl oyster *Margaritifera vulgaris* in Sri Lanka (Herdman and Hornell 1906; Shipley and Hornell 1906; Jameson 1912; Southwell 1924), the eastern oyster in Hawaii (Sparks 1963) and elsewhere, including the east coast of the USA, and the clam *Tapes semidecussatus* in Hawaii (Cheng and Rifkin 1968) must have metamorphosed from an earlier stage (a coracidium) after the latter had entered the bivalve host. Although Cheng (1966) described what was thought to be the coracidium of *Tetragonocephalum* sp., Freeman (1982) pointed out that this was probably an error.

Hyman (1951) reported that the plerocercoid of the tetraphyllidean *Echeneibothrium* sp. encysted in the foot of a clam collected in Puget Sound. This larva is probably the same as that reported by Sparks and Chew (1966) encysted throughout the tissues of the Pacific littleneck *Protothaca staminea* collected in Humboldt Bay, California. In this case, the plerocercoid must have metamorphosed from a procercoid that is probably the infective stage for Pacific littlenecks.

What host factor(s) triggers the coracidium of *Tetragonocephalum* sp. and the procercoid of *Echeneibothrium* sp. to metamorphose into the subsequent life cycle stage? Based on what is known of exogenously stimulated differentiation, the unknown stimulatory host factor(s) must be recognized at the molecular level at the cell surfaces of the parasites to activate the genomes responsible for differentiation. Thus, the possession of a surface recognition site and a sequence of molecular events that results in metamorphosis must be considered adaptive features that permit successful establishment of the parasites.

A series of asexually reproducing stages usually occurs in larval digeneans parasitic in molluscs. In some species, primary sporocysts differentiate from miracidia and give rise to secondary sporocysts, which then give rise to cercariae. In other species, primary sporocysts give rise to rediae which then give rise to cercariae. In still other species, only redial generations occur prior to the development of cercariae. What host factor(s) stimulates the sequence of asexually derived larval generations? Szidat (1959) suggested that sex hormones of the molluscan host may act as a stimulatory factor that also controls growth of the parasite; however, I have not been able to duplicate his results (unpublished data).

The ability of one asexually reproducing generation to give rise to another suggests that the exogenous host factor is recognized by the target cells and results in embryogenesis of the subsequent generation. This development must also be considered an adaptive feature that promotes successful parasitism.

Escape of Parasite

To perpetuate a parasitic species, the parasites or their germ cell-bearing progeny must escape from their hosts to contact and invade new hosts. Cheng (1967) suggested the terms "active escape" and "involuntary escape" to describe how the parasite or its germ cells escape from the host. In active escape, either the parasite or its progeny actively effects the escape; in involuntary escape, the parasite or its progeny are egested, as with intestinal parasites. The discharge of gymnospores of *Nematopsis*, *Porospora*, and related genera from the hindgut of the crustacean host represents involuntary escape.

A third category, passive escape, involves no active behavior either by the escaping stage or the parental stage. Examples include the encysted metacestodes of *Tetragonocephalum* sp. in marine bivalves and the encysted metacercariae of such digeneans as *Himasthla* spp. in marine bivalves.

The ability of a parasite to escape actively, involuntarily, or passively represents other adaptations that enhance the perpetuation of the parasitic relationship.

Some Conclusions and Suggestions

The intent of this article is to emphasize that when the essential components of parasitism are analyzed, successful parasites display a variety of strategies that enhance their ability to survive. Although parasite avoidance of host immunological reactions is being investigated with rigor, there are additional structural, physiological, biochemical, and even ethological strategies that represent important adaptive mechanisms. Very little is known about the strategies employed by parasites of marine bivalves that enhance their establishment within the host. This area of marine parasitology is essentially unexplored. Studies to elucidate such strategies and the mechanisms underlying them will help explain the intricacies of parasitism in the marine environment and may also be of practical value. Thus, the adaptive strategies could be interrupted mechanically, physically, chemically, or genetically to reduce or eliminate those parasitisms that lead to morbidity and mortality among economically important marine bivalves. The use of marine organisms as models to answer questions of basic biomedical importance is the main objective of marine biomedicine (Cheng 1982).

Acknowledgments

I acknowledge the input of many persons over the years with whom I have had stimulating discussions about the subject. Special appreciation is due to the late Leslie A. Stauber and to Harold H. Haskin, both of Rutgers University, and to S. Y. Feng of the University of Connecticut, M. R. Tripp of the University of Delaware, Burton J. Bogitsh of Vanderbilt University, Timothy P. Yoshino of the University of Oklahoma, Gary E. Rodrick of the University of Florida, and Claude Combes of the University of Perpignan, France. I am also most grateful to Albert K. Sparks of the National Marine Fisheries Service for critically reviewing the original draft of this contribution and making many excellent suggestions.

References

Axmann, M. C. 1947. Morphological studies on glycogen deposition in schistosomes and other flukes. Journal of Morphology 80:321–334.

Baer, J. G. 1952. Ecology of animal parasites. University of Illinois Press, Urbana.

Barrett, J. 1981. Biochemistry of parasitic helminths. University Park Press, Baltimore, Maryland.

Bartoli, P., and C. Combes. 1986. Stratégies de dissémination des cercaires des trématodes dans un écosysteme marine littoral. Acta Ecologica 7:101–114.

Becker, C. D., and G. B. Pauley. 1968. An ovarian parasite (Protista incertae sedis) from the Pacific oyster, Crassostrea gigas. Journal of Invertebrate Pathology 12:425–437.

Bishop, S. H., L. L. Ellis, and J. M. Burcham. 1983. Amino acid metabolism in molluscs. Pages 243–327 in K. M. Wilbur and P. W. Hochachka, editors. The Mollusca, volume 1. Academic Press, New York.

Bloom, B. R. 1979. Games parasites play: how parasites evade immune surveillance. Nature (London) 279: 21–26.

Borst, T. P., and G. A. M. Cross. 1982. Molecular basis for trypanosome antigenic variation. Cell 29:291–303.

Boss, K. J., and A. S. Merrill. 1965. Degree of host specificity in two species of Odostomia (Pyramidellidae: Gastropoda). Proceedings of the Malacological Society of London 36:349–355.

Campbell, W. C., editor. 1983. Trichinella and trichinosis. Plenum, New York.

Chang, K. P., and D. M. Dwyer. 1978. Leishmania donovani hamster macrophage interactions in vitro: cell entry, intracellular survival, and multiplication of amastigotes. Journal of Experimental Medicine 147:515–529.

Cheng, T. C. 1963a. Biochemical requirements of larval trematodes. Annals of the New York Academy of Sciences 113:289–321.

Cheng, T. C. 1963b. Histological and histochemical studies on the effects of parasitism of Musculium partumeium (Say) by the larvae of Gorgodera amplicara Looss. Proceedings of the Helminthological Society of Washington 30:101–107.

Cheng, T. C. 1963c. The effects of Echinoparyphium larvae on the structure and glycogen deposition in the hepatopancreas of Helisoma trivolvis and glycogenesis in the parasite larvae. Malacologia 1:291–303.

Cheng, T. C. 1964. Studies on phosphatase systems in hepatopancreatic cells of the molluscan host of Echinoparyphium sp. and in the rediae and cercariae of this trematode. Parasitology 54:73–79.

Cheng, T. C. 1965. Histochemical observations on changes in the lipid composition of Crassostrea virginica parasitized by the trematode Bucephalus sp. Journal of Invertebrate Pathology 7:398–407.

Cheng, T. C. 1966. The coracidium of the cestode Tylocephalum and the migration and fate of this parasite in the American oyster, Crassostrea virginica. Transactions of the American Microscopical Society 85:246–255.

Cheng, T. C. 1967. Marine molluscs as hosts for symbioses: with a review of known parasites of commercially important species. Advances in Marine Biology 5:1–424.

Cheng, T. C. 1968. The compatibility and incompatibility concept as related to trematodes and molluscs. Pacific Science 22:141–160.

Cheng, T. C. 1970. Hartmannella tahitiensis sp. n., an amoeba associated with mass mortalities of Crassostrea commercialis in Tahiti, French Polynesia. Journal of Invertebrate Pathology 15:405–419.

Cheng, T. C. 1978. Larval nematodes parasitic in shell-fish. U.S. National Marine Fisheries Service Marine Fisheries Review 40(10):39–42.

Cheng, T. C. 1982. The (re-)emerging area of marine biomedicine. Journal of Invertebrate Pathology 40: 155–158.

Cheng, T. C. 1986a. General parasitology, 2nd edition. Academic Press, Orlando, Florida.

Cheng, T. C. 1986b. Specificity and the role of lysosomal hydrolases in molluscan inflammation. International Journal of Tissue Reactions 8:439–445.

Cheng, T. C. 1987. Some cellular mechanisms governing self and nonself recognition and pathogenicity in vertebrates and invertebrates relative to protistan parasites. Aquaculture 67:1–14.

Cheng, T. C., and R. W. Burton. 1966. Relationships between *Bucephalus* sp. and *Crassostrea virginica*: a histochemical study of some carbohydrates and carbohydrate complexes occurring in the host and parasite. Parasitology 56:111–122.

Cheng, T. C., and C. Combes. In press. Influence of environmental factors on the invasion of molluscs by parasites: with special reference to Europe. *In* F. de Castri and A. J. Hansen, editors. Biological invasions. Dr. W. Junk, Dordrecht, The Netherlands.

Cheng, T. C., and K. H. Howland. 1979. Chemotactic attraction between hemocytes of the oyster, *Crassostrea virginica*, and bacteria. Journal of Invertebrate Pathology 33:204–210.

Cheng, T. C., and J. Jourdane. 1987. Transient cellular reaction in *Biomphalaria glabrata* (Mollusca) to heterotopic isografts. Journal of Invertebrate Pathology 49:273–278.

Cheng, T. C., J. Jourdane, and C. Combes. 1988. Stratégies de survie des parasites chez leurs hôtes. Année Biologique 26:73–92.

Cheng, T. C., and E. Rifkin. 1968. The occurrence and resorption of *Tylocephalum* metacestodes in the clam *Tapes semidecussata*. Journal of Invertebrate Pathology 19:65–69.

Cheng, T. C., C. N. Schuster, Jr., and A. H. Anderson. 1966. A comparative study of the susceptibility and response of eight species of marine pelecypods to the trematode *Himasthla quissetensis*. Transactions of the American Microscopical Society 85: 284–295.

Cheng, T. C., and R. W. Snyder, Jr. 1962a. Studies on host–parasite relationships between larval trematodes and their hosts. I. A review. II. Host glycogen utilization by the intramolluscan larvae of *Glypthelmins pennsylvaniensis* Cheng and associated phenomena. Transactions of the American Microscopical Society 81:209–228.

Cheng, T. C., and R. W. Snyder, Jr. 1962b. Studies of host–parasite relationships between larval trematodes and their hosts. III. Certain aspects of lipid metabolism in *Helisoma trivolvis* (Say) infected with the larvae of *Glypthelmins pennsylvaniensis* Cheng and related phenomena. Transactions of the American Microscopical Society 81:327–331.

Cheng, T. C., and R. W. Snyder, Jr. 1963. Studies on host–parasite relationships between larval trema-todes and their hosts. IV. A histochemical determination of glucose and its role in the metabolism of molluscan host and parasites. Transactions of the American Microscopical Society 83:343–349.

Chernin, E. 1970. Behavioral responses of miracidia of *Schistosoma mansoni* and other trematodes to substances emitted by snails. Journal of Parasitology 56:287–296.

Christensen, A. M., and J. J. McDermott. 1958. Life-history and biology of the oyster crab, *Pinnotheres ostreum* Say. Biological Bulletin (Woods Hole) 114: 146–179.

Collin, W. K. 1970. Electron microscopy of postembryonic stages of the tapeworm, *Hymenolepis citelli*. Journal of Parasitology 56:1159–1170.

Corliss, J. O. 1979. The ciliated protozoa: characterization, classification and guide to the literature, 2nd edition. Pergamon Press, Oxford, England.

Couch, J. A., C. A. Farley, and A. Rosenfield. 1966. Sporulation of *Minchinia nelsoni* (Haplosporida, Haplosporidae) in *Crassostrea virginica* (Gmelin). Science (Washington, D.C.) 153:1529–1531.

Cross, G. A. M. 1978. Antigenic variation in trypanosomes. Proceedings of the Royal Society of London B, Biological Sciences 202:55–72.

Damian, R. T. 1964. Molecular mimicry: antigen sharing by parasite and host and its consequences. American Naturalist 98:129–149.

Damian, R. T. 1979. Molecular mimicry in biological adaptation. Pages 103–126 *in* B. B. Nichol, editor. Host–parasite interfaces. Academic Press, New York.

Davey, J. T., and J. M. Gee. 1976. The occurrence of *Mytilicola intestinalis* Steuer, an intestinal copepod parasite of *Mytilus*, in the south-west of England. Journal of the Marine Biological Association of the United Kingdom 56:85–94.

Dawes, B. 1960. A study of the *Fasciola hepatica* and an account of the mode of penetration of the sporocyst into *Lymnaea trunculata*. Pages 95–111 *in* Libro Homen-je al Dr. Eduardo Caballero y Caballero, Jubileo 1940–1960. Escuela Nacional de Ciencias Biológicas, Mexico City.

Despommier, D. D. 1981. Partial purification and characterization of protection-inducing antigens from the muscle larva of *Trichinella spiralis* by molecular sizing chromatography and preparative flatbed iso-electric focusing. Parasite Immunology (Oxford) 3: 261–272.

De Zwaan, A. 1983. Carbohydrate catabolism in bivalves. Pages 137–175 *in* K. M. Wilbur and P. W. Hochachka, editors. The Mollusca, volume 1. Academic Press, New York.

Erasmus, D. A. 1972. The biology of trematodes. Edward Arnold, London.

Farley, C. A. 1968. *Minchinia nelsoni* (Haplosporida, Haplosporidiidae) disease syndrome in the American oyster, *Crassostrea virginica*. Journal of Protozoology 15:585–599.

Fenchel, T. 1965. Ciliates from Scandinavian molluscs. Ophelia 2:71–174.

Fenchel, T. 1966. On the ciliated protozoa inhabiting

the mantle cavity of lamellibranchs. Malacologia 5:35–36.

Feng, S. Y. 1958. Observations on distribution and elimination of spores of *Nematopsis ostrearum* in oysters. Proceedings National Shellfisheries Association 48:162–173.

Feng, S. Y., and W. J. Canzonier. 1970. Humoral responses in the American oyster (*Crassostrea virginica*) infected with *Bucephalus* sp. and *Minchinia nelsoni*. American Fisheries Society Special Publication 5:497–510.

Feng, S. Y., and L. A. Stauber. 1968. Experimental hexamitiasis in the oyster *Crassostrea virginica*. Journal of Invertebrate Pathology 10:94–110.

Freeman, R. S. 1982. Do any *Anonchotaenia, Cyathocephalus, Echeneibothrium*, or *Tetragonocephalum* (= *Tylocephalum*) (Eucestoda) have hookless oncospheres or coracidia? Journal of Parasitology 68:737–843.

Fretter, V., and A. Graham. 1949. The structure and mode of life of the Pyramidellidae, parasitic opisthobranchs. Journal of the Marine Biological Association of the United Kingdom 28:493–532.

Ginecinskij, T. A. 1960. [Glycogen in the body of cercariae and the dependence of its distribution upon the peculiarities of their biology.] Doklady Akademii Nauk SSSR 135:1012–1015. (In Russian.)

Ginecinskij, T. A. 1961. [The dynamics of the storage of fat in the course of the life cycle of trematodes.] Doklady Akademii Nauk SSSR 139:1016–1019. (In Russian.)

Ginecinskij, T. A., and A. A. Dobrovalskij. 1962. [Glycogen and fat in the various phases of the life cycle of trematodes.] Vestnik Leningradskogo Universiteta, Seriya Biologii 9(2):67–81. (In Russian.)

Green, N. M. 1957. Protease inhibitors from *Ascaris lumbricoides*. Biochemical Journal 66:416–419.

Gutteridge, W. E., and G. H. Coombs. 1977. Biochemistry of parasitic protozoa. University Park Press, Baltimore, Maryland.

Hepper, B. T. 1956. The European flat oyster, *Ostrea edulis* L., as a host for *Mytilicola intestinalis* Steuer. Journal of Animal Ecology 25:144–147.

Herdman, W. A., and J. Hornell. 1906. Pearl production. Pages 1–42 *in* W. A. Herdman, editor. Report to the Government of Ceylon on the pearl oyster fisheries of the Gulf of Manaar. Royal Society, London.

Hogue, M. J. 1915. Studies on the life history of an amoeba of the Limax group, *Vahlkampfia calkensi*. Archiv für Protistenkunde 35:154–163.

Hogue, M. J. 1921. Studies of the life history of *Vahlkampfia patuxent* n. sp., parasitic in the oyster, with experiments regarding its pathogenicity. American Journal of Hygiene 1:321–343.

Hopkins, S. H. 1956. *Odostomia impressa* parasitizing southern oysters. Science (Washington, D.C.) 124: 628–629.

Huehner, M. K., and F. J. Etges. 1977. The life cycle and development of *Aspidogaster conchicola* in the snails, *Viviparus malleatus* and *Goniobasis livescens*. Journal of Parasitology 63:669–674.

Huehner, M. K., and F. J. Etges. 1981. Encapsulation of *Aspidogaster conchicola* (Trematoda: Aspidogastrea) by unionid mussels. Journal of Invertebrate Pathology 37:123–128.

Hyman, L. H. 1951. The invertebrates, part 2. The Platyhelminthes and Rhynchocoela. McGraw-Hill, New York.

Jameson, H. L. 1912. Studies on pearl-oysters and pearls. I. The structure of the shell and pearls of the Ceylon pearl-oyster (*Margaritifera vulgaris* Schumacher): with an examination of the cestode theory of pearl production. Proceedings of the Zoological Society of London 11:266–358.

Jones, T. C., S. Yeh, and J. G. Hirsch. 1972. The interaction between *Toxoplasma gondii* and mammalian cells. II. The absence of lysosomal fusion with phagocytic vacuoles containing living parasites. Journal of Experimental Medicine 136:1173–1194.

Jourdane, J., and T. C. Cheng. 1987. The two-phase recognition process of allografts in a Brazilian strain of *Biomphalaria glabrata*. Journal of Invertebrate Pathology 49:145–158.

Jourdane, J., and J. B. Mounkassa. 1986. Topographic shifting of primary sporocysts of *Schistosoma mansoni* in *Biomphalaria pfeifferi* as a result of coinfection with *Echinostoma caproni*. Journal of Invertebrate Pathology 48:269–274.

Kent, M. L., R. A. Elston, T. A. Herad, and T. K. Sawyer. 1987. An *Isonema*-like flagellate (Protozoa: Mastigophora) infection in larval geoduck clams, *Panopea abrupta*. Journal of Invertebrate Pathology 50:221–229.

Kozloff, E. N. 1946a. Studies on cilates of the family Ancistrocomidae Chatton and Lwoff (order Holotricha, suborder Thigmotricha). I. *Hypocomina tegularum* sp. nov. and *Crebricoma* gen. nov. Biological Bulletin (Woods Hole) 90:1–7.

Kozloff, E. N. 1946b. Studies of the family Ancistrocomidae Chatton and Lwoff (order Holotricha, suborder Thigmotricha). III. *Ancistrocoma pelseneeri* Chatton and Lwoff, *Ancistrocoma dissimilis* sp. nov. and *Hypocomagalma pholadidis* sp. nov. Biological Bulletin (Woods Hole) 91:189–199.

Laird, M. 1961. Microecological factors in oyster epizootics. Canadian Journal of Zoology 39:449–485.

Lauckner, G. 1983. Diseases of Mollusca: Bivalvia. Pages 447–961 *in* O. Kinne, editor. Diseases of marine animals, volume 2. Biologische Anstalt Helgoland, Hamburg, West Germany.

Lee, D. L. 1966. The structure and composition of the helminth cuticle. Advances in Parasitology 4:187–254.

Lee, D. L. 1972. The structure of the helminth cuticle. Advances in Parasitology 10:347–379.

Léger, L., and A. C. Hollande. 1917. Sur un nouveau protiste à facies de *Chytridiopsis*, parasite des ovules de l'huître. Comptes Rendus des Séances de Societé de Biologie et de ses Filiales 80:61–64.

Livingstone, D. R., and A. De Zwaan. 1983. Carbohydrate metabolism in gastropods. Pages 177–242 *in* K. W. Wilbur and P. W. Hochachka, editors. The Mollusca, volume 1. Academic Press, New York.

Love, R. J., B. M. Ogilvie, and D. J. McLaren. 1976. The immune mechanism which expels the intestinal stage of *Trichinella spiralis* from rats. Immunology 30:7–15.

MacInnis, A. J. 1976. How parasites find hosts: some thoughts on the inception of host–parasite integration. Pages 3–20 in C. R. Kennedy, editor. Ecological aspects of parasitology. Elsevier, Amsterdam.

Mackin, J. G., P. Korringa, and S. H. Hopkins. 1952. Hexamitiasis of *Ostrea edulis* L. and *Crassostrea virginica* (Gmelin). Bulletin of Marine Science of the Gulf and Caribbean 1:266–277.

Mackinnon, D. L., and R. S. J. Hawes. 1961. An introduction to the study of protozoa. Clarendon Press, Oxford, England.

Müller, M. 1967. Digestion. Chemical Zoology 1:351–393.

Nelson, T. C. 1938. The feeding mechanism of the oyster. I. On the pallium and the branchial chambers of *Ostrea virginica*, *O. edulis* and *O. angulata*, with comparisons with other species of the genus. Journal of Morphology 63:1–61.

Nelson, T. C. 1960. The feeding mechanism of the oyster. II. On the gills and palps of *Ostrea edulis*, *Crassostrea virginica* and *C. angulata*. Journal of Morphology 107:163–202.

Ortega-Pierres, G., C. D. Mackenzie, and R. M. E. Parkhouse. 1984. Protection against *Trichinella spiralis* induced by a monoclonal antibody that promotes killing of newborn larvae by granulocytes. Parasite Immunology (Oxford) 6:275–284.

Pauley, G. B., and C. D. Becker. 1968. *Aspidogaster conchicola* in mollusks of the Columbia River system with comments on the host's pathological response. Journal of Parasitology 54:917–920.

Peanasky, R. J., and M. Laskowski. 1960. Chymotrypsin inhibitor from *Ascaris*. Biochimica et Biophysica Acta 37:167–169.

Perkins, F. O., and R. W. Menzel. 1967. Ultrastructure of sporulation in the oyster pathogen *Dermocystidium marinum*. Journal of Invertebrate Pathology 9:205–229.

Prytherch, H. F. 1940. The life cycle and morphology of *Nematopsis ostrearum* sp. nov., a gregarine parasite of the mud crabs and oysters. Journal of Morphology 66:39–65.

Ratcliffe, N. A., A. F. Rowley, S. W. Fitzgerald, and C. P. Rhodes. 1985. Invertebrate immunity: basic concepts and recent advances. International Review of Cytology 97:183–350.

Richards, C. S. 1968. Two new species of *Hartmannella* amebae infecting freshwater molluscs. Journal of Protozoology 15:651–656.

Richards, C. S. 1973. Susceptibility of adult *Biomphalaria glabrata* to *Schistosoma mansoni* infection. American Journal of Tropical Medicine and Hygiene 22:748–756.

Roder, J. C., T. K. R. Bourns, and S. K. Singhal. 1977. *Trichobilharzia ocellata*: cercariae masked by antigens of the snail, *Lymnaea stagnalis*. Experimental Parasitology 41:206–212.

Sastry, A. N., and R. W. Menzel. 1962. Influence of hosts on the behavior of the commensal crab, *Pinnotheres maculatus* Say. Biological Bulletin (Woods Hole) 123:388–395.

Sawyer, T. K. 1966. Observations on the taxonomic status of amoeboid organisms from the American oyster, *Crassostrea virginica*. Journal of Protozoology 13 (supplement):23.

Schad, G. A. 1977. The role of arrested development in the regulation of nematode populations. Pages 111–167 in G. W. Esch, editor. Regulations of parasite populations. Academic Press, New York.

Schaeffer, A. A. 1926. Taxonomy of the amebas. Carnegie Institution of Washington Publication 345:1–116.

Scheltema, R. S. 1962. The relationship between the flagellate protozoon *Hexamita* and the oyster *Crassostrea virginica*. Journal of Parasitology 48:137–141.

Shipley, A. E., and J. Hornell. 1906. Cestode and nematode parasites from the marine fishes of Ceylon. Pages 43–96 in W. A. Herdman, editor. Report to the Government of Ceylon on the pearl oyster fisheries of the Gulf of Manaar. Royal Society, London.

Sibley, L. D., J. L. Krahenbuhl, G. M. W. Adams, and E. Weidner. 1986. *Toxoplasma* modifies macrophage phagosomes by secretion of a vesicular network rich in surface proteins. Journal of Cell Biology 103:867–874.

Sindermann, C. J. 1970. Principal diseases of marine fish and shellfish. Academic Press, New York.

Smithers, S. R., and R. J. Terry. 1969. The immunology of schistosomiasis. Advances in Parasitology 7:41–93.

Soulsby, E. J. L. 1982. Evasion of the immune response by parasites. Revista Ibérica de Parasitologia 1982 (volumen extra):43–47.

Southwell, T. 1924. The pearl-inducing worm in the Ceylon pearl oyster. Annals of Tropical Medicine and Parasitology 18:37–53.

Sparks, A. K. 1963. Infection of *Crassostrea virginica* (Gmelin) from Hawaii with a larval tapeworm, *Tylocephalum*. Journal of Insect Pathology 5:284–288.

Sparks, A. K. 1985. Synopsis of invertebrate pathology exclusive of insects. Elsevier Scientific, Amsterdam.

Sparks, A. K., and K. K. Chew. 1966. Gross infestation of the littleneck clam (*Venerupis staminea*) with a larval cestode (*Echeneibothrium* sp.). Journal of Invertebrate Pathology 8:413–416.

Sponholtz, G. M., and R. B. Short. 1976. *Schistosoma mansoni* miracidia: stimulation by calcium and magnesium. Journal of Parasitology 62:155–157.

Sprague, V. 1963. Revision of genus *Haplosporidium* and restoration of genus *Minchinia* (Haplosporida, Haplosporidiidae). Journal of Protozoology 10:263–266.

Sprague, V. 1965. Observations of *Chytridiopsis mytilovum* (Field), formerly *Haplosporidium mytilovum* Field, (Microsporida?). Journal of Protozoology 12:385–389.

Stauber, L. A. 1945. *Pinnotheres ostreum*, parasitic on the American oyster, *Ostrea* (*Gryphaea*) *virginica*. Biological Bulletin (Woods Hole) 88:269–291.

Stein, J. E., J. G. Denison, and J. G. Mackin. 1962. *Hexamita* sp. and an infectious disease in the commercial oyster, *Ostrea lurida*. Proceedings National Shellfisheries Association 50:67–81.

Stibbs, H. H., E. Chernin, S. Ward, and M. L. Karnovsky. 1976. Magnesium emitted by snails alters swimming behavior of *Schistosoma mansoni* miracidia. Nature (London) 260:702–703.

Szidat, L. 1959. Hormonale Beeinflussung von Parasiten durch ihren Wirt. Zeitschrift für Parasitenkunde 19:503–524.

Trager, W. 1986. Living together: the biology of animal parasitism. Plenum, New York.

Ubelaker, J. E., B. Cooper, and V. F. Allison. 1970. Possible defense mechanisms of *Hymenolepis diminuta* cysticercoids to hemocytes of the beetle *Tribolium confusum*. Journal of Invertebrate Pathology 16:310–312.

Ulmer, M. J. 1971. Site-finding behaviour in helminths in intermediate and definitive hosts. Pages 123–159 *in* A. M. Fallis, editor. Ecology and physiology of parasites. University of Toronto Press, Toronto.

van der Knaap, W. P. W., A. M. H. Boots, E. A. Meuleman, and T. Sminia. 1985. Search for shared antigens in the schistosome–snail combination *Trichobilharzia ocellata–Lymnaea stagnalis*. Zeitschrift für Parasitenkunde 71:219–226.

Vickerman, K. 1978. Antigenic variation in trypanosomes. Nature (London) 273:613–620.

Voogt, P. A. 1983. Lipids: their distribution and metabolism. Pages 329–370 *in* K. M. Wilbur and P. W. Hochachka, editors. The Mollusca, volume 1. Academic Press, New York.

Warhurst, D. C. 1985. Pathogenic free-living amoebae. Parasitology Today 1:24–28.

White, M. E., E. N. Powell, S. M. Ray, and E. A. Wilson. 1987. Host-to-host transmission of *Perkinsus marinus* in oyster (*Crassostrea virginica*) populations by the ectoparasitic snail *Boonea impressa* (Pyramidellidae). Journal of Shellfish Research 6:1–5.

Wolf, P. H. 1977. An unidentified protistan parasite in the ova of the blacklipped oyster, *Crassostrea echinata*, from northern Australia. Journal of Invertebrate Pathology 29:244–246.

Yoshino, T. P., and C. P. Bayne. 1983. Mimicry of snail host antigens by miracidia and primary sporocysts of *Schistosoma mansoni*. Parasite Immunology (Oxford) 5:317–328.

Yoshino, T. P., and T. C. Cheng. 1978. Snail host-like antigens associated with the surface membranes of *Schistosoma mansoni* miracidia. Journal of Parasitology 64:752–754.

American Fisheries Society Special Publication 18:130–137, 1988

Ecology and Evolution of Bivalve Parasites

Antonio J. Figueras

Instituto de Investigaciones Marinas, Muelle de Bouzas
Vigo 36208, Spain

William S. Fisher[1]

University of Maryland, Horn Point Environmental Laboratories
Center for Environmental and Estuarine Studies
Cambridge, Maryland 21613, USA

Abstract.—Host–parasite relationships are dynamic, complex, and central to understanding various aspects of disease. Due to extreme ecological and environmental constraints, bivalve parasites may have developed complex life cycles for nutrient utilization, reproduction, or dispersal. Parasites have high speciation rates due to rapid generation times and high fecundity and can normally adapt more quickly than their hosts. Pathogenicity can develop from a commensal relationship or from encounter of a nonenzootic with a vulnerable host population. Introduced parasites in oyster fisheries, where there is large-scale transplanting of seed oysters, have caused the most serious bivalve epizootics. Hosts may respond in a variety of ways to the presence of a parasite, but this response can be diminished by antigenic naivete or molecular mimicry. Physical interactions elicit responses in host and parasite that are adaptive over time, and a more benign relationship can develop. This may be the case for Malpeque Bay, *Haplosporidium nelsoni*, and *Perkinsus marinus* diseases of the eastern oyster *Crassostrea virginica*. Breeding programs to select oysters resistant to disease offer a useful tool to study the mechanisms of susceptibility and defense. Success of such programs in producing disease-resistant oysters may depend on whether the parasite can evolve away from damaging the host. Implementation of new technology will aid in the development of theoretical concepts of host–parasite interactions.

Much of bivalve molluscan pathology has been restricted to the description of individual occurrences of parasites and diseases; interest or support has been insufficient to develop an understanding of the parasite–host interrelationships. There have been, however, a few epizootiological studies on parasites and diseases of bivalve molluscs, such as *Perkinsus marinus*, *Haplosporidium nelsoni*, and Malpeque Bay diseases of eastern oyster *Crassostrea virginica*, and *Bonamia ostreae* and *Marteilia refringens* diseases of edible oyster *Ostrea edulis*. We must draw upon past studies and theoretical concepts to elucidate the dynamic aspects of the sometimes complex interactions that lie at the core of basic questions concerning spread of disease, epizootic mortalities, and selection of resistant stock. In this contribution we will consider the known aspects of bivalve parasite ecology and evolution in the context of general theories.

Parasite Ecology

Life cycles of parasites are often quite complex and difficult to trace. Because the organization of a life cycle is a product of natural selection, understanding it requires some knowledge of parasite ecology and population dynamics. General characteristics of parasitic life have been summarized by Price (1980) as follows.

(1) Parasites exploit small, discontinuous environments. They exist in small homogeneous populations with little gene flow between them but develop means to ensure their reproduction and dispersal. Dispersal in time is often achieved by the longevity of resting stages.

(2) Parasites represent the extreme in the exploitation of specialized resources. Predators and grazers may use 10 or more species as food sources, but it is rare to find parasites that utilize more than two.

(3) Parasites exist in nonequilibrium conditions. Their world is a small patch of resources surrounded by other small patches where the probability of colonization is very low. Resources can be temporary, so tenure on a patch is brief. Alternate species and environmental conditions

---

[1]Present address: Marine Biomedical Institute, University of Texas Medical Branch, 200 University Boulevard, Galveston, Texas 77550, USA.

must overlap in time and space before new resources become exploitable.

The major bivalve parasites (species of *Bonamia, Haplosporidium, Marteilia*, and *Perkinsus*) are internal parasites, which means that the host must have a vulnerable site, usually along the gills, mantle, mouth, or digestive system, and the parasite must have a means to penetrate or chemically erode the epithelial layer. Some trematodes and cestodes have penetration glands (Lauckner 1983), and *P. marinus* has an apical complex that may allow the parasite to attach and burrow into the epithelium (Perkins 1976).

The first site of infection, although meeting requirements for parasite entry, may not suffice for continued parasite survival. Throughout their existence, parasites must derive metabolic energy from the host or from the environment provided by the host. Many parasites appear in the digestive tubules and stomach, always with relatively easy access to energy liberated by the digestive process. In an extreme case, *Mytilicola intestinalis* resides in the stomach and intestinal lumen of its host to take advantage of host food-gathering processes (Davey and Gee 1988, this volume).

Bonamia ostreae and *H. nelsoni* are found in the interstitial fluid or in cells where there is no gross liberation of nutrients by host digestive processes. The hemolymph does, however, transport nutrients throughout the organism, and infection can cause a depression of biochemical components in the hemolymph (Feng et al. 1970; Ford 1986). Extracellular parasites probably derive their nutrients directly from the hemolymph. Probably, parasites gain some nutrients from the breakdown of host tissue. *Perkinsus marinus* infections destroy epithelial tissue and lyse the basal membrane of the stomach (Mackin 1951). Chlamydiae and rickettsias are found in the cells of digestive tubules but also in the gills (Harshbarger et al. 1977; Comps 1980), where they must obtain nutrients from the hemolymph or from nearby tissue. *Steinhausia mytilovum* appears to take advantage of increased nutrients available in the mature oocytes of female blue mussels *Mytilus edulis* (A. J. Figueras, unpublished data). Degradation of tissue is perhaps the most obvious way that parasites, in fulfilling their own requirements for life, can cause damage to the host.

Nutrition is the most conspicuous, but not the only, parasite requirement. Various stages of a parasite life cycle can have different requirements, so infection sites may change as the parasite develops. *Marteilia refringens*, for example,

works its way through the digestive system of edible oysters, beginning in the stomach epithelium and finally establishing itself in the digestive tubules where sporulation occurs (Grizel 1977). *Haplosporidium nelsoni* sporulates only in the digestive tubules of eastern oyster (Kern 1976), whereas the closely related species *H. costale* sporulates in all the connective tissues (Andrews and Castagna 1978). Parasites may be limited in the number of host species they can utilize, but they may vary their resource utilization by movement within a single host species.

Due to the rigorous ecological constraints outlined by Price (1980), surviving parasite strains may have developed complex life cycles to accomplish reproduction, dispersal, and exploitation of new resources. Among the most important bivalve molluscan disease agents, only the life cycle of *Perkinsus marinus* is clear (Perkins and Menzel 1966). The vegetative prezoosporangia multiply within the oyster and produce large numbers of biflagellated zoospores. Both the prezoosporangia and zoospores are released when diseased oysters die. Infections of *P. marinus* have been transmitted in the laboratory by proximity and by feeding infected tissue to healthy oysters. Thus, there is no requisite for intermediate or reservoir hosts. Nonetheless, *Perkinsus*-like cells have been found in a variety of other benthic organisms (including oyster predators) which may serve as reservoirs (Andrews 1988, this volume). *Bonamia ostreae* has also been transmitted by proximity in the laboratory (Poder et al. 1982; Tige and Grizel 1984), but the infective stage has not yet been described (Elston et al. 1986).

Because direct transmission of *M. refringens* has not been demonstrated, intermediate hosts may exist. But sporulation has been observed regularly in oysters, and a reservoir host is not necessary to account for observed prevalences. Presumably, spores are released into the water to infect a new host, but dispersion may also occur when predators eat moribund or dead oysters opened (gaping) from the disease (Perkins 1988, this volume).

Although a life cycle for *H. nelsoni* has been proposed (Farley 1967), transmission has not been demonstrated, and sporulation in oysters is rarely seen. An intermediate host has been suggested for *H. nelsoni* (Farley 1967; Andrews 1984), and Burreson (1988) hypothesized intermediate hosts for four *Haplosporidium* species found in marine invertebrates. Perkins (1988) has found *Haplosporidium* spores to survive at least 2 weeks in seawater and suggested that spores in the sediment and seawater could be a reservoir completely independent of the biota.

Host Defense Responses

Once an invading organism contacts and enters a host, the cells of the host usually recognize them as foreign material and attempt to destroy the invader or at least restrict it to a single site. Defense mechanisms used by the host have been reviewed elsewhere in this volume (Chu 1988; Feng 1988). If the host response is successful, the invader is eliminated; if not, the invader becomes a parasite or a pathogen. The immune system of invertebrates is not considered as complex as that of vertebrates, especially in terms of specificity and memory (Klein 1977). Yet, recognition of nonself, or foreignness, is a property shared among all animals and may be a prerequisite for multicellular existence (Bodmer 1972). Cheng (1986) has recently argued that there is molecular specificity to most phases of molluscan cellular responses.

Nonetheless, bivalve responses to parasitic infection can vary widely. Douglass (1976) reported an intense cellular reaction to *Bucephalus* sp. infection in some oysters, whereas the response to the most severe infections usually is limited. S. Y. Feng (Marine Sciences Institute, University of Connecticut, personal communication) and Figueras (unpublished data) found that blue mussels infected with *Perkinsus maculatus* produced a reaction only with certain parasite developmental stages. Some bivalves infected with parasite sporocysts and cercariae rejected the metacercariae of the same parasite species (Dolgikh 1968).

Lack of host response to a newly introduced, nonenzootic parasite is probably due to a lack of molecular experience (naivete) with the parasite as an antigen. This nonresponsiveness can cause diseases of epizootic proportions (see below). Lack of host response to a long-standing, enzootic parasite, on the other hand, may result from antigen sharing between parasite and host (Damian 1979, 1987). Antigen sharing, or molecular mimicry, has been recorded between platyhelminths and their intermediate molluscan hosts (Capron et al. 1968; Jackson 1976; Roder et al. 1977; Yoshino and Cheng 1978; Yoshino and Bayne 1983). Parasites might acquire or synthesize hostlike properties, but Bradley (1965) suggested they are preadapted to certain hosts with similar cell surface molecules. Other parasite strategies to evade recognition by the host are reviewed in this volume by Cheng (1988). Hemo-

cytes of edible oyster may not be able to effectively recognize *B. ostreae* parasites. In vitro, they were able to recognize and bind latex particles as well as hemocytes from an insusceptible oyster species, Pacific oyster *Crassostrea gigas*, but they were able to bind only half the number of isolated *B. ostreae* particles (Fisher 1988).

Parasitism and Pathogenicity

Parasitism was defined by Cheng (1970) as an intimate and obligatory relationship between two heterospecific organisms during which the parasite, usually the smaller of the two, is metabolically dependent on the host. Parasitism is said to be obligatory because the parasite cannot normally survive if it does not contact the host during some part of its life cycle. Anderson and May (1978) felt that a clear negative effect on the growth rate of the host population must exist to distinguish parasitism from commensalism where there is no net effect on either organism. Noble and Noble (1976), however, suggested that even a symbiont could adversely affect its host, but in ways more subtle than physical damage. Thus, if host damage is removed from consideration, a continuum exists between commensalism and parasitism, the equilibrium depending on a complex of factors such as the physiological status of the host and parasites, host defenses, season, and environmental stress.

MacInnis (1974) provided the following distinctions. "A parasite is defined as one partner of a pair of interacting species that have integrated their genomes so that the parasite is dependent upon the minimum of one gene of the other interactive species, called the host, for survival. Mutualism is defined as a case in which each of the interacting species functions as host and parasite. In commensalism, the interacting species have not evolved genetic dependencies, but may do so in the future." Vickerman (1987) believed that this definition of parasite, based on genetic dependency, addresses the essence of parasitism better than definitions that imply exploitation and damage to the host.

This definition also emphasizes that a parasite can exist in a host without causing a disease; indeed, a parasite that damages or kills its host can jeopardize its own survival. Cheng (1970) felt that host–parasite relationships evolved from symbioses and that a deleterious parasitic relationship was the exception. Haskin and Ford (1982) suggested that adverse effects of *H. nelsoni* on selected resistant stocks of eastern oyster may be due more to increased parasite intensity than simply to the presence of the parasite.

Certainly, there are parasites that cause disease. The major pathogens of commercially important bivalve stocks, those that cause epizootic mortalities, are found in oysters. This occurrence may be due to the large-scale transport of stock in the international aquaculture community (Farley and Durfort 1987). There is great potential for introducing nonenzootic parasites to a vulnerable population with transport of stock, and the high densities of potential hosts in culture and fisheries areas abets the spread of a pathogen. Rapid onset of disease and high mortalities generated from a geographical focal point characterize diseases caused by newly introduced, nonenzootic pathogens (Haskin et al. 1966). Most of the major epizootics in commercial bivalve fisheries follow this pattern, but in only one case have the details of the introduction been adequately reconstructed: *Bonamia ostreae* was probably brought into susceptible edible oyster stocks in France and the northwestern USA by a contaminated shipment of seed from California (Elston et al. 1986). Circumstantial evidence links the occurrence of *H. nelsoni* in eastern oysters of Delaware Bay and lower Chesapeake Bay. Both regions showed rapid onset of disease and high mortalities in the late 1950s. As Andrews (1968) stated, "It is hard to believe that this close parallel of events two years apart could be accidental or caused by natural changes or cycles." Transplanted seed could have been the common factor since both sites received seed from the seaside region of Virginia (Chincoteague Bay) in the early 1950s. Chincoteague Bay may not have been the original source of the parasite, but it could have served as a site of transfer because of the traffic of oysters to and from different locales.

Pathogenicity in epizootics caused by newly introduced parasites is probably due to the inability of the host defenses to recognize the new parasite as a foreign entity or perhaps to the hosts' inability to effectively defend against a novel parasite strategy. As parasites and surviving hosts live together longer, they could coevolve so that damaging effects of the parasite are diminished. This topic will be discussed in a later section.

A variety of organs are affected by pathogens and, in each case, the mechanism of pathogenicity may be different, whether through tissue destruction, nutrient competition, toxic metabolites, or interruption of biological processes and biosynthetic pathways. Pathogen virulence is often considered in

terms of dosage, mortalities, or the severity of host damage or response, but several factors are involved in an animal's health. Environmental, physiological, and seasonal (reproductive) variables play major roles in the overall effect of parasites on their hosts, and these must be considered in a comprehensive picture of any disease situation.

Parasite Evolution

Rates of evolution, until recently, were largely measured in terms of morphological change (Simpson 1953). Those taxa with considerable morphological complexity were believed to be evolving more rapidly than simpler organisms. Thus, the small and morphologically simple parasites were said to evolve slowly. Schopf et al. (1975) demonstrated convincingly, however, that morphological change was a poor estimator of evolutionary change and reflected only apparent rates of evolution.

Some investigators view parasites as slowly evolving species that represent phylogenetic dead ends. Noble and Noble (1976) stated that parasites are worthy examples of the inexorable march of evolution into blind alleys. Mayr (1963) expressed the same sentiment for any highly specialized species. Even though these views are widely held, more specialized species do not seem to have disappeared any faster than more generalized species over the course of evolutionary time (Price 1980).

The concept of slow evolution by parasites may result in part from the knowledge that asexual reproduction and inbreeding are common among parasites and that these traits lead to reduced variability and slower changes in the gene pool. However, there appears to be ample variability on which natural selection can act; parasites account for over half the animal species in the world and as many as 60,000 species in a single family (Price 1980). Speciation of parasites must rely on extensive adaptive radiation promoted by short generation times, high fecundity, and the fractionation of gene pools by dissimilar selective pressures (Price 1980). Subtle ecological or temporal isolation can promote independent differentiation of parasite populations.

Documentation of evolution by parasites of bivalve molluscs is limited, but there are good examples of speciation. According to Perkins (1988), presumptive prezoosporangia of *Perkinsus* have been observed in 34 species of bivalves with such wide geographic and host isolation that it is unlikely the cells came from a single parasite species. Perkins documented at least five species of *Marteilia* found in a variety of bivalves and six described species of *Haplosporidium*; eight other *Haplosporidium* species have been reported. *Bonamia ostreae* may belong to the Haplosporidia (Perkins 1988). It appears, within bivalve hosts at least, that parasites are quite capable of speciation and a wide host range. Host, geographic, and temporal isolation may actually promote speciation due to the influence of ecological and environmental factors.

Host and Parasite Coevolution

Members of host populations that survive an epizootic disease usually have some characteristic in their biology, behavior, or environment that prevents contact with or limits the parasites. If these characteristics are heritable, future generations are selected for reduced susceptibility. Obviously, parasite evolution also plays a role in this selection process; with fast generation times, parasites can greatly accelerate coevolution toward a more benign relationship. This coevolution can be very specific. Flor (1971) and Day (1974), working with plants, suggested that a gene-for-gene relationship develops between host and parasite. For example, a change in the virulence of the parasite can be followed by a change in an associated defense mechanism of the host. After prolonged coevolution, during which this stepwise process has occurred several times, many complementary genes can exist between the pair.

Malpeque Bay disease of oysters is an example of reduced severity of disease after long-term selective pressure. Large-scale mortalities of eastern oyster occurred at Prince Edward Island (eastern Canada) in 1915 following the importation of seed oysters from New England. Although the causative agent has never been clearly demonstrated, oysters died at an astonishing rate, and only 2% of the population survived the first 6 years of the epizootic. However, these survivors apparently had some heritable form of resistance to the disease (Needler 1931; Needler and Logie 1947) because the populations returned to previous levels of abundance after 20 years. Oysters imported to Prince Edward Island from disease-free areas contract the disease, and 90% die within 2 years (Logie et al. 1960). Other examples of naturally acquired resistance to disease mortalities by eastern oyster include resistance to *H. nelsoni* in Delaware Bay (Haskin and Ford 1979) and in the lower York River and Mobjack Bay of Virginia (Andrews 1988), and to *P. marinus* in seaside bays of the eastern shores of Virginia and Maryland and in oysters from South Carolina (Andrews 1988).

As a consequence of epizootic diseases in economically valuable oyster fisheries, selective breeding programs have been initiated to enhance the more benign aspects of the parasite–host relationship. Since little is understood of parasite selection, attempts are being made to optimize physiological and defensive traits of the oysters that increase tolerance or resistance. These programs include breeding of eastern oysters tolerant of *H. nelsoni* (Andrews 1968; Haskin and Ford 1979; Ford and Haskin 1987) and breeding of Pacific oysters less susceptible to "summer mortality," which has no confirmed etiological agent (Beattie et al. 1980, 1988). The potential also exists for breeding edible oysters resistant to mortalities caused by *B. ostreae* (Elston et al. 1987). These breeding efforts bring a ray of hope to a battered commercial industry and can provide a useful research tool for determining the mechanisms of defense to specific parasites. Pacific oysters selected for survival against summer mortality have higher glycogen levels than unselected oysters; high energy reserves may be vital during the stressful period of gonadal maturation and spawning in this highly fecund species (Beattie et al. 1988). Eastern oysters resistant to Malpeque Bay disease are serologically unique (Li et al. 1967), which may be a factor in their ability to withstand exposure to the disease. Ford (1988, this volume) has outlined several potential mechanisms of resistance against *H. nelsoni*. In order to efficiently capitalize on hatchery technology, the mechanisms responsible for resistance or tolerance must be singled out, optimized, and applied to specific cases.

The success of selective breeding programs for disease resistance may paradoxically depend on the parasite. If the pathogen evolves in the direction of a benign relationship with its host, then the selection of coevolving stock could be simple and efficient. But it is possible that the life requirements of the parasite for nutrition or dispersal are best met by damaging the host. Dispersal could be a driving factor in diseases caused by *B. ostreae* and *P. marinus*, for which dead and gaping oysters may be an effective means to spread the infective propagules to new hosts. If so, the task of selective breeding programs will be formidable. Parasites, with fast generation times and high fecundity, might easily adopt new strategies faster than the host.

This phenomenon does not, however, negate application of selective breeding, which can focus on the breeding of traits that slow the disease or increase the rate of bivalve growth so that animals may be harvested before they succumb. Perhaps a particular defense mechanism or physiological trait can overshadow the adaptive abilities of the parasite.

Discussion

Price (1980) stated that much of science is biased because our senses and minds are eclectic. Large and colorful organisms command our attention while the small and secretive ones are largely ignored. We see predators kill, but rarely do we see parasites kill, and the visual impact is subliminally converted to a ranking in ecological importance. Yet, is is hard to imagine life without parasites and harder still to estimate the full impact of parasites on the existing biota. Because every species has a set of parasites that exerts selective pressure in a nonrandom manner, the repercussions are enormous. Parasites influence every aspect of the host population, including size, structure, temporal and spatial dynamics, coexistence, and competition. There is a great deal to learn and develop in our concept of parasites and bivalve hosts, but understanding these concepts is essential to management of bivalve fisheries and resources. The occurrence of epizootic diseases has had a major economic impact on bivalve fisheries in several parts of the world.

In order to make the needed progress in understanding parasite relationships with bivalve molluscan hosts, several approaches are necessary. Until now, much of our understanding has been based on histological sections; yet parasitism and disease are dynamic activities between two organisms moderated by the environment. New efforts must be made to understand the transmission and life cycles of parasites as well as the defense mechanisms of the host and coevolution. It is very important to utilize new technology to supplement the snapshot approach that has been so successfully applied to describing diseases and epizootiology. These technologies (see subsequent papers in this volume) include in vitro cell culture of host and parasite tissue, flow cytometry, enzyme immunoassays involving monoclonal antibodies, and electrophoresis. Electrophoretic studies, for example, could illuminate the correlation between genotypic variability of a host species and its parasites, the diversity of host species against genotypic diversity of a parasite species, the population structure of a host compared to that of the parasites, the genetic divergence between parasite populations, and modes of speciation.

Acknowledgments

We thank S. E. Ford and T. C. Cheng for their constructive comments on this manuscript. This is contribution 1908 from the Center for Environmental and Estuarine Studies, University of Maryland.

References

Anderson, R. M., and R. M. May. 1978. Regulation and stability of host–parasite population interactions. I. Regulatory processes. Journal of Animal Ecology 47:219–247.

Andrews, J. D. 1968. Oyster mortality studies in Virginia. VII. Review of epizootiology and origin of *Minchinia nelsoni*. Proceedings National Shellfisheries Association 58:23–36.

Andrews, J. D. 1984. Epizootiology of diseases of oysters (*Crassostrea virginica*) and parasites of associated organisms in eastern North America. Helgoländer Meeresuntersuchungen 37:149–166.

Andrews, J. D. 1988. Epizootiology of the disease caused by the oyster pathogen *Perkinsus marinus* and its effects on the oyster industry. American Fisheries Society Special Publication 18:47–63.

Andrews, J. D., and M. Castagna. 1978. Epizootiology of *Minchinia costalis* in susceptible oysters in seaside bays of Virginia's eastern shore, 1959–1976. Journal of Invertebrate Pathology 24:127–140.

Beattie, J. H., K. K. Chew, and W. K. Hershberger. 1980. Differential survival of selected strains of Pacific oysters (*Crassostrea gigas*). Proceedings National Shellfisheries Association 70:184–189.

Beattie, J. H., J. P. Davis, S. L. Downing, and K. K. Chew. 1988. Summer mortality of Pacific oysters. American Fisheries Society Special Publication 18:265–268.

Bodmer, W. F. 1972. Evolutionary significance of the HL–A system. Nature (London) 237:139–145.

Bradley, D. J. 1965. Analysis of larval forms in the trematoda and marine bottom invertebrates. Proceedings of the Central African Science and Medicine Congress 1:343–350. (Lusaka, Northern Rhodesia.) (Not seen; cited in Damian 1979.)

Burreson, E. M. 1988. Use of immunoassays in haplosporidan life cycle studies. American Fisheries Society Special Publication 18:298–303.

Capron, A., J. Biguet, A. Vernes, and D. Afchain. 1968. Structure antigénique des helminthes. Aspects immunologiques des relations hôte–parasite. Pathologie et Biologie 16:121–138.

Cheng, T. C. 1970. Symbioses. Organisms living together. Pegasus, New York.

Cheng, T. C. 1986. Specificity and the role of lysosomal hydrolases in molluscan inflammation. International Journal of Tissue Reactions 8:439–445.

Cheng, T. C. 1988. Strategies employed by parasites of marine bivalves to effect successful establishment in hosts. American Fisheries Society Special Publication 18:112–129.

Chu, F.-L. E. 1988. Humoral defense factors in marine bivalves. American Fisheries Society Special Publication 18:178–188.

Comps, M. 1980. Infections rickettsiennes chez les mollusques bivalves marins. Haliotis 10(2):39. (Société Française de Malacologie, Paris.)

Damian, R. T. 1979. Molecular mimicry in biological adaptation. Pages 103–126 in B. B. Nickoll, editor. Host parasite interfaces. Academic Press, New York.

Damian, R. T. 1987. The exploitation of host immune responses by parasites. Journal of Parasitology 73: 3–13.

Davey, J. T., and J. M. Gee. 1988. *Mytilicola intestinalis*, a copepod parasite of blue mussels. American Fisheries Society Special Publication 18:64–73.

Day, P. R. 1974. Genetics of host–parasite interaction. Freeman, San Francisco.

Dolgikh, A. V. 1968. The life cycle of *Parvatrema isostoma* Belopolskaja, 1966 (Trematoda, Gymnophallidae). Doklady Akademii Nauk SSSR 183:1229–1231.

Douglass, W. R. 1976. Host response to infection with *Bucephelas* in *Crassostrea virginica*. Proceedings National Shellfisheries Association 65:1.

Elston, R. A., C. A. Farley, and M. L. Kent. 1986. Occurrence and significance of bonamiasis in European flat oysters *Ostrea edulis* in North America. Diseases of Aquatic Organisms 2:49–54.

Elston, R. A., M. L. Kent, and M. T. Wilkinson. 1987. Resistance of *Ostrea edulis* to *Bonamia ostreae* infection. Aquaculture 64:237–242.

Farley, C. A. 1967. A proposed life cycle of *Minchinia nelsoni* (Haplosporida, Haplosporidiidae) in the American oyster *Crassostrea virginica*. Journal of Protozoology 14:616–625.

Farley, C. A., and M. Durfort. 1987. Parasites and diseases of marine bivalves. Pages 9–13 in W. S. Fisher and A. J. Figueras, editors. Marine bivalve pathology. University of Maryland Sea Grant Programs, UM-SG-TS-87-02, College Park, Maryland.

Feng, S. Y. 1988. Cellular defense mechanisms of oysters and mussels. American Fisheries Society Special Publication 18:153–168.

Feng, S. Y., E. A. Khairallah, and W. J. Canzonier. 1970. Hemolymph-free amino acids and related nitrogenous compounds of *Crassostrea virginica* infected with *Bucephelas* sp. and *Minchinia nelsoni*. Comparative Biochemistry and Physiology B, Comparative Biochemistry 34:547–556.

Fisher, W. S. 1988. In vitro binding of parasites (*Bonamia ostreae*) and latex particles by hemocytes of susceptible and insusceptible oysters. Developmental and Comparative Immunology 12:43–53.

Flor, H. H. 1971. Current status of the gene-for-gene concept. Annual Review of Phytopathology 9: 272–296.

Ford, S. E. 1986. Comparison of hemolymph proteins from resistant and susceptible oysters, *Crassostrea virginica*, exposed to the parasite *Haplosporidium nelsoni* (MSX). Journal of Invertebrate Pathology 47:283–294.

Ford, S. E. 1988. Host–parasite interactions in eastern oysters, selected for resistance to *Haplosporidium nelsoni* (MSX) disease: survival mechanisms against a natural pathogen. American Fisheries Society Special Publication 18:206–224.

Ford, S. E., and H. H. Haskin. 1987. Infection and mortality patterns in strains of oysters *Crassostrea virginica* selected for resistance to the parasite *Haplosporidium nelsoni* (MSX). Journal of Parasitology 73:368–376.

Grizel, H. 1977. Parasitologie des mollusques. Oceanus 3:190–207.

Harshbarger, J. C., S. C. Chang, and S. V. Otto. 1977. Chlamydiae (with phages), mycoplasmas, and rickettsiae in Chesapeake Bay bivalves. Sciences (New York) 196:666–668.

Haskin, H. H., and S. E. Ford. 1979. Development of resistance to *Minchinia nelsoni* (MSX) mortality in laboratory-reared and native oyster stocks in Delaware Bay. U.S. National Marine Fisheries Service Marine Fisheries Review 41(1–2):54–63.

Haskin, H. H., and S. E. Ford. 1982. *Haplosporidium nelsoni* (MSX) on Delaware Bay seed oyster beds: a host–parasite relationship along a salinity gradient. Journal of Invertebrate Pathology 40:388–405.

Haskin, H. H., L. A. Stauber, and J. G. Mackin. 1966. *Minchinia nelsoni* n. sp. (Haplosporida, Haplosporidiidae): causative agent of the Delaware Bay oyster epizootic. Science (Washington, D.C.) 153: 1414–1416.

Jackson, T. F. H. G. 1976. Intermediate host antigens associated with the cercariae of *Schistosoma haemotobium*. Journal of Helminthology 50:45–47.

Kern, F. G. 1976. *Minchinia nelsoni* (MSX) disease of the American oyster. U.S. National Marine Fisheries Service Marine Fisheries Review 38(10):22–24.

Klein, J. 1977. Evolution and function of the major histocompatibility system: facts and speculations. Pages 339–378 in D. Gotze, editor. The major histocompatibility system in man and animals. Springer-Verlag, Berlin.

Lauckner, G. 1983. Diseases of Mollusca: Bivalvia. Pages 477–961 in O. Kinne, editor. Diseases of marine animals, volume 2. Biologische Anstalt Helgoland, Hamburg, West Germany.

Li, M. F., C. Flemming, and J. E. Stewart. 1967. Serological differences between two populations of oysters (*Crassostrea virginica*) from the Atlantic coast of Canada. Journal of the Fisheries Research Board of Canada 24:443–446.

Logie, R. R., R. E. Drinnan, and E. B. Henderson. 1960. Rehabilitation of disease-depleted oyster populations in eastern Canada. Proceedings of the Gulf and Caribbean Fisheries Institute 13:109–113.

MacInnis, A. J. 1974. A general theory of parasitism. Pages 1511–1512 in Proceedings of the 3rd International Congress of Parasitology. Munich, West Germany. (Not seen; cited in Vickerman 1987.)

Mackin, J. G. 1951. Histopathology of infection of *Crassostrea virginica* (Gmelin) by *Dermocystidium marinum* Mackin Owen Collier. Bulletin of Marine Science of the Gulf and Caribbean 1:72–87.

Mayr, E. 1963. Animal species in evolution. Harvard University Press, Cambridge, Massachusetts.

Needler, A. W. H. 1931. The oysters of Malpeque Bay. Biological Board of Canada Bulletin 22:1–30.

Needler, A. W. H., and R. R. Logie. 1947. Serious

mortalities in Prince Edward Island oysters caused by a contagious disease. Transactions of the Royal Society of Canada 41(5):73–89.

Noble, E. R., and G. A. Noble. 1976. Parasitology. The biology of animal parasites, 4th edition. Lea and Febiger, Philadelphia.

Perkins, F. O. 1976. *Dermocystidium marinum* infection in oysters. U.S. National Marine Fisheries Service Marine Fisheries Review 38(10):19–21.

Perkins, F. O. 1988. Structure of protistan parasites found in bivalve molluscs. American Fisheries Society Special Publication 18:93–111.

Perkins, F. O., and R. W. Menzel. 1966. Morphological and cultural studies of a motile stage in the life cycle of *Dermocystidium marinum*. Proceedings National Shellfisheries Association 56:23–30.

Poder, M., A. Cahour, and G. Balouet. 1982. Hemocytic parasitosis in European oyster *Ostrea edulis* L: pathology and contamination. Proceedings International Colloquium on Invertebrate Pathology 3:254–257.

Price, P. W. 1980. Evolutionary biology of parasites. Princeton University Press, Princeton, New Jersey.

Roder, J. C., T. K. R. Bourns, and S. K. Singhal. 1977. *Trichobilharzia ocellata*: cercariae masked by antigens of the snail, *Lymnaea stagnalis*. Experimental Parasitology 41:206–212.

Schopf, T. J. M., D. M. Raup, S. J. Gould, and D. S. Simberloff. 1975. Genomic versus morphologic rates of evolution: influence of morphologic complexity. Paleobiology 1:63–70.

Simpson, G. G. 1953. The major features of evolution. Columbia University Press, New York.

Tige, G., and H. Grizel. 1984. Essai de contamination d'*Ostrea edulis* L. par *Bonamia ostreae* (Pichot et al., 1979) en riviére de Crach (Morbihan). Revue des Travaux de l'Institut des Pêches Maritimes 46(4):1–8.

Vickerman, K. 1987. The impact of future research on our understanding of parasitism. International Journal of Parasitology 17:731–735.

Yoshino, T. P., and C. P. Bayne. 1983. Mimicry of snail host antigens by miracidia and primary sporocysts of *Schistosoma mansoni*. Parasite Immunology (Oxford) 5:317–328.

Yoshino, T. P., and T. C. Cheng. 1978. Snail host-like antigens associated with the surface membranes of *Schistosoma mansoni* miracidia. Journal of Parasitology 64:752–754.

American Fisheries Society Special Publication 18:139–152, 1988

HOST DEFENSES

Inflammation and Wound Repair in Bivalve Molluscs

ALBERT K. SPARKS AND J. FRANK MORADO

Northwest and Alaska Fisheries Center, National Marine Fisheries Service
7600 Sand Point Way, NE, Seattle, Washington 98115, USA

Abstract.—Injury to marine bivalve molluscs typically evokes an inflammatory response in which leukocytes infiltrate the area of injury and phagocytose or sequester the injurious agent, the inflammatory exudate, and necrotic tissue debris. Successful termination of the inflammatory response is wound repair; dead or damaged tissue is replaced by healthy tissue and the original architecture is completely restored.

Injury to marine bivalve molluscs typically evokes a protective response that destroys, dilutes, or sequesters the injurious agent and damaged or dead tissue (Sparks 1972, 1985). The pattern of response is basically similar irrespective of the nature of the injurious agent, site of injury, or the species of bivalve sustaining the injury. However, some disease organisms are so virulent that infected bivalves succumb before an inflammatory response can be initiated, and some parasites are so compatible with the host that they are not recognized as nonself and, therefore, they elicit no response.

Successful termination of the inflammatory response is wound repair, with dead or damaged tissue replaced by healthy, functional tissue. The major features of the reparative process are removal of the inflammatory exudate, necrotic cellular debris, and the noxious agent, and restoration of the original architecture. When repair is less than successful, outcomes range from death of the host at some stage of the inflammatory reaction or reparative process to the in situ isolation of the injurious agent and tissue debris by some form of encapsulation.

Differentiation between the inflammatory response and postmortem changes is important because autolytic changes may easily be confused with inflammatory responses or histopathological effects of an injurious agent. This distinction is particularly germane because death in molluscs is difficult to define; somatic death, or death of the entire organism, typically occurs only after necrosis, or cell death, has progressed over large areas of the body (Sparks 1972). Detailed descriptions of the normal postmortem changes are available only for the Pacific oyster *Crassostrea gigas* (Sparks

and Pauley 1964). Reaction to injury and wound repair have been investigated more thoroughly for molluscs than for any other major invertebrate group except insects, but, as Tripp (1970) noted, virtually all available information is restricted to relatively few species of gastropods and pelecypods. Although experienced invertebrate pathologists are able, almost intuitively, to recognize differences between autolysis and either the inflammatory response or histopathological effects of injurious agents, detailed studies of the rate and pattern of normal postmortem changes in other molluscs are needed.

Inflammation and Wound Repair in Oysters

Intravascular injection of india ink particles into the eastern oyster *Crassostrea virginica* results in virtual embolism of the arterial system, including all its major branches (Plate 1, Figure 1), but adjacent hemal sinuses are initially unaffected (Stauber 1950). Although masked by the ink particles, accumulating leukocytes begin to phagocytose ink particles between 2 and 4 h postinjection. By the end of the first day, large numbers of leukocytes, most of which contain ink particles, are associated with the ink mass and they persist for some 8 d (Plate 1, Figure 2).

Between 8 and 17 d, most phagocytes migrate from the arterial system and are distributed throughout the oyster (Plate 2, Figure 3). The ink particles are eventually eliminated from the oyster by migration of ink-laden phagocytes through various epithelial surfaces (diapedesis). Diapedesis begins at approximately 8 d postinjection, the chief sites being the epithelia of the entire digestive system (Plate 2, Figure 4). Discharge into the

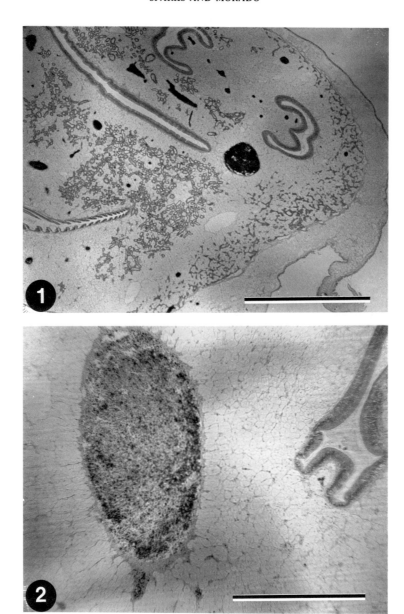

PLATE 1.—Eastern oyster after intracardial injection with india ink. (Reprinted, by permission, from Stauber 1950. Photographs by L. A. Stauber. © 1950 by Marine Biological Laboratory.) FIGURE 1 (above). Cross section shortly after injection. Note the virtual embolism of the arterial system. Bar = 2,000 μm. FIGURE 2 (below). Arterial emboli several days after injection. Note increased numbers of leukocytes present in the artery. Bar = 550 μm.

pericardium and through the epithelium of the mantle and palps occurs rarely, but epithelia of the external, shell-secreting portion of the mantle, excretory tubules, gonaducts, and gill are almost never involved in diapedesis.

Intracardial injection of vertebrate erythrocytes also causes virtual arterial occulsion, but phagocytosis begins within 10 min and is essentially complete within 6 h (Tripp 1958). Diapedesis begins as early as 48 h postinjection, reaches a peak at 8 d, and involves the same epithelial surfaces reported by Stauber (1950). Intracellular digestion of the erythrocytes also occurs, as demonstrated particularly well by Tripp's injection of avian erythrocytes infected with a malarial parasite. Parasitized cells contain a pigment (hematin),

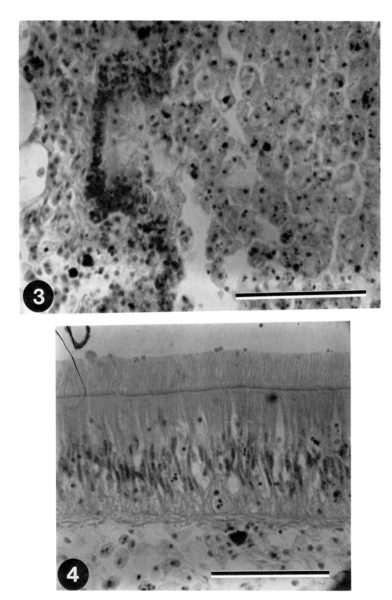

PLATE 2.—Ink-laden leukocytes in eastern oyster. Bars = 50 μm. (Reprinted, by permission, from Stauber 1950. Photographs by L. A. Stauber. © 1950 by Marine Biological Laboratory.) FIGURE 3 (above). Leukocytes throughout tissue subsequent to phagocytosis and migration from the arterial system. FIGURE 4 (below). Migration of leukocytes to and through the intestinal epithelium.

resulting from the destruction of hemoglobin, that is extremely resistant to degradation. Twelve days after injection of the parasitized erythrocytes, leukocytes containing malarial pigment were numerous in the tissues, but intact phagocytosed erythrocytes were rare.

Stauber (1950) noted that activity of leukocytes is correlated with temperature. Numerous studies have shown that physiological processes, including leukocyte activity, virtually cease when temperatures are depressed below 5°C and are slowed substantially at temperatures of 5–10°C. Additionally, some foreign materials are not recognized as nonself and are not phagocytosed, and others are resistant to degradation after being incorporated into the leukocyte.

Pauley and Sparks (1965, 1966) studied the acute inflammatory reaction in Pacific oysters. In

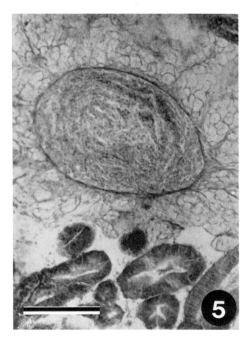

PLATE 3.—FIGURE 5. Blood vessel of Pacific oyster completely filled with leukocytes and talc particles 32 h after the animal was injected with talc. Hematoxylin and eosin stain; bar = 200 μm. (Reprinted, by permission, from Pauley and Sparks 1967. © 1967 by Academic Press.)

separate studies, turpentine and talc were each injected into the connective tissue above the palps. Injection of talc into connective tissue results in a greenish discoloration around the injection site within 16 h that persists for about 72 h, then gradually becomes yellowish-green and disappears by 96 h postinjection. This greenish discoloration is comparable to the rubor or redness charcteristic of vertebrate inflammation.

Leukocytic infiltration accompanied by edema is the first histological response to talc injection, beginning at about 16 h postinjection. By 24 h, blood sinuses adjacent to the injected talc become congested and contain talc particles. Congestion also occurs in many large blood vessels some distance from the talc (Plate 3, Figure 5), and by 24 h, postinjection granulomatous tissue that consists of elongate cells with round nuclei begins to form around masses of talc in blood vessels (Plate 4, Figure 6). By 160 h, the granulomatous tissue is well organized into thrombi that partially occlude the vessels.

Lesions that consist of talc particles and loosely aggregated leukocytes surrounded by a compacted peripheral band of leukocytes (Plate 4, Figure 7) begin to form in the connective tissue as early as 24 h after talc injection. The peripheral leukocytes become elongated and arranged parallel to one another by 40 h. At 88 h, the lesion consists of necrotic leukocytes in the center and aggregations of compact viable leukocytes between the center and the elongated peripheral cells. Nuclei in the peripheral band begin to elongate and the central area is invaded by elongate cells to produce, by approximately 300 h, well-formed granulomas consisting of a band of elongated cells 25–100 cells thick surrounding a mass of talc particles and necrotic leukocytes heavily infiltrated by elongated leukocytes. The lesions also contain numerous brown pigment cells after 56 h, whereas adjacent normal tissue lacks such cells. These cells are typically more abundant in diseased than in normal oyster tissue (Stein and Mackin 1955).

The injection of turpentine into the Leydig tissue results in edema and general leukocytic infiltration into the area of injury within 8 h, accompanied by congestion of adjacent small blood vessels and sinusoids. Marked vascular dilation develops by 16 h, and the large vessel walls are pavemented by leukocytes in the vicinity of the injury. Increased numbers of leukocytes appear in blood vessels and sinuses and begin to migrate toward the injury site.

Edematous areas in mantle and gonadal tissues are present at 24 h, and the gonadal ducts are markedly distended. Concomitantly, digestive tubules and Leydig cells in the injection site undergo massive necrosis that is characterized by pyknotic nuclei and faded cytoplasm. The mantle epithelium over the area of injection is necrotic, and only shadowy outlines of cell membranes remain.

The lesion is heavily infiltrated by leukocytes at 40 h, and 8 h later a conspicuous band of leukocytes surrounds the necrotic area. Multinucleate giant cells that are normal components of postmortem change (Sparks and Pauley 1964) appear at about 64 h and are common thereafter. They appear to be macrophages, phagocytosing necrotic cells and tissue debris. Mass migration of leukocytes across epithelial borders is a constant feature after 64 h. Although it is not certain that the diapedesing leukocytes contain turpentine diluted by the cytoplasm of destroyed cells, coagulation necrosis of the epithelia of the gut, stomach, and digestive tubules is conspicuous at 64 h postinjection.

By 72 h, necrosis of the gonadal area is initiated, and it is characterized by edema, heavy leukocytic infiltration, distention of gonadal

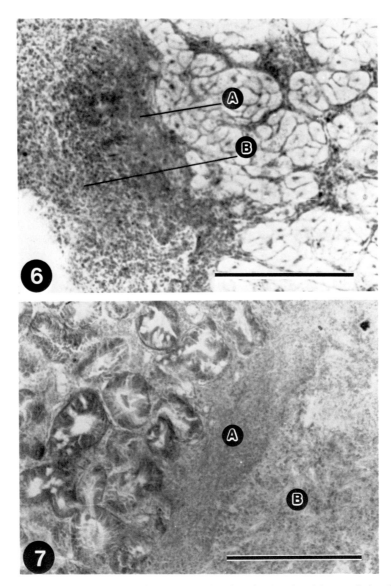

PLATE 4.—Inflammation in Pacific oyster. Hematoxylin and eosin stain. (Reprinted, by permission, from Pauley and Sparks 1967. © 1967 by Academic Press.) FIGURE 6 (above). Blood vessel 48 h after talc injection. A = darker-staining formative granulomatous tissue. B = lighter-staining adherent leukocytes. Lumen of the blood vessel is at lower left. Bar = 200 μm. FIGURE 7 (below). Talc lesion in connective tissue 200 h after talc injection. A = thick bank of leukocytes encapsulating entire lesion. B = central area of lesion consisting of talc particles and necrotic and viable-appearing leukocytes. Bar = 300 μm.

ducts, and formation of walls of leukocytes around the necrotic gonadal area. Concomitantly, conspicuous edema develops in adjacent mantle areas, and the several types of epithelial cells in the digestive tubules are reduced to a uniform low cuboidal epithelium with loss of normal crypt structure (Plate 5, Figure 8). However, sloughing or fragmentation of the epithelium, characteristic of postmortem changes, does not occur.

Multiple abscesses develop in the digestive tubule area after 88 h. Each abscess consists of a central area of liquefaction necrosis, surrounded by a band of necrotic leukocytes and a peripheral band of viable leukocytes. The abscesses appear to

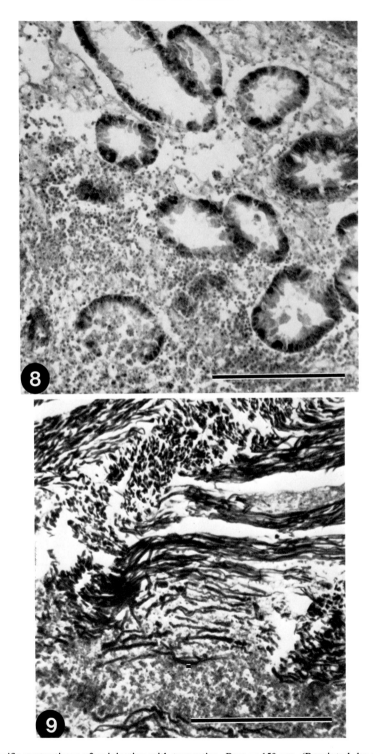

PLATE 5.—Pacific oyster tissue after injection with turpentine. Bars = 150 μm. (Reprinted, by permission, from Pauley and Sparks 1965. © 1965 by Academic Press.) FIGURE 8 (above). Periphery of an abscess in the digestive tubule area 88 h postinjection. Note the progressive necrosis of leukocytes and digestive tubule epithelial cells from the periphery of the abscess inward. Hematoxylin and eosin stain. FIGURE 9 (below). Adductor muscle 168 h postinjection. Note edema, heavy leukocytic infiltrate, and the disorganized hypertrophied appearance of the muscle fibers. Mallory's trichrome stain.

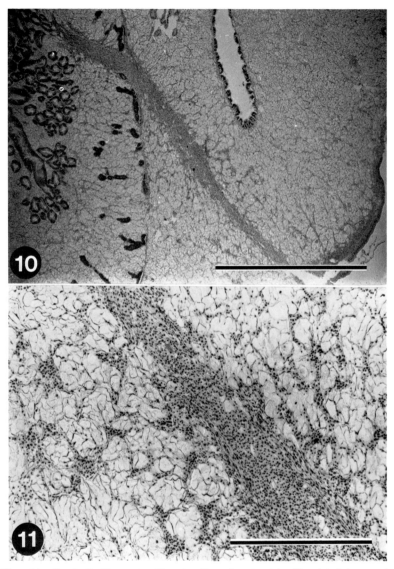

PLATE 6.—Wound healing in Pacific oyster. Hematoxylin and eosin stain. (Reprinted, by permission, from Des Voigne and Sparks 1968. © 1968 by Academic Press.) FIGURE 10 (above). Infiltration of leukocytes into the wound channel, 96 h postwounding. Bar = 1,500 μm. FIGURE 11 (below). Wound channel in 94-h wound, higher magnification. Note the elongate leukocytes at the periphery of the lesion and round leukocytes in the center. Bar = 350 μm.

be well confined, but there is no indication of fibrous encapsulation. Despite histological evidence that the turpentine is confined in abscesses, some turpentine spreads or is carried to adjacent tissues. The swollen adductor muscle at 32 h postinjection is highly edematous; blood channels are congested, leukocytic infiltration is light, and muscle fibers are necrotic. Subsequently, small localized areas of coagulation necrosis develop, and pus, characterized by aggregations of necrotic leukocytes and tissue debris, is present throughout the muscle. An extensive, spreading cellulitis commonly develops after 160 h, and most muscle tissue becomes hypertrophied, necrotic, and heavily infiltrated by leukocytes (Plate 5, Figure 9).

The turpentine apparently spreads to the visceral ganglion, where it causes marked liquefaction necrosis by 300 h. The epithelium of the kidney becomes hypertrophied, and the nuclei become karyolytic; the epithelium begins to slough

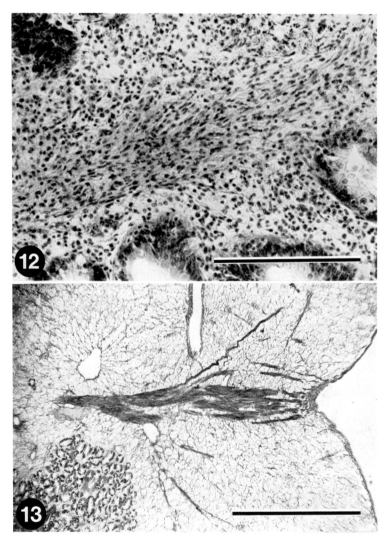

PLATE 7.—Wound healing in Pacific oyster. (Reprinted, by permission, from Des Voigne and Sparks 1968. © 1968 by Academic Press.) FIGURE 12 (above). Leukocyte plug in 144-h wound. Note increased elongation of leukocytes and beginning of collagen deposition. Hematoxylin and eosin stain; bar = 150 μm. FIGURE 13 (below). Increased collagen deposition along wound channel in 144-h wound. Mallory's trichrome stain; bar = 1,500 μm.

from the basement membrane at approximately 40 h and may be liquified a short time later. Ulceration of the epithelium of the gut, style sac, digestive tubules, and rectum is common after 200 h.

Casual observations on the healing of surface wounds (Pauley and Sparks 1967) during the inflammation study led to a detailed investigation of the wound repair process in Pacific oysters (Des Voigne and Sparks 1968). A small portion of the shell was removed, and the animals were incised with a cataract knife to produce a more pronounced and persistent wound than one caused by a hypodermic needle. Grossly, the tissues that surround a surface wound begin to darken approximately 16 h after injury, a phenomenon analogous to the redness of vertebrate inflammation. The area is first yellowish green in color; it gradually turns dark green and then fades to a yellow green that persists for at least 28 d.

Histologically, the first recognizable response is the pavementing by leukocytes of small blood vessels near the lesion, and infiltrating leukocytes begin forming a band beneath the mantle epithelium adjacent to the wound. By 16 h postinjury, the subepidermal band of leukocytes thickens, and masses of leukocytes migrate toward the wound

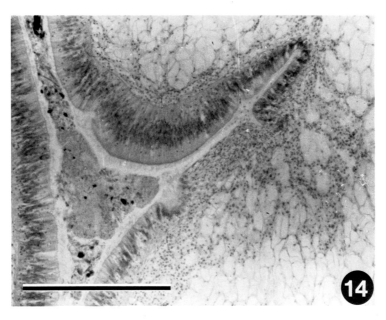

PLATE 8.—FIGURE 14. Wound healing in Pacific oyster. Band of leukocytes replacing necrotic epithelium in the intestine, 88-h wound. Hematoxylin and eosin stain; bar = 350 μm. (Reprinted, by permission, from Des Voigne and Sparks 1968. © 1968 by Academic Press.)

channel. Leukocytic infiltration of the wound area continues and, by 24–48 h, a thick band of normal round leukocytes surrounds the entire lesion. Blood vessels adjacent to the wound are packed with leukocytes, and a heavy infiltration into the region of the wound is well established.

Healing proceeds from the interior of the lesion toward the surface. The round leukocytes, after delineating the margin of the wound, become fusiform and line up parallel to the wound channel. After approximately 160 h postinjury, the nuclei also become fusiform. As the band of fusiform leukocytes thickens at the periphery, the wound channel is infiltrated by masses of round leukocytes and fibroblasts (Plate 6, Figures 10, 11) that effectively plug the wound channel by 144 h. After this infiltration, the leukocytes also elongate and align themselves along the axis of the lesion; collagen deposition also increases markedly (Plate 7, Figures 12, 13). Subsequently, groups of the fusiform leukocytes that fill the wound channel in parallel rows form randomly arranged whorls. At this stage, varying in time from 88 to 448 h postwounding, the lesion resembles a vertebrate scar. However, the original architecture is eventually restored; the whorls of leukocytes and collagenous fibers are replaced by Leydig cells indistinguishable from normal tissue. It is not known if this replacement is accomplished by

invasion of adjacent Leydig cells into the lesion or by differentiation of the infiltrated fusiform leukocytes into Leydig cells.

Digestive epithelium traumatized by the incision causes a leukocytic infiltration and phagocytosis of the necrotic cell debris. Other leukocytes, not involved in phagocytosis, form a band along the edge of the digestive tissue (Plate 8, Figure 14) and elongate. Des Voigne and Sparks (1968) believed that the leukocytes were totipotent and that they differentiated into digestive epithelial cells to replace the destroyed epithelium. However, Mix and Sparks (1971) showed that digestive epithelium in oysters traumatized by ionizing radiation is repaired by mitosis and migration of adjacent uninjured epithelial cells.

At the surface of the wound, the band of fusiform leukocytes that underlies the mantle epithelium closes the lesion within 24–32 h and gradually thickens and forms a laminated border. The external portion is composed of fusiform leukocytes arranged parallel to the surface, the middle portion consists of cells perpendicular to the surface and is infiltrated with round leukocytes, and the inner layer contains normal, round leukocytes and randomly arranged elongate cells. Des Voigne and Sparks (1968) reported that fusiform cells in the outer lamina gradually turn perpendicular to the surface and appear to form a

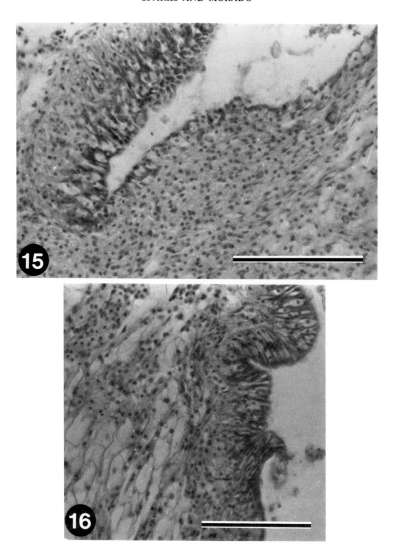

PLATE 9.—Wound healing in Pacific oyster, 88-h wound. Hematoxylin and eosin stain. (Reprinted, by permission, from Des Voigne and Sparks 1968. © 1968 by Academic Press.) FIGURE 15 (above). Newly formed epithelial covering of the mantle over the surface of the wound. Note the vacuolated nature. Bar = 130 μm. FIGURE 16 (below). Newly formed epithelial covering over the surface of the wound. Note the more highly organized nature of the mantle covering and the irregular, convoluted border. Bar = 100 μm.

loose, vacuolated pseudoepithelium (Plate 9, Figures 15, 16). At this stage, 40–120 h after wounding, the underlying lamina is heavily infiltrated with round leukocytes that eventually become fusiform and align themselves parallel to the surface. The surface layer frequently becomes stratified and is always deeply convoluted. By 168 h, the surface is covered by a loose cuboidal epithelium that gradually becomes a tall columnar ciliated epithelium similar in appearance to the adjacent normal epithelium. Des Voigne and Sparks were unable to demonstrate mitosis and migration

in adjacent normal epithelium and were led to believe that reepithelialization of the wound surface was by differentiation of the surface fusiform leukocytes into epithelium. However, Ruddell (1969) reported that adjacent epithelial cells proliferate and migrate across the surface of the wound to restore the epithelial covering.

Ruddell (1969) utilized histochemical techniques and electron microscopy to confirm the observations of Des Voigne and Sparks, except for the mechanism of reepithelialization, and he provided additional information on the precise

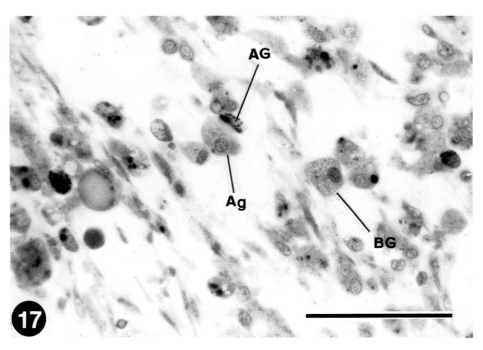

PLATE 10.—FIGURE 17. Light micrograph of an area adjacent to a 144-h wound in Pacific oyster showing an agranular amoebocyte-type cell containing a large glycogen deposit. Granular and agranular amoebocytes are also seen in this section. AG = acidophilic granular amoebocyte; BG = basophilic granular amoebocyte; Ag = agranular amoebocyte. Aldehyde-osmium tetroxide fixation; embedded in epon; stained in 0.5% toluidine blue O, pH 9.0; bar = 40 μm. (From Ruddell 1969. Photograph by C. L. Ruddell.)

mechanisms by which the process is accomplished. Ruddell identified three types of leukocytes (Plate 10, Figure 17) in the inflammatory response and wound repair process, agranular, acidophilic granular, and basophilic granular.

Two to four hours after wounding, basophilic granular leukocytes in the wound swell to three to six times their normal size and burst. Concurrently, acidophilic granular leukocytes begin to release copper and a diazotized p-nitroaniline-positive substance into the wound area. By 18 h after wounding, the copper disseminates and can be demonstrated bound to the cytoplasm and nucleus of Leydig, muscle, and nerve cells and to remnants of exploded basophilic granular leukocytes (Plate 11, Figure 18). By 24 h, copper is bound to most of the cells in the wound area, including the epithelial cells. The period of copper release appears to be brief; maximum dispersal is attained within 48 h and is largely confined to the immediate area of trauma.

The agranular leukocytes perform an important role in the early inflammatory period by infiltrating the wound and phagocytosing and sequestering cellular debris and particulate foreign materials. A late inflammatory period, from 48 to 72 h postwounding, is characterized by an influx of both granular and agranular leukocytes into the wound and the beginning of epithelial migration over the external surface.

In the early wound-healing stage, extending from 72 to 144 h, electron microscopy reveals that the plug filling the wound is composed of typical, undifferentiated agranular leukocytes that have elongated and produced numerous small fibrils beneath and parallel to the cell membranes. Ruddell (1969) demonstrated that agranular leukocytes travel at least 2 cm to infiltrate the wound and form the plug. Meanwhile, epithelial cells continue to migrate across the leukocyte plug at the external surface, and epithelial cells adjacent to the wound undergo mitotic division to add new cells to the migrating epithelium.

During the late wound-healing period, extending from 120 h postinjury to 2–3 weeks, fibrous material is deposited by many of the leukocytes in the wound plug. This material is composed of poorly defined fuzzy fibrils 100–250 nm in diameter. The cells that produce the fibrils resemble agranular leukocytes in every respect except that

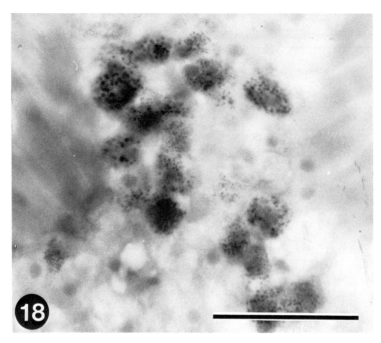

PLATE 11.—FIGURE 18. Section of an area adjacent to an 18-h wound in Pacific oyster. Note copper bound to virtually every cell in the field. Aldehyde fixation; embedded in hydroxyethyl methacrylate; stained with 0.05% aqueous hematoxylin for 20 min; bar = 40 μm. (From Ruddell 1969. Photograph by C. L. Ruddell.)

the cytoplasmic matrix, organelles, and membranes are markedly granular in appearance.

Migrating epithelial cells completely cover the wound between 144 and 200 h after wounding. On completion of reepithelialization, the torpedo-shaped cells of the actively migrating epithelium round up and become oriented perpendicular to the wound surface. During this terminal phase of wound repair, large numbers of both acidophilic and basophilic granular leukocytes become incorporated into the newly formed epithelium (Plate 12, Figures 19, 20), and they often outnumber the epithelial cells. This massive infiltration of leukocytes into the newly formed epithelium may have masked the mitotic activity and epithelial migration. The masking may have caused Des Voigne and Sparks (1968) to assume that differentiation of leukocytes occurred.

Inflammation and Wound Repair in Other Marine Bivalves

Drew (1910) reported that small incisions in the foot of *Cardium norvegicus* healed within 3 weeks by a process remarkably similar to that of mammals. In 1910, Drew and de Morgan described the early defense reactions of the scallop *Pecten maximus* to insertion of foreign substances into the adductor muscle, but they were unable to maintain wounded animals for more than 6 d. However, by that time, fibrous encapsulation of the foreign bodies had occurred.

Mikhailova and Prazdnikov (1961, 1962) reported that the main component of the inflammatory process in the edible blue mussel *Mytilus edulis* is phagocytosis, but the reaction varies depending on the nature of the foreign body. The first stage of inflammation occurs during the first 12 h postinjury and is characterized by accumulation of damaged connective tissue and homogenized collagenous material around the foreign material and is accompanied by emigration of numerous leukocytes to nearby blood spaces.

The second stage begins after 12–24 h and continues for 1–10 d postinjury. It is characterized by intensive phagocytosis and both intra- and extracellular digestion of biotic foreign materials. After approximately 48 h, eosinophilic amoebocytes infiltrate between the fibers of an introduced thread, and the amoebocytes infiltrate the periphery of the wound channel to form layers surrounding the foreign body.

The third stage of inflammation is characterized by the removal of foreign particles from the wound area. Leukocytes that contain engulfed

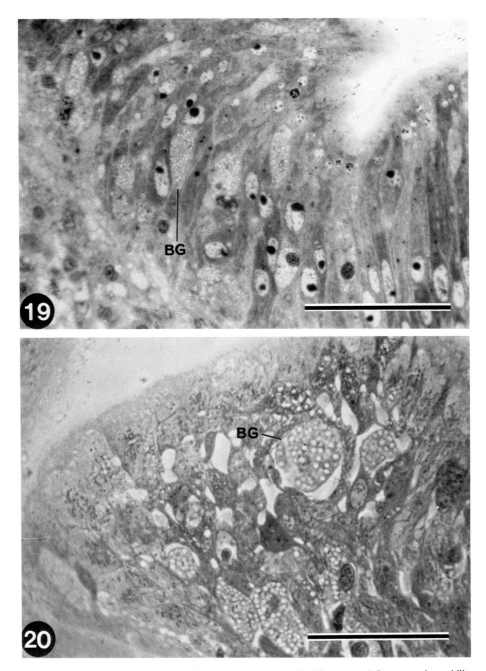

PLATE 12.—Cross sections of epithelial cells covering wounds in Pacific oyster. BG = many basophilic granular amoebocytes incorporated into wound epithelium. Aldehyde-osmium tetroxide fixation; embedded in epon; stained with 0.05% toluidine blue O, pH 9.0. (From Ruddell 1969. Photograph by C. L. Ruddell.) FIGURE 19 (above). Section of 144-h wound. Bar = 40 μm. FIGURE 20 (below). Section of a 400-h wound. Bar = 25 μm.

particles migrate to the surface epithelium and are discharged. Basophilic leukocytes also participate in phagocytosis and diapedesis of foreign particles. The third stage may be superim-posed on the second stage and occurs between 5 and 10 d postinjury.

Cheney (1969) described the response to and repair of incision wounds in the Manila clam

Tapes semidecussata. A light leukocytic infiltration into the wound develops within 14 h. Infiltrated cells and their nuclei elongate and become oriented parallel to the axis of the wound. At 1 d postwounding, the infiltrate is much heavier but similar in appearance to that occurring at 14 h. Epithelium adjacent to the incision begins to migrate down the wound channel. The leukocytic infiltrate begins to recede and, at 5 d after wounding, remains only at the base of the wound. The migrating epithelium continues to move along the surface of the wound channel and covers most of the cut surface by 5 d postinjury; epithelial pavementing of the wound channel is almost always complete 13 d after wounding. Within 20–30 d, the wound channel is bridged by extension of the connective tissue, and the outer surface of the wound is covered by cuboidal epithelium, which isolates the epithelium lining the wound channel as in the Oregon floater *Anodonta oregonesis*, a freshwater mussel (Pauley and Heaton 1969). Cheney (1969) did not follow the fate of the epithelium that migrates down the surface of the wound channel, but it probably follows the same sequence of necrosis and subsequent removal by phagocytosis that occurs in Oregon floaters.

References

Cheney, D. P. 1969. The morphology, morphogenesis, and reactive response of 3H-thymidine labeled leucocytes in the Manila clam, *Tapes semidecussata* (Reeve). Doctoral dissertation. University of Washington, Seattle.

Des Voigne, D. M., and A. K. Sparks. 1968. The process of wound healing in the Pacific oyster, *Crassostrea gigas.* Journal of Invertebrate Pathology 12:53–65.

Drew, G. H. 1910. Some points in the physiology of lamellibranch blood corpuscles. Quarterly Journal of Microscopical Science 54:605–623.

Drew, G. H., and W. de Morgan. 1910. The origin and formation of fibrous tissue produced as a reaction to injury in *Pecten maximus*, as a type of the Lamellibranchiata. Quarterly Journal of Microscopical Science 55:595–610.

Makhailova, I. G., and E. V. Prazdnikov. 1961. Two questions on the morphological reactivity of mantle tissues in *Mytilus edulis* L. Trudy Murmanskogo Morskogo Biologicheskogo Instituta 3:125–130.

Mikhailova, I. G., and E. V. Prazdnikov. 1962. Inflammatory reactions in the Barents Sea mussel. (*Mytilus edulis* L.) Trudy Murmanskogo Morskogo Biologicheskogo Instituta 4:208–220.

Mix, M. C., and A. K. Sparks. 1971. Repair of digestive tubule tissue of the Pacific oyster, *Crassostrea gigas*, damaged by ionizing radiation. Journal of Invertebrate Pathology 16:14–37.

Pauley, G. B., and L. H. Heaton. 1969. Experimental wound repair in the freshwater mussel *Anodonta oregonensis*. Journal of Invertebrate Pathology 13: 241–249.

Pauley, G. B., and A. K. Sparks. 1965. Preliminary observations on the acute inflammatory reaction in the Pacific oyster, *Crassostrea gigas* (Thunberg). Journal of Invertebrate Pathology 7:248–256.

Pauley, G. B., and A. K. Sparks. 1966. The acute inflammatory reaction in two different tissues of the Pacific oyster, *Crassostrea gigas.* Journal of the Fisheries Research Board of Canada 23:1913–1921.

Pauley, G. B., and A. K. Sparks. 1967. Observations on experimental wound repair in the adductor muscle and the Leydig cells of the oyster, *Crassostrea gigas.* Journal of Invertebrate Pathology 9:298–309.

Ruddell, C. L. 1969. A cytological and histochemical study of wound repair in the Pacific oyster, *Crassostrea gigas.* Doctoral dissertation. University of Washington, Seattle.

Sparks, A. K. 1972. Invertebrate pathology—noncommunicable diseases. Academic Press, New York.

Sparks, A. K. 1985. Synopsis of invertebrate pathology—exclusive of insects. Elsevier, Amsterdam.

Sparks, A. K., and G. B. Pauley. 1964. Studies of the normal postmortem changes in the oyster, *Crassostrea gigas* (Thunberg). Journal of Insect Pathology 6:78–101.

Stauber, L. A. 1950. The fate of India ink injected intracardially into the oyster, *Ostrea virginica.* Biological Bulletin (Woods Hole) 98:227–241.

Stein, J. E., and J. G. Mackin. 1955. A study of the nature of pigment cells of oysters and the relation of their numbers to the fungus disease caused by *Dermocystidium marinum.* Texas Journal of Science 7:422–429.

Tripp, M. R. 1958. Disposal by the oyster of intracardially injected red blood cells of vertebrates. Proceedings National Shellfisheries Association 48: 143–147.

Tripp, M. R. 1970. Defense mechanisms of mollusks. Journal of the Reticuloendothelial Society 7:173–182.

American Fisheries Society Special Publication 18:153–168, 1988
© Copyright by the American Fisheries Society 1988

Cellular Defense Mechanisms of Oysters and Mussels[1]

SUNG Y. FENG

Department of Marine Sciences, The University of Connecticut
Groton, Connecticut 06340, USA

Abstract.—From the 1950s to the 1970s, research on molluscan immunobiology emphasized descriptive aspects of cellular responses as well as morphological and functional studies of the formed hemolymph elements. These early studies revealed that the cellular defense mechanisms of oysters and mussels elicited by natural and experimental infections are characterized by an immediate commencement of cellular infiltration at the affected site, phagocytosis, and the eventual removal or isolation of foreign entities. Whether the ingested particles are eliminated by intracellular digestion, attrition, or both, and isolation by encapsulation depends on the size and susceptibility of the particles to host lysosomal enzymes. Repair and regeneration of injured tissues signifies the final episode of host–parasite interplay. However, there are cases in which host cellular responses are either absent or ineffective. The variability in host cellular responses could be a consequence of parasites avoiding recognition or interfering with host defenses. Recently, we have sought to understand the molecular basis of cellular and humoral defense, of self–nonself recognition mechanisms (proposed in the 1960s), and of molluscan host–parasite interactions. The more recent important findings include the discovery of membrane-bound and hemolymph-associated lectins, the release of lysosomal enzymes and cytotoxic molecules from hemocytes as a possible means of extracellular destruction of particulate antigens, and the detection of killing agents such as hydrogen peroxide in hemocytes of certain bivalve molluscs. Nevertheless, critical issues remain unresolved, e.g., identification of hemopoietic tissues, ontogeny of hemocytes, significance of cell membrane and hemocyte-associated lectins in self–nonself recognition and phagocytosis, cellular involvement in allograft acceptance and specificity of xenograft rejection, and cellular origins of reported humoral factors and their role in defense and disease resistance.

The definitions of susceptibility, insusceptibility, and resistance advanced by Read (1958) have served as conceptual frameworks for the investigation of host–parasite interactions in the last three decades. He defined the terms susceptibility and insusceptibility on the basis of whether or not a host can fulfill the nutritional needs of a parasite. He further reasoned that resistance could be independent of susceptibility.

Under natural conditions, animals are probably constantly explored as potential hosts by a wide spectrum of parasitic microorganisms and metazoa. But the success of these parasites appears to be limited because the suite of parasites associated with a given host species is more or less constant. The host is probably capable of destroying most, if not all, potential parasites immediately or shortly after they enter the host. The mechanism of destruction may be attributable to the existence of passive or active factors, e.g., the lack of proper nutrients to sustain the growth of the parasites or the presence of cytotoxic factors. Susceptibility and resistance determine host–parasite compatibil-

ity and are the result of sundry innate and acquired resistance mechanisms in the host.

The factors influencing host resistance can be divided into extrasystemic or extrinsic factors and systemic or intrinsic factors. The extrasystemic factors are fortuitous and epizootiological (ecological) and are associated with climatic conditions, density-dependent factors, and food preference of the host. The systemic factors are host-specific and consist of innate and acquired resistances. Innate resistance is manifested by the inherited anatomical, physiological, biochemical, and behavioral traits of a host that effectively prevent the host from acquiring infections. Acquired resistance, on the other hand, includes resistance mechanisms that are enhanced as a result of previous exposure to an agent. Acquired resistance differs from innate resistance both in the intensity and specificity of the response. Moreover, the manifestation of innate resistance is rapid, whereas the appearance of acquired resistance is delayed. In invertebrates, most observed resistance mechanisms are considered to be innate because acquired resistance mechanisms are not well developed. However, several reports suggest that quasi-immune responses are functionally and biochemically analo-

---

[1]Contribution 201 from Marine Sciences Institute, The University of Connecticut.

gous to vertebrate responses (Cheng 1981, 1985; Fries 1984).

Both innate and acquired resistance may consist of cellular and humoral factors. This scheme, however, is basically an artificial one because probably all humoral principles are ultimately of cellular origin. It is these cellular factors which I shall discuss.

Our knowledge of bivalve immunobiology has been derived from observations on the reaction of the host to a variety of pathogens and from experimental investigations of the fate of parenterally inoculated inanimate substances and particulate and soluble antigens. Over the last three decades, much research effort has been devoted to the oyster genus *Crassostrea* and to the eastern oyster *C. virginica* in particular. Furthermore, most of the studies have emphasized cell-mediated responses expressed as clearance of inoculated foreign substances.

The search for inducible humoral response has been less successful, and the results have been more equivocal. In a few instances, where elevated levels of lysozyme and hemagglutinin were demonstrated in eastern oysters infected with *Bucephalus* sp. (Feng and Canzonier 1970) and in Pacific oysters *Crassostrea gigas* exposed to *Vibrio anguillarum* (Hardy et al. 1977a), the enhancement of these nonspecific humoral factors was achieved by infecting the hosts under natural conditions or by administering *Vibrio anguillarum* to the oysters orally. These approaches differed basically from the parenteral route employed by Acton et al. (1969) and Tripp and Kent (1967), who met with limited success. In the hemolymph of the northern quahog *Mercenaria mercenaria*, a clam, hemolytic activity was enhanced by injecting rabbit erythrocytes or by experimental wounding (Anderson 1981).

Cellular Defense Mechanisms

Cell Types

The unit of cellular defense in oysters and mussels is the hemocyte. Two types of hemocytes, agranulocytes and granulocytes, are demonstrable by light and electron microscopy and by cytomchemical studies (Feng et al. 1971, 1977; Ruddell 1971a, 1971b, 1971c; Cheng and Cali 1974; Cheng et al. 1974; Moore and Lowe 1977; Rasmussen et al. 1985). Cheng (1981) reviewed the literature on the functional morphology of bivalve hemocytes and proposed a scheme that placed the hemocytes in three categories, granu-

locytes, hyalinocytes, and serous cells. He eliminated fibrocytes as a viable class of hemocytes because some cells designated as fibrocytes in the study of the northern quahog by light microscopy (Foley and Cheng 1974) were later revealed to be degranulated granulocytes by electron microscopy (Cheng and Foley 1975; Cheng et al. 1975). However, this important cell type is known to participate in wound healing (Pauley and Sparks 1967; Des Voigne and Sparks 1968), in response to helminth parasitism (Rifkin and Cheng 1968; Cheng and Rifkin 1970), and in response to inoculated avian erythrocytes (Feng and Feng, 1974). Also, fibroblastlike cells have been seen in oyster leukocyte populations maintained in vitro (Tripp et al. 1966). Thus, there is considerable evidence to support the view that fibrocytes or fibroblastlike cells are normal components of formed hemolymph elements of the oyster and probably of other bivalves and that they constitute an integral part of the cellular defense.

Fisher (1986) comprehensively reviewed the structure and function of oyster hemocytes. The reviews of Stauber (1961), Tripp (1963, 1969, 1970, 1975), Cheng (1967), Feng (1967), Sindermann (1970), and Bayne (1983) contain detailed historic information on the development of molluscan immunity. Auffret (1988, this volume) has synthesized the observations of several researchers.

Inflammatory Process

Inflammation is a process that involves both cellular and humoral elements of the host. The process begins following tissue injury alone or accompanied by invasion with a biological, chemical, or physical agent; it terminates with three possible consequences: (1) destruction of tissues, (2) isolation of invading organisms, and (3) complete repair of damaged tissues. Failure to contain the invading organism can lead to the destruction or death of host tissues (necrosis). Even in cases where containment of a pathogen is achieved, the reparative process could result in the loss of tissue and structural integrity. For example, fibrotic repair of hepatocytes damaged by hepatitis B virus in humans will lead to liver cirrhosis and eventual loss of liver functions. Therefore, the inflammatory process cannot always be interpreted as being protective to the host. In mammals, the process involves increased permeability of the local capillary wall, resulting in hyperemia, stasis, characteristic cellular infiltration, leukocytosis, exudations, and deposition of fibrin. Externally, symptoms are characterized by redness, pain, heat, and swelling, the well known cardinal signs. Generally, the inflammatory response of invertebrates is similar to that of vertebrates, but the two responses may differ greatly in detail. For example, increased permeability of the capillary wall in oysters and mussels cannot be determined readily because these animals possess a circulatory system that is only partially closed.

Wound healing and the inflammatory response was first investigated by Sparks (1972, 1985) in the 1960s using the Pacific oyster. Hemocytes play a major role in wound repair (Des Voigne and Sparks 1968, 1969; Ruddell 1971c). The following sequence of events leads to healing (Sparks and Morado 1988, this volume): (1) infiltration of the wound site by numerous hemocytes, (2) formation of a plug by the aggregated hemocytes 144 h postwounding, (3) replacement of damaged tissues with elongated hemocytes from the interior of the lesion toward the surface, (4) deposition of collagen by fibrolasts (Ruddel 1971c), (5) removal of the necrotic tissue debris by phagocytic granulocytes, and (6) restoration of the normal tissue architecture. Similar wound-healing events follow xenograft implantation and shell damage in the California mussel *Mytilus californianus* and the blue mussel *M. edulis* (Bubel et al. 1977; Bayne et al. 1979); apparently, diapedesis of hemocytes through the mantle epithelium is required in the process.

Cellular elements that participate in wound repair are similar in the three species of bivalve molluscs studied, and they are of three types, agranular amoebocytes, basophilic granulocytes, and acidophilic granulocytes. During the initial phase, the infiltrating cell population is dominated by phagocytic agranular amoebocytes in Pacific oysters (Ruddell 1971b) and by small basophilic hyalinocytes in California mussels (Bayne et al. 1979). In California mussels, large basophilic hyalinocytes (macrophages) are highly phagocytic and contain a number of lysosomal enzymes. As healing progresses in Pacific oysters, acidophilic granulocytes begin to accumulate and apparently release copper into the lesion, but the importance of this event remains to be elucidated (Ruddell 1971a). Finally, the wound is closed by new epithelium which originates in the adjacent undamaged tissue by mitoses (Ruddell 1971b). Such mitotic regeneration of the epithelium has also been observed in eastern oysters (Hillman 1963) and California mussels (Bayne et al. 1979). Phagocytosis and its associated events of aggregation, hyperplasia, and encapsulation are integral parts of the inflammatory response (Metchnikoff 1891).

Aggregation

The behavior of hemocytes to form aggregates or clumps is well known to investigators of molluscan hematology. In oysters, this phenomenon can be demonstrated readily both in vitro and in vivo. Hemocyte aggregates are frequently encountered during the preparation of blood smears on glass slides (Bang 1961; Foley and Cheng 1972). Upon standing, oyster hemocyte clumps exhibit exomigration of cells at the periphery that results in a monolayer of concentrically arranged cells.

In vivo hemocyte aggregation in the form of a plug after wounding of the oyster soft tissue was discussed above (Des Voigne and Sparks 1968; Ruddell 1971c). Bang (1961) described the development of intracellular clotting or thrombosis (i.e., hemocyte aggregates) in the circumpallial artery of eastern oysters following intracardial injection of tissue extracts. He further reported that the clotting resolves spontaneously 2 h after injection. Feng (1965a) demonstrated that hemocyte aggregates or clotting can be formed by placing eastern oysters at 5°C, and that these aggregates can be resolved by gradual warming to 23°C. Hemocyte aggregates, which persisted for at least 22 d, were also found in eastern oysters inoculated with avian erythrocytes (Feng and Feng 1974). Histological studies revealed three recognizable host cell types in such cell aggregates: hemocytes with or without ingested erythrocytes, fibroblastlike cells, and lymphocytelike cells. Therefore, these hemocyte aggregates differ basically from the transient forms described by Bang (1961) and Feng (1965a).

The factors which govern the formation of cell aggregates are not known. Morphologically, the hemocytes first link together by their filopodia, which then shorten and thicken to draw the cells closer into a cell aggregate (Drew 1910). To explain the formation of hemocyte aggregates purely on a physical basis, two factors must be resolved: the probability of collision among hemocytes and the force which binds hemocytes by their filopodia. These factors are in turn determined by the force generated by the pulsating and accessory hearts in the circulatory system. One may be able to predict optimal conditions that enhance the chance of contact among cells but do not disrupt the formed aggregates by the agitating force of the hearts. It also appears reasonable to assume that the membrane properties of the circulating hemocytes have changed after the injection of foreign materials (Feng and Feng 1974) such that binding among hemocytes is facilitated and strengthened.

Available evidence suggests that the reversible type of hemocyte aggregates of Bang (1961) and Feng (1965a) could be explained by the physical forces involved, whereas the more permanent type of hemocyte aggregates of Feng and Feng (1974) favors the assumption of membrane alteration mediated by macromolecules. However, it is possible that both physical and chemical elements are involved in the formation of hemocyte aggregates. Attempts to elucidate the chemical basis of bivalve hemocyte aggregation are few. Thus, Fisher (1986) reported that filopodia of eastern oyster hemocytes are not inhibited by EDTA which inhibits aggregation of amoebocytes of the horseshoe crab *Limulus polyphemus* amoebocytes (Kenney et al. 1972) and of coelomocytes of the echinoderms *Asteria forbesi* (Kanungo 1982) and *Cucumaria frondosa* (Noble 1970). Further studies of this phenomenon of mulluscan hemocytes are needed. The functional significance of hemocyte aggregation in wound healing has been discussed above.

Phagocytosis

Phagocytosis occurs throughout the animal kingdom from protozoa to metazoa. It is the means by which single-celled animals ingest food, and it also serves as an integral part of the inflammatory process in higher animals. Metchnikoff's (1891) extensive studies on a variety of animals were the first approach to this basic fact in an evolutionary context; he described phagocytes as cells having the capacity to ingest and sometimes absorb food particles. Thus, phagocytosis is a basic biological process of great antiquity that possibly antedates the development of immune responses by evolutionary eons.

Phagocytosis involves five events: foreign-particle recognition, adherence, uptake, destruction, and disposal. The first two events could be brought about either passively, by random collisions, or actively, by chemotaxis. Foreign-particle recognition is probably determined by the surface properties of the particle and the receptors on the hemocyte. The degree of intracellular destruction of ingested particles varies; it depends on the susceptibility of the particles to lysosomal enzymes. In extreme cases, an ingested living agent not only survives but also multiples within the phagocyte; it may inflict injury on the host cell (intracellular parasitism), or it may benefit both parties (intracellular mutualism).

Since the publication of Metchnikoff's 1884 paper on fungal disease of *Daphnia* spp., many investigators have described phagocytosis as a defense mechanism of invertebrates. In molluscs, particularly oysters, most of the experimentally inoculated particles are cleared quickly from the recipient. The mechanisms of clearance were delineated in an experiment in which eastern oysters were injected intracardially with india ink (Stauber 1950). The first prominent event was occlusion of major blood vessels by the ink particles. Then, hemocytes began to infiltrate the occluded lumen of the vessels, and phagocytosis of the ink particles ensued. By the eighth day postinjection, emigration of hemocytes laden with ink particles through arterial walls and various epithelia was well underway. Arterial emboli were completely resolved in 17 d. Exit of ink-laden hemocytes was seldom seen in the gills, gonoducts, renal tubules, and the pallial epithelium of the mantle. Tripp (1958, 1960) made similar observations in eastern oysters injected with bacteria, yeasts, and erythrocytes. However, erythrocyte-laden hemocytes were observed in the gonoducts, and they were seen to migrate through the mantel epithelium into the pallial space (Feng and Feng 1974). Moreover, phagocytosis, intracellular digestion, and migration of erythrocyte-laden hemocytes were shown to be temperature-dependent processes. Apparently, carmine-laden phagocytes of the edible oyster *Ostrea edulis* exited through the gonoduct and the renal tubules as well (Takatsuki 1934).

Most experimentally introduced vegetative bacteria were eventually rendered nonviable by intracellular digestion (Tripp 1960; S. Y. Feng 1966). However, the nondigestible *Bacillus mycoides* spores were found in the fecal material for 60 d (Tripp 1960), and a *Pseudomonas*-like bacterium was isolated from morbid eastern oysters that exhibited a relapsing course of infection at 22–27°C but not at 9°C (S. Y. Feng 1966). Thus, the elimination of inoculated bacteria was determined by the digestibility of the microbe and by the ambient temperature of the seawater in which the oysters were maintained. Intracellular digestion is more prevalent than attrition of particle-laden hemocytes as a means of final disposal. This preference probably reflects the original food-acquisition role of the ancestral form of hemocytes (Yonge 1926; Cheng and Rudo 1976; Feng et al. 1977). Intracellular degradation of phagocytosed bacteria and eukaryotic cells were accomplished by an array of hemocyte lysosomal enzymes: acid and alkaline phosphatases, nonspecific esterases, indoxyl esterase, β-glucuronidase, lipase, aminopeptidase, lysozyme, and β-hexosaminidase (McDade and

Tripp 1967; Eble and Tripp 1969; Feng et al. 1971, 1977; Rodrick and Cheng 1974; Yoshino and Cheng 1976; Moore and Lowe 1977). However, extracellular digestion of *Bacillus megaterium* by elevated levels of hemolymph lysozyme in northern quahogs was observed by Cheng et al. (1975), Foley and Cheng (1977) and Mohandas et al. (1985).

In mammals, stimulation of phagocytic activities in polymorphonuclear leukocytes, monocytes, and macrophages elicits respiratory bursts and a concomitant production of superoxide anions (O_2^-), hydrogen peroxide (H_2O_2), hydroxyl radical ($\cdot OH$), and singlet oxygen ($^1O_2$). These four oxygen species are potent killing agents of microbes and known as the myeloperoxidase-peroxide-halide system; they can function independently or in concert with lysosomal enzymes. Only three published papers deal with the biochemistry of phagocytosing molluscan hemocytes. Cheng (1976) could not detect increased oxygen consumption or tetrazolium blue reduction in phagocytosing hemocytes of northern quahogs; this result indicated either the absence of the myeloperoxidase-peroxide-halide system or the existence of some yet unidentified microbicidal systems. However, hemocytes of *Patinopecten yessoensis*, which were stimulated with concanavalin A or bacteria, produced and released peroxide in vitro (Nakamura et al. 1985). Dikkeboom et al. (1987) demonstrated the generation of superoxide and peroxide in vitro in hemocytes of the swamp lymnaea *Lymnaea stagnalis* (Gastropoda) stimulated with latex, *Escherichia coli, Staphylococcus saprophyticus*, zymosan, and phorbol myristate acetate. More comprehensive studies of this killing mechanism in marine bivalve molluscs are needed to resolve the apparent discrepancies between hemocytes of northern quahog (Cheng 1976) and those of *Patinopectin yessoensis* (Nakamura et al. 1985).

The elimination of injected *Staphylococcus aureus* phage 80 from eastern oysters depended on temperature (J. S. Feng 1966); hemocytes probably did not play a major role in the expulsion of the viral particles from the hosts because the recovery of phage 80 particles from the hemocytes was less than one per hemocyte. However, this observation could also indicate intracellular destruction of the virus. In vitro uptake of the blue-green algal virus LPP-1 by oyster leukocytes was reported by Fries and Tripp (1970), but the viability of the ingested virus was not determined. Secondary clearance of the T2 bacteriophage from eastern oysters was enhanced by a "not

entirely specific" cell-mediated mechanism (Acton and Evans 1968; Acton et al. 1969). These results vary markedly from those obtained in the naive and "immune" animal experiment of J. S. Feng (1966). However, Acton's experiments were conducted under less than ideal conditions, i.e., water temperatures varied between 25 and 30°C, and results were derived from oysters with one valve removed.

The response of molluscan hemocytes to soluble antigens needs to be investigated more fully. Bang (1961) reported that the intracardiac injection of eastern oysters with tissue extract stimulated the transient appearance of "intravascular clotting or thrombosis," which consisted of leukocytes. In vitro and in vivo experiments showed that oyster hemocytes pinocytose rhodamine-labeled mammalian and *Limulus polyphemus* serum proteins and that protein-laden hemocytes are first sequestered in the periintestinal region, later migrate through the epithelia of the intestine and other organs, and are eventually incorporated in the ejecta and rejecta (Feng 1965b). Pinocytosis, like phagocytosis, was temperature-sensitive but proceeded more quickly. When concanavalin A, a lectin, was presented to oyster hemocytes (Yoshino et al. 1979), the hemocytes capped and internalized this membrane-bound glycoprotein; these processes are analogous to those manifested in vertebrate immunocytes. Whether a similar mechanism is involved in the pinocytosis of serum proteins is yet unknown. It was suggested that, in mammalian systems, the redistribution of surface membrane receptors plays a role in antigenic modulation, immune expression in lymphocytes, or inter- and intracellular recognition and communication (Singer 1974; Edelman 1976).

Chemotaxis, random collisions, or both result in initial adherence and subsequent endocytosis of particulate matters by hemocytes. There are few publications on this subject. Oyster hemocytes exhibit chemotactic responses to a variety of biological agents: metacercarial cysts of *Himasthla quissetensis* (Cheng et al. 1974), *Micrococcus varians* (=*Staphylococcus lactus*), *Bacillus megaterium*, and *Escherichia coli* (Cheng and Rudo 1976; Cheng and Howland 1979). But heat-killed bacteria and both live and heat-killed *Vibrio parahaemolyticus* do not elicit such responses. Howland and Cheng (1982) identified the agents from *B. megaterium* and *E. coli* that initiate chemotaxis by oyster hemocytes as proteins of about 10,000 daltons.

The importance of understanding the link between phagocytic cells and humoral components of the blood or hemolymph was recognized early in the study of endocytosis. Metchnikoff (1891) sought the link to defend his phagocytosis theory against humoralists who stated that chemical substances present in the blood account for immunity. The link was found in 1903 when Wright and Douglas discovered opsonin, a humoral component that could render bacteria more susceptible to phagocytosis. Henceforth, the two theories of immunity (cellular versus humoral) were reconciled. In the oyster, the discovery of hemagglutinin and its functional role as an opsonin (Tripp 1966) and the characterization of lysozyme (McDade and Tripp 1967) are historically important to invertebrate immunologists. These humoral principles also occur in blue and California mussels and in Pacific oysters (Hardy et al. 1976, 1977b; Bayne et al. 1979; Renwrantz and Stahmer 1983).

These discoveries have raised questions as to the source(s) of the humoral principles and their roles in defense. As mentioned earlier, certain types of eastern oyster hemocytes patch, cap, and internalize concanavalin A (Yoshino et al. 1979). Vasta et al. (1982) identified two serologically distinct oyster hemolymph lectins and one hemocyte membrane-associated lectin that has identical specificity with one of the two hemolymph lectins. They suggested that the hemocyte membrane-associated lectin may serve as a membrane receptor in nonself recognition by oyster hemocytes. Similar observations on blue mussel hemocytes were reported by Renwrantz and Stahmer (1983). They showed that the heightened phagocytosis of yeast is attributable to two hemolymph factors: opsonizing properties of the purified agglutinin, and Ca^{++} ions. Apparently, the phagocytosis of yeast in the absence of agglutinin is enhanced by the presence at the hemocyte surface of recognition molecules that depend on divalet cations. Because these recognition sites can be blocked by a purified specific antiagglutinin, immunoglobulin G, derived from blue mussel hemolymph proteins, the hemolymph and membrane-associated lectins may share similar antigenic structures. These important findings have clearly elevated the investigation of invertebrate immunological recognition systems to a new level of understanding.

Lysozyme, a lysosomal enzyme, was detected in both the hemolymph and hemocytes of eastern oysters (McDade and Tripp 1967; Feng et al. 1971; Rodrick and Cheng 1974; Cheng and Rodrick 1975). Its increased release from hemocytes of northern quahogs was stimulated by phagocytosis of *Bacillus megaterium* (Cheng et al. 1975; Foley

and Cheng 1977). Yoshino and Cheng (1976) reported similar elevation of aminopeptidase activities in eastern oyster hemocytes undergoing phagocytosis. Additional lysosomal enzymes, alkaline and acid phosphatases, nonspecific esterases, β-glucuronidase, amylase (in hemolymph only), and lipase were demonstrated in the hemolymph and hemocytes of eastern oysters and northern quahogs (Feng et al. 1971; Cheng and Rodrick 1975). Wittke and Renwrantz (1984) reported that certain blue mussel hemocytes secrete cytotoxic substances against human group-A erythrocytes; it is not known whether or not the lytic factor is of lysosomal origin. The secretion of cytotoxic substances by eastern oyster hemocytes, however, requires further confirmation (Wittke and Renwrantz 1985).

The morphological basis for the release of hemocyte lysosomes into the bivalve hemolymph has been elucidated recently. In a scanning electron microscope study, Mohandas et al. (1985) showed that bacteria stimulated hemocytes of northern quahog to extrude more and larger intact lysosomes into the hemolymph than those of the control group and that contact between lysosomes and bacteria caused the degradation of bacterial cell walls, an indication of extracellular digestion. They referred to this process as degranulation. Degranulation originally described the fusion of lysosomes with phagosomes (to form secondary phagosomes) in polymorphonuclear leukocytes (Hirsch and Cohn 1960; Hirsch 1962). Intracellular coalescence or fusion of the lysosomes with phagosomes during phagocytosis of *Escherichia coli* by the hemocytes of Pacific oyster and *Mytilus coruscus* was demonstrated by electron microscopy and enzyme cytochemical procedures (Feng et al. 1977). Therefore, degranulation of bivalve hemocytes could be associated with the intracellular fusion of lysosomes with phagosomes and with the extrusion of intact lysosomes from hemocytes into the extracellular medium.

The presence of lysosomal enzymes in bivalve hemolymph led Cheng (1983, 1985) to conclude that "hypersynthesized enzymes released into the serum play a protective role in destroying susceptible infectious, biotic agents," that "elevated levels of serum lysosomal hydrolases may initiate autolysis in inflamed areas," and that the cellular inflammatory response "is specific at the molecular level rather than what has been designated as nonspecific."

Thus far, I have dealt only with the interaction between oysters and experimentally introduced inanimate and living agents. But oysters are also naturally invaded by a variety of microorganisms and metazoan parasites. Oyster hemocytes invaded by intracellular parasites, such as *Nematopsis ostrearum* (Prytherch 1940), *Bonamia ostreae* (Pichot et al. 1980), and *Perkinsus marinus* (Mackin et al. 1950; Perkins and Menzel 1967) manifest varying degrees of cytopathology. Mackin et al. (1950) reported the invasion of eastern oyster hemocytes by *P. marinus*, which reproduces intracellularly and eventually destroys the host cell. *Bonamia ostreae* is a small haplosporidianlike parasite of granulocytes; infected cells show extensive degenerative changes, pyknotic nuclei, and disruption of nuclear membranes, cytoplasmic organelles, and cell membranes leading to release of the parasite (Balouet et al. 1983). *Nematopsis ostrearum*, on the other hand, is a benign sporozoan parasite (Sprague and Orr 1955; Feng 1958) that develops into mature sporozoites within the host hemocytes, where it causes hypertrophy. Under favorable conditions, *P. marinus* and *N. ostrearum* were eliminated from the host (Ray 1954; Feng 1958). Hemocytes laden with *Perkinsus marinus* can be seen to traverse the intestinal epithelium of the oysters (Figure 1). Farley (1968) reported phagocytosis of *Haplosporidium nelsoni*, a highly virulent oyster pathogen, by eastern oyster hemocytes. Oyster hemocytes were also found to ingest moribund *H. nelsoni* (Kern 1976). These observations could represent either unsuccessful attempts to control the pathogen or oysters manifesting some degree of resistance. Certain selected strains of oysters that were substantially more resistant to *H. nelsoni* could localize the infection (Ford and Haskin 1982; Ford 1988, this volume). Although this suppression may involve hemocyte activity, the mechanism is not yet known. The infection of eastern oysters by *Hexamita nelsoni*, a facultative flagellate, elicits a heightened hemocytosis but little or no phagocytosis (Feng and Stauber 1968). The inconsistencies found in these examples emphasize questions of nonself recognition, resistance, and coexistence.

Hyperplasia

Hyperplasia is defined as "unusual enlargement of an organ or tissue due to controlled, increased proliferation of normal cells" (Kinne, 1980). In oysters, an increase in the number of circulating hemocytes in response to an infection can be considered as hemocytic hyperplasia or hemocytosis. Migration of available hemocytes to the in-

flamed site can result in a transient shortage of circulating hemocytes before the host has an opportunity to replenish them. In histologic sections, foci of hemocytic infiltrations are clearly discernible; circulating hemocytes at this time are often reduced in number but may return to a normal or heightened level depending on the pathogen present. Feng (1965a) pointed out that one should interpret the apparent hemopenia and hemocytosis with care because the number of circulating hemocytes increases as the heart rate increases, and the heart rate is influenced by temperature.

Encapsulation

Encapsulation occurs when the host is successful in isolating or segregating foreign bodies. Tripp (1961) observed that particles too large to be ingested by a single hemocyte are often enveloped in concentric layers of fibroblastlike cells.

The type of cellular response of the oyster varies with the type of metazoan parasite present, e.g., trematode or cestode. Little or no cellular reaction occurs in eastern oysters infected with the sporocysts of *Bucephalus* sp. from New England waters (Cheng and Burton 1965) or with the larval *B. cuculus* from the Gulf of Mexico and mid-Atlantic coastal regions (Tennent 1906; Hopkins 1954). However, intense cellular responses are induced in oysters by *Bucephalus* sp. sporocysts infected with a haplosporidian hyperparasite (Mackin and Loesch 1955). *Tylocephalum* sp., a metacestode found in eastern oysters (Sparks 1963; Cheng 1966; Rifkin and Cheng 1968), in *Tapes semidecussata* (Cheng and Rifkin 1968), and in *Venerupis staminea* (Sparks and Chew 1966), is completely encapsulated. According to Cheng and Rifkin (1970), the encapsulation complex consists of an inner layer of fibroblastlike cells with acid mucopolysaccharides and an outer layer of thick concentric fibrous materials which are reticular and contain glycoproteins or mucoproteins (or both) and neutral mucopolysaccharides. This thick fibrous layer is also infiltrated by hemocytes. Cheng and Rifkin (1986) noted that 56% of the metacestodes in *Tapes semidecussata* were resorbed.

The cellular response of blue mussels to *Proctoeces maculatus* infection changes according to the developmental stage of the trematode; both larval sporocysts and adults could be found in the same host individual (Uzmann 1953; Stunkard and Uzmann, 1959). Although developing sporocysts do not induce host responses, spent sporocysts, free cercariae and adult worms elicit intense hemocytic

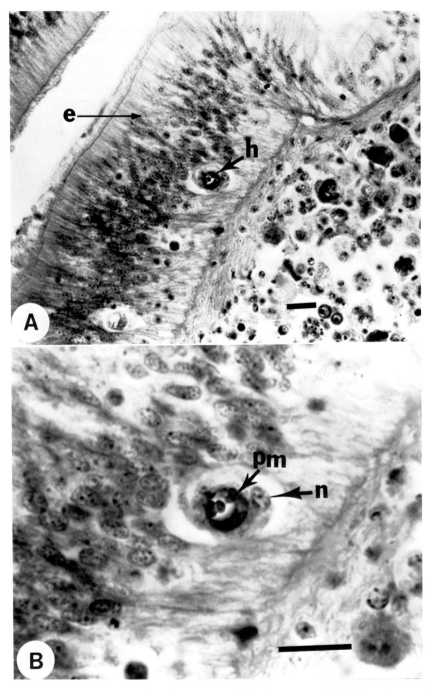

FIGURE 1.—Transepithelial migration of *Perkinsus marinus*-laden hemocytes (PMLH) in eastern oysters. Bars = 10 μm. Giemsa stain. **A.** A PMLH (h) moving through the intestinal epithelium (e). **B.** An enlarged view of the intestinal epithelium. pm = hemocyte with *P. marinus*; n = host hemocyte nucleus.

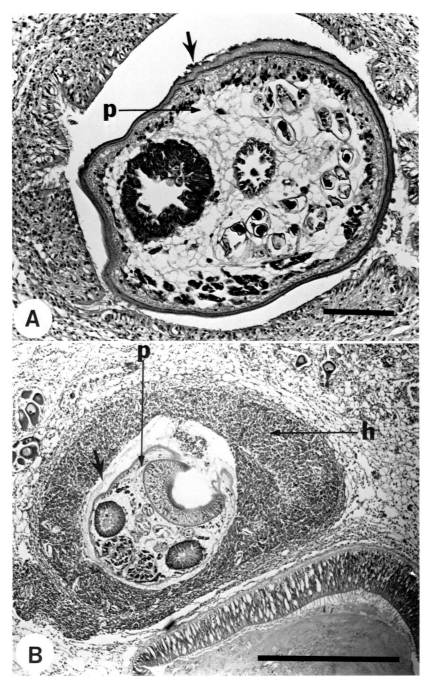

FIGURE 2.—Destruction of adult *Proctoeces maculatus* worms in blue mussels. Hematoxylin and eosin stain. **A.** The integument of an adult worm (p) begins to lose its integrity. Large arrow points to pits and irregularities. Bar = 100 μm. **B.** An adult worm (p) is surrounded by a dense layer of hemocytes (h). Large arrow points to the delamellated outer layer of its integument. Bar = 500 μm.

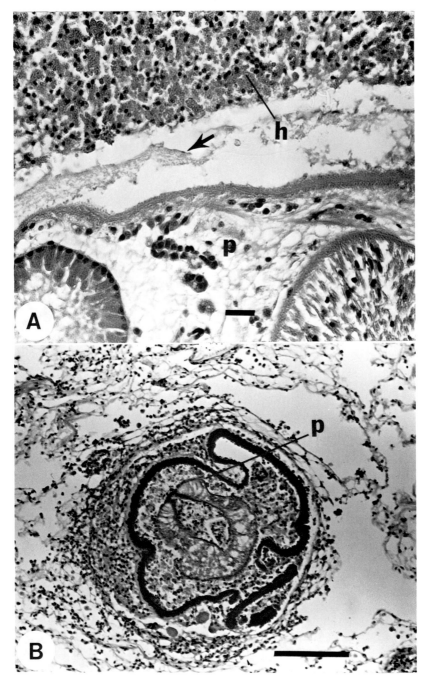

FIGURE 3.—Destruction of adult *Proctoeces maculatus* worms in blue mussels. Hematoxylin and eosin stain. **A.** A portion of Figure 2B enlarged. Arrow points to the delamellated integument. h = mass of hemocytes. p = adult worm. Bar = 20 μm. **B.** Advanced stage of destruction. Internal structures of the worm (p) have been largely destroyed, and only remnants of the anterior sucker are still recognizable. Also shown is the collapse of densely stained integument. Bar = 100 μm.

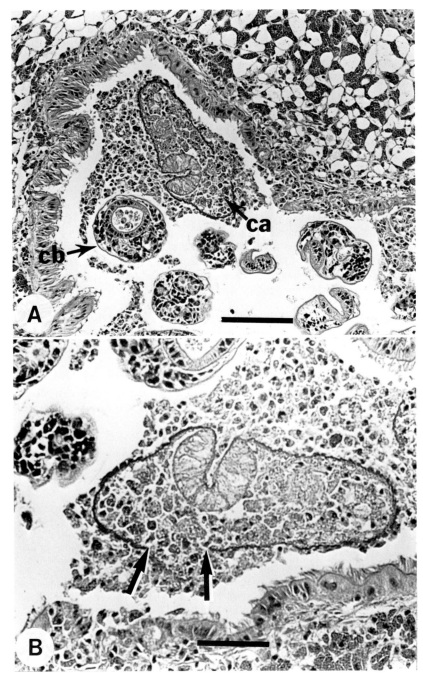

FIGURE 4.—Destruction of *Proctoeces maculatus* cercariae in blue mussels. Hematoxylin and eosin stain. **A.** A cercaria (ca) encircled by hemocytes is being destroyed. cb = cercaria with ingested hemocytes. Bar = 100 μm. **B.** An enlarged view of encircled cercaria showing the breach of its integument (at arrow heads) and invasion of hemocytes through the breach. Bar = 50 μm.

infiltrations (Tripp and Turner 1978). Adult *P. maculatus* are encapsulated by a thick layer of hemocytes (Figure 2) upon which adult worms and free cercariae feed. A lysosomal enzyme released from hemocytes, β-glucoronidase, may be responsible for the destruction of the adult worms and cercariae (Figures 3, 4) because it could hydrolyze acid mucopolysaccharides, i.e., structural components of the helminth integument (Lee 1966).

Future Research

It has been 103 years since Metchnikoff first announced the theory of phagocytosis. Since then, we have witnessed spectacular advances in vertebrate immunology. Although considerable information on cellular and humoral defense mechanisms has been obtained from some commercially imporant molluscs, the data base is incomplete. Some of the most pressing gaps in our knowledge of oyster and mussel immunology include the following.

Is there a hemopoietic center in these molluscs? Are hemocytes derived from a single- or a multiple-cell lineage? Much of the current disagreement regarding the cell types of oysters and mussels arise from the lack of uniformity in methodologies. Recent separation and identification of three subpopulations of oyster hemocytes by lectins represent a novel approach to the study of molluscan hemocytes (Renwrantz et al. 1979; Cheng et al. 1980). Can we use this approach to shed light on the ontogenesis of hemocytes?

Why are certain protozoan parasites recognized as nonself and phagocytosed, while others are perceived as self and tolerated or not destroyed? Why do certain ingested parasites resist digestion or coexist with the host hemocytes, whereas others eventually kill the host hemocytes? Are there specific recognition receptors on the membrane of hemocytes (Vasta et al. 1982)? Are some parasites "invisible" to the host due to shared or "eclipsed" antigens (Damian 1964)? Smyth (1973) hypothesized that such hostlike antigens could be acquired by parasites through natural selection, antigen induction, and antigen masking. In the oyster, Tripp (1975) proposed an "all purpose" hemolymph protein which coats and tags nonself entities. However, coating with host proteins conceivably could either facilitate or inhibit their recognition. Aside from the above considerations, Feng and Barja (1987) speculated that lack of host responses, e.g., phagocytosis, could be independent of self–nonself recognition and could result from the presence of factors that inhibit membrane synthesis and of receptor-blocking substances of parasite origin. In experiment studies, the rapid clearance by oysters of injected particulate and soluble antigens conveyed the sense that the defense mechanism was omnipotent. There is a real need to rectify such biased views and to understand the host–parasite interaction. The eastern oyster, which is naturally infected by parasites of varying virulence, offers good host–parasite systems in which to seek answers to the questions outlined above.

Early approaches to bivalve immunobiological research stressed the search for inducible principles and may have limited our perspective of investigation. One could argue that, in the absence of a highly specific antigen–antibody system, the oysters and mussels appear to be at least as well guarded as vertebrates. Therefore, we should seek all principles (specific, quasi specific, or nonspecific) that free the host from infections. More efforts should be directed towards investigating humoral factors and their cellular origins as well as the roles of the cell membrane- and hemolymph-associated lectins in self and nonself recognition.

Acknowledgments

I thank two anonymous reviewers for their constructive criticism of the draft manuscript. I am grateful to Joyce Rodriguez for typing the manuscript, and to Robert DeGoursey for collecting the mussels and preparing the photomicrographs.

References

Acton, R. T., and E. E. Evans. 1968. Bacteriophage clearance in the oyster *Crassostrea virginica*. Journal of Bacteriology 96:1260–1266.

Acton, R. T., E. C. Evans, and J. C. Bennett. 1969. Immunobiological capabilities of the oyster *Crassostrea virginica*. Comparative Biochemistry and Physiology 29:149–160.

Anderson, R. S. 1981. Inducible hemolytic activity in *Mercenaria mercenaria* hemolymph. Development and Comparative Immunology 5:575–585.

Auffret, M. 1988. Bivalve hemocyte morphology. American Fisheries Society Special Publication 18:169–177.

Balouet, G., M. Poder, and A. Cahour. 1983. Haemocytic parasitosis: morphology and pathology of lesions in the French flat oysters, *Ostrea edulis* L. Acquaculture 34:1–14.

Bang, F. B. 1961. Reaction to injury in the oyster (*Crassostrea virginica*). Biological Bulletin (Woods Hole) 121:57–68.

Bayne, C. J. 1983. Molluscan immunobiology. Pages 407–486 *in* A. S. M. Saleuddin and K. M. Wilbur, editors. The Mollusca, volume 5. Academic Press, New York.

Bayne, C. J., M. N. Moore, T. H. Carefoot, and R. J. Thompson. 1979. Hemolymph functions in *Mytilus californianus*: the cytochemistry of hemocytes and their responses to foreign implants and hemolymph factors in phagocytosis. Journal of Invertebrate Pathology 34:1–20.

Bubel, A., M. N. Moore, and D. Lowe. 1977. Cellular responses to shell damage in *Mytilus edulis* L. Journal of Experimental Marine Biology and Ecology 30:1–27.

Cheng, T. C. 1966. The coracidium of the cestode *Tylocephalum* and the migration and fate of this parasite in the American oyster, *Crassostrea virginica*. Transactions of the American Microscopical Society 85:256–255.

Cheng, T. C. 1967. Marine molluscs as hosts for symbioses: with a review of known parasites of commercially important species. Advances in Marine Biology 5:1–424.

Cheng, T. C. 1976. Aspects of substrate utilization and energy requirement during molluscan phagocytosis. Journal of Invertebrate Pathology 27:263–268.

Cheng, T. C. 1981. Bivalves. Pages 233–300 *in* N. A. Ratcliffe and A. F. Rowley, editors. Invertebrate blood cells, volume 1. Academic Press, New York.

Cheng, T. C. 1983. The role of lysosomes in molluscan inflammation. American Zoologist 23:129–144.

Cheng, T. C. 1985. Evidence for molecular specificities involved in molluscan inflammation. Pages 129–142 *in* T. C. Cheng, editor. Comparative pathobiology, volume 8. Plenum, New York.

Cheng, T. C., and R. W. Burton. 1965. Relationships between *Bucephalus* sp. and *Crassostrea virginica*: histopathology and sites of infection. Chesapeake Science 6:3–16.

Cheng, T. C., and A. Cali. 1974. An electron microscope study of the fate of bacteria phagocytosed by granulocytes of *Crassostrea virginica*. Contemporary Topics in Immunobiology 4:25–35.

Cheng, T. C., A. Cali, and D. A. Foley. 1974. Cellular reaction in marine pelecypods as a factor influencing endosymbiosis. Pages 61–91 *in* W. B. Vernberg, editor. Symbiosis in the sea. University of South Carolina Press, Columbia.

Cheng, T. C., and D. A. Foley. 1975. Hemolymph cells of the bivalve mollusc *Mercenaria mercenaria*: an electron microscopical study. Journal of Invertebrate Pathology 26:341–351.

Cheng, T. C., and K. H. Howland. 1979. Chemotactic attraction between hemocytes of the oyster, *Crassostrea virginica*, and bacteria. Journal of Invertebrate Pathology 33:204–210.

Cheng, T. C., J. W. Huang, H. Karadogan, L. R. Renwrantz, and T. P. Yoshino. 1980. Separation of oyster hemocytes by density gradient centrifugation and identification of their surface receptors. Journal of Invertebrate Pathology 36:35–40.

Cheng, T. C., and E. Rifkin. 1968. The occurrence and resorption of *Tylocephalum* metacestodes in the clam *Tapes semidecussata*. Journal of Invertebrate Pathology 10:65–69.

Cheng, T. C., and E. Rifkin. 1970. Cellular reactions in marine molluscs in response to helminth parasitism. American Fisheries Society Special Publication 5: 443–496.

Cheng, T. C., and G. E. Rodrick. 1975. Lysosomal and other enzymes in the hemolymph of *Crassostrea virginica* and *Mercenaria mercenaria*. Comparative Biochemistry and Physiology 52:443–447.

Cheng, T. C., G. E. Rodrick, D. A. Foley, and S. A. Koehler. 1975. Release of lysozyme from hemolymph cells of *Mercenaria mercenaria* during phagocytosis. Journal of Invertebrate Pathology 25: 261–265.

Cheng, T. C., and B. M. Rudo. 1976. Distribution of glycogen resulting from degradation of ^{14}C-labelled bacteria in the American oyster, *Crassostrea virginica*. Journal of Invertebrate Pathology 27:259–262.

Damian, R. T. 1964. Molecular mimicry: antigen sharing by parasite and host and its consequences. American Naturalist 98:129–149.

Des Voigne, D. M., and A. K. Sparks. 1968. The process of wound healing in the Pacific oyster *Crassostrea gigas* (Thunberg). Journal of Invertebrate Pathology 12:23–65.

Des Voigne, D. M., and A. K. Sparks. 1969. The reaction of the Pacific oyster, *Crassostrea gigas*, to homologous tissue implants. Journal of Invertebrate Pathology 14:293–300.

Dikkeboom, R., J. M. G. H. Tijnagel, E. C. Mulder, and W. P. W. Van der Knaap. 1987. Hemocytes of the pond snail *Lymnaea stagnalis* generate reactive forms of oxygen. Journal of Invertebrate Pathology 49:321–331.

Drew, G. H. 1910. Some points in the physiology of lamellibranch blood corpuscles. Quarterly Journal of Microscopical Science 54:605–623.

Eble, A. F., and M. R. Tripp. 1969. Oyster leucocytes in tissue culture: a functional study. Proceedings National Shellfisheries Association 59:5.

Edelman, G. M. 1976. Surface modulation in cell recognition and cell growth. Science (Washington, D.C.) 192:218–226.

Farley, C. A. 1968. *Minchinia nelsoni* (Haplosporidia) disease syndrome in the American osyter, *Crassostrea virginica*. Journal of Protozoology 15:585–599.

Feng, J. S. 1966. The fate of a virus, *Staphylococcus aureus* phage 80, injected into the oyster, *Crassostrea virginica*. Journal of Invertebrate Pathology 8: 496–504.

Feng, S. Y. 1958. Observations on the distribution and elimination of spores of *Nematopsis ostrearum* in oysters. Proceedings National Shellfisheries Association 48:162–173.

Feng, S. Y. 1965a. Heart rate and leucocyte circulation in *Crassostrea virginica* (Gmelin). Biological Bulletin (Woods Hole) 128:198–210.

Feng, S. Y. 1965b. Pinocytosis of proteins by oyster leucocytes. Biological Bulletin (Woods Hole) 129: 95–105.

Feng, S. Y. 1966. Experimental bacterial infections in the oyster *Crassostrea virginica*. Journal of Invertebrate Pathology 8:505–511.

Feng, S. Y. 1967. Responses of molluscs to foreign bodies, with special reference to the oyster. Federation Proceedings 26:1685–1692.

Feng, S. Y., and J. L. Barja. 1987. Summary of cellular defense mechanisms of oysters and mussels. Pages 29–41 in W. S. Fisher and A. J. Figueras, editors. Marine bivalve pathology. University of Maryland Sea Grant Publication UM-SG-TS-87-02, College Park.

Feng, S. Y., and W. J. Canzonier. 1970. Humoral responses in the American oyster (Crassostrea virginica) infected with Bucephalus sp. and Minchinia nelsoni. American Fisheries Society Special Publication 5:497–510.

Feng, S. Y., and J. S. Feng. 1974. The effect of temperature on cellular reactions of Crassostrea virginica to the injection of avian erythrocytes. Journal of Invertebrate Pathology 23:22–37.

Feng, S. Y., J. S. Feng, C. N. Burke, and L. H. Khairallah. 1971. Light and electron microscopy of the leucocytes of Crassostrea virginica (Mollusca: Pelecypoda). Zeitschrift für Zellforschung und mikroskopische Anatomie 120:222–245.

Feng, S. Y., J. S. Feng, and T. Yamasu. 1977. Roles of Mytilus coruscus and Crassostrea gigas blood cells in defense and nutrition. Pages 31–67 in L. A. Bulla and T. C. Cheng, editors. Comparative pathobiology, volume 3. Plenum, New York.

Feng, S. Y., and L. A. Stauber. 1968. Experimental hexamitiasis in the oyster, Crassostrea virginica. Journal of Invertebrate Pathology 10:94–110.

Fisher, W. S. 1986. Structure and function of oyster hemocytes. Pages 25–35 in M. Brehelin, editor. Immunity in invertebrates. Springer-Verlag, Berlin.

Foley, D. A., and T. C. Cheng. 1972. Interaction of molluscs and foreign substances: the morphology and behavior of hemolymph cells of the oyster, Crassostrea virginica, in vitro. Journal of Invertebrate Pathology 19:383–394.

Foley, D. A., and T. C. Cheng. 1974. Morphology, hematologic parameters, and behavior of hemolymph cells of the quahaug clam, Mercenaria mercenaria. Biological Bulletin (Woods Hole) 146:343–356.

Foley, D. A., and T. C. Cheng. 1977. Degranulation and other changes of molluscan granulocytes associated with phagocytosis. Journal of Invertebrate Pathology 29:321–325.

Ford, S. E. 1988. Host–parasite interactions in eastern oysters selected for resistance to Haplosporidium nelsoni (MSX) disease: survival mechanisms against a natural pathogen. American Fisheries Society Special Publication 18:206–224.

Ford, S. E., and H. H. Haskin. 1982. History and epizootiology of Haplosporidium nelsoni (MSX), an oyster pathogen in Delaware Bay, 1957–1980. Journal of Invertebrate Pathology 40:118–141.

Fries, C. R. 1984. Protein hemolymph factors and their roles in invertebrate defense mechanisms: a review. Pages 49–109 in T. C. Cheng, editor. Comparative pathobiology, volume 6. Plenum, New York.

Fries, C. R., and M. R. Tripp. 1970. Uptake of viral particles by oyster leucocytes in vitro. Journal of Invertebrate Pathology 15:136–137.

Hardy, S. W., T. C. Fletcher, and L. M. Gerrie. 1976. Factors in haemolymph of the mussel, Mytilus edulis L., of possible significance as defense mechanisms. Biochemical Society Transactions 4:473–475.

Hardy, S. W., T. C. Fletcher, and J. A. Olafsen. 1977a. Aspects of cellular and humoral defense mechanisms in the Pacific oyster, Crassostrea gigas. Pages 59–66 in J. B. Solomon and J. D. Horton, editors. Developmental immunobiology. Elsevier, New York.

Hardy, S. W., P. T. Grant, and T. C. Fletcher. 1977b. A haemagglutinin in the tissue fluid of the Pacific oyster, Crassostrea gigas, with specificity for sialic acid residues in glycoproteins. Experientia (Basel) 33:767–769.

Hillman, R. E. 1963. An observation of the occurrence of mitosis in regenerating mantle epithelium of the eastern oyster, Crassostrea virginica. Chesapeake Science 4:172–174.

Hirsch, J. C. 1962. Cinematographic observations of granule lysis in polymorphonuclear leucocytes during phagocytosis. Journal of Experimental Medicine 116:827–834.

Hirsch, J. C., and Z. A. Cohn. 1960. Degranulation of polymorphonuclear leucocytes following phagocytosis of microorganisms. Journal of Experimental Medicine 112:1005–1014.

Hopkins, S. H. 1954. The American species of trematode confused with Bucephalus (Bucephalopsis) haimeanus. Journal of Parasitology 44:353–370.

Howland, K. H., and T. C. Cheng. 1982. Identification of bacterial chemoattractants for oyster (Crassostrea virginica) hemocytes. Journal of Invertebrate Pathology 39:123–132.

Kanungo, K. 1982. In vitro studies on the effects of cell-free coelomic fluid, calcium, and/or magnesium on clumping of coelomocytes of the sea star Asteria forbesi (Echinodermata: Asteroidea). Biological Bulletin (Woods Hole) 163:438–452.

Kenney, D. M., F. A. Belamarich, and D. Shepro. 1972. Aggregation of horseshoe crab (Limulus polyphemus) amebocytes and reversible inhibition of aggregation by EDTA. Biological Bulletin (Woods Hole) 143:548–567.

Kern, F. G. 1976. Minchinia nelsoni (MSX) disease of the American oyster. U.S. National Marine Fisheries Service Marine Fisheries Review 38(10):22–24.

Kinne, O. 1980. Diseases of marine animals: general aspects. Pages 13–73 in O. Kinne, editor. Diseases of marine animals, volume 1. Wiley, New York,

Lee, D. L. 1966. The structure and composition of the helminth cuticle. Advances in Parasitology 5:187–254.

Mackin, J. G., and H. Loesch. 1955. A haplosporidian hyperparasite of oysters. Proceedings National Shellfisheries Association 45:182–183.

Mackin, J. G., H. M. Owen, and A. Collier. 1950. Preliminary note on the occurrence of a new protistan parasite, Dermocystidium marinum n. sp., in

Crassostrea virginica (Gmelin). Bulletin of Marine Science of the Gulf and Caribbean 1:266–277.

McDade, J. E., and M. R. Tripp. 1967. Lysozyme in the hemolymph of the oyster, *Crassostrea virginica*. *Journal of Invertebrate Pathology 9:531–535*.

Metchnikoff, E. 1884. Uber eine Sprosspilzkrankheit der Daphnien: Beitrag zur Lehre über den Kampf der Phagocyten gegen Krankheitserreger. Virchows Archiv für pathologische Anatomie and Physiologie und für klinische Medizin 96:178–193.

Metchnikoff, E. 1891. Lectures on the comparative pathology of inflammation. Dover, New York.

Mohandas, A., T. C. Cheng, and J. B. Cheng. 1985. Mechanism of lysosomal enzyme release from *Mercenaria mercenaria* granulocytes: a scanning electron microscope study. Journal of Invertebrate Pathology 46:189–197.

Moore, M. N., and D. M. Lowe. 1977. The cytology and cytochemistry of the hemocytes of *Mytilus edulis* and their responses to experimentally injected carbon particles. Journal of Invertebrate Pathology 29:18–30.

Nakamura, M., K. Mori, S. Inocka, and T. Nomura. 1985. In vitro production of hydrogen peroxide by the amoebocytes of the scallop, *Patinopecten yessoensis* (Jay). Developmental and Comparative Immunology 9:407–417.

Noble, P. B. 1970. Coelomocyte aggregation in *Cucumaria frondosa*: effect of ethylenediaminetetracetate, adenosine, and adenosine nucleotides. Biological Bulletin (Woods Hole) 139:549–556.

Pauley, G. B., and A. K. Sparks. 1967. Observations of experimental wound repair in the adductor muscle and the leydig cells of the oyster *Crassostrea gigas*. Journal of Invertebrate Pathology 9:298–309.

Perkins, F. O., and R. W. Menzel. 1967. Ultrastructure of sporulation in the oyster pathogen *Dermocystidium marinum*. Journal of Invertebrate Pathology 9: 205–229.

Pichot, Y., M. Comps, G. Tigé, H. Grizel, and M. A. Rabouin. 1980. Recherche sur *Bonami ostreae* gen. n., sp. n., parasite nouveau de l'huître plate *Ostrea edulis* L. Revue des Travaux de l'Institut des Pêches Maritimes 43(1):131–140.

Prytherch, H. F. 1940. The life cycle and morphology of *Nematopsis ostrearum* sp. nov., a gregarine parasite of the mud crab and oyster. Journal of Morphology 66:39–65.

Rasmussen, L. P. D., E. Hage, and O. Karlog. 1985. An electron microscope study of the circulating leucocytes of the marine mussel, *Mytilus edulis*. Journal of Invertebrate Pathology 45:158–167.

Ray, S. M. 1954. Biological studies of *Dermocystidium*. Rice Institute Pamphlet 41 (special issue). (The Rice Institute, Houston, Texas.)

Read, C. P. 1958. Status of behavioral and physiological "resistance." Rice Institute Pamphlet 45:36–54. (The Rice Institute, Houston, Texas.)

Renwrantz, L., and A. Stahmer. 1983. Opsonizing properties of an isolated hemolymph agglutinin and demonstration of lectin-like recognition molecules at the surface of hemocytes from *Mytilus edulis*.

Journal of Comparative Physiology 149:535–546.

Renwrantz, L., T. Yoshino, T. Cheng, and K. Auld. 1979. Size determination of hemocytes from the American oyster, *Crassostrea virginica*, and the description of a phagocytosis mechanism. Zoologische Jahrbücher Abteilung für allgemeine Zoologie und Physiologie der Tiere 83:1–12.

Rifkin, E., and T. C. Cheng. 1968. The origin, structure, and histochemical characterization of encapsulating cysts in the oyster *Crassostrea virginica* parasitized by the cestode *Tylocephalum* sp. Journal of Invertebrate Pathology 10:54–64.

Rodrick, G. E., and T. C. Cheng. 1974. Kinetic properties of lysozyme from *Crassostrea virginica* hemolymph. Journal of Invertebrate Pathology 24:41–48.

Ruddell, C. L. 1971a. Elucidation of the nature and function of the granular oyster amoebocytes through histochemical studies of normal and traumatized oyster tissues. Histochemie 26:98–112.

Ruddell, C. L. 1971b. The fine structure of oyster agranular amoebocytes from regenerating mantle wounds in the Pacific oyster, *Crassostrea gigas*. Journal of Invertebrate Pathology 18:260–268.

Ruddell, C. L. 1971c. The fine structure of the granular amoebocytes of the Pacific oyster, *Crassostrea virginica*. Journal of Invertebrate Pathology 18: 269–275.

Sindermann, C. J. 1970. Principal diseases of marine fish and shellfish. Academic Press, New York.

Singer, S. J. 1974. Molecular biology of cellular membranes with applications to immunology. Advances in Immunology 19:1–66.

Smyth, J. D. 1973. Some interface phenomena in parasitic protozoa and platyhelminths. Canadian Journal of Zoology 51:367–377.

Sparks, A. K. 1963. Infection of *Crassostrea virginica* (Gmelin) from Hawaii with a larval tapeworm, *Tylocephalum*. Journal of Insect Pathology 5:284–288.

Sparks, A. K. 1972. Invertebrate pathology: noncommunicable diseases. Academic Press, New York.

Sparks, A. K. 1985. Synopsis of invertebrate pathology. Elsevier, New York.

Sparks, A. K., and K. K. Chew. 1966. Gross infestation of the littleneck clam, *Venerupis staminea*, with a larval cestode (*Echeneibothrium* sp.). Journal of Invertebrate Pathology 8:413–416.

Sparks, A. K., and J. F. Morado. 1988. Inflammation and wound repair in bivalve molluscs. American Fisheries Society Special Publication 18:139–152.

Sprague, V., and P. E. Orr, Jr. 1955. *Nematopsis ostrearum* and *N. prytherchi* (Eugregarinina: Porosporidae) with special reference to the host–parasite relations. Journal of Parasitology 40:89–104.

Stauber, L. A. 1950. The fate of India ink injected intracardially into the oyster, *Ostrea virginica* Gmelin. Biological Bulletin (Woods Hole) 98:227–241.

Stauber, L. A. 1961. Immunity in invertebrates, with special reference to the oyster. Proceedings National Shellfisheries Association 50:7–20.

Stunkard, H. W., and J. R. Uzmann. 1959. The life cycle of the digenetic trematode, *Proctoeces maculatus* (Looss, 1901) Odhener, 1911 [syn. *P. sub-*

tenuis (Linton, 1907) Hanson, 1950], and description of *Cercaria adranocerca* n. sp. Biological Bulletin (Woods Hole) 116:184–193.

Takatsuki, S. 1934. On the nature and functions of the amoebocytes of *O. edulis*. Quarterly Journal of Microscopical Science 76:374–431.

Tennent, D. H. 1906. A study of the life-history of *Bucephalus haimeanus*, a parasite of the oyster. Quarterly Journal of Microscopical Science 49:536–690.

Tripp, M. R. 1958. Disposal by the oyster of intracardially injected red blood cells of vertebrates. Proceedings National Shellfisheries Association 48: 143–147.

Tripp, M. R. 1960. Mechanisms of removal of injected microorganisms from the American oyster, *Crassostrea virginica* (Gmelin). Biological Bulletin (Woods Hole) 119:273–282.

Tripp, M. R. 1961. The fate of foreign materials experimentally introduced into the snail *Australorbis glabratus*. Journal of Parasitology 47:745–751.

Tripp, M. R. 1963. Cellular responses of mollusks. Annals of the New York Academy of Sciences 113: 467–474.

Tripp, M. R. 1966. Hemagglutinin in the blood of the oyster. Journal of Invertebrate Pathology 8:478–484.

Tripp, M. R. 1969. Mechanisms and general principles of invertebrate immunity. Pages 111–128 *in* G. J. Jackson, R. Herman, and I. Singer, editors. Immunity to parasitic animals, volume 1. Appleton-Century-Crofts, New York.

Tripp, M. R. 1970. Defense mechanisms of mollusks. Journal of the Reticuloendothelial Society 7:173–182.

Tripp, M. R. 1975. Humoral factors and molluscan immunity. Pages 201–223 *in* K. Maramorosch and R. E. Shope, editors. Invertebrate immunity. Academic Press, New York.

Tripp, M. R., L. A. Bisignani, and M. T. Kenny. 1966. Oyster amoebocytes in vitro. Journal of Invertebrate Pathology 8:137–140.

Tripp, M. R., and V. E. Kent. 1967. Studies on oyster cellular immunity. In Vitro (Rockville) 3:129–135.

Tripp, M. R., and R. M. Turner 1978. Effects of the trematode *Proctoeces maculatus* on the mussel *Mytilus edulis*. Pages 73–84 *in* L. A. Bulla, and T. C. Cheng, editors. Comparative immunobiology: invertebrate models for biomedical research, volume 4. Plenum, New York.

Uzmann, J. R. 1953. *Cercaria milfordensis* nov. sp. a microcerous trematode larva from the marine bivalve, *Mytilus edulis* L., with special reference to its effect on the host. Journal of Parasitology 39: 445–451.

Vasta, G. R., J. T. Sullivan, T. C. Cheng, J. J. Marchalonis, and G. W. Warr. 1982. A cell-associated lectin of the oyster hemocyte. Journal of Invertebrate Pathology 40:367–377.

Wittke, M., and L. Renwrantz. 1984. Quantification of cytotoxic hemocytes of *Mytilus edulis* using a cytotoxic assay in agar. Journal of Invertebrate Pathology 43:248–253.

Wittke, M., and L. Renwrantz. 1985. On the capability of bivalve and gastropod hemocytes to secrete cytotoxic molecules. Journal of Invertebrate Pathology 46:209–210.

Wright, A., and S. Douglas. 1903. An experimental investigation of the role of the blood fluids in connection with phagocytosis. Proceedings of the Royal Society of London B, Biological Sciences 72: 357–370.

Yonge, C. M. 1926. Structure and physiology of the organs of feeding and digestion in *Ostrea edulis*. Journal of the Marine Biological Association of the United Kingdom 14:295–388.

Yoshino, T. P., and T. C. Cheng. 1976. Fine structural localization of acid phosphatase in granulocytes of the pelecypod *Mercenaria mercenaria*. Transactions of the American Microscopical Society 95: 215–220.

Yoshino, T. P., L. R. Renwrantz, and T. C. Cheng. 1979. Binding and redistribution of surface membrane receptors for concanavalin A on oyster hemocytes. Journal of Experimental Zoology 207: 439–450.

American Fisheries Society Special Publication 18:169–177, 1988

Bivalve Hemocyte Morphology

MICHEL AUFFRET

Laboratory of Zoology, University of Bretagne Occidentale
Avenue Le Gorgeu, 29287 Brest Cedex, France

Abstract.—Blood cells (or hemocytes) of marine bivalve molluscs play a prominent role in the internal defense of the animals from pathogens as well as in other biological and physiological functions. This paper poses questions of present interest concerning hemocyte morphology, occurrence of cell types, and hemocytogenesis. In hemocyte nomenclature, a working scheme has recently emerged in which mature hemocytes are either granular hemocytes (granulocytes) or nongranular hemocytes (hyalinocytes). There are conspicuous similarities and differences among bivalve species in the relative numbers and fine morphology of hemocytes, but the two general classes of hemocytes appear to be common features of bivalve families. The aim of morphological and cytochemical studies is to reach a better understanding of the essential functions of hemocytes.

The term hemocyte, proposed by Farley (1968), is now broadly accepted to designate the blood cells of bivalve molluscs. Molluscs possess an open circulatory system, as the hemolymph is believed to have a homogenous composition with hemocytes distributed throughout the vascular system and tissues. Bivalve hemocytes are involved in a wide range of physiological functions, including digestion and excretion (Narain 1973). It has become increasingly evident, however, that hemocytes are also predominantly responsible for internal defense against pathogens (Bayne et al. 1980; Feng 1988, this volume). A better understanding of the morphology and functions of hemocytes will help us understand the internal defense mechanisms and their role in the diseases that affect bivalve species. In addition, there are disease conditions that act directly on the hemocytes. These include parasitic diseases by protozoans such as *Perkinsus marinus* (Andrews 1988, this volume) and *Bonamia ostreae* (Grizel et al. 1988, this volume), as well as hemocytosarcomas (Peters 1988, this volume).

This paper examines some of the questions that presently concern the morphology of hemocytes, the occurrence of cell types, and cytogenetic lines. Also, unpublished results from a comparative morphological study of hemocytes from several European bivalve species are presented here as a vehicle for discussion. Reviews of the literature on bivalve hemocytes by Narain (1973), Cheng (1981), and Fisher (1986) are available for background information.

Hemocyte Types in Bivalve Molluscs

In spite of numerous studies published during the last 15 years, there is no widely accepted nomenclature for bivalve hemocyte types. Although relatively few species have been investigated, attempts have been made to establish a general comprehensive scheme that could be applied to a variety of species. Such a scheme would provide a useful tool for future research. In the past, investigators have based their classification of hemocytes on the nature of cytoplasmic organelles (Feng et al. 1971), the staining affinity and morphology of the nucleus (Feng et al. 1977), or cytogenesis and cell function (Moore and Lowe 1977). Such criteria, however, have not led to an acceptable comprehensive model. Cheng (1981) pointed out that classification based purely on morphological differences, especially those visible by light microscopy, has been unsatisfactory. Transmission electron microscopy, now broadly applied to hemocyte research, has been very valuable and eventually will yield objective results that can be used to compare observations by different authors on a variety of species.

Despite his reservations, Cheng (1981) presented a morphological scheme based on two cell types: (1) hyalinocytes, which are cells with nearly clear cytoplasm containing few or no granules; and (2) granular hemocytes (GH), or granulocytes, which are cells with well-developed cytosplasm containing granules that range from scarce to numerous. "Agranular" hemocytes are from the same hematopoietic line as the granular hemocytes but have no visible granules. "Nongranular" describes cytoplasm with few or no granules, regardless of cell lineage. In the Ostreidae, for example, nongranular hemocytes include both hyalinocytes and agranular hemocytes. The intent of this scheme was to place all distinguishable hemocyte types into the two classes and

TABLE 1.—Hemocyte types[a] in the original nomenclature of the authors (1–8 in parentheses)[b] in four families of bivalve molluscs. Cell types are arrayed horizontally according to their ultrastructural similarities.

Ostreidae							Veneridae		Mytilidae Blue mussel		Pectinidae	
Edible oyster (1)	Pacific oyster			Eastern oyster			Japanese littleneck (1)	Northern quahog (5)	(7)	(8)	*Chlamys varia* (1)	*Pecten maximus* (1)
	(1)	(2)	(3)	(4)	(5)	(6)						
Nongranular cell types[a]												
Hy	Hy	aA	aA	aHeI	Hy	HeI	HyI	Hy	Ly	aLe	HeI	HeI
		aA	aA	aHeII	Gr	HeII						
aHe	aHe	aA	aA	aHeIII	Hy							
							HyII				HeII	HeII
											HeIII	HeIII
										Ma		
Granular cell types[a]												
gHe	bgHe	bgHe	gHe	gHe	Gr	HeIV						
	cgHe	cgHe	gHe									
							gHe	Gr	Gr	gLe		

[a]Authors terms have been abbreviated by the following conventions.
   Cell types: A = amoebocyte; Gr = granulocyte; He = hemocyte; Hy = hyalinocyte; Le = leukocyte; Ly = lymphocyte; Ma = macrophage.
   Prefixes: a = agranular; b = basophilic; c = acidophilic; g = granular.
   Suffixes: I, . . ., IV = type I, . . ., IV.
[b]References: (1) Auffret (unpublished)   (4) Feng et al. (1971)   (7) Moore and Lowe (1977)
                  (2) Ruddell (1971a, 1971b)   (5) Cheng (1975)   (8) Rasmussen et al. (1985)
                  (3) Feng et al. (1977)   (6) Hawkins and Howse (1982)

describe any variants as cytogenetic or physiological stages. This scheme is relatively successful for most, but not all, of the bivalves that have been investigated. Recent comparative studies indicate that there may be notable exceptions.

Cytological and ultrastructural criteria were used to compare the hemocytes of five bivalve species (from the families Ostreidae, Veneridae, and Pectinidae) and to classify them into 2–4 morphological types for each species (my unpublished data). These results, compared with descriptions of hemocytes from other species, reveal general similarities among species and a relative homogeneity in the ultrastructure of hemocytes, especially within the Ostreidae (Table 1). Some differences exist between species, usually the addition of one or more types such as the acidophilic GH of Pacific oyster *Crassostrea gigas* and the type-II hyalinocytes of Japanese littleneck *Tapes philippinarum*.

Nongranular hemocytes were more widely distributed than granulocytes, each species possessing one or more types. One of these (Table 1, first row) has been variously designated as hyalinocyte, type-I hyalinocyte, type-I hemocyte, type-I agranular hemocyte, or lymphocyte. Such cells have been reported to be phagocytic by Ruddell (1971b) and Reade and Reade (1976). Observed by transmission electron microscopy, this lymphocyte- or monocytelike cell has a large nucleus with a rather irregular shape and relatively reduced cytoplasm containing endoplasmic reticulum but few granules. This cell type may be called hyalinocyte, although it differs slightly in morphological features among the species. Other nongranular types appear more variable, and cell typing was more difficult among the species (Table 1). In the Ostreidae, the other nongranular cell was the agranular hemocyte (my unpublished data). This cell was similar to the GH in general morphology and nuclear shape and size, but the cytoplasm lacked granules, both by light and electron microscopy. It was also different from a degranulated GH. The relationship of agranular hemocytes to GHs is discussed in a subsequent section.

There is greater agreement among the types of GH because these cells are more easily characterized by their numerous cytoplasmic granules. In the Ostreidae, there is one type of GH that is morphologically similar throughout the family, except Pacific oyster possesses an additional type (Table 1). The Veneridae and Mytilidae also have a single GH that is morphologically similar within these two families.

The Pectinidae differ from other groups by the absence of GHs. Surprisingly, their hemocytes resemble blood cells of gastropod molluscs in

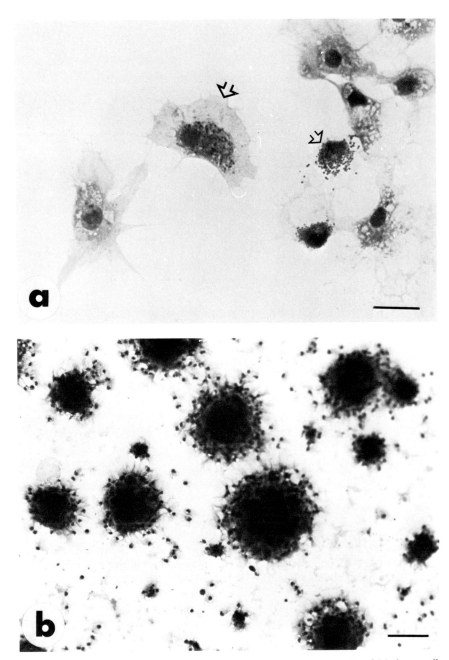

FIGURE 1.—Hemolymph smears from (**a**) Pacific oyster and (**b**) *Pecten maximus*. Panel (**a**) shows cells spreading on a glass slide with basophilic granules in basophilic granular hemocyte (GH) endoplasm (large arrow) and acidophilic granules in acidophilic GH endoplasm (small arrow; bar = 10 μm). Panel (**b**) shows hemocyte aggregates of various sizes (May Grünwald–Giemsa stain; bar = 50 μm).

terms of cell morphology and classification (as described by Joky et al. 1983). It is also interesting that the type-II hyalinocyte from Japanese little-neck and the type-III hemocyte of *Chlamys varia* so closely resemble one other ultrastructurally.

This could represent a cytogenetic identity and possibly similar cellular functions.

Thus, there is diversity in bivalve hemocyte types, both among families and among species of the same family. Generally, however, species

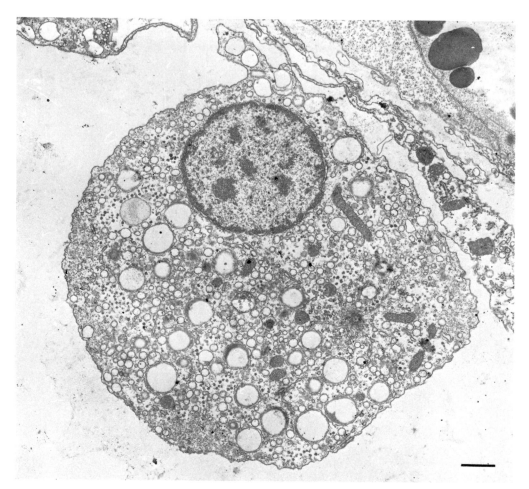

FIGURE 2.—Transmission electron micrograph of a basophilic granular hemocyte of Pacific oyster in the lumen of a hemolymph sinus (bar = 1 μm).

appear to have both granular and nongranular hemocytes. The possibility of "intermediate" cell types is considered in a subsequent section.

Elements of Hemocyte Morphology

When examined in histological section, hemocytes in tissue sinuses are spherical; when they are found in tissues, they are deformed by neighboring structures. Deformation commonly occurs during diapedesis, a phenomenon often associated with inflammation. Some hemocytes spontaneously produce thick cytoplasmic projections (type-II hyalinocytes of Japanese littleneck) or thin projections (granular and agranular hemocytes of the Ostreidae), which are believed to be involved in endocytosis (Mohandas 1985). Hemocytes in vitro have a distinctive behavior. In the Ostreidae, they adhere quickly to glass slides, and

migrate by amoeboid movements (Figure 1a) and form aggregates in a second step. In other species, the hemocytes form aggregates as soon as hemolymph is withdrawn from an animal (or there may be some formed in the circulating hemolymph) and begin to migrate away when they are spread on the glass (Figure 1b). It is believed by some that the ability of hemocytes to spread and locomote in vitro indicates their abilities and capacity in internal defense (Fisher 1988, this volume).

When examined by transmission electron microscopy, hemocytes of bivalve molluscs are typically eukaryotic with mitochondria, endoplasmic reticulum, Golgi apparatus, ribosomes, and various inclusion bodies (Figure 2). The nucleus of a hyalinocyte is large and irregular in shape, whereas that of the GH is small and round (or

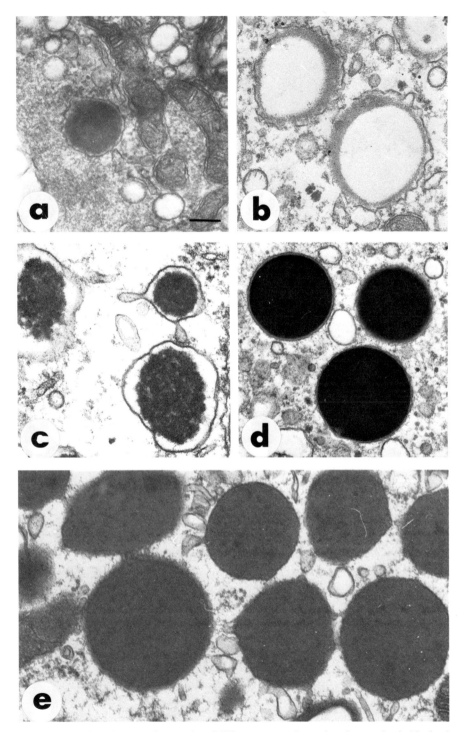

FIGURE 3.—Transmission electron micrographs of different types of cytoplasmic granules in bivalve hemocytes (bar = 1 μm). (**a**) Type-I hemocyte of *Pecten maximus*. (**b**) Basophilic granular hemocyte of Pacific oyster. (**c**) Granular hemocyte of Japanese littleneck. (**d**) Acidophilic granular hemocyte of Pacific oyster. (**e**) Granular hemocyte of Japanese littleneck.

COMPARTMENTS

```
stem cell      proliferating      maturing        functional       cell loss

                                                      basophilic
                                         granular    granulocytes          senescence
                                         hyalinocytes
                                                      acidophilic
                                                      granulocytes

( ¯ )          ?    ( ¯ )
 \ /          ⇌     \ /                                fibroblasts       defence
leucoblast         hyalinocytes                                         reactions
                                         intermediates
                                                      myoblasts

                                         intermediates ──► pigment cells      diapedesis
```

FIGURE 4.—General model of hemocyte renewal in bivalve molluscs according to Mix (1976).

oval). Chromatin is abundant and sometimes forms clumps. Hyalinocytes possess one or two large nucleoli, which are less prominent in the agranular hemocytes and in the GH. Mitochondria are often quite numerous and may form a complex, called the juxtanuclear body, whose function is not yet known. This structure was first observed by Hawkins and Howse (1982) in hyalinocytes of eastern oyster *Crassostrea virginica*, and I have seen it frequently in hyalinocytes from the Ostreidae and in the type-I hemocytes from Pectinidae (unpublished data).

Almost all hemocyte types that have been described possess cytoplasmic granules; they are scarce in hyalinocytes but numerous in GHs where they might be considered specific organelles. By light microscopy, the granules are easily demonstrated on smears with appropriate cytological staining. The staining affinity of the granules sometimes allows specific cell types to be identified, but this is not generally possible over all the bivalves, nor should it be considered a basis for functional classification until research can verify individual cases. At least three different granule-staining patterns have been described in cell populations treated with standard cytological stains.

(1) Some granules have a single staining characteristic, such as those in edible oyster *Ostrea edulis* (Takatsuki 1934; Franc 1975; Auffret, unpublished data) and in Japanese littleneck (Auffret, unpublished data). These most likely constitute a single cell type.

(2) Other granules have several staining characteristics, such as those in eastern oyster with basophilic and neutrophilic granules (Feng et al. 1971) or basophilic and acidophilic granules (Cheng 1975), those in the softshell *Mya arenaria* with basophilic and neutrophilc granules (Huffman and Tripp 1982), and those in the blue mussel *Mytilus edulis* with neutrophilic and acidophilic granules (Moore and Lowe 1977). As emphasized by Cheng (1975), these variations probably reflect different physiological stages and probably do not represent different GH subpopulations.

(3) Still other granules have a specific staining property for each morphological cell type. In Pacific oyster, for example, the basophilic and acidophilic granules occur in two distinct cell types, the basophilic and acidophilic GHs (Auffret, unpublished data).

Transmission electron microscopy has revealed some basic morphological information about the granules. The most striking aspect is that the granulations in hyalinocytes appear sufficiently similar to be classified in the same category (see below), whereas the granules in GHs exhibit such a varied morphology that they define at least four additional categories.

(1) Small (0.2–0.4 μm), round, electron-dense vesicles (Figure 3a) occur in small numbers in most of the nongranular hemocytes, for example, in hyalinocytes of edible oyster, in type I and II hemocytes of Japanese littleneck and of *Pecten maximus*, and in type-I of hemocytes of *Chlamys varia* (my unpublished data).

(2) Some granules have electron-dense contents (matrix) condensed at the periphery, leaving a lucid core (Figure 3b). These are variable in size (0.5–1.0 μm) and occur in edible oyster GHs (my

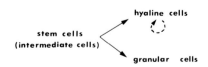

FIGURE 5.—Model for hemocyte maturation in bivalve molluscs according to Balouet and Poder (1979).

granuloblast → progranulocyte → granulocyte I → granulocyte II

hyalinoblast → prohyalinocyte → hyalinocyte

FIGURE 6.—Simplified scheme of the model for hemocyte maturation in bivalve molluscs according to Cheng (1981).

unpublished data), in Pacific oyster basophilic GHs (Ruddell 1971c), and in eastern oyster GHs (Feng et al. 1971).

(3) Some round or elongated granules have the matrix condensed in the center, sometimes leaving a lucid periphery (Figure 3c). These are also variable in size (0.3–0.6 μm) and occur in most Japanese littleneck GHs (my unpublished data) and in GHs of the northern quahog *Mercenaria mercenaria* (Cheng and Foley 1975).

(4) Round granules (about 0.6 μm) with a homogenous matrix (Figure 3d) occur in Pacific oyster acidophilic GHs (Ruddell 1971c).

(5) Polymorphous granules (Figure 3e) with a homogenous matrix, highly variable in size (0.5–2.0 μm), occur in rare GHs of Japanese littleneck (my unpublished data).

Neither the biochemical nor the functional nature of the GH granules is precisely known in any species, although the granules evidently play a fundamental role in biological processes (Yoshino and Cheng 1976). Using histochemistry, Ruddell (1971c) demonstrated the release of ions from Pacific oyster GHs and different roles for the two GH types. However, the granules are believed to be predominantly involved in mechanisms of intracellular digestion (Cheng 1975; Auffret 1986). Cheng (1975) considered the granules in eastern oyster GHs as secondary lysosomes. However, fusion of hydrolase-containing vesicles and phagosomes in *Mytilus coruscus* was reported by Feng et al. (1977). Undoubtedly, some of the granules belong to the cellular lysosomal system and function in the intracellular degradation of phagocytized material. But even with these insights, I emphasize that the functions of the GH

and the role of the cytoplasmic granules are not well understood. Cytochemical techniques should be employed to complement morphological studies in the search for hemocyte function. Biochemical investigation of the granules and isolation of organelles from cell suspensions may soon be possible with the recent development of techniques in cell molecular biology.

Hemocytogenesis

Research on bivalve mollusc hemocytes has not provided a great deal of information on cytogenesis. A better knowledge of stem cells and maturation processes would help us to understand the phylogenetic relationships between different morphological hemocyte types and to elucidate many of the problems linked to the occurrence of "intermediate" cells that do not fit well the criteria of any single category.

Although there have been several attempts to establish a general model of hemocyte renewal and maturation (e.g., Mix 1976; Balouet and Poder 1979; Cheng 1981), hemocyte histogenesis remains a poorly understood topic. One of the most crucial questions is whether granular cells arise from hyaline cells or represent a distinct histogenetic type. Mix (1976) proposed a general model for a "cell renewal system" that included compartments for hemocytes from the stem cell level to the functional level (Figure 4). Hyalinocytes, whose function is the least clear, were considered undifferentiated cells that would proliferate into other cell types. Balouet and Poder (1979) and Cheng (1981), however, hypothesized two distinct cell lines, hyalinocytes and GHs, each originating from a unique stem cell (Figures

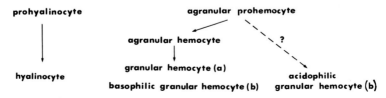

FIGURE 7.—Hypothesis for hemocyte maturation in the Ostreidae, leading to the cells found in (**a**) edible oyster and (**b**) Pacific oyster.

5, 6). Moore and Eble (1977) offered a third scheme, suggesting that different hemocyte types are maturing stages within a single line.

Recent unpublished data on Ostreidae (Figure 7) appeared to support the model of Cheng (1981). In addition to two different stages of maturation, two types of prohemocytes were found, one for hyalinocytes and one for agranular hemocytes. The latter is considered an early stage of GH. This indicates an early cellular differentiation, which supports the hypothesis of two distinct cytogenetic lines. Also, a similar model, based on the parallel evolution of two cell lines (i.e., beyond the consideration of granular or nongranular cells), could be applied to other groups of bivalve molluscs—even to the Pectinidae, which do not possess granular hemocytes.

It is quite likely that morphological studies will not provide satisfactory answers to the questions of hemocytogenesis. Progress more likely will come from the development of new cellular techniques to label and trace hemocytes from genesis and differentiation through maturation and senescence.

Conclusions

A general scheme for the nomenclature of bivalve hematology is needed to more easily describe and identify the functions of blood cells. The present working outline includes two types of hemocytes: (1) nongranular, lymphocytelike cells, or hyalinocytes; and (2) granular cells or granulocytes. In the latter group, there may be several different cell populations or subpopulations.

A major problem in the classification of blood cells is the occurrence of intermediate types, which may be differentiating cells from phylogenetic lines or morphological variations in the same functional cell line. To elucidate these distinctions, it will be necessary to either separate the subpopulations or to label the hemocytes. Separation of oyster hemocytes by density-gradient centrifugation (Cheng et al. 1980) yielded subpopulations that differed in the surface receptors available to selected lectins. Immunolabeling of hemocytes has been accomplished by Yoshino and Granath (1985), who studied the functional heterogeneity of gastropod hemocyte subpopulations. Monoclonal antibody techniques also have been applied to bivalve mollusc hemocytes (Smolowitz and Reinisch 1986), and we can expect that immunologic techniques will be of great help to bivalve hematologists in the near future.

Considering the important role of hemocytes in the internal defense of bivalves, I emphasize that identification of morphological types is essential if we are to learn the functions of these cells. Precise cell types must be identified for in vitro studies of phagocytosis, nonself recognition, chemotaxis, and other phenomena that may account for the ability or inability of bivalves to clear foreign agents.

References

Andrews, J. D. 1988. Epizootiology of the disease caused by the oyster pathogen *Perkinsus marinus* and its effects on the oyster industry. American Fisheries Society Special Publication 18:47–63.

Auffret, M. 1986. Internal defence in bivalve molluscs: ultrastructural observations on the fate of experimentally injected bacteria in *Ostrea edulis* granular hemocytes. Pages 351–356 *in* C. P. Vivares, J.-R. Bonami, and E. Jaspers, editors. Pathology in marine aquaculture. European Aquaculture Society Special Publication 9, Bredene, Belgium.

Balouet, G., and M. Poder. 1979. A proposal for classification of normal and neoplastic types of blood cells in molluscs. Pages 205–208 *in* D. S. Yohn, B. Lapin, and J. Blakeslee, editors. Advances in comparative leukemia research. Elsevier, Amsterdam.

Bayne, C. J., T. Sminia, and W. P. W. Van der Knaap. 1980. Immunological memory: status of molluscan studies. Pages 57–64 *in* M. J. Manning, editor. Phylogeny of immunological memory. Elsevier, Amsterdam.

Cheng, T. C. 1975. Functional morphology and biochemistry of molluscan phagocytes. Annals of the New York Academy of Sciences 266:343–379.

Cheng, T. C. 1981. Bivalves. Pages 231–300 *in* N. A. Ratcliffe and A. F. Rowley, editors. Invertebrate blood cells, volume 2. Academic Press, London.

Cheng, T. C., and D. A. Foley. 1975. Hemolymph cells of the bivalve mollusc *Mercenaria mercenaria*: an electron microscopical study. Journal of Invertebrate Pathology 26:341–351.

Cheng, T. C., J. W. Huang, H. Karadogan, L. R. Renwrantz, and T. P. Yoshino. 1980. Separation of oyster hemocytes by density gradient centrifugation and identification of their surface receptors. Journal of Invertebrate Pathology 36:35–40.

Farley, C. A. 1968. *Minchinia nelsoni* (Haplosporida) disease syndrome in the American oyster, *Crassostrea virginica*. Journal of Protozoology 15:585–599.

Feng, S. Y. 1988. Cellular defense mechanisms of oysters and mussels. American Fisheries Society Special Publication 18:153–168.

Feng, S. Y., J. S. Feng, C. N. Burke, and L. H. Khairallah. 1971. Light and electron microscopy of the hemocytes of *Crassostrea virginica* (Mollusca–Pelecypoda). Zeitschrift für Zellforschung und mikroskopische Anatomie 120:225–243.

Feng, S. Y., J. S. Feng, and T. Yamasu. 1977. Roles of *Mytilus coruscus* and *Crassostrea gigas* blood cells in defense and nutrition. Pages 31–67 *in* L. A. Bulla, Jr., and T. C. Cheng, editors. Comparative pathobiology, volume 3. Plenum, London.

Fisher, W. S. 1986. Structure and function of oyster hemocytes. Pages 25–35 *in* M. Brehelin, editor. Immunity in invertebrates. Springer-Verlag, Berlin.

Fisher, W. S. 1988. Environmental influence on bivalve hemocyte function. American Fisheries Society Special Publication 18:225–237.

Franc, A. 1975. Hyperplasie hémocytaire et lésions chez l'huître plate (*Ostrea edulis* L.). Comptes Rendus Hebdomadaires des Séances de l'Académie des Sciences, Série D, Sciences Naturelles 280:495–497.

Grizel, H., E. Mialhe, D. Chagot, V. Boulo, and E. Bachère. 1988. Bonamiasis: a model study of diseases in marine molluscs. American Fisheries Society Special Publication 18:1–4.

Hawkins, W. E., and H. D. Howse. 1982. Ultrastructure of cardiac hemocytes and related cells in the oyster *Crassostrea virginica*. Transactions of the American Microscopical Society 101:241–252.

Huffman, J. E., and M. R. Tripp. 1982. Cell types and hydrolytic enzymes of soft shell clam (*Mya arenaria*) hemocytes. Journal of Invertebrate Pathology 40:68–74.

Joky, A., M. Matricon-Gondran, and J. Benex. 1983. Fine structural differences in the amoebocytes of *Biomphalaria glabrata*. Developmental and Comparative Immunology 7:669–672.

Mix, M. C. 1976. A general model for leucocyte cell renewal in bivalve mollusks. U.S. National Marine Fisheries Service Marine Fisheries Review 38(10):37–41.

Mohandas, A. 1985. An electron microscope study of endocytosis and subsequent events in *Mercenaria mercenaria* granulocytes. Pages 143–161 *in* T. C. Cheng, editor. Comparative pathobiology, volume 8. Plenum, New York.

Moore, C. A., and A. F. Eble. 1977. Cytochemical aspects of *Mercenaria mercenaria* hemocytes. Biological Bulletin (Woods Hole) 152:105–119.

Moore, M. N., and D. M. Lowe. 1977. The cytology and cytochemistry of the hemocytes of *Mytilus edulis* and their responses to experimentally injected carbon particles. Journal of Invertebrate Pathology 29:18–30.

Narain, A. S. 1973. The amoebocytes of lamellibranch molluscs, with special reference to the circulating amoebocytes. Malacological Review 6:1–12.

Peters, E. C. 1988. Recent investigations on the disseminated sarcomas of marine bivalve molluscs. American Fisheries Society Special Publication 18:74–92.

Rasmussen, L. P. D., E. Hage, and O. Karlog. 1985. An electron microscope study of the circulating leucocytes of the marine mussel, *Mytilus edulis*. Journal of Invertebrate Pathology 45:158–167.

Reade, P., and E. Reade. 1976. Phagocytosis in invertebrates: studies on the hemocytes of the clam *Tridacna maxima*. Journal of Invertebrate Pathology 28:281–290.

Ruddell, C. L. 1971a. Elucidation of the nature and function of the granular oyster amoebocytes through histochemical studies of normal and traumatized oyster tissues. Histochemie 26:98–112.

Ruddell, C. L. 1971b. The fine structure of oyster agranular amoebocytes from regenerating mantle wounds in the Pacific oyster, *Crassostrea gigas*. Journal of Invertebrate Pathology 18:260–268.

Ruddell, C. L. 1971c. The fine structure of the granular amoebocytes of the Pacific oyster, *Crassostrea gigas*. Journal of Invertebrate Pathology 18:268–275.

Smolowitz, R. M., and C. L. Reinisch. 1986. Indirect peroxidase staining using monoclonal antibodies specific for *Mya arenaria* neoplastic cells. Journal of Invertebrate Pathology 48:139–145.

Takatsuki, S. 1934. On the nature and functions of the amoebocytes of *Ostrea edulis*. Quarterly Journal of Microscopical Science 76:379–431.

Yoshino, T. P., and T. C. Cheng. 1976. Fine structural localization of acid phosphatase in granulocytes of the pelecypod *Mercenaria mercenaria*. Transactions of the American Microscopical Society 95:215–220.

Yoshino, T. P., and W. O. Granath, Jr. 1985. Surface antigens of *Biomphalaria glabrata* (Gastropoda) hemocytes: functional heterogeneity in cell subpopulations recognized by a monoclonal antibody. Journal of Invertebrate Pathology 45:176–186.

American Fisheries Society Special Publication 18:178–188, 1988

Humoral Defense Factors in Marine Bivalves

Fu-Lin E. Chu

Virginia Institute of Marine Science, School of Marine Science
The College of William and Mary, Gloucester Point, Virginia 23062, USA

Abstract.—Natural humoral components have been discovered and described in hemolymph from several marine bivalve species including eastern oyster *Crassostrea virginica*, blue mussel *Mitilus edulis*, northern quahog *Mercenaria mercenaria*, softshell *Mya arenaria*, and Pacific oyster *Crassostrea gigas*. These hemolymph components are enzymes of lysosomal origin, agglutinins, lectins, hemolysin, and antimicrobial substances. These components are proteins or glycoproteins found in the serum, hemocytes, or both. The exact relationship of these substances to the internal defense of marine bivalves against parasites and pathogenic microorganisms is not known. Lysosomal enzymes seem to have a double role, defense and nutrition. The free- and cell-bound lectins and agglutinins are believed to serve as recognition factors for the attachment of nonself particles to the phagocytes. The nature of bivalve hemolymph components appears to be innate and nonspecific. It has been suggested that the elevation in titer of enzymes of lysosomal origin in bivalve hemocytes and hemolymph after antigenic challenge is the acquired "humoral" protection produced by the animal. As yet, no experimental evidence has been obtained to support this idea. Whether humoral factors can be acquired in oysters and other marine bivalves needs further investigation. The specificity and the function of humoral factors relating to internal defense remain to be determined. Several mechanisms have been hypothesized for humoral defense in invertebrates. In these hypothesized mechanisms, recognition sites presented on the hemocytes are proposed for the events of phagocytosis, encapsulation, hypersynthesis and release of lysosomal enzymes.

In recent years, the question of immune defense mechanisms in some economically important bivalve molluscs has attracted much attention and generated substantial controversy. Though humoral defense of marine bivalves has not been studied as extensively as cellular defense (see Feng 1988, this volume), the humoral factors in bivalves and their exact relationship to the internal defense of the animals against parasites and pathogenic microorganisms are of interest. The understanding of bivalve humoral factors and their function in defense might allow us to manipulate their resistance or tolerance to diseases. In this chapter, the literature concerning humoral defense factors in marine bivalves and the possible roles of humoral factors in defense against diseases is reviewed and discussed.

Lysosomal Enzymes

Lysozyme and Lysosomal Enzymes

Lysozyme activity has been documented in many invertebrate species (e.g., McDade and Tripp 1967a, 1967b; Feng 1974; Cheng et al. 1975). The bacteriolytic enzyme, lysozyme, is a basic protein with a molecular weight of approximately 15,000 daltons; its characteristics are (1) ability to lyse the bacterium *Micrococcus lysodeikticus* with the release of amino sugars and (2) stability at low pH and high temperature but becoming inactive at high pH especially at high temperatures (Jolly 1967). Lysozyme activity was first reported in the hemolymph and mantle mucus of the eastern oyster *Crassostrea virginica* by McDade and Tripp (1967a, 1967b). Lysozyme or lysozymelike substances have subsequently been described in softshell *Mya arenaria* (Cheng and Rodrick 1974), northern quahog *Mercenaria mercenaria* (Cheng et al. 1975), and blue mussel *Mytilus edulis* (Hardy et al. 1976; McHenery and Birkbeck 1982).

The enzymatic activity of lysozymes found in these bivalve species resembles avian lysozyme in their ability to lyse bacteria and in sharing some biochemical properties (Cheng and Rodrick 1974; Rodrick and Cheng 1974) As with egg-white lysozyme, the lytic activity on *Micrococcus lysodeikticus* is salt-dependent, relatively heat-stable, and very sensitive to changes in ionic concentration. However, Feng (1974) reported that the oyster lysozyme differs from egg-white lysozyme in electrophoretic mobility, molecular weight, isoelectric point, and association with acidic proteins.

Bivalve lysozymes are active not only against *M. lysodeikticus* but also against other species of bacteria including *Bacillus subtilis*, *B. megaterium*, *Escherichia coli*, *Gaffkya tetragina*, *Proteus vulgaris*, *Salmonella pullorum*, and *Shigella sonnei* (Cheng and Rodrick 1974; Rodrick and Cheng 1974).

Lysozyme activities were found in both the serum and hemocytes of eastern oyster, softshell, and northern quahog. Except for northern quahog, lysozyme activity was greater in serum than in cells (Cheng and Rodrick 1974; Rodrick and Cheng 1974; Cheng et al. 1975).

Other lysosomal enzymes were also reported to be present in hemolymph of eastern oyster and northern quahog (Cheng and Rodrick 1975; Cheng 1976). These lysosomal enzymes are β-glucuronidase, alkaline and acid phosphatases, lipase, aminopeptidase, and amylase. All of these enzymes, except amylase, were also found in hemocytes (Yoshino and Cheng 1976). Antimicrobial substances have also been detected in tissue extracts of oysters and clams (Li 1960). These substances were able to inhibit growth of bacteria. The nature and origin of these substances are unknown.

Function and Origin of Lysosomal Enzymes

The biological roles of lysosomal enzymes in molluscan bivalves are not completely understood. They are generally presumed to be involved with host defense and digestion. Lysozymes from eastern oyster, softshell, and northern quahog are active against several species of bacteria. The β-glucuronidase from northern quahog was found to hydrolyse mucopolysaccharides which are found in bacteria cell walls. It has been speculated (Cheng 1983a, 1983b, 1983c) that lysosomal enzymes were inducible and served as a form of "acquired" humoral immunity in the animal. The elevated level of lysosomal enzymes in the serum fraction, when oyster and clam hemocytes were challenged with bacteria, is thought to represent a type of humoral defense mechanism against invading microorganisms (Cheng et al. 1975; Cheng et al. 1978; Cheng and Butler 1979).

Beside the suggested role of lysozyme in defense, evidence provided by McHenery et al. (1979) indicates that its primary function is digestion. The tissue distribution of lysozyme in more than 30 species of bivalve molluscs was studied, including the marine species *Chlamys opercularis, Tellina tenuis,* softshell, and blue mussel. In most cases, the style and the digestive gland contained more enzyme than the blood and body fluid. McHenery et al. (1979), therefore, concluded that (1) the function of lysozyme is associated with digestion and degradation of bacteria for nourishment of the animal, (2) lysozyme concentration is related to the proportion of bacteria utilized as food, and (3) host defense is secondary to that of nutrition.

Lysosomal enzymes from different marine bivalves share similar biochemical and bacteriolytic properties. The roles of lysosomal enzymes as agents of defense and digestion are probably equally important. Digestion and defense could actually occur at the same time. Because bacteria are one of the foods of filter feeders such as marine bivalves, a substance that is able to hydrolyze bacteria would fulfill the purposes of both defense and digestion. Perhaps digestion and defense are more distinct in the minds of investigators than in the life of the animal.

The origin of lysosomal enzymes is postulated to be in the lysosomes of granular hemocytes (Rodrick and Cheng 1974). The release of lysosomal enzymes into the serum is a consequence of hemocyte "degranulation" during phagocytosis (Rodrick and Cheng 1974; Cheng and Yoshino 1976a, 1976b; Foley and Cheng 1977; Cheng and Butler 1979). Cheng and his associates have demonstrated that lysosomal enzymes could be induced experimentally by in vitro exposure of molluscan hemocytes to bacteria or in vivo challenge with bacteria or bacteria lipids (Cheng et al. 1975; Cheng and Yoshino 1976a, 1976b; Cheng et al. 1977, 1978; Cheng and Butler 1979). For example, when northern quahog hemocytes were exposed to *Bacillus megaterium,* the level of lysozyme in the serum fraction rose. It was postulated that the elevated levels of serum lysosomal enzymes (e.g., lysozyme, aminopeptidase, acid phosphatase) are due to the hypersynthesis and release of these enzymes in the phagocytosing hemocytes in response to the bacterial challenge. The experimental induction of lysosomal enzymes will be discussed further in the section, Induction of Humoral Factors.

Agglutinins and Lectins

Serum agglutinins are common components of bivalve hemolymph. Among the various humoral factors, agglutinins have received much attention because of their ability to agglutinate vertebrate erythrocytes and bacteria in vitro.

Hemagglutinin and Bacterial Agglutinin

Naturally occurring substances that agglutinate red blood cells (hemagglutinins) are known to be widely distributed among plants and animals. Such substances in plants are called lectins (carbohydrate-binding proteins), and they agglutinate cells or materials by glycosyl moieties (Sharon and Lis 1972). The specificity of plant lectins varies with the plant species (Boyd 1962). Natural hemagglutinins have been found in the body and seminal fluids of numerous invertebrates, including several marine bivalves (Tyler 1946; Cheng and Sanders 1962; Cushing et al. 1963; Johnson 1964; Boyd and Brown 1965; Cohen et al. 1965; Tripp 1966; Li and Fleming 1967; Brown et al. 1968; Hardy et al. 1976, 1977b).

Hemagglutinins were found in tissue extracts and in the hemolymph of blue mussel (Brown et al. 1968; Hardy et al. 1976), eastern oyster (Tripp 1966; McDade and Tripp 1967a, 1967b; Acton et al. 1969), and Pacific oyster *Crassostrea gigas* (Hardy et al. 1977a, 1977b). These hemagglutinins could agglutinate more than one type of human erythrocyte and the erythrocytes of many vertebrate species. Some marine bivalve agglutinins are heterogenous, i.e., they are not group specific (e.g., agglutinins from Pacific oyster and northern quahog). They agglutinated cells other than red blood cells (e.g., algae, vertebrate sperm, and bacteria). Bacterial agglutination has been observed only in Pacific oyster and northern quahog (Arimoto and Tripp 1977; Hardy et al. 1977a, 1977b); hemolymph from eastern oyster did not agglutinate bacteria (Tripp 1966). Because marine bivalves live in an environment full of bacteria, it is not known why other bivalves lack bacterial agglutinins. Perhaps investigation to date has been too limited to detect them.

Lectins

Two serum and one cell-bound lectin have been discovered in the eastern oyster (Vasta et al. 1982; Cheng et al. 1984). Each of these serum lectins was shown to have a distinct serological agglutination specificity. The hemocyte-bound lectin was able to agglutinate certain vertebrate erythrocytes (Vasta et al. 1982; Cheng et al. 1984). It showed only one of the two serological activities found in the serum lectins and did not appear to be associated with a specific subpopulation of oyster hemocytes (Cheng et al. 1984).

Nature and Biochemical Properties of Agglutinins

Hemagglutinins and bacterial agglutinins from marine bivalves showed strong biochemical and biological similarity. Hemagglutinins and bacterial agglutinins are all lectinlike glycoproteins, bearing divalent or multivalent receptors for certain specific carbohydrate determinants. They appear to have a common structure consisting of small identical subunits with molecular weights of 20,000–21,000 daltons. The bacterial agglutinin of northern quahog was found to be composed of subunits, each with a molecular weight of approx-

imately 21,000 daltons (Arimoto and Tripp 1977). The oyster hemagglutinin was composed of identical subunits with molecular weights of approximately 20,000 daltons (Acton et al. 1969) and contained fewer protein components than human immunoglobulin. These subunits are usually linked by noncovalent bonds, and each subunit has a carbohydrate binding site. Hemagglutination and bacterial agglutination can be inhibited by saccharides. Bacterial agglutination of northern quahog hemolymph was partially inhibited by N-acetyl-D-glucosamine, D-glucosamine, N-acetyl-D-galactosamine, and D-fucose (Arimoto and Tripp 1977). Agglutination of human red cells by eastern oyster hemolymph was completely inhibited by galactosamine, N-acetylglucosamine, and N-acetylgalactosamine, and partially inhibited by glucosamine. N-Acetyl-D-galactosamine and N-acetylglucosamine inhibited the agglutination of human erythrocytes A by an extract from butter clam *Saxidomus giganteus* (Johnson 1964). Bovine salivary gland glycoprotein inhibited hemagglutination by the hemolymph of Pacific oyster (Hardy et al. 1977b).

Marine bivalve agglutinins are reported to be nondialysable and are inactivated by temperatures of 65–70°C. Calcium ions are required for agglutination activity and contribute to the heat stability of the molecule (Johnson 1964; Tripp 1966; McDade and Tripp 1967a, 1967b; Arimoto and Tripp 1977). A more detailed account of biochemical properties of agglutinins is presented elsewhere in this volume (Olafsen 1988).

Functional Role of Agglutinins and Lectins

There is confusion among the terms, lectin, agglutinin, and opsonin, even though they are different by definition. But in reality, we still cannot distinguish agglutinins from lectins in many invertebrate species, including bivalves. Agglutinins have some characteristics of lectins. It is likely that many of the agglutinins found in bivalves are lectins, and both can function as opsonins. The terminology of agglutinins and opsonins may only represent various functions of the same protein.

The role of agglutinins and lectins in molluscs has been discussed in an immunological context (Hardy et al. 1977b; Lackie 1980; Warr 1981; Vasta et al. 1982; Cheng et al. 1984; Coombe et al. 1984). The following possible roles of agglutinins and lectins in defense mechanisms are suggested.

(1) Agglutinins and lectins inactivate bacteria or parasites by agglutination. The inactivation

may lead to two defense events: lysis of bacteria by extracellular enzymes (e.g., lysozyme) and subsequent phagocytosis and encapsulation of agglutinated particles. As described earlier, hemagglutination and bacteria agglutination by hemolymph, tissue extract, or both of marine bivalves have been well documented.

(2) Serum agglutinins and lectins serve as opsonins to link receptors with similar glycosyl moieties on the surface of nonself particles and hemocytes. There is evidence for the opsonic effect of purified agglutinins from the Pacific oyster on the phagocytosis of bacteria (Hardy et al. 1977b). Various studies have also demonstrated opsonic activity of molluscan serum both in vitro (Tripp and Kent 1967; Prowse and Tait 1969; Anderson and Good 1976; Arimoto and Tripp 1977; Hardy et al. 1977a; Sminia et al. 1979; Van der Knapp et al. 1982) and in vivo (Renwrantz and Mohr 1978, Renwrantz 1981), although in some cases the evidence could be considered circumstantial (Coombe et al. 1984).

(3) Agglutinins and lectins attach to the hemocytes and function as cell surface recognition factors to bind nonself particle bearing appropriate glycosyl moieties. A cell membrane-associated lectin has been found in hemocytes of eastern oyster (Vasta et al. 1982). Lectin-binding receptors have also been detected at the hemocyte surface of the marine bivalves eastern oyster (Yoshino et al. 1979; Cheng et al. 1980) and blue mussel (Renwrantz et al. 1985), and of gastropods bloodfluke planorb *Biomphalaria glabrata* (Schoenberg and Cheng 1980; Yoshino 1981), swamp lymnaea *Lymnaea stagnalis* (Sminia et al. 1981), and escargot *Helix pomatia* (Renwrantz and Cheng 1977). These animals possess binding sites on the hemocytes for various lectins.

Sialic acids are suggested as the binding sites of serum lectins in Pacific oyster and of membrane-associated lectins in eastern oyster, although the configuration of lectin molecules should be taken into consideration (Hardy et al. 1977b; Cheng et al. 1984). Lectin-binding activity of hemocytes was found to be temperature-dependent (Yoshino et al. 1979) and relied on the presence of carbohydrates such as glucose, fructose, mannose, galactose, N-acetylneuraminic acid, and N-acetylgalactosamine (Renwrantz and Cheng 1977).

The second and third suggested roles are the ones best supported by the evidence. They describe adhesion of nonself particles and hemocyte through the binding of serum or cell-associated agglutinins–lectins to nonself particles and to phagocytes. The serum and cell-bound lectins–agglutinins mediate the processes of phagocytosis and encapsulation of foreign materials.

Hemolysin

Hemolytic factor (hemolysin) was found in northern quahogs (Graham 1968; Anderson 1981) and blue mussels (Hardy et al. 1976; Feng and Barja 1986). The lytic factor described by Graham (1968) from hemolymph and shell liquor lysed erythrocytes of many vertebrate species. This lysin was heat-labile (inactivated at 47°C for 30 min), nondialysable, could not be absorbed by erythrocytes, and its activity was dependent on calcium ions. Hemolysin from northern quahog was further investigated by Anderson (1981). Hemolytic activity was detected in both sera and in hemocytes and hemolymph of northern quahog and could be induced experimentally. Anderson's results differed from those of Graham in erythrocyte specificity, reactivity at temperature optima for hemolytic reaction kinetics, and absorption and inactivation of hemolytic activity by homologous erythrocytes. The hemolymph of blue mussels was examined by Hardy et al. (1976); human erythrocytes of type O were lysed at pH 7–9 and an optimum temperature of 4°C.

The exact role of hemolysin is not defined, and the mechanism for hemolysis is unknown. The capability to lyse erythrocytes from species of distant phylogenetic origin does not have practical importance for bivalve defense. If hemolysin is a rudimentary complement system in bivalves, why is it restricted only to northern quahogs and blue mussels?

Induction of Humoral Factors

In marine bivalves, the humoral factors, agglutinins, lectins, and enzymes of lysosomal origin appear to be innate and nonspecific. Attempts have been made to induce lysosomal enzymes in bivalves. An increase of lipase activity has been demonstrated in the serum and hemolymph cells of softshells that were injected with heat-killed *Bacillus megaterium* (Cheng and Yoshino 1976b).

Similarly, levels of aminopeptidase and lysozyme were increased in eastern oyster hemocytes and northern quahog serum after in vitro exposure to bacteria (Cheng et al. 1975; Yoshino and Cheng 1976). No data exist on the induction of lysozyme in oysters or clams in vivo. Messner and Mohrig (1969) failed to produce this effect in the freshwater mussel *Anodonta anatina*.

Like other invertebrates, molluscs are incapable of synthesizing antibodies (immunoglobulins). Cheng (1981, 1983b) hypothesized that the elevated lysosomal enzymes found in marine bivalves are inducible "protective" humoral factors. The increased levels of lysosomal enzymes in the cell and serum after in vivo or in vitro bacterial exposure are believed to be the consequence of hypersynthesis of enzymes in the cells and their subsequent release to the sera. These soluble molecules may play a role in acquired resistance. For example, the elevated level of serum aminopeptidase may alter the surface protein of secondarily introduced parasites and thus act as a form of acquired humoral immunity. Acquired immunity, however, is characterized by its specificity, adaptive nature (i.e., capacity to learn), and irreversibility. The specificity of lysosomal enzymes has not been elucidated, and thus far, response to antigenic challenge appears to be nonspecific. The production of lysosomal enzymes can be stimulated, but their formation cannot be "induced" since these lysosomal enzymes are innate. Moreover, as stated by Feng and Barja (1986), elevated levels of hemolymph enzymes could also be interpreted as pathological manifestations rather than defense responses. Feng and Canzonier (1970) found that hemolymph lysozyme levels decreased in eastern oysters infected with *Haplosporidium nelsoni* (MSX), but it increased in oysters infected with *Bucephalus* sp.

Lysozyme activity has been reported higher in winter than in summer, and it exhibits a great variation among individual oysters (S. Y. Feng, Marine Sciences Institute, University of Connecticut, personal communication; Chu, unpublished data). Lysozyme activity was detected only in a few (0–10%) of the eastern oysters maintained in ambient estuarine water (York River, Virginia) during the summer (Chu, unpublished data).

All of these data leave some questions unanswered. What is the real role of lysozyme in diseases oysters? If lysozyme is an inducible "protective" agent, why did its level drop when the oysters were infected with MSX (Feng and Canzonier 1970), and why is it absent during the summer? Why does the hemolymph lysozyme level vary so much individually within the same population? Is this variation related to age or to the physiological and nutritional status of the animal? Is seasonal variation due to changes of environmental temperature and salinity?

Attempts to increase oyster hemagglutinin titers by challenging with erythrocytes were unsuc- cessful (Acton et al. 1969; Feng 1974). Hardy et al. (1977a) found that exposure of Pacific oysters to bacteria stimulated an increase in the titer of human hemagglutinin. Anderson (1981) also reported that hemolysin could be induced by both red blood cell injection and experimental wounding.

In studying the chemical and physical properties of cell-free hemolymph of "normal" eastern oysters and those infected with *Bucephalus* sp. and *H. nelsoni* (MSX), Feng and Canzonier (1970) found quantitative changes in some of the hemolymph proteins of infected oysters. They suggested that these changes were evidence of host humoral responses to infection. However, the study was performed on naturally infected oysters. The observed changes in hemolymph protein could be caused by environmental effects rather than by infection.

Hypothesized Humoral Defense Mechanisms

There are several humoral defense mechanisms proposed for invertebrates. These hypothesized mechanisms are applicable also to molluscs.

(1) Lysosomal enzymes act as defense molecules against invading microorganisms and abiotic foreign particles. These lysosomal enzymes are inducible and serve as a form of acquired humoral immunity in the animal. This mechanism was hypothesized and discussed by Cheng (1983a, 1983b, 1983c). Cheng (1983b) proposed three recognition sites for the occurrence of hypersynthesis and release of lysosomal enzymes, the cell surface, the nuclear membrane, and the lysosomal surface. The operational sequence of this mechanism is shown schematically in Figure 1. Sites on the surfaces of hemocytes recognize the challenging agent. Signals pass from the cell membrane recognition sites to those on the nuclear membrane that receive the transcytoplasmic messenger. This sequence of events results in the synthesis of enzymes on ribosomes and their eventual release from the lysosomes as a consequence of degranulation.

(2) Agglutinins and lectins are humoral recognition factors which may also act as opsonins. They may form a molecular link between hemocytes and the foreign object or attach to the hemocytes and act as cell-bound recognition molecules. This hypothesized mechanism explains the phenomena of self–nonself discrimination in invertebrates very well, although the exact nature of this discrimination is not yet known. Cheng et al. (1984) postulated that, during phagocytosis and encapsulation, lectins facilitate attachment be-

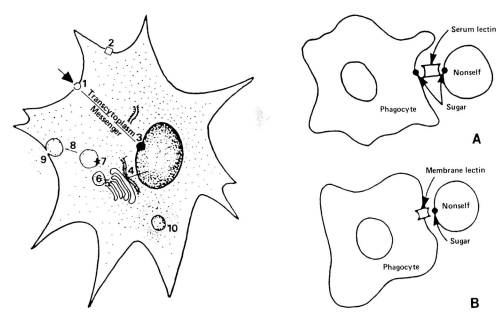

FIGURE 1.—Diagram explaining hypersynthesis of lysosomal enzymes and their subsequent release to serum. (After Cheng et al. 1984.) Surface sites (e.g., 1 and 2) on the cell membrane recognize stimulating agents and send a transcytoplasmic messenger to recognition sites (3) on the nuclear membrane. Enzyme synthesis on ribosomes (4) is directed by a deoxyribonucleic acid–messenger ribonucleic acid–transfer ribonucleic acid sequence, and lysosomal enzymes are packaged by the Golgi apparatus (5) into lysosomes (6). Activated recognition sites (7) on lysosomal membranes trigger lysosome migration (8). Degranulation of lysosomes (9) releases lysosomal enzymes to the serum. Lysosomes may also be destabilized to release lysosomal enzymes into the cytoplasm (10).

FIGURE 2.—Diagrams showing positions of lectins and sugars that would facilitate attachment between nonself particles and a phagocyte. (After Cheng et al. 1984.) **A.** A serum lectin links the phagocyte and nonself particle by appropriate sugar moieties. **B.** A nonself particle binds to a membrane lectin with compatible sugar moieties.

tween cell and nonself material through the binding of appropriate sugar moieties (Figures 2, 3). Coombe et al. (1984) proposed a different model for the mechanism of self–nonself discrimination (Figure 4) based simply on a specific recognition of self. Self-reactivity is controlled by the recognition of self through a self-marker (H) and a recognition structure on the phagocyte (anti-H). Self-recognition is hypothesized to control phagocytosis in the absence or presence of an opsonin. In the absence of opsonin, the selection of nonself particles could occur either by means of a cell-surface receptor or through nonspecific means, e.g., a physical attraction. In the presence of opsonin, phagocytosis is initiated by the binding of opsonized nonself cells or particles. In both cases, any cell or particle to which the phagocyte becomes attached would be ingested unless self is

specifically recognized. Olafsen (1988) suggested the presence of multiple heterogenous receptor sites on lectins, some for nonself particles and some for potential phagocytes (self). He contends that self-receptors must be masked to avoid premature binding of the agglutinin with hemocytes, but these are unmasked after a nonself particle is bound to the agglutinin.

(3) Self–nonself discrimination in invertebrates is based on recognition of carbohydrate determinants by dissolved or cell-bound oligomers of glycosyl-transferases. The model (Figure 5) hypothesized by Parish (1977) is also of interest. Five proteins (transferases) with different sugar binding properties are present in the organism. These five transferases act as the subunits of the recognition factors and are synthesized and secreted by hemocytes into the hemolymph. They randomly associate into hexamers by a catalyzed protein which has an acceptor site on the hemocyte surface. The inclusion of this additional protein gives cytophilic and opsonic properties to the recognition factors.

All of these hypothesized defense mechanisms propose the occurrence of recognition (or binding)

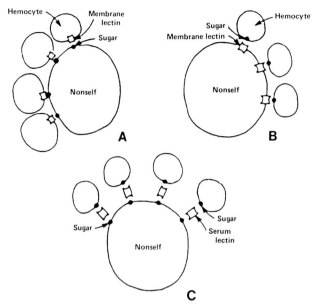

FIGURE 3.—Diagrams showing positions of lectins and sugars that would facilitate encapsulation of nonself particles. (After Cheng et al. 1984.) **A.** Hemocytes attach to nonself particles through the binding of membrane-associated lectins and compatible sugars on the surface of nonself particles. **B.** Nonself particles bind to hemocytes bearing lectin-binding receptors. **C.** Serum lectin links hemocytes and nonself particles together with appropriate sugar moieties.

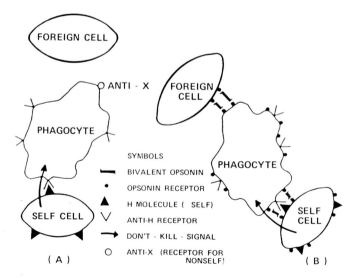

FIGURE 4.—A model proposed to explain phagocytosis by invertebrate cells. Self-reactivity is controlled by the recognition of self through a self-marker (H) and a recognition structure on the phagocyte (anti-H). Self-recognition is proposed to control phagocytosis in the absence of an opsonin (**A**) or when phagocytosis is initiated by the binding of opsonized cells (**B**). In both cases it is envisaged that any cell or particle to which the phagocyte becomes attached would be ingested unless the self is specifically recognized. (After Coombe et al. 1984.)

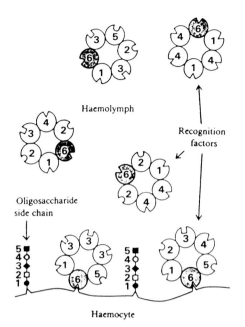

Haemolymph

Recognition
factors

Oligosaccharide
side chain

Haemocyte

FIGURE 5.—Possible model for the recognition factors in invertebrates which augment the phagocytosis of foreign substances. Various monosaccharides on the hemocyte surface, which form oligosaccharide side chains, are represented by small circles, diamonds and squares numbered 1–5. Each monosaccharide is inserted in the oligosaccharide by a specific glycosyltransferase (notched circles 1–5). Thus, transferase 2 binds to monosaccharide 1 and catalyses the attachment of monosaccharide 2, transferase 3 binds to monosaccharide 2 and facilitates the attachment of monosaccharide 3, and so on. The transferases randomly associate into recognition factors (hexamers), a process of polymerization that is initiated by an additional protein (6). Protein 6 has an acceptor site on the hemocyte surface which enables the recognition factors to be cytophilic for hemocytes. (After Parish 1977.)

sites on the hemocytes. As mentioned previously, recent studies have revealed that cell-bound lectins and sites (receptors) that are able to bind more than one kind of lectin are present on molluscan hemocytes. Hemocyte lectin receptors of eastern oysters include concanavalin A, albumin gland extracts of escargot and grovesnail *Cepaea nemoralis,* and wheat germ agglutinin (Yoshino et al. 1979; Cheng et al. 1980). Those of blue mussels include wheat germ agglutinin, *Ricinus* 60 and 120, and agglutinin from albumin gland extracts of escargot (Renwrantz et al. 1985). The second mechanism implies the direct interaction of agglutinins–lectins and cellular components. However, very little is known about the interaction mode of agglutinin–lectin and the phago-

cytic cells at the molecular level. Although there is still no evidence for secretion of glycosyltransferases into serum, Parish (1977) made an important point that enzymes of several specificities should be available to extend the range of recognition possible.

Summary

Humoral substances with agglutinating, opsonic, lytic, and antimicrobial activities have been discovered and described in several bivalve species, including eastern oyster, Pacific oyster, northern quahog, and softshell (Table 1). These humoral components may play a significant role in host defense. Some of the humoral factors (e.g., lysozyme, β-glucuronidase, hemolysin) can lyse bacteria and erythrocytes of several vertebrate species. Some (e.g., agglutinins and lectins) are thought to mediate cellular defense mechanisms. The nature of these humoral components appear to be innate and nonspecific rather than acquired. Although there are indications in the literature that directly and indirectly point to induction of humoral response in oysters and other bivalve species, firm evidence that the inducible humoral response is both specific to challenged antigen and memorized has yet to be presented.

Mechanisms hypothesized to interpret the phenomena of self–nonself discrimination in invertebrates involve the interaction of humoral components and phagocytic cells. Agglutinins and lectins have been found in molluscs, including oysters, clams, and mussels and have been designated "recognition factors" (or receptors) and "opsonic factors" for foreign particles. These substances are both free in the hemolymph and bound to the cell membrane. Although the dissolved recognition factor(s) in the serum and the cell-bound molecule(s) are distinct, they are serologically related. It is suggested that a specific recognition of more than one kind is involved in phagocytosis. The molecular mode of these humoral components is not clear, but existing hypotheses are discussed by Olafsen (1988). Similarly, the origin, specificity, and precise role of the lysosomal enzymes in humoral defense need further investigation. Lack of reliable information makes it difficult to construct a general scheme for humoral defense in marine bivalves. Nevertheless, the discovery of cell-associated lectins in marine bivalves may open the door for studies of host and parasite interaction at the molecular level. The hypothesized humoral mechanisms presented in this chapter offer a basis for experimentation as well as direction for future research.

TABLE 1.—Humoral components of host defense in some marine bivalves.

Humoral component	Species	Source	Activity	Reference[a]
Lysozyme	Eastern oyster	Hemocytes and serum	Lyse bacteria, inducible	(4, 10)
	Northern quahog	Hemocytes and serum	Lyse bacteria, inducible	(5)
	Softshell	Hemocytes and serum	Lyse bacteria	(3)
	Blue mussel	Body fluid and hemolymph	Lyse bacteria	(7)
Aminopeptidase	Eastern oyster	Hemocytes and serum	Exopeptidase	(4)
	Northern quahog	Hemocytes and serum	Exopeptidase	(4)
β-Glucuronidase	Eastern oyster	Hemocytes and serum	Hydrolyze mucopolysaccharides	(13)
	Northern quahog	Hemocytes and serum	Hydrolyze mucopolysaccharides	(13)
Antimicrobial factors	Eastern oyster	Tissue extract and juice	Inhibit bacteria growth	(9)
	Northern quahog	Tissue extract and juice	Inhibit bacteria growth	(9)
Hemagglutinin	Pacific oyster	Serum	Agglutinate vertebrate erythrocytes	(10)
	Eastern oyster	Hemocytes and serum	Agglutinate vertebrate erythrocytes	(10, 11, 12)
Bacterial agglutinin	Pacific oyster	Serum	Agglutinate bacteria, fish sperm, and algae	(8)
	Northern quahog	Serum	Agglutinate bacteria and algae	(2)
Hemolysin	Blue mussel	Hemocytes and serum	Lyse mammalian erythrocytes and human lymphocytes	(7)
	Northern quahog	Serum and shell liquor	Lyse mammalian and avian erythrocytes, inducible	(1, 6)

[a](1) Anderson (1981)
(2) Arimoto and Tripp (1977)
(3) Cheng and Rodrick (1974)
(4) Cheng and Rodrick (1975)
(5) Cheng et al. (1975)
(6) Graham (1968)
(7) Hardy et al. (1976)
(8) Hardy et al. (1977a, 1977b)
(9) Li (1960)
(10) McDade and Tripp (1967a, 1967b)
(11) Tripp (1966)
(12) Vasta et al. (1982)
(13) Yoshino and Cheng (1976)

Acknowledgments

The author thanks J. Walker for typing and preparing manuscripts, M. Shackelford for graphic work, and M. Bender and E. Burreson for critically reviewing the manuscript. This paper is contribution 1441 from the Virginia Institute of Marine Science.

References

Acton, R. T., E. E. Evans, and J. C. Bennett. 1969. Immunological capabilities of the oyster *Crassostrea virginica*. Comparative Biochemistry and Physiology 29:149–160.

Anderson, R. S. 1981. Inducible hemolytic activity in *Mercenaria mercenaria* hemolymph. Developmental and Comparative Immunology 5:575–585.

Anderson, R. S., and R. A. Good. 1976. Opsonic involvement in phagocytosis by mollusk hemocytes. Journal of Invertebrate Pathology 27:57–64.

Arimoto, R., and M. R. Tripp. 1977. Characterization of a bacterial agglutinin in the hemolymph of the hard clam *Mercenaria mercenaria*. Journal of Invertebrate Pathology 30:406–413.

Boyd, W. C. 1962. Introduction to immunochemical specificity. Wiley, New York.

Boyd, W. C., and R. Brown. 1965. A specific agglutinin in the snail *Octala lactea* (*Helix*). Nature (London) 208:593–594.

Brown, R., L. R. Almondovar, H. M. Bhatia, and W. C. Boyd. 1968. Blood group specific agglutinins in invertebrates. Journal of Immunology 100:214–216.

Cheng, T. C. 1976. Beta-glucuronidase in the serum and hemolymph cells of *Mercenaria mercenaria* and *Crassostrea virginica* (Mollusca: Pelecypoda). Journal of Invertebrate Pathology 27:125–128.

Cheng, T. C. 1981. Bivalves. Pages 233–300 *in* N. A. Ratcliffe and A. F. Rowley, editors. Invertebrate blood cells, volume 1. Academic Press, New York.

Cheng, T. C. 1983a. Internal defense mechanisms of molluscs against invading microorganisms: personal reminiscences. Transaction of the American Microscopical Society 102:185–193.

Cheng, T. C. 1983b. The role of lysosomes in molluscan inflammation. American Zoologist 23:129–144.

Cheng, T. C. 1983c. Triggering of immunologic defense mechanisms of molluscan shellfish by biotic and abiotic challenge and its applications. Marine Technology Society Journal 17:18–25.

Cheng, T. C., and M. S. Butler. 1979. Experimentally induced elevations of acid phosphatase activity in hemolymph of *Biomphalaria glabrata* (Mollusca). Journal of Invertebrate Pathology 34:119–124.

Cheng, T. C., M. J. Chorney, and T. P. Yoshino. 1977. Lysozymelike activity in the hemolymph of *Biomphalaria glabrata* challenged with bacteria. Journal of Invertebrate Pathology 29:170–174.

Cheng, T. C., V. G. Guida, and P. L. Gerjhart. 1978. Aminopeptidase and lysozyme activity levels and serum protein concentrations in *Biomphalaria glabrata* (Mollusca) challenged with bacteria. Journal of Invertebrate Pathology 32:297–302.

Cheng, T. C., J. W. Huang, H. Karadogan, L. R. Renwrantz, and T. P. Yoshino. 1980. Separation of oyster hemocytes by density gradient centrifugation and identification of their surface receptors. Journal of Invertebrate Pathology 36:35–40.

Cheng, T. C., J. J. Marchalonis, and G. R. Vasta. 1984. Recognition proteins, receptors, and probes: invertebrates. Progress in Clinical and Biological Research 157:1–15.

Cheng, T. C., and G. E. Rodrick. 1974. Identification and characterization of lysozyme from the hemolymph of the soft-shelled clam *Mya arenaria*. Biological Bulletin (Woods Hole) 147:311–320.

Cheng, T. C., and G. E. Rodrick. 1975. Lysosomal and other enzymes in the hemolymph of *Crassostrea virginica* and *Mercenaria mercenaria*. Comparative Biochemistry and Physiology B, Comparative Biochemistry 52:443–447.

Cheng, T. C., G. E. Rodrick, D. A. Foley, and S. A. Koehler. 1975. Release of lysozyme from hemolymph cells of *Mercenaria mercenaria* during phagocytosis. Journal of Invertebrate Pathology 25:261–265.

Cheng, T. C., and B. G. Sanders. 1962. Internal defense mechanisms in molluscs and an electrophoretic analysis of a naturally occurring serum hemagglutinin in *Viviparus malleatus* Reeve. Proceedings of the Pennsylvania Academy of Science 36:72–83.

Cheng, T. C., and T. P. Yoshino. 1976a. Lipase activity in the hemolymph of *Biomphalaria glabrata* (Mollusca) challenged with bacterial lipids. Journal of Invertebrate Pathology 28:143–146.

Cheng, T. C., and T. P. Yoshino. 1976b. Lipase activity in the serum and hemolymph cells of the soft-shelled clam, *Mya arenaria*, during phagocytosis. Journal of Invertebrate Pathology 27:243–245.

Cohen, E., A. W. Rowe, and F. C. Wissler. 1965 Heteroagglutinins of the horseshoe crab, *Limulus polyphemus*. Life Sciences 4:2009–2016.

Coombe, D. R., P. L. Ey, and C. R. Jenkin. 1984. Self/non-self recognition in invertebrates. Quarterly Review of Biology 59:231–254.

Cushing, J. E., N. L. Calaprice, and G. Trump. Blood group reactive substances in some marine invertebrates. Biological Bulletin (Woods Hole) 125:69–80.

Feng, S. Y. 1974. Lysozymelike activities in the hemolymph of *Crassostrea virginica*. Contemporary Topics in Immunobiology 4:225–231.

Feng, S. Y. 1988. Cellular defense mechanisms of oysters and mussels. American Fisheries Society Special Publication 18:153–168.

Feng, S. Y., and J. L. Barja. 1986. Summary of cellular defense mechanisms of oysters and mussels. Pages 29–41 *in* W. S. Fisher and A. J. Figueras, editors. Marine bivalve pathology. University of Maryland Sea Grant Publications UM-SG-TS-87-02, College Park.

Feng, S. Y., and W. J. Canzonier. 1970. Humoral responses in the American oyster *(Crassostrea virginica)* infected with *Bucephalus* sp. and *Minchinia nelsoni*. American Fisheries Society Special Publication 5:497–510.

Foley, D. A., and T. C. Cheng. 1977. Degranulation and other changes of molluscan granulocytes associated phagocytosis. Journal of Invertebrate Pathology 29:321–325.

Graham, M. A. 1968. A hemolytic enzyme in the hemolymph of the clam *Mercenaria mercenaria*. Master's thesis. University of Delaware, Newark.

Hardy, S. W., T. C. Fletcher, and L. M. Gerrie. 1976. Factors in hemolymph of the mussel, *Mytilus edulis* L., of possible significance as defense mechanisms. Biochemical Society Transactions 4:473–475.

Hardy, S. W., T. C. Fletcher, and J. A. Olafsen. 1977a. Aspects of cellular and humoral defense mechanisms in the Pacific oyster, *Crassostrea gigas*. Pages 59–66 *in* J. B. Solomon and J. D. Horton, editors. Developmental immunobiology. Elsevier, Amsterdam.

Hardy, S. W., P. T. Grant, and T. C. Fletcher. 1977b. A haemagglutinin in the tissue fluid of the Pacific oyster *Crassostrea gigas*, with specificity for sialic acid residue in glycoproteins. Experientia (Basel) 33:767–769.

Johnson, H. 1964. Human blood A, specific agglutinin of the butter clam *Saxidomus giganteus*. Science (Washington, D.C.) 146:548–549.

Jolly, P. 1967. Relationship between chemical structure and biological activity of hen egg-white lysozyme and lysozymes of different species. Proceedings of Royal Society Serial Bulletin 167:350–364.

Lackie, A. M. 1980. Invertebrate immunity. Parasitology 80:393–412.

Li, C. P. 1960. Antimicrobial activity of certain marine fauna. Proceedings of the Society of Experimental Biology and Medicine 104:366–368.

Li, M. F., and C. Fleming. 1967. Hemagglutinins from oyster hemolymph. Canadian Journal of Zoology 45:1225–1234.

McDade, J. E., and M. R. Tripp. 1967a. Lysozyme in oyster mantle mucus. Journal of Invertebrate Pathology 9:581–582.

McDade, J. E., and M. R. Tripp. 1967b. Lysozyme in the hemolymph of the oyster, *Crassostrea virginica*. Journal of Invertebrate Pathology 9:531–535.

McHenery, J. G., T. H. Birkbeck, and J. A. Allen. 1979. The occurrence of lysozyme in marine bivalves. Comparative Biochemistry and Physiology B, Comparative Biochemistry 63:25–28.

McHenery, J. G., and T. H. Birkbeck. 1982. Characterization of the lysozyme of *Mytilus edulis* (L.). Comparative Biochemistry and Physiology B, Comparative Biochemistry 71:583–589.

Messner, B., and W. Mohrig. 1969. Zum Lysozym-

Vorkommen bei Muschln, *Anadonta anatina*. Zoologische Jahrbücher Abteilung für allgemeine Zoologie und Physiologie der Tiere 74:427–435.

Olafsen, J. A. 1988. Role of lectins in invertebrate humoral defense. American Fisheries Society Special Publication 18:189–205.

Parish, C. R. 1977. Simple model for self/non-self discrimination in invertebrates. Nature (London) 267: 711–713.

Prowse, R. E., and N. N. Tait. 1969. In vitro phagocytosis and amoebocytes from the hemolymph of *Helix aspersa* (Muller). Immunology 17:437–443.

Renwrantz, L. 1981. *Helix pomatia*: recognition and clearance of bacteria and foreign cells in an invertebrate. Pages 133–138 in J. B. Solomon, editor. Aspects of development and comparative immunology, volume 1. Pergamon Press, New York.

Renwrantz, L. R., and T. C. Cheng. 1977. Identification of agglutinin receptors on hemocytes of *Helix pomatia*. Journal of Invertebrate Pathology 29:88–96.

Renwrantz, L., J. Daniels, and P. Hansen. 1985. Lectin-binding hemocytes of *Mytilus edulis*. Developmental and Comparative Immunology 9:203–210.

Renwrantz, L., and W. Mohr. 1978. Opsonizing effect of serum and albumin gland extracts on the elimination of human erythrocytes from the circulation of *Helix pomatia*. Journal of Invertebrate Pathology 31:164–170.

Rodrick, G. E., and T. C. Cheng. 1974. Kinetic properties of lysozyme from the hemolymph of *Crassostrea virginica*. Journal of Invertebrate Pathology 24:41–48.

Schoenberg, D. A., and T. C. Cheng. 1980. Lectin-binding specificities of hemocytes from two strains of *Biomphalaria glabrata* as determined by microhemadsorption assays. Development and Comparative Immunology 4:617–628.

Sharon, N., and H. Lis. 1972. Lectins: cell agglutinating and sugar-specific proteins. Science (Washington, D.C.) 177:949–959.

Sminia, T., W. P. W. Van der Knapp, and P. Edelenbosch. 1979. The role of serum factors in phagocytosis of foreign particles by blood cells of the freshwater snail *Lymnaea stagnalis*. Developmental and Comparative Immunology 3:37–44.

Sminia, T., A. A. Winsemius, and W. P. W. Van der Knapp. 1981. Recognition of foreignness of blood cells of the freshwater snail *Lymnaea stagnalis*, with special reference to the role and structure of the cell coat. Journal of Invertebrate Pathology 38: 175–183.

Tripp, M. R. 1966. Hemagglutinin in the blood of the oyster. Journal of Invertebrate Pathology 8:478–484.

Tripp, M. R., and V. E. Kent. 1967. Studies on oyster cellular immunity. In Vitro (Rockville) 3:129–135.

Tyler, A. 1946. Natural hemagglutinins in the body fluids and seminal fluids of various invertebrates. Biological Bulletin (Woods Hole) 90:213–219.

Van der Knapp, W. P. W., L. H. Boerrigter-Barendsen, D. S. P. Van der Hoeven, and T. Sminia. 1982. Immunocytochemical demonstration of humoral defense factor in blood cells (amoebocytes) of the pond snail, *Lymnaea stagnalis*. Cell and Tissue Research 213:291–296.

Vasta, G. R., J. T. Sullivan, T. C. Cheng, J. J. Marchalonis, and G. W. Warr. 1982. A cell membrane-associated lectin of the oyster hemocyte. Journal of Invertebrate Pathology 40:367–377.

Warr, G. W. 1981. Immunity in invertebrates. Journal of Invertebrate Pathology 38:311–314.

Yoshino, T. P. 1981. Comparison of concanavalin A-reactive determinants on hemocytes of two *Biomphalaria glabrata* snail stocks: receptor binding and redistribution. Developmental and Comparative Immunology 5:229–239.

Yoshino, T. P., and T. C. Cheng. 1976. Experimental induced elevation of aminopeptidase activity in hemolymph cells of the American oyster, *Crassostrea virginica*. Journal of Invertebrate Pathology 27:267–270.

Yoshino, T. P., L. R. Renwrantz, and T. C. Cheng. 1979. Binding and redistribution of surface membrane receptors for concanavalin A on oyster hemocytes. Journal of Experimental Zoology 207: 439–450.

American Fisheries Society Special Publication 18:189–205, 1988

Role of Lectins in Invertebrate Humoral Defense

Jan A. Olafsen[1]

Department of Marine Biochemistry, College of Fisheries
University of Tromsø, Tromsø, Norway

Abstract.—Marine filter-feeders such as oysters are exposed to a microflora that harbors multiple pathogens and must have evolved effective protection to distinguish between self and nonself on a molecular level. This ability is fundamental to all multicellular organisms in internal defense, symbiosis, and the control of cell growth and proliferation. The evidence that lectins are involved in recognition and defense in invertebrates is considerable. Even though biochemical information on lectins is voluminous and detailed, the molecular mechanisms for discrimination between self and nonself by invertebrates are still not known. Extensive homologies that have been observed between the primary sequences of lectins from taxonomically related sources demonstrate that these proteins have been conserved through evolution. The discussion presented here focuses on the relationships between molecular architecture and biological function of invertebrate lectins, with emphasis on heterogeneity. The evidence of lectin induction through antigenic stimulation also is discussed, as are lectins as opsonins in invertebrate humoral defense; the Pacific oyster *Crassostrea gigas* is used as a molluscan example. A lectin needs at least two different binding sites to facilitate the recognition of nonself and then the reaction with hemocyte surfaces. Multimeric lectins, such as the lectins of Pacific oyster, can bind through multipoint attachment between antigen and ligand. Recognition or binding could also be modulated by molecular rearrangement and increase of available noncryptic binding sites. Reaction of Pacific oyster lectins with ligands is modulated by divalent cations and possibly also by serum lipid. The discussion is focused on different theories and molecular models for binding or recognition by multimeric lectin molecules in this invertebrate system.

Chemical mediation of fundamental biological reactions is a well-documented phenomenon in all living organisms. In the marine environment, the sensory systems of taste, smell, and touch form a continuum where chemoreception and recognition are fundamental (Grant and Mackie 1974). Müller et al. (1984) established that a lectin was the biochemical key for a symbiotic relationship between algae and colonial ascidians. The symbiosis involved cytostatic compounds produced by the animal and transfer of nutriments from algae to host, and this relationship was maintained by a lectin. Müller et al. (1981) demonstrated that a sponge lectin was the molecular cue to a symbiotic relationship between the sponge and a bacterium. In contrast, attachment and ingestion of certain bacteria by pathogenic amoebae were facilitated by fimbrial lectins of the associated bacteria. The pathogenicity of these amoebae was related to the association of the amoebae with specific bacteria.

Settlement and metamorphosis of marine invertebrate larvae can be regulated by molecules such as lectins (Maki and Mitchell 1985), and there is increasing evidence that interactions inside vertebrate bodies have external analogs that function at the interface between marine animals and the environment. For a recent review of macromolecular cues in marine systems, see Rittschof and Bonaventura (1986).

The most fundamental chemical cues are related to feeding, reproduction, and defense (discrimination between self and nonself). For marine invertebrates, the following compartments and interfaces can be visualized; seawater carrying food particles, pathogens and predators, hemolymph with circulating hemocytes (and invading pathogens), and the intracellular fluid. The outer shell and soft tissues, the surface of the circulating cells, walls of blood vessels, and the soft connective tissues constitute primary and secondary interfaces.

The invertebrate protective system appears to be primitive relative to that of mammals because of a seeming lack of diversity in chemical specificity and reactions with epitopes, absence of immunoglobulins, and relatively short-term protective memory. Marine invertebrates frequently occupy areas with a rich microflora that harbors multiple pathogens, and their filter-feeding habits greatly increase the risk for infection; yet their

---

[1]Present address: Institute of Fishery Technology Research, Post Office Box 677, N-9001, Tromsø, Norway.

body fluids are normally sterile. They must have evolved effective protection over millions of years to distinguish between self and nonself, and, in this context, the invertebrate immune system must be considered highly successful.

Invertebrate immunity has been extensively reviewed (Tripp 1970; Bang 1973; Jenkin and Hardy 1975; Chorney and Cheng 1980; Lackie 1980, 1981; Fletcher and Cooper-Willis 1982; Renwrantz 1983, 1986; Solomon 1986); for a recent comprehensive treatise the reader is referred to Brehelin 1986. The role of lectins in invertebrate immunity has also been extensively discussed by Yeaton (1981a, 1981b), Ey and Jenkin (1982), Renwrantz (1983, 1986), Vasta and Marchalonis (1983), Fries (1984), and Olafsen (1986).

Role of Hemocytes in Humoral Defense

Marine bivalves possess hemolymph blood cells (hemocytes) that play a role in internal defense, in parallel with mammalian phagocytes (Foley and Cheng 1975). Hemocytes are multifunctional cells that take part in phagocytosis and encapsulation of foreign material, tissue and shell repair, and transport of shell precursor material. They originate possibly from the endothelial lining (Rhodes and Ratcliffe 1983), and many invertebrates possess organs containing depots of phagocytes. The escargot *Helix pomatia* possesses fixed phagocytes associated with the digestive diverticula, and snail hemocytes display chemotaxis toward bacteria (Schmied 1975). In molluscs, hemocytes are distributed throughout the circulatory system and are also transported across the epithelium to the alimentary canal (Feng et al. 1977). Specificity of hemocyte populations has also been demonstrated by selective phagocytosis of Gram-positive or Gram-negative bacteria effected by different subpopulations of hemocytes from the polychaete *Arenicola marina* (Fitzgerald and Ratcliffe 1982).

ZoBell and Feltham (1938) established the importance of bacteria as food for marine invertebrates; they argued that bacteria might play a role for marine animals by furnishing cell substances and micronutrients. Marine fish larvae ingest bacteria which eventually become endocytosed by epithelial cells in the hindgut of the alimentary canal (Olafsen 1984; Olafsen, unpublished data). Phagocytosis is a major attribute of hemocytes, and the reaction with nonself and the disposal of unwanted material are fundamental processes to all living organisms. In filter-feeding invertebrates, the process has been related to nutrition;

for example, the amoebocytes of the oyster move freely through the gut epithelium and engulf partially digested particulate food (Takatsuki 1934). Intracellular digestion follows, and the products of digestion are made available to all the cells of the body. In molluscs with more complex digestive organs, intracellular digestion plays a minor role (Yonge 1937; Tripp 1960). An antigalactan-specific agglutinin–precipitin system important to the symbiosis between bivalves and algae was described from the hemolymph of tridacnid clams by Uhlenbruck and Steinhausen (1977). The agglutinating molecules were essential for elimination and utilization of algae, which had become senescent and, during degeneration, had exposed galactan structures on their surfaces. Evidence was presented that the lectin system of tridacnids served mainly for digesting their symbionts and their photosynthetic products. The role of hemocytes in cellular feeding and digestion and their importance in defense has been discussed by Jenkin (1976), Elston (1980), and Fisher (1986).

The phagocytic response is very similar for all invertebrates studied; small particles are phagocytosed, but particles greater than 10 μm tend to be encapsulated rather than phagocytosed. Rodrick and Ulrich (1984) reported that the concentration of hemolymph glycogen increased during in vivo phagocytosis of *Escherichia coli* and *Vibrio anguillarum* by the eastern oyster *Crassostrea virginica*. The study of phagocytosed material by transmission electron microscopy revealed the presence of numerous glycogen rosettes in endophagocytic vesicles. Glycogen concentration increased significantly in the hemocytes after 60 min of exposure, and high activity of hydrolytic enzymes was demonstrated. The results agreed with previous investigations by Cheng and Cali (1974) and by Cheng and Rudo (1976).

Phagocytosis of foreign particles, heterologous proteins, and allogeneic or xenogeneic cells implanted in the body cavity; the specificity of phagocytic reactions in invertebrates; and the roles of substances that act as opsonins have been described in several review articles (Chorney and Cheng 1980; Cheng 1981, 1984; Ey and Jenkin 1982; Sminia et al. 1983; Robohm 1984; Cooper 1985; Solomon 1986). Hemocytes and molecules in molluscan immunity were reviewed by Fisher (1986) and Sminia and van der Knapp (1987).

Hemolymph Factors—Lectins

There is strong and increasing evidence that the hemocyte–foreign body interactions are receptor-mediated; thus, these interactions depend on reactions between hemocyte surface components and foreign molecular structures. Glycoproteins are involved in binding bacteria to macrophages (Sharon 1984), and it is conceivable that lectins reacting with glycoprotein recognition determinants also function in invertebrate phagocytosis (for recent discussions see Vasta and Marchalonis 1983; Cheng et al. 1984; Olafsen 1986; Renwrantz 1986, Sminia and van der Knapp 1987). In a similar manner, lectins in reproductive organs may function to protect eggs (Vasta and Marchalonis 1983; Rögener and Uhlenbruck 1984) or prevent polyspermy (Yeaton 1981b), and may be involved in other functions such as cell–cell interactions or organization (Gold and Balding 1975; Barondes 1981).

Even though specific factors and recognition of defined determinants are probably required for phagocytosis by hemocytes, other forces, such as hydrophobicity and charge interaction, are also important (van Oss and Gillman 1972). Bacteria more hydrophobic or with a greater negative charge than the phagocyte are readily taken up, whereas those more hydrophilic are not easily engulfed. Surface charge and wettability of invertebrate hemocytes and the physical properties of foreign substances have been reported to influence phagocytosis (Lackie 1983).

Lectins are carbohydrate-binding, nonimmunoglobulin proteins that agglutinate cells or precipitate glycoconjugates through interaction with glycoproteins or glycolipids (Boyd and Shapleigh 1954). Although lectins, by definition, bind carbohydrates, forces such as hydrophobicity, charge interaction, and multivalent binding may stabilize the bond.

Lectins have been described from viruses, bacteria, yeasts, plants, invertebrates, vertebrates, and as endogenous membrane components of cells (Sharon and Lis 1988). Carbohydrate binding by lectins has been implicated in fertilization of sea urchin and mammalian eggs, in gamete fusion in algae, in sexual compatibility in yeasts, in cell–cell communication and adhesions of slime molds (Barondes 1981), and in symbiosis between bivalve clams (tridacnids) and algae (Uhlenbruck and Steinhausen 1977). Müller et al. (1984) determined that lectins are the biochemical basis for the symbiotic relationship between *Prochloron* sp. cells and the colonial ascidians *Didemnum fulle* and *Didemnum molle*. The symbiosis was based on cytostatic compounds produced by the animal and transfer of nutriments from algae to host, and this relationship was maintained by a lectin. Colonies rich in these algae produced large amounts of lectin, whereas colonies lacking *Prochloron* did not contain any lectin activity. Müller et al. (1981) demonstrated that a lectin from the sponge *Halichondria panicea* was the molecular cue of a symbiotic relationship between the sponge and the bacterium *Pseudomonas insolita*. In contrast, attachment and ingestion of certain bacteria by the pathogenic amoeba *Entamoeba histolytica* was facilitated by fimbrial lectins of the associated bacteria *Escherichia coli* and *Serratia marcescens*. It is notable that the pathogenicity of these amoebae is related to the association of the amoebae with bacteria.

Apart from playing a role in self–nonself recognition, lectins may organize macromolecules or multienzyme complexes, take part in transport of calcium or sugars, or promote signals across biological membranes. The occurrence, structure, specificity, and possible function of lectins have been extensively reviewed (see references in Olafsen 1986). The function and applications of lectins (primarily noninvertebrate) in biology and medicine have recently been comprehensively treated (Liener et al. 1986). Invertebrate lectins have been reviewed by Yeaton (1981a, 1981b), Ey and Jenkin (1982), Renwrantz (1983, 1986), Cheng et al. (1984), and Fries (1984), and, with specific reference to heterogeneity, by Olafsen (1986). For detailed descriptions the reader is referred to these reviews; only recent examples are presented here.

Many lectins are oligomers or multimers of homogeneous subunits with uniform carbohydrate-binding specificity. To accomplish recognition of nonself and subsequent opsonization by phagocytic cells, a lectin would need at least two different binding sites, one to bind the antigen and the other to react with the hemocyte surface. To avoid reaction with self when not bound to antigens, the hemocyte-directed site should be masked, inactive, or cryptic. Reaction with nonself could possibly invoke molecular rearrangement to expose self-determinants (Olafsen 1986). For a detailed discussion of hemocyte-bound lectins the reader is referred to Coombe et al. (1982, 1984) or Renwrantz (1986).

Recently, a sialic acid-binding lectin (achatinin$_H$) has been purified in a single step from hemolymph of the giant African small *Achatina fulica* by affinity chromatography of sheep submaxillary mucin coupled to Sepharose 4B (Basu et al. 1986). The molecular weight (M_r) of the native lectin was 242,000 with identical subunits of M_r 15,000; it agglutinated rabbit erythrocytes in the presence of

Ca^{2+}, and fetuin was the best inhibitor of agglutination. Cassels et al. (1986) isolated heterogeneous humoral and hemocyte-associated lectins with N-acylaminosugar specificities from the blue crab *Callinectus sapidus*. The lectins required divalent cations and agglutinated free sugars as well as sialoglycoproteins. It was inferred from the binding characteristics that the lectins would also agglutinate a number of *Vibrio parahemolyticus* serotypes. Flower et al. (1985) reported two lectins in the hemolymph of the oyster *Pinctada maxima*; the major lectin was a macromolecular aggregate (M_r 8,000,000) and agglutinated human type-A erythrocytes with a specificity for N-acetylgalactosamine. The minor lectin (M_r 400,000) bound to α-galactose and agglutinated all human blood groups.

Lectins of the Pacific Oyster

Hardy et al. (1977b) described agglutinating activity of a variety of erythrocytes in the cell-free hemolymph of the Pacific oyster *C. gigas*. The active agents against human and horse erythrocytes were separately named gigalin H (human) and gigalin E (equine), respectively. They were purified by affinity chromatography on bovine submaxillary mucin (BSM) coupled to Sepharose 4B. Gigalin H was specific for N-acetylneuraminic acid residues. The inhibitory effect of BSM, expressed as sialic acid residues, was dramatically greater than that of free N-acetylneuraminic acid, and this greater inhibition indicated multiple binding sites with affinity for sialic acid residues.

Further characteristics of these agglutinins were described by Olafsen (1986). Both lectins contained lipid and carbohydrate (about 20% of each) and were extremely hydrophobic with a tendency to self-aggregate, especially in the presence of calcium. They were high-molecular-weight aggregates in the native state as well as in the presence of either EDTA or calcium. They appeared as macromolecular aggregates at distinct stages of aggregation with M_r from 500,000 to about 1,600,000. These properties were demonstrated both by electrophoresis and gel filtration of purified oyster lectins and also of crude fractions and cell-free hemolymph in different buffer systems. That these aggregates were native was also confirmed by the fact that their state of aggregation was unaffected by the addition of ligand, other proteins, or cell-free hemolymph (Olafsen 1986; Olafsen, unpublished data). The aggregates dissociated in 8 M urea; the resulting subunits could not agglutinate cells, but addition of excess subunits

inhibited agglutination of human erythrocytes. The subunits reaggregated easily but never to form active complexes. Sodium dodecyl sulfate–polyacrylamide gel electrophoresis (SDS–PAGE) under reducing conditions demonstrated at least three subunits, α_1 (M_r 21,000), α_2 (M_r 22,500) and β(M_r 33,000). Gigalin H activity was predominated by $\alpha_1(\alpha_2)$, whereas gigalin E activity was $\alpha_2(\beta)$. Although calcium did not appear necessary for binding to BSM, gigalin E, but not gigalin H, demonstrated a calcium requirement for the agglutination of erythrocytes (Olafsen, unpublished data). By measuring the effect of metal substitution on intrinsic protein fluorescence, it was demonstrated that Ca^{2+} affected protein conformation but was not directly involved in ligand binding (R. Hovik and Olafsen, unpublished data).

The native lectins had binding sites both for human and horse erythrocytes on the same polymeric molecule. Of further importance was the finding that, in affinity-purified lectin preparations, one binding site could be masked (or hidden) and reexposed by chemical treatment or interaction with a ligand (Olafsen, unpublished data). When cell-free hemolymph was filtered on Biogel P-300 in 1 mM EDTA and 2.5 M sodium chloride, the activity of the recovered gigalin E was five times that applied. Even though gigalin E had no affinity for sialic acid or BSM, complete separation of gigalin H and gigalin E by affinity chromatography was impossible. Absorption of all detectable gigalin E with a great excess of glutaraldehyde-treated horse erythrocytes prior to purification resulted in the two activities reappearing when the bound fractions were eluted and dialyzed. Likewise, rechromatography of the nonbound fraction from BSM–Sepharose (devoid of gigalin H activity) always resulted in mixed activities. This result indicates that gigalin multimers contained masked or cryptic binding sites that could be activated by reaction with a ligand or exposure to high ionic strength.

The lectins from Pacific oyster were extremely hydrophobic and easily precipitated from aqueous solutions. Extraction with chloroform–methanol reduced lectin activity against human erythrocytes but did not affect activity against horse erythrocytes (Olafsen, unpublished data). Also, extraction to remove lipid from cell-free hemolymph and purified preparations resulted in increased gigalin E activity. The results thus indicate that serum- or hemocyte-associated lipid may inhibit gigalin E. In comparison, embryos of the frog *Xenopus laevis* at cleavage, gastrula, and neurula stages contained a galactose-specific lectin (Harris and Zalik 1985). In order to identify a single lectin band in SDS–PAGE, it was necessary to extract aqueous suspensions of the purified lectin with chloroform–methanol. The lectin remained in the aqueous layer and gave rise on SDS–PAGE to a distinct band of M_r 65,500. Aqueous suspensions of the purified lectin that were not subjected to chloroform–methanol extractive gave rise to several bands. In addition the lectin showed a high degree of aggregation under aqueous conditions when gel-filtered.

Induction of Lectin Activity

Lectins were induced in the earthworm *Lumbricus terrestris* (Stein and Cooper 1982) by injection of foreign erythrocytes, and the response reached a maximum titer at 24 h. Later it was shown that coelomic fluid of *L. terrestris* contained agglutinins that bound with and agglutinated Gram-positive and Gram-negative bacteria (Stein et al. 1986). When worms were injected with five bacterial strains, three of them isolated from *Lumbricus* coelomic fluid, four of five strains induced higher levels of coelomic fluid agglutinins within 24 h. Each bacterial strain reacted with different agglutinins, and 2-keto-3-deoxyoctanate was the only sugar to consistently inhibit agglutination of all five bacterial strains. Faye et al. (1975) demonstrated that the injection of viable bacteria into larvae of the insects *Hyalophora cecropia* and *Samia cynthia* induced hemolymph proteins within 6 h. Although these proteins were not lectins, the ability to respond to bacterial challenge was demonstrated. Also, bacteria injected into the hemocoel of larval tobacco hornworm *Manduca sexta* induced a number of proteins in the hemolymph (Minnick et al. 1986). One of these proteins was bifunctional with glucose-specific lectin activity which also initiated coagulation in certain populations of hemocytes.

Qu et al. (1987) isolated and purified a lectin with affinity to galactose from the hemolymph of Chinese oak silk moth pupae *Antheraea pernyi*. The lectin had a molecular weight of 380,000 and formed oligomeric structures of a subunit of M_r 38,000. The agglutinating activity increased with time following immunization with *Escherichia coli*, and it was thus inferred that de novo synthesis of the lectin was promoted by injection of *E. coli*.

Clearance of bacteria was studied in eastern oysters and northern quahogs *Mercenaria mercenaria* experimentally contaminated with *E. coli, Salmonella typhimurium*, and *Shigella flexneri* either by intracardial injection or via natural ingestion (Hartland and Timoney 1979). Bacterial inactivation in the hemolymph was monitored for 72 h at 20° and 6°C after exposure to these enteric pathogens. At 6°C, both mean bacterial uptake by ingestion and subsequent clearance was significantly lower than at 20°C, but substantial clearance of bacteria from the hemolymph occurred at both temperatures for both shellfish. After 24 h at 20°C, viable bacteria were no longer detectable in hemolymph of either host species after exposure to contaminated water containing 4×10^3 bacteria·mL^{-1}.

An antigalactan-specific agglutinin–precipitin system, described from the hemolymph of tridacnids, may be important for symbiosis between the clams and algae (Uhlenbruck and Steinhausen 1977). The agglutinating and precipitating tridacnid molecules were regarded as essential for the elimination and utilization of senescent algal guests, which, during degeneration, had exposed galactan structures on their surfaces. Evidence was presented that the lectins of tridacnids served to digest the symbionts and their photosynthetic products.

Acton and Evans (1968) reported enhanced secondary clearance of bacteriophage from eastern oysters. The California seahare *Aplysia californica* cleared bacteria more rapidly after second exposure than after first exposure (Pauley et al. 1971b); however, titers were not increased by prior challenge with any of the bacteria. Following second and third bacterial injections of *Staphylococcus aureus* into the freshwater snail *Planobarius corneus*, clearance rates were faster and clearance patterns were markedly different from responses to the first injection (Ottaviana et al. 1986). Clearance correlated with increases in serum lactate dehydrogenase and hemocyte acid phosphatase activities; serum acid and alkaline phosphatases decreased and α-amylase remained constant.

Hardy et al. (1977a) demonstrated that in vivo exposure of Pacific oysters to bacteria in the water resulted in increased lectin activity in oyster hemolymph; the increased activity suggested the involvement of lectins in a defense reaction. Later, this result was verified by in vivo exposure of oysters to the pathogen *Vibrio anguillarum* NCMB 6 at different time intervals, temperatures, and seasons (Olafsen, unpublished data). Increased activity was demonstrated 6 h after bacteria were added to the water (about 10^7 bacteria·mL^{-1}), indicating activation or release of oyster lectin rather than de novo synthesis. The response affected primarily gigalin E, and the lectin activity returned to background values after a week. By then, the bacterial count in the water was down to normal (about 10^3–10^4 bacteria·mL^{-1}).

The failure of other workers to induce lectin activity in molluscs by challenge may be due to several factors. Acton and Evans (1968) injected sheep erythrocytes into eastern oysters without detectable response. However, injection may not have provoked a response, or possibly sheep erythrocytes were an unnatural challenge. Pauley et al. (1971b) observed an initial decrease in agglutinating activity following injection of bacteria into the California seahare. They interpreted the decrease as rapid depletion of recognition factors resulting from the massive injection of bacteria.

For filter feeders such as oysters, exposure to enhanced levels of a pathogenic bacterium in seawater represents a natural challenge. *Vibrio anguillarum* is a facultative pathogen responsible for the fish disease vibriosis (Hendrie et al. 1971; Olafsen et al. 1981). It is one of the preponderant organisms associated with Pacific oysters (Colwell and Sparks 1967), and is pathogenic to oysters (Grischkowsky and Liston 1974). We (Hardy et al. 1977a) demonstrated that Pacific oysters responded to bacterial challenge with increased lectin activity in cell-free hemolymph and thus were capable of induced humoral responses. This demonstration was the first to show an induced lectin response in an invertebrate.

Lectin activation by proteolytic cleavage was discussed by Olafsen (1986). Komano et al. (1980, 1981) described an inducible lectin in the larvae of the flesh fly *Sarcophaga peregrina*; the lectin consisted of four α and two β subunits. Hemolymph from normal larvae contained only the α subunit (Komano et al. 1981). Upon injury to the larval body wall or pupation, some of the α subunits were converted to β subunits by a protease, resulting in formation of active $\alpha_4\beta_2$ lectins. Evidence that this lectin stimulated elimination of erythrocytes introduced into the abdominal cavity of the *S. peregrina* larvae has since been presented (Komano and Natori 1985).

A lectin in *Giardia lamblia* was activated by secretions from the human duodenum, the environment where the parasite lives (Lev et al. 1986). Because trypsin inhibitors prevented activation, proteases were implicated as the activating agent. Activation by crystalline trypsin and pronase was demonstrated; other proteases tested were ineffective. When activated, the lectin agglutinated intestinal cells to which the parasite adhered in vivo. Activation of a parasite lectin by a host protease represents a new mechanism of host–parasite interaction and may contribute to the affinity of *Giardia lamblia* for the infection site.

Lectins and Reaction with Bacteria; Opsonization

The term opsonin is used to describe agents in serum which act on particles or cells to increase their ingestion by phagocytic cells. Several reports on the opsonic properties of hemolymph factors from different invertebrates are available (Ratcliffe and Rowley 1984; Olafsen 1986; Renwrantz 1986; Solomon 1986). However, most of these reports are concerned with crude or partly purified hemolymph factors, and only rarely with purified lectins.

Opsonization provides an important link between cellular and humoral defense mechanisms. Tripp (1966) demonstrated that eastern oyster hemolymph, which agglutinated rabbit erythrocytes, also enhanced the in vitro uptake of these erythrocytes by oyster hemocytes. Prowse and Tait (1969) showed that hemolymph promoted in vitro phagocytosis of yeast cells and sheep erythrocytes by the brown gardensnail *Helix aspersa*. Phagocytosis of erythrocytes by lobster hemocytes occurred only in the presence of purified lobster lectins (Hall and Rowlands 1974b).

Pistole (1976) described a natural agglutinin for bacteria (*Salmonella* sp.) in the hemolymph of the horseshoe crab *Limulus polyphemus*, and Pauley et al. (1971a) detected agglutinins for marine bacteria in hemolymph from the California seahare. Zipris et al. (1986) demonstrated that the gonads and hemolymph of two Mediterranean seahares, *Aplysia depilans* and *A. fasciata*, contained lectins that were specific for D-galacturonic acid and D-galactosides. Whereas the hemolymph lectins exhibited heterogeneic specificity, they bound N-acetylated sugars. Both *Aplysia* gonad lectin and hemolymph lectin interacted with bacteria, including certain *E. coli* strains, *Bacillus subtilis*, *Pseudomonas aeruginosa* strains, marine bacteria such as the light-producing *Vibrio harveyi* and *Photobacterium leiognathi*, and marine bacteria cultivated from the close environment of *Aplysia*.

In slime molds *Dictyostelium discoideum* that were fed bacteria with abundant discoidin-binding glycoconjugates, the bacteria and slime mold lectin accumulated in special subcellular, multilamellar bodies (Cooper et al. 1986). In contrast, in cells fed bacteria that had been treated so as to thoroughly deplete them of discoidin-binding glycoconjugates, neither endogenous discoidin nor complementary glycoconjugates were found in the multilamellar bodies. The results indicate that the function of the carbohydrate-binding site of discoidin is to interact with bacterial glycoconjugates, which the slime mold does not degrade. This interaction directs compartmentalization of the lectin in multilamellar bodies and its externalization from the cell in these structures.

Renwrantz and his colleagues (Renwrantz and Mohr 1978; Renwrantz et al. 1981; Renwrantz 1983; Renwrantz and Stahmer 1983) reported increased clearance from circulation of erythrocytes preincubated with hemolymph. Secondary injections were cleared less effectively unless erythrocytes were pretreated with hemolymph; this reduced clearance indicated that recognition molecules had been depleted during the first injection. Evidence was presented to indicate that *Helix* lectin and opsonin were the same.

Renwrantz and Stahmer (1983) concluded that a calcium-dependent, membrane-associated recognition factor was present on hemocytes of blue mussel *Mytilus edulis*. This presence was demonstrated by the fact that Ca^{2+} alone could stimulate in vitro phagocytosis of yeast and erythrocytes by blue mussel hemocytes. The presence of identical lectins in the hemolymph and on the plasma membrane of hemocytes has also been reported for the eastern oyster (Vasta et al. 1982, 1984a) and the blue crab (Cassels et al. 1986). Mullainadhan and Renwrantz (1986) recently demonstrated that induction of phagocytosis by four heterologous lectins only occurred in blue mussels when the lectin could bind to carbohydrate determinants on both the hemocyte surface and the target cell. Lectins that bound to only one of the cell surfaces failed to stimulate phagocytosis.

Cell-free hemolymph from Pacific oysters agglutinated the marine bacterium *Vibrio anguillarum* in vitro. The reaction of oyster lectins with the bacterial surface was demonstrated by the observation that resting or dead *V. anguillarum* cells absorbed lectin activity from oyster hemolymph (Olafsen, unpublished data). Hardy et al. (1977a) demonstrated that affinity-purified oyster lectins from oyster cell-free hemolymph acted as opsonins, i.e., they increased the phagocytosis of bacteria (*V. anguillarum* NCMB 6 and *E. coli* K235) by oyster hemocytes in vitro. A 3- to 5-fold increase in phagocytosis was demonstrated in all cases when bacteria were precoated with lectin or when cell-free hemolymph (plasma) was used as the incubation medium. This report was the first of opsonization by highly purified lectins in an autologous system. Moreover, limulin, a sialic acid-specific

lectin from the horseshoe crab failed to increase phagocytosis by oyster hemocytes.

Thus, increased in vitro phagocytosis could be demonstrated when bacteria were precoated with hemolymph factors or purified lectin. However, purified lectins had to be added to bacteria prior to addition of hemocytes to obtain this effect. When bacteria and lectins were incubated simultaneously with hemocytes in saline solution, no increase in phagocytosis above the baseline value could be observed; however, uptake was stimulated when cell-free hemolymph was the incubation medium. This result indicated that either affinity-purified lectins exposed self-binding sites (high affinity for hemocytes) prior to the reaction with bacteria, or some factor in the cell-free hemolymph was preventing the reaction of lectins with self prior to recognition of nonself.

The subcellular events in the phagocytosis of *Vibrio vulnificus* and *V. anguillarum* by eastern oyster hemocytes were studied by scanning and transmission electron microscopy by Rodrick and Ulrich (1984). Phagocytosis of bacteria was influenced by the presence of serum in the medium. With serum present, phagocytosis was increased 3–10 times, depending on the bacterial strain employed. For hemocytes reconstituted with heat-treated serum, phagocytosis was depressed, indicating that the effect was most likely due to a heat-labile serum factor.

Discussion

The understanding of invertebrate immune protection has, to a great extent, been influenced by the search for primitive forms of mammalianlike immunocompetent molecules. Even though the invertebrates lack inducible immunoglobulins or a complement system, the presence of invertebrates in such great numbers in marine and terrestrial ecosystems indicates a defense system of great efficiency.

Although invertebrates, in some aspects, demonstrate induced immune responses following antigen challenge, induction of lectin activity remains a disputed phenomenon. However, increased lectin activity following exposure to bacteria has been demonstrated (see Olafsen 1986 and previous section). Bacteria may be regarded as food for filter-feeding bivalves. In an evolutionary sense, "effective eating" could be related to defense, even though the coelomic vasculature normally remains sterile. The same type of amoeboid cells are involved both in intracellular digestion and as defense. In filter-feeding bivalves like oysters, amoebocytes move freely through the gut epithelium and engulf food particles which are digested intracellularly (Tripp 1960).

Recognition Theories

Recognition phenomena have been described for more than 100 years. Important observations were made as early as 1891, when Metchnikoff recognized the ability of macrophages to differentiate between self cells and intruding pathogens. All multicellular organisms must have acquired, at an early evolutionary stage, the ability to discriminate between self and nonself to prevent pathogenic organisms from invading and multiplying in their tissues, to maintain control of cell growth and proliferation, and to prevent cancerous cells from multiplying without restriction. During evolution from Protozoa to Metazoa, cells have acquired the ability to adhere to cells of the same species. Such adherence is mediated by specific recognition factors (receptors for self) even in primitive organisms such as sponges (Barondes 1981). The ability to discriminate between self and nonself must have evolved early in evolution to allow development of multicellular organisms and to enable phagocytes to free the organisms from pathogens; some of the fundamental mechanisms of self–nonself discrimination may thus be the same in animals and in plants.

The evidence that lectins are involved in recognition and defense in invertebrates is considerable. However, even though comprehensive and detailed biochemical information on lectins are now available, the molecular mechanisms for discrimination between self and nonself by invertebrates are still not known. Recognition theories have recently been discussed by Solomon (1986); some of them are briefly described here.

Theories dealing with self–nonself recognition have developed along two themes, recognition of self ("dealing with the expected") and recognition of nonself ("dealing with the unexpected") as in the vertebrate immune system. This was discussed by Kolb (1977), who proposed that nonspecific phagocytosis was due to indirect recognition of foreign structures; particles that failed to fit receptors for self would be rejected as foreign. Lack of specific self-recognition might be either due to lack of self-determinants on foreign cells, or differences in electrical charge patterns, surface tension, or other physical variables.

Langman (1978) proposed a "switch-off" mechanism concerned with the preservation of colony identity and feeding groups as well as defense against infectious agents. Kolb (1977) suggested

that proliferative cells were unable to establish cell contact via self-receptors and thus were detected and destroyed as foreign cells. Indirect recognition of foreign structures (lack of self-determinants) would be a drawback if foreign cells with surface structures mimicking the host's self-determinants were accepted as self; pathogens with self-determinants would not be recognized as nonself. It was proposed that "invariant," or unchanging, antiforeign cells of invertebrates would recognize structures or pathogenic microbes that were also invariant; otherwise microbes would soon avoid detection by the host. Invariancy would be conserved only if mutation was disadvantageous, as would be the case with those structures which mimicked host self-determinants; variation of such structures would impair mimicry and reduce virulence. To avoid cancerous growth, it would be imperative to detect cells expressing altered self-determinants. Kolb (1977) proposed that self-determinants were carbohydrates, whereas receptors for self were proteins or lectins. The invertebrate immune system thus could exist with 20–100 receptors directed against invariable antigenic determinants of environmental pathogens.

Involvement of carbohydrates of the glycocalyx of eukaryotic cells in recognition was proposed by Roseman (1970, 1974). In recent years, it has been proposed that enzymes could function in such a way (Parish 1977); the theory involved glycosyltransferases that recognized their sugar acceptor and were randomly polymerized into recognition factors. This hypothesis was not supported by experimental evidence, and most lectins are devoid of enzymatic activity.

The ability to express self-markers of the host on the cell surface is a form of immune evasion used by successful parasites (see Cheng 1988, this volume). Such mimicry of host self is most likely driven by increased variety and sophistication of the host's ability to recognize specific structures. The idea of antigen sharing based on similarities in antigen determinants was proposed by Damian (1964), and if lectins are important as defense molecules, perhaps selection has favored sharing of their carbohydrate determinants–receptors between parasite and host. Aspects of such mechanisms have more recently been discussed by Maramorosch and Shope (1975) and Lackie (1980). However, caution was warranted in conclusions about molecular mimicry based on serological tests with mammalian antisera if host and parasite were both invertebrates. Distantly related immunogens could bind epitopes which are widespread phylogenetically and yield spurious evidence of homology. According to Bayne et al. (1987), this was probably the case when *Schistosoma mansoni* was first interpreted to mimic the host bloodfluke planorb *Biomphalaria glabrata*. Boswell et al. (1987) showed that common carbohydrate epitopes could account for cross-reactivity among host and parasite components. Parasite sporocysts expressed a heterogeneous population of glycosylated surface proteins. Common and widespread carbohydrate epitopes were associated with a variety of parasite surface glycoconjugates, suggesting that they might play a role in cellular interaction with host hemocytes. Results indicated that the surface epitopes were protease-resistant or had an internal source. Although natural resistance of molluscs to trematode infection is known to be mediated by circulating hemocytes (Bayne 1983), the molecular mechanisms underlying host immune recognition or parasite immune evasion strategies are not known in detail (Boswell et al. 1987).

Jenkin and Hardy (1975) proposed a hypothesis for "the generation of diversity" from their experience with the multispecific crayfish lectins. Erythrocyte agglutinating activity was removed by absorption but could be restored by first treating hemolymph with EDTA and then with tris–calcium buffer. It was inferred that this treatment caused the association of monomers into new functional lectins. Six different subunits combined in a random fashion could form nearly a thousand different specificities, which could perhaps suffice for defense in invertebrates.

A lectin needs two binding sites to facilitate first the reaction with nonself and then ingestion by self-hemocytes. In a previous review, Olafsen (1986) suggested that, for multimeric lectins, recognition and binding could operate through multipoint attachments between antigen and ligand. If each bond at attachment was weak, opsonization could be facilitated through cooperative or induced forces. Thus, recognition or binding could be the result of an increase (above a threshold) of available (noncryptic) binding sites reacting with nonself. This could explain the much reported failure of monomers to agglutinate cells or even bind firmly to ligands, and also the frequent potent inhibition of lectins by "multivalence" glycoproteins (mucins).

Renwrantz (1986) independently proposed a model that would support Olafsen's (1986) suggestions. If soluble serum lectin molecules have only low affinity for the complementary receptor site on the hemocyte surface, dissociation of the hemocyte-receptor–lectin complex would be much greater than association between receptors and lectin. According to the law of mass action, accumulation of serum lectin molecules on the surface of a microorganism would greatly increase binding of the target to the hemocyte surface. The effects described by Renwrantz would be even greater for multimeric lectins because the interaction would benefit from induced or cooperative forces.

Specificity and Heterogeneity in Ligand Binding

The blood-group specificity of some lectins is so sharply defined that it can distinguish between blood subgroups (Sharon and Lis 1988), but inhibition of lectins can usually be achieved by relatively few and simple monosaccharides or oligosaccharides (discussed by Olafsen 1986). A universal role of lectins as recognition molecules could be disputed because of this lack of diversity in sugar specificity. However, other factors, apart from sugar specificity, also may determine the successful agglutination of particles. Strong binding usually involves large ligands, which may possess multipoint attachment sites or induce positive cooperation. Many lectins bind glycolipids or agglutinate liposomes, and bacteria may be agglutinated by lectins specific for carbohydrates not usually found on the bacterial surface, as is the case when sialic acid-specific lectins (like limulin and gigalin) agglutinate bacterial cells devoid of this receptor (Olafsen, unpublished data). Thus, other forces beside the known carbohydrate specificity may determine agglutination, or there may be multiple undetected lectin activities in invertebrates (for recent discussion, see Renwrantz 1986).

We do not know how lectins can function as opsonins. After combination of a lectin with bacteria, the lectin may undergo a molecular change so that a different recognition site on the molecule can combine with receptors on the hemocyte surface. This receptor must be cryptic under "normal" conditions to prevent reaction with the hemocyte surface. The significance of heterogeneic binding sites has been treated elsewhere (Olafsen 1986), and multiple lectins, isolectins, and lectins with heterogeneous binding sites have been described in the hemolymph of different invertebrates. Some of these lectins are present as aggregated forms.

Finstad et al. (1974) reported amino acid heterogeneity of limulin and speculated that this heterogeneity could introduce functional variability. Evidence for heterogeneity of oyster agglutinins and difference in calcium activation was presented by McDade and Tripp (1967). They speculated that oyster hemolymph contained a family of polymeric agglutinins with different subunits reacting with structurally related sugars. Binding-site heterogeneity of lobster lectins was demonstrated by Hall and Rowlands (1974a, 1974b). Lectins from the Japanese horseshoe crab *Tachypleus tridentatus* showed heterogeneity in molecular weight and immunological properties (Shishikura and Sekiguchi 1983). Lectins of the tunicate *Halocynthia pyriformis* also demonstrated heterogeneity in calcium requirement and molecular weight (Form 1979; Form et al. 1979). Nguyen et al. (1986a, 1986b) reported evidence of polymorphism when they studied amino acid sequences of *Limulus* C-reactive protein (CRP) and nucleotide sequence of genomic DNA encoding the protein.

Function of Metals in Lectin Activation

Divalent cations, particularly calcium, frequently are a prerequisite for the agglutination of cells by lectins and also for stabilization of the native molecule (see Olafsen 1986). Binding sites or subunits with differing cation requirements in the same molecule have been reported for a number of lectins (Form 1979; Schluter et al. 1981; Ingram et al. 1983). Calcium was needed for the agglutination of horse erythrocytes but not human erythrocytes by oyster lectins (Olafsen 1986); agglutination of horse erythrocytes was inhibited by magnesium and zinc; 100% inhibition was achieved at zinc concentrations found in hemolymph. The demonstration that oyster hemocytes accumulate zinc (George et al. 1978; Simkiss and Mason 1983) led Olafsen (1986) to speculate that the opsonin could be influenced by the intracellular–extracellular ratio of divalent cations.

Manen and Comte (1986) demonstrated that each *Phaseolus vulgaris* isolectin exists in five electrophoretic forms related to the number of metal ions bound per molecule of this plant lectin. By measuring the effect of metal substitution on the intrinsic protein fluorescence for the oyster lectins, we could demonstrate that metals affected protein conformation but were not directly involved in ligand binding (Hovik and Olafsen, unpublished data). This makes it possible to speculate further that the different aggregation states of the oyster lectins could be stabilized and regulated by cations.

Relationship between Molecular Architecture and Function

The lectin of the yellow gardenslug *Limax flavus* behaved as a rapidly associating–dissociating polymeric system at high protein concentration (Miller et al. 1982), and the *Pinctada* lectins were macromolecular aggregates. Thus, these lectins resembled those of Pacific oyster, which were high-molecular-weight aggregates of three different subunits existing in a series of distinct aggregation states that expressed two functionally different activities. The association of subunits into biologically active multimers were probably complex since monomers could never be aggregated into active lectins.

The lectins from Pacific oyster were extremely hydrophobic (Olafsen 1986) and easily precipitated from aqueous solutions. Moreover, dialysis against different buffers always resulted in a 2- to 4-fold increase in the activity directed towards horse erythrocytes. The results so far indicate that lipid may be involved in inhibition of gigalin E (Olafsen et al., unpublished data). The increase of gigalin E activity by lipid removal leads to the presumption that lipid as well as cations could be modulators of lectin activity in invertebrates.

For multimeric lectins, hemocyanin could serve as a model for the relationship between molecular architecture and biological activity (Olafsen 1986). Similarities in specificity, serology, or molecular architecture inferred from chemical data or electron-microscopical observations between invertebrate lectins and hemocyanins have been observed. Arthropod hemocyanins are assembled from one, two, four, or eight hexameric units; thus *Limulus* hemocyanin contains 48 subunits (M_r 70,000), whereas the lectin, limulin, is a double-stacked hexamer (Gilbride and Pistole 1981). Hemolymph from common octopus *Octopus vulgaris* contained a lectin which revealed striking similarities with octopus hemocyanin (Rögener et al. 1986); i.e., its M_r (260,000–280,000) was similar to that of the minimal functional subunit of hemocyanin, an antiserum against octopus lectin cross-reacted with agglutinin hemocyanin.

Crayfish hemocyanin is a random association of three types of subunits into hexamers or dodecamers, and the molecule occurs in a number of distinct forms stabilized by calcium (Ey and Jenkin 1982). Richelli et al. (1984) described the importance of Ca^{2+} ions for the molecular conformation and the interaction between subunits in hemocyanin. Salvato et al. (1979) proposed that hemocyanin subunits might be connected through their carbo-hydrate moiety. Rögener et al. (1985) also found a strong calcium dependence among the binding properties of common octopus lectin. They implied that the lectin may have evolved from hemocyanin or perhaps from a tyrosinaselike molecule, a presumed precursor of hemocyanin in evolution (Van Holde and Miller 1982), without being functionally related to respiratory protein.

Mollusc hemocyanins are assembled from 10–20 subunits (M_r 400,000), and the aggregates are cylinders with a typical M_r of about 9×10^6 (Bonaventura and Bonaventura 1983). Markl et al. (1978) demonstrated that heterogeneity of subunits is of great importance to the formation of high-molecular-weight hemocyanin molecules. The stability of the aggregated hemocyanins depends on external conditions such as pH, ionic strength, and divalent cations; changes may bring about dissociation into several distinct states of aggregation. The parallel molecular architecture of multimeric lectins and hemocyanin is striking, and the similarity is supported by experimental evidence of the biochemical "behavior" of some invertebrate lectins.

What Lectins Do; Phylogenetic and Structural Aspects

The exact function of invertebrate lectins is still unknown, but the findings that lectins occur both in the serum and on the surface of hemocytes (Vasta et al. 1982, 1984a) suggest that these molecules play an essential role in self–nonself discrimination. Perhaps invertebrate lectins are related to early precursors of vertebrate immunoglobulins, but little information has yet been obtained to support this hypothesis. Kaplan et al. (1977) reported that the amino acid sequence of limulin showed no sequence homology with vertebrate immunoglobulins.

The acute phase of infection in most mammals is characterized by rapid increase in the concentration of certain plasma proteins, such as C-reactive protein (CRP), named for the unique property of reacting with pneumococcal C polysaccharide (and phosphorylcholine) in the presence of calcium and precipitating it (Tillett and Francis 1930). Limulus lectin bound phosphorylcholine as well as sialic acid (Robey and Liu 1981) and showed a number of similarities to CRP. Vasta et al. (1984b) used monoclonal antibodies to estimate the minimal structural requirements for idiotypic determinants and found cross-reaction between limulin, CRP, and the phosphorylcholine-binding myeloma protein TEPC-15. C-reac-tive protein, limulin, and the TEPC-15 variable-region heavy chain shared short stretches of homology (8–10 amino acids), which might account for the cross-reactivity. Nguyen et al. (1986a, 1986b) reported evidence of polymorphism from the amino acid sequence of Limulus CRP and the nucleotide sequence of genomic DNA encoding the protein. The amino acid sequence deduced from nucleotide analysis of the Limulus CRP gene also helped to provide insight into the evolutionary origin of CRP.

Among the invertebrates, tunicates are considered closest to vertebrates in evolution, and the study of recognition molecules in this subphylum might provide evidence for relatedness to vertebrate immunomolecules. Vasta et al. (1986) demonstrated that the plasma of the ascidian Didemnum candidum possessed lectin activity (DCL) toward galactosyl moieties. Isoelectric focusing revealed that DCL-I focused as a family of bands at pH 3.8–5.2, whereas DCL-II focused at pH 9.2–10.2. Gas chromatographic analysis indicated that carbohydrate was not associated with the lectins. Studies of amino acid composition revealed similarities to lectins from the tunicate Halocynthia pyriformis, the sea lamprey Petromyzon marinus, and the horseshoe crab Carcinoscorpius rotunda cauda, as well as to rabbit CRP and immunoglobulin μ-chains from sea lamprey and common carp Cyprinus carpio. The results were supported by serological evidence: enzyme-linked immunosorbent experiments showed that antibodies made against human CRP cross-reacted with DCL-I.

Lectins from some invertebrates have been highly purified, and though such work is highly warranted, caution should be applied to any biological interpretation based on lectins subjected to harsh chemical treatment. Even treatment such as biospecific affinity chromatography could theoretically induce self-determinants, and it is imperative to verify results under natural conditions. In a review of the assembly of blood clotting complexes on membranes, Mann (1987) described a 1,000-fold increase in the catalytic rate constant k_{cat} when activation factors acted on prothrombin in the presence of the cell membrane. Thus, activities and action of purified invertebrate lectins should also be tested in complete hemolymph.

Lectins from different invertebrate sources are diverse molecules of different shapes and specificities. The lectins described so far presumably represent only "the tip of an iceberg" (Renwrantz 1986), so the complexity of structure–function relationships of such molecules from one species should be investigated before general assumptions are made for all invertebrates.

If only two subunits or valences are required to facilitate agglutination or binding of a bacterial cell to a hemocyte, why are some invertebrate lectins large, multimeric proteins (Olafsen 1986)? One would expect nature to economize better. Possibly, binding depends on multisubunit attachment, or the multiple subunits of each molecule bear cryptic recognition sites. Induction of lectin activity could come about by molecular rearrangement to expose such cryptic binding sites or by deaggregation and release in the hemolymph of multiple, activated lectin oligomers. Binding of a polymeric multisubunit protein to an invading pathogen could activate and release multiple functional oligomers and thus increase the concentration of active circulating recognition molecules of a given specificity. If the subunit size is about M_r 2×10^4, and polymers as large as M_r 2×10^6 exist in the hemolymph, deaggregation to lectin oligomers could cause the concentration of active oligomers to increase at least 20-fold. Upon reaction with a hemocyte, such molecules, free or coupled to antigen, could become deactivated by aggregation to passive (latent) defense molecules. This reaction would make available a battery of recognition molecules which could be mobilized and activated, then deaggregated and released as active oligomers in the hemolymph. This would be consistent with the low-affinity, multisubunit-binding theory discussed by Olafsen (1986) and Renwrantz (1986).

The results and discussion presented here support the following hypothesis of the influence of molecular architecture on biological function of those invertebrate lectins which exist as polymers or multimers of monovalent subunits.

• Multimeric lectins may facilitate recognition and binding through multipoint attachment. Even if each bond is weak, recognition could result from reaction of binding sites above a threshold. The existence of such multimeric lectins at distinct stages of aggregation, such as in Pacific oysters, could make possible the following alternative hypotheses on induction of lectin activity by antigens and their function as opsonins.

• Interaction of a lectin multimer with nonself could induce deaggregation and release of oligomers with resulting increase in serum nonself activity.

• Reaction with nonself could invoke molecular rearrangement and exposure of cryptic binding sites with self-directed activity.

• Interaction of an oligomer with ligand (nonself) could induce polymerization to multimers with resulting increase in self-affinity.

In all cases, the reactions of Pacific oyster lectins with ligands are most likely modulated by divalent cations and possibly also by serum lipid.

Important tasks for future research would be to gain insight into the complexity of lectin-combining sites and to relate this to biological activity. It will be imperative to understand if differences exist in the combining sites of lectins of identical chemical specificity. The contribution of cooperativity and molecular rearrangement to the interaction of lectins with ligands and with cells is another area for future research.

In the past, emphasis has been placed on the chemistry of carbohydrate binding and the phylogenetic relationships of lectins. This approach is important if we are to understand the evolution of immunological reactions, but effort should also be devoted to understanding the involvement of molecular architecture in activation and recognition of lectins—and to relate this to other molecular events involved in immune protection of this successful and abundant group of animals.

Acknowledgments

The author gratefully acknowledges financial support from the Norwegian Research Council for Science and the Humanities and the Norwegian Fisheries Research Council.

References

Acton, R. T., and E. E. Evans. 1968. Bacteriophage clearance in the oyster (*Crassostrea virginica*). Journal of Bacteriology 96:1260–1266.

Barondes, S. H. 1981. Lectins: their multiple endogeneous cellular functions. Annual Review of Biochemistry 50:207–231.

Basu, S., M. Sarkar, and C. Mandal. 1986. A single step purification of a sialic acid binding lectin (achatinin$_H$) from *Achatina fulica* snail. Molecular and Cellular Biochemistry 71:149–157.

Bayne, C. J. 1983. Molluscan immunobiology. Pages 407–486 *in* A.S.M. Saleuddin and K. M. Wilbur, editors. The Mollusca, volume 5. Academic Press, New York.

Bayne, C. J., C. A. Boswell, and M. A. Yui. 1987. Widespread antigenic crossreactivity between plasma proteins of a gastropod, and its trematode parasite. Developmental and Comparative Immunology 11:321–329.

Bonaventura, C., and J. Bonaventura. 1983. Respiratory pigments: structure and function. Pages 1–50 *in* P. W. Hochachka, editor. The Mollusca, volume 2. Academic Press, London.

Boswell, C. A., T. P. Yoshino, and T. S. Dunn. 1987. Analysis of the tegumental surface of *Schistosoma mansoni* primary sporocysts. Journal of Parasitology 73:778–786.

Boyd, W. C., and E. Shapleigh. 1954. Antigenic relations of blood group antigens as suggested by tests with lectins. Journal of Immunology 73:226–231.

Brehelin, M., editor. 1986. Immunity in invertebrates. Cells, molecules, and defense reactions. Springer-Verlag, Berlin.

Cassels, F. J., J. J. Marchalonis, and G. R. Vasta. 1986. Heterogeneous humoral and hemocyte associated lectins with *N*-acylaminosugar specificities from the blue crab, *Callinectus sapidus* Rathbun. Comparative Biochemistry and Physiology B, Comparative Biochemistry 85:23–30.

Cheng, T. C. 1981. Bivalves. Pages 233–300 *in* N. A. Ratcliffe and A. F. Rowley, editors. Invertebrate blood cells. Academic Press, London.

Cheng, T. C. 1984. Classification of molluscan hemocytes based on functional evidences. Pages 111–146 *in* T. C. Cheng, editor. Invertebrate blood. Cells and serum factors, volume 6. Plenum, New York.

Cheng, T. C. 1988. Strategies employed by parasites of marine bivalves to effect successful establishment in hosts. American Fisheries Society Special Publication 18:112–129.

Cheng, T. C., and A. Cali. 1974. An electron microscope study of the fate of bacteria phagocytized by granulocytes of *Crassostrea virginica*. Contemporary Topics in Immunobiology 4:25–35.

Cheng, T. C., J. J. Marchalonis, and G. R. Vasta. 1984. Role of molluscan lectins in recognition processes. Progress in Clinical and Biological Research 157:1–15.

Cheng, T. C., and B. M. Rudo. 1976. Distribution of glycogen resulting from degradation of ^{14}C-labelled bacteria in the American oyster *Crassostrea virginica*. Journal of Invertebrate Pathology 27:259–262.

Chorney, M. J., and T. C. Cheng. 1980. Discrimination of self and non-self in invertebrates. Contemporary Topics in Immunobiology 9:37–54.

Colwell, R. R., and A. K. Sparks. 1967. Properties of *Pseudomonas enalia*, a marine bacterium pathogenic for the invertebrate *Crassostrea gigas*. Applied Microbiology 15:980–986.

Coombe, D. R., P. L. Ey, and C. R. Jenkin. 1982. Haemagglutinin levels in haemolymph from the colonial ascidian *Botrylloides leachii* following injection with sheep or chicken erythrocytes. Australian Journal of Experimental Biology and Medical Science 60:359–368.

Coombe, D. R., P. L. Ey, and C. R. Jenkin. 1984. Self/non-self recognition in invertebrates. Quarterly Review of Biology 59:231–255.

Cooper, E. L. 1985. Comparative immunology. American Zoologist 25:649–664.

Cooper, D. N. W., P. L. Haywood-Reid, W. R. Springer, and S. H. Barondes. 1986. Bacterial glycoconjugates are natural ligands for the carbohydrate binding site of discoidin I and influence on its cellular compartmentalization. Developmental Biology 114:416–425.

Damian, R. T. 1964. Molecular mimicry: antigen sharing by parasite and host and its consequences. American Naturalist 98:129–149.

Elston, R. 1980. Functional morphology of the coelomocytes of the larval oysters (*Crassostrea virginica* and *Crassostrea gigas*). Journal of the Biological Association of the United Kingdom 60:947–957.

Ey, P. L., and C. R. Jenkin. 1982. Molecular basis of self/non-self discrimination in the Invertebrata. Pages 321–391 *in* N. Cohen and N. Sigel, editors. The reticuloendothelial system, volume 3. Plenum, New York.

Faye, I., A. Pye, T. Rasmuson, H. G. Boman, and I. A. Boman. 1975. Insect immunity. II. Simultaneous induction of antibacterial activity and selective synthesis of some hemolymph proteins in diapausing pupae of *Hyalophora cecropia* and *Samia cynthia*. Infection and Immunity 12:1426–1438.

Feng, S. Y., J. S. Feng, and T. Yamasu. 1977. Roles of *Mytilus coruscus* and *Crassostrea gigas* blood cells in defense and nutrition. Pages 31–67 *in* L. A. Bulla, Jr., and T. C. Cheng, editors. Invertebrate immune responses, volume 3. Plenum, London.

Finstad, C. L., R. A. Good, and G. W. Litman. 1974. The erythrocyte agglutinin from *Limulus polyphemus* hemolymph: molecular structure and biological function. Annals of the New York Academy of Sciences 234:170–182

Fisher, W. S. 1986. Structure and functions of oyster hemocytes. Pages 25–35 *in* M. Brehelin, editor. Immunity in invertebrates. Springer-Verlag, Berlin.

Fitzgerald, S. W., and N. A. Ratcliffe. 1982. Evidence for the presence of subpopulations of *Arenicula marina* coelomocytes identified by their selective response towards Gram+ve and Gram−ve bacteria. Developmental and Comparative Immunology 6:23–34.

Fletcher, T. C., and C. A. Cooper-Willis. 1982. Cellular defence systems in the Mollusca. Pages 141–166 *in* N. Cohen and N. N. Sigel, editors. The reticuloendothelial system, volume 3. Plenum, London.

Flower, R. L. P., G. E. Wilcox, and D. A. Pass. 1985. Detection of two lectins in hemolymph from the oyster *Pinctada maxima*. Australian Journal of Experimental Biology and Medical Science 63:703–707.

Foley, D. A., and T. C. Cheng. 1975. A quantitative study of phagocytosis by hemolymph cells of the pelecypods *Crassostrea virginica* and *Mercenaria mercenaria*. Journal of Invertebrate Pathology 25: 189–197.

Form, D. M. 1979. Isolation and characterization of a lectin from the hemolymph of a tunicate, *Halocynthia pyriformis*. Doctoral dissertation. Yale University, New Haven, Connecticut.

Form, D. M., G. W. Warr, and J. J. Marchalonis. 1979. Isolation and characterization of a lectin from the hemolymph of a tunicate, *Halocynthia pyriformis*. Federation Proceedings 38:934.

Fries, C. R. 1984. Protein hemolymph factors and their roles in invertebrate defense mechanisms: a review. Pages 49–109 *in* T. C. Cheng, editor. Invertebrate blood. Cells and serum factors, volume 6. Plenum, New York.

George, S. G., B. J. S. Pirie, A. R. Cheyne, T. L. Coombs, and P. T. Grant. 1978. Detoxification of metals by marine bivalves: an ultrastructural study of the compartmentation of copper and zinc in the oyster *Ostrea edulis*. Marine Biology (Berlin) 45: 147–156.

Gilbride, K. J., and T. G. Pistole. 1981. The presence of copper in a purified lectin from *Limulus polyphemus*: possible new role for hemocyanin. Developmental and Comparative Immunology 5:347–352.

Gold, E. R., and P. Balding, editors. 1975. Receptor specific proteins: plant and animal lectins. Elsevier, New York.

Grant, P. T., and A. M. Mackie. 1974. Interspecies and intraspecies chemoreception by marine invertebrates. Pages 105–141 *in* P. T. Grant and A. M. Mackie, editors. Chemoreception in marine organisms. Academic Press, London.

Grischkowski, R. S., and J. Liston. 1974. Bacterial pathogenicity in laboratory-induced mortality of the Pacific oyster (*Crassostrea gigas*, Thunberg). Proceedings National Shellfisheries Association 64:82–91.

Hall, J. L., and D. T. Rowlands. 1974a. Heterogeneity of lobster agglutinins. I. Purification and physiochemical characterization. Biochemistry 13:821–827.

Hall, J. L., and D. T. Rowlands. 1974b. Heterogeneity of lobster agglutinins. II. Specificity of agglutinin-erythrocyte binding. Biochemistry 13:828–832.

Hardy, S. W., T. C. Fletcher, and J. A. Olafsen. 1977a. Aspects of cellular and humoral defence mechanisms in the Pacific oyster, *Crassostrea gigas*. Pages 59–66 *in* J. B. Solomon and J. D. Horton, editors. Developmental immunobiology. Elsevier, Amsterdam.

Hardy, S. W., P. T. Grant, and T. C. Fletcher. 1977b. A hemagglutinin in the tissue fluid of the Pacific oyster, *Crassostrea gigas*, with specificity for sialic acid residues in glycoproteins. Experientia (Basel) 33:767–768.

Harris, H., and S. E. Zalik. 1985. Studies on the endogenous galactose-binding lectin during early development of the embryo of *Xenopus laevis*. Journal of Cell Science 79:105–117.

Hartland, B., and J. F. Timoney. 1979. In vivo clearance of enteric bacteria from the hemolymph of the hard clam and the American oyster. Applied and Environmental Microbiology 37:517–520.

Hendrie, M. S., W. Hodgkiss, and J. M. Shewan. 1971. Proposal that *Vibrio marinus* (Russel 1891) Ford 1927 be amalgamated with *Vibrio fischeri* (Beijerinck 1889) Lehmann and Neumann 1896. International Journal of Systematic Bacteriology 21:217–221.

Ingram, G. A., J. East, and D. H. Molyneux. 1983. Agglutinins of *Trypanosoma, Leishmania* and *Crithidia* in insect hemolymph. Developmental and Comparative Immunology 7:649–652.

Jenkin, C. R. 1976. Factors involved in the recognition of foreign material by phagocytic cells from invertebrates. Pages 80–97 *in* J. J. Marchalonis, editor. Comparative immunology. Blackwell, Oxford, England.

Jenkin, C. R., and D. Hardy. 1975. Recognition factors of the crayfish and the generation of diversity. Pages 55–64 *in* W. H. Hildeman and A. A. Benedict, editors. Immunologic phylogeny. Plenum, New York.

Kaplan, R., S. S.-L. Li, and J. M. Kehoe. 1977. Molecular characterization of limulin, a sialic acid binding lectin from the horseshoe crab, *Limulus polyphemus*. Biochemistry 16:4297–4303.

Kolb, H. 1977. On the phylogenetic origin of the immune system. A hypothesis. Developmental and Comparative Immunology 1:193–206.

Komano, H., D. Mizuno, and S. Natori. 1980. Purification of a lectin induced in the hemolymph of *Sarcophaga peregrina* larvae on injury. Journal of Biological Chemistry 255:2919–2924.

Komano, H., D. Mizuno, and S. Natori. 1981. A possible mechanism of induction of insect lectin. Journal of Biological Chemistry 256:7087–7089.

Komano, H., and S. Natori. 1985. Participation of *Sarcophage peregrina* humoral lectin in the lysis of sheep red blood cells injected into the abdominal cavity of larvae. Developmental and Comparative Immunology 9:31–40.

Lackie, A. M. 1980. Invertebrate immunity. Parasitology 80:393–412.

Lackie, A. M. 1981. Immune recognition in insects. Developmental and Comparative Immunology 5: 191–204.

Lackie, A. M. 1983. Effect of substratum wettability and charge on adhesion in vitro and encapsulation in vivo by insect hemocytes. Journal of Cell Science 63:181–190.

Langmann, R. E. 1978. Cell-mediated immunity and the major histocompatibility complex. Reviews of Physiology, Biochemistry and Pharmacology 81:1–37.

Lev, B., H. Ward, G. T. Keusch, and M. E. A. Pereira. 1986. Lectin activation in *Giardia lamblia* by host protease: a novel host–parasite interaction. Science (Washington, D.C.) 232:71–73.

Liener, I. E., N. Sharon, and I. J. Goldstein, editors. 1986. The lectins. Properties, functions, and applications in biology and medicine. Academic Press, London.

Maki, J. S., and R. Mitchell. 1985. Involvement of lectins in the settlement and metamorphosis of marine invertebrate larvae. Bulletin of Marine Science 37:675–683.

Manen, J. F., and M. Comte. 1986. Each *Phaseolus vulgaris* isolectin exist in five electrophoretic forms related to the number of metal ions bound per molecule. Plant Science (Shannon) 43:51–56.

Mann, K. G. 1987. The assembly of blood clotting complexes on membranes. Trends in Biochemical Sciences 12:229–233.

Maramorosh, K., and R. E. Shope, editors. 1975. Invertebrate immunity: mechanisms of invertebrate vector parasite relations. Academic Press, New York.

Markl, J., A. Markl, A. Hofer, W. Schartau, and B. Linzen. 1978. Subunit heterogeneity in arthropod hemocyanins. Verhandlungen deutschen zoologischen Gesellschaft 71:265.

McDade, J. E., and M. R. Tripp. 1967. Mechanism of agglutination of red blood cells by oyster hemolymph. Journal of Invertebrate Pathology 9:523–530.

Metchnikoff, E. 1891. Lectures on the comparative pathology of inflammation. Reprint 1968. Dover, New York.

Miller, R. L., J. F. Collawn, Jr., and W. W. Fish. 1982. Purification and macromolecular properties of a sialic acid-specific lectin from the slug *Limax flavus*. Journal of Biological Chemistry 257:7574–7580.

Minnick, M. F., R. A. Rupp, and K. D. Spence. 1986. A bacterial-induced lectin which triggers hemocyte coagulation in *Manduca sexta*. Biochemical and Biophysical Research Communications 137:729–735.

Mullainadhan, P., and L. Renwrantz. 1986. Lectin-dependent recognition of foreign cells by hemocytes of the mussel *Mytilus edulis*. Immunobiology 171:263–273.

Müller, W. E. G., and eight coauthors. 1984. Biochemical basis for the symbiotic relationship *Didemnum–Prochloron* (Prochlorophyta). Biology of the Cell 51:381–388.

Müller, W. E. G., R. K. Zahn, B. Kurelec, C. Lucu, I. Müller, and G. Uhlenbruck. 1981. Lectin, a possible basis for symbiosis between bacteria and sponges. Journal of Bacteriology 145:548–558.

Nguyen, N. Y., A. Suzuki, R. A. Boykins, and T.-Y. Liu. 1986a. The amino acid sequence of *Limulus* C-reactive protein. Evidence of polymorphism. Journal of Biological Chemistry 261:10456–10465.

Nguyen, N. Y., A. Suzuki, S.-M. Cheng, G. Zon, and T.-Y. Liu. 1986b. Isolation and characterization of *Limulus* C-reactive protein genes. Journal of Biological Chemistry 261:10450–10455.

Olafsen, J. A. 1984. Ingestion of bacteria by cod (*Gadus morhua* L.) larvae. Pages 627–643 *in* E. Dahl, D. S. Danielsen, E. Moksness, and P. Solemdal, editors. The propagation of cod, *Gadus morhua* L. Institute of Marine Research, Flødevigen Biological Station, Flødevigen rapportserie 1, Arendal, Norway.

Olafsen, J. A. 1986. Invertebrate lectins: biochemical heterogeneity as a possible key to their biological function. Pages 94–111 *in* M. Brehelin, editor. Immunity in invertebrates. Springer-Verlag, Berlin.

Olafsen, J. A., M. Christie, and J. Raa. 1981. Biochemical ecology of psychrotrophic strains of *Vibrio anguillarum* isolated from outbreaks of vibriosis at low temperature. Systematic and Applied Microbiology 2:339–348.

Ottaviani, E., G. Aggazotti, and S. Tricoli. 1986. Kinetics of bacterial clearance and selected enzyme activities in serum and haemocytes of the freshwater snail *Planorbarius corneus* (L.) (Gastropoda, Pulmonata) during the primary and secondary response to *Staphylococcus aureus*. Comparative Biochemistry and Physiology A, Comparative Physiology 85:91–95.

Parish, C. R. 1977. Simple model for self–non-self-discrimination in invertebrates. Nature (London) 267:711–713.

Pauley, G. B., G. A. Granger, and S. M. Krassner. 1971a. Characterization of a natural agglutinin present in the hemolymph of the California sea hare, *Aplysia californica*. Journal of Invertebrate Pathology 18:207–218.

Pauley, G. B., S. M. Krassner, and F. A. Chapman. 1971b. Bacterial clearance in the California sea hare, *Aplysia californica*. Journal of Invertebrate Pathology 18:227–239.

Pistole, T. G. 1976. Naturally occurring bacterial agglutinin in the serum of the horseshoe crab, *Limulus polyphemus*. Journal of Invertebrate Pathology 28:153–154.

Prowse, R. H., and N. N. Tait. 1969. In vitro phagocytosis by amoebocytes from the hemolymph of *Helix aspersa* (Müller). I. Evidence for opsonic factor(s) in serum. Immunology 17:437–443.

Qu, X.-M., C.-F. Zhang, H. Komano, and S. Natori. 1987. Purification of a lectin from the hemolymph of Chinese oak silk moth (*Antheraea pernyl*) pupae. Journal of Biochemistry 101:545–551.

Ratcliffe, N. A., and A. F. Rowley. 1984. Opsonic activity of insect hemolymph. Pages 187–204 *in* T. C. Cheng, editor. Invertebrate blood. Cells and serum factors, volume 6. Plenum, New York.

Renwrantz, L. 1983. Involvement of agglutinins (lectins) in invertebrate defense reactions: the immunobiological importance of carbohydrate-specific binding molecules. Developmental and Comparative Immunology 7:603–608.

Renwrantz, L. 1986. Lectins in molluscs and arthropods: their occurrence, origin and roles in immunity. Symposia of the Zoological Society of London 56:81–93.

Renwrantz, L., and W. Mohr. 1978. Opsonizing effect of serum and albumin gland extract on the elimination of human erythrocytes from the circulation of *Helix pomatia*. Journal of Invertebrate Pathology 31:164–170.

Renwrantz, L., W. Schancke, H. Harm, H. Erl, H. Leibsch, and J. Gercken. 1981. Discriminative ability and function of the immunobiological recognition system of the snail *Helix pomatia*. Journal of Comparative Physiology 141:477–488.

Renwrantz, L., and A. Stahmer. 1983. Opsonizing properties of an isolated hemolymph agglutinin and demonstration of lectin-like recognition molecules at the surface of hemocytes from *Mytilus edulis*. Journal of Comparative Physiology 149:535–546.

Rhodes, C. P., and N. A. Ratcliffe. 1983. Coelomocytes and defence reactions of the primitive chordartes, *Brachiostoma lanceolatum* and *Saccoglossus horsti*. Developmental and Comparative Immunology 7:695–698.

Ricchelli, F., G. Jori, L. Tallandini, P. Zatta, M. Beltramini, and B. Salvato. 1984. The role of copper and the quarternary structure on the conformational properties of *Octopus vulgaris* hemocyanin. Archives of Biochemistry and Biophysics 235:461–469.

Rittschof, D., and J. Bonaventura. 1986. Macromolecular cues in marine systems. Journal of Chemical Ecology 12:1013–1023.

Robey, F. A., and T.-Y. Liu. 1981. Limulin: a C-reactive protein from *Limulus polyphemus*. Journal of Biological Chemistry 256:969–975.

Robohm, R. A. 1984. In vitro phagocytosis by molluscan hemocytes: a survey and critique of methods. Pages 147–172 *in* T. C. Cheng, editor. Invertebrate blood. Cells and serum factors, volume 6. Plenum, New York.

Rodrick, G. E., and S. A. Ulrich. 1984. Microscopical studies on the haemocytes of bivalves and their phagocytic interaction with selected bacteria. Helgoländer Meeresuntersuchungen 37:167–176.

Rögener, W., L. Renwrantz, and G. Uhlenbruck. 1985. Isolation and characterization of a hemolymph lectin from the cephalopod *Octopus vulgaris* inhibited by α-D-lactose and *N*-acetyl-lactosamine. Developmental and Comparative Immunology 9:605–616.

Rögener, W., L. Renwrantz, and G. Uhlenbruck. 1986. Comparison of a hemolymph lectin from *Octopus vulgaris* with haemocyanin. Comparative Biochemistry and Physiology B, Comparative Biochemistry 85:119–123.

Rögener, W., and G. Uhlenbruck. 1984. Invertebrate lectins: the biological role of a biological rule. Developmental and Comparative Immunology 8:159–164.

Roseman, S. 1970. The synthesis of complex carbohydrates by multiglycosyltransferase systems and their potential function in intercellular adhesion. Chemistry and Physics of Lipids 5:270–297.

Roseman, S. 1974. Complex carbohydrates and intercellular adhesion. Pages 255–272 *in* A. A. Moscona, editor. The cell surface in development. Wiley, New York.

Salvato, B., A. Ghiretti-Magaldi, and F. Ghiretti. 1979. Hemocyanin of *Octopus vulgaris*. The molecular weight of the minimal functional subunit in 3 M urea. Biochemistry 18:2731–2736.

Schluter, S. F., P. L. Ey, D. R. Keough, and C. R. Jenkin. 1981. Identification of two carbohydrate-specific erythrocyte agglutinins in the haemolymph of the protochordate *Botrylloides leachii*. Immunology 42:241–250.

Schmied, L. S. 1975. Chemotaxis of hemocytes from the snail *Viviparus malleatus*. Journal of Invertebrate Pathology 25:125–131.

Sharon, N. 1984. Surface carbohydrates and surface lectins are recognition determinants in phagocytosis. Immunology Today 5:143–147.

Sharon, H., and H. Lis. 1988. A century of lectin research (1888–1988). Trends in Biochemical Sciences 12:488–491.

Shishikura, F., and K. Sekiguchi. 1983. Agglutinins in the horseshoe crab hemolymph: purification of a potent agglutinin of horse erythrocytes from the hemolymph of *Tachypleus tridentatus*, the Japanese horseshoe crab. Journal of Biochemistry 93:1539–1546.

Simkiss, K., and A. Z. Mason. 1983. Metal ions: metabolic and toxic effects. Pages 101–164 *in* P. W. Hochachka, editor. The Mollusca, volume 2. Academic Press, New York.

Sminia, T., W. P. U. van der Knapp, and L. A. van Asselt. 1983. Blood cell types and blood cell formation in gastropod molluscs. Developmental and Comparative Immunology 7:665–668.

Sminia, T., and W. P. W. van der Knapp. 1987. Cells and molecules in molluscan immunology. Developmental and Comparative Immunology 11:17–28.

Solomon, J. B. 1986. Invertebrate receptors and recognition molecules involved in immunity and determination of self and non-self. Pages 9–43 *in* R. M. Gorczynski, editor. Receptors in cellular recognition and developmental processes. Academic Press, London.

Stein, A. A., and E. L. Cooper. 1982. Agglutinins as receptor molecules: A phylogenetic approach. Pages 85–98 *in* E. L. Cooper and M. A. B. Brazier, editors. Developmental immunology: clinical problems and aging. Academic Press, New York.

Stein, E. A., Y. Soheil, and E. L. Cooper. 1986. Bacterial agglutinins of the earthworm *Lumbricus terrestris*. Comparative Biochemistry and Physiology B, Comparative Biochemistry 84:409–415.

Takatsuki, S. 1934. On the nature and functions of the amoebocytes of *Ostrea edulis*. Quarterly Journal of Microbiological Science 76:379–431.

Tillett, W. S., and T. Francis, Jr. 1930. Serological reactions in pneumonia with a non-protein somatic fraction of *Pneumococcus*. Journal of Experimental Medicine 52:561–571.

Tripp, M. R. 1960. Mechanisms of removal of injected microorganisms from the American oyster, *Crassostrea virginica* (Gmelin). Biological Bulletin (Woods Hole) 119:210–223.

Tripp, M. R. 1966. Haemagglutinin in the blood of the oyster *Crassostrea virginica*. Journal of Invertebrate Pathology 8:478–484.

Tripp, M. R. 1970. Defense mechanisms of molluscs. Journal of the Reticuloendothelial Society 7:173–182.

Uhlenbruck, G., and G. Steinhausen. 1977. Tridacnins: symbiosis profit or defense purpose? Developmental and Comparative Immunology 1:183–192.

van Holde, K. E., and K. I. Miller. 1982. Haemocyanins. Quarterly Reviews of Biophysics 15:1–29.

van Oss, C. J., and C. F. Gillman. 1972. Phagocytosis as a surface phenomenon. II. Contact angles and phagocytosis of encapsulated bacteria before and after opsonization by specific antiserum and complement. Journal of the Reticuloendothelial Society 12:497–502.

Vasta, G. R., T. C. Cheng, and J. J. Marchalonis. 1984a. A lectin on the hemocyte membrane of the oyster (*Crassostrea virginica*). Cellular Immunology 88:475–488.

Vasta, G. R., J. C. Hunt, J. J. Marchalonis, and W. W. Fish. 1986. Galactosyl-binding lectin from the tunicate *Didemnum candidum*. Purification and physicochemical properties. Journal of Biological Chemistry 261:9174–9181.

Vasta, G. R., and J. J. Marchalonis. 1983. Humoral recognition factors in the Arthropoda. The specificity of Chelicerata serum lectins. American Zoologist 23:157–171.

Vasta, G. R., J. J. Marchalonis, and H. Kohler. 1984b. Invertebrate recognition protein cross-reacts with an immunoglobulin idiotype. Journal of Experimental Medicine 159:1270–1276.

Vasta, G. R., J. T. Sullivan, T. C. Cheng, J. J. Marchalonis, and G. W. Warr. 1982. A cell membrane-associated lectin of the oyster hemocyte. Journal of Invertebrate Pathology 40:367–377.

Yeaton, R. W. 1981a. Invertebrate lectins: I. Occurrence. Developmental and Comparative Immunology 5:391–402.

Yeaton, R. W. 1981b. Invertebrate lectins: II. Diversity of specificity, biological synthesis and function in recognition. Developmental and Comparative Immunology 5:535–545.

Yonge, C. M. 1937. Evolution and adaption of the digestive system of Metazoa. Biological Reviews of the Cambridge Philosophical Society 12:87–115.

Zipris, D., N. Gilboa-Garber, and A. J. Susswein. 1986. Interaction of lectins from gonads and haemolymph of the sea-hare *Aplysia* with bacteria. Microbios 46:193–198.

ZoBell, C. E., and C. B. Feltham. 1938. Bacteria as food for certain marine invertebrates. Journal of Marine Research 1:312–327.

American Fisheries Society Special Publication 18:206–224, 1988

Host–Parasite Interactions in Eastern Oysters Selected for Resistance to *Haplosporidium nelsoni* (MSX) Disease: Survival Mechanisms against a Natural Pathogen

SUSAN E. FORD

Shellfish Research Laboratory, New Jersey Agricultural Experiment Station, Cook College
Rutgers University, Port Norris, New Jersey 08349, USA

Abstract.—Strains of eastern oysters *Crassostrea virginica* have been selectively bred for improved survival when exposed to the pathogen *Haplosporidium nelsoni* (MSX). Survival mechanisms involve restriction of parasite development and tolerance of infection. The most consistent and obvious host response to infection is an infiltration of tissues with large numbers of hemocytes, mostly hyalinocytes. However, there is no good evidence to date that such a response is correlated with improved survival. Phagocytosis of *H. nelsoni* plasmodia, both in vivo and in vitro, is low compared to that reported for many injected inert particles as well as for certain natural parasites. No protective humoral responses have been identified. A large percentage of the parasites found in localized lesions in eastern oysters selected for survival are uni- and binucleated plasmodia, in contrast to the rapidly proliferating multinucleated forms found in unselected individuals. This finding suggests that the parasites face conditions in the epithelium of selected oysters that inhibit their ability to survive, proliferate, or both. An hypothesis is presented that restriction of parasites involves decreasing susceptibility (provision of life needs to the parasite) rather than increasing resistance (active response against the parasite).

The disease-causing organisms described in this volume have caused widespread mortalities of affected bivalve populations. Losses may exceed 90% within a year in newly exposed stocks (Haskin et al. 1965; Andrews 1968; Grizel 1983; Elston 1986). Such extreme selective pressure should eventually result in the development of resistance in the hosts, a situation reported for populations of the eastern oyster *Crassostrea virginica* afflicted with Malpeque disease (Needler and Logie 1947), *Perkinsus marinus* (Andrews and Hewatt 1957), *Haplosporidium costale* (Andrews and Castagna 1978), and *H. nelsoni* (Haskin and Ford 1979). Early indications that native eastern oysters in Delaware Bay were experiencing lower mortalities after the initial epizootics of *H. nelsoni* (MSX) in the late 1950s stimulated a program of selective breeding at Rutgers University to create highly resistant oyster strains. The specific objectives of the project were (1) to determine whether improved survival of native stocks was heritable and, if so, whether a rigorous selection and breeding regime could produce strains with better survival than the natural population, (2) to provide experimental animals of known lineage for the study of defense mechanisms against a natural pathogen and of the genetic basis for disease resistance, and (3) to provide brood stock for hatcheries in the event that natural seed supplies became insufficient to meet demand in areas enzootic for *H. nelsoni*.

Several recent papers from this laboratory have discussed results of this project and compared characteristics of selected eastern oyster strains with those of unselected control groups (Ford 1986; Haskin and Ford 1986; Ford and Haskin 1987). This review will (1) summarize methods used to select, breed, and test eastern oysters for improved survival when exposed to *H. nelsoni*, (2) examine in detail the characteristics of the apparent survival mechanism(s), and (3) propose an hypothesis about "resistance" to *H. nelsoni* disease that may be applicable to other pathogens of bivalves.

Much of the information examined in this review is taken from the unpublished theses of three graduate students, J. L. Myhre, G. A. Valiulis, and W. R. Douglass, who investigated mechanisms of defense against *H. nelsoni* under the direction of H. H. Haskin, Rutgers University.

Selection, Breeding, and Testing for Improved Survival

The original selected strains were produced from eastern oysters that had survived the initial epizootic (1957–1959) on the planting grounds of lower Delaware Bay. To properly evaluate their performance, control groups were bred from unselected parents imported from regions where *H. nelsoni* was absent or at least had not caused measurable mortality. Over the years, imported stocks originated mostly in the James River in

Virginia, upper Chesapeake Bay, the Navesink River in northern New Jersey, the Connecticut shore of Long Island Sound, and Great South Bay on the south shore of Long Island, New York. Occasional imports were made from Maine, New Hampshire, the Potomac River, and Florida. All stocks experienced heavy mortalities when exposed to the parasite in Delaware Bay.

Spawning and larval rearing procedures were described by Haskin and Ford (1979). To minimize adverse effects of inbreeding described for pair matings (Longwell and Stiles 1973; Andrews 1979b), we employed group spawnings involving about 30 eastern oysters. The number of individuals actually discharging gametes ranged between 2 and 20, averaging about 12, approximately equally divided between males and females. Comparison of allozyme frequencies at several loci between a number of highly inbred strains and the wild stocks from which they originated showed that rare alleles have been lost in the breeding process but that heterozygosity has been maintained (Vrijenhoek and Ford 1987). Further, we experienced none of the problems associated with pair matings at other laboratories.

The testing procedure involved exposure of strains at a location in lower Delaware Bay, which regularly received very heavy natural infection pressure. Exposure to natural infections was necessary because we have been unable to transmit the parasite experimentally (Canzonier 1968, 1973). The test site was a tidal flat so that eastern oysters, which were kept in replicate trays, could be conveniently checked at low tide. A standard test period was established, beginning when the oysters were 3–4 months old and continuing until they were 3 years old, when most individuals reached market size. Survivors of the test exposure then became parents of the next generation. To date, we have reared several strains through six generations. The survival of each selected group was measured against the offspring of unselected (imported) parents produced in the same year and tested under the same conditions. This procedure provided assurance that improved performance in experimental strains was not due to lower parasite abundance or virulence. Control-group survivors were then used to begin selected lines from areas outside of Delaware Bay. Early comparisons of outcrossed with inbred strains showed no consistent advantage to the former (Haskin and Ford 1986); therefore, our efforts have concentrated on the production of inbred lines.

TABLE 1.—Survival of six generations of eastern oysters during selection for disease resistance following 33 months of exposure to *Haplosporidium nelsoni*.

Generation	Number of lines	Percent survival		
			Range	
		Mean	Median	Overall
Unselected	39	8	1–10	<1–47
F_1	12	33	31–40	16–43
F_2	15	38	31–40	17–90
F_3	11	53	51–60	25–89
F_4	8	37	51–60	9–61
F_5	6	49	61–70	<1–81
F_6	3	40	21–30	29–61

Characteristics of Selected Oysters

Mortality Patterns

When results for all groups were pooled by generation, it was evident that increased selection improved survival, a trend best observed in the median percentile range for survival (Table 1). For instance, median survival of 39 unselected groups was 1–10%, whereas it was 61–70% for six 5th-generation strains. Considerable variation existed among individual strains: Several 4th-, 5th- and 6th-generation groups have experienced unexpectedly high *H. nelsoni*-caused mortalities. The reasons for such reversions are not clear; however, initial results of comparing allozyme frequencies in several highly selected inbred lines with very different performance records indicated that loss of heterozygosity was not a factor (R. C. Vrijenhoek and Ford, Rutgers University, unpublished data). Overall, the breeding project has demonstrated that resistance to *H. nelsoni*-caused mortality is heritable, that it most probably has a multigenic basis, and that on the average, increased selection improves performance. By the time they reach market size, survivors in the best groups outnumber controls by 10 to 1.

Infection Patterns

Restriction and elimination of parasites.—An understanding of resistance follows from the knowledge that the earliest infections consist of multinucleated plasmodia situated between epithelial cells of the eastern oyster's gills and palps, where they multiply rapidly (Farley 1965b, 1968; Haskin et al. 1965). Myhre and Haskin (1968) estimated a generation time of 24 h for parasites in early gill lesions. Invasion of subepithelial tissues follows, and the parasite eventually spreads throughout the oyster via the circulatory system. Approximately 90% of dead or dying oysters

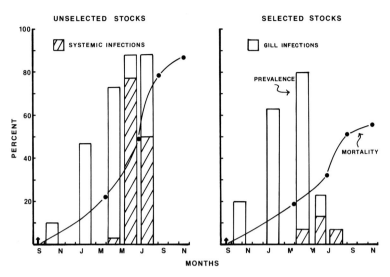

FIGURE 1.—Infection and mortality patterns in strains of eastern oysters selected (right) for resistance to *Haplosporidium nelsoni*-caused mortality and in unselected (left) control strains. Prevalence is shown by bars that are hatched for systemic infections or open for gill infections. Mortality is shown by line curves. Each point represents the mean of three strains. Histological samples consisted of 20 oysters per strain on each sampling date. Arrow on abscissa indicates the time of initial exposure. (From Myhre 1973.)

infected with *H. nelsoni* have systemic parasitism, a considerably higher proportion than that in living oysters (Ford and Haskin 1982, 1987). Conversely, the proportion of dead oysters with gill infections is never greater than that in live oysters, indicating that gill lesions are rarely lethal.

Myhre (1973) provided the first experimental evidence for a defense mechanism in mortality-resistant eastern oysters (see also Myhre and Haskin 1970). He studied in detail the early development of infection in three selected (F_1 and F_2) and three unselected (control) laboratory-reared strains of the same age. All oysters were initially exposed to natural infections, as spat, in early to mid-September 1967. (Earlier experiments with timed imports of susceptible oysters had demonstrated that the infection period in Delaware Bay extends from June through October.) Some individuals in both groups had patent gill infections as early as 27 October and by the end of April 1968, 70–80% of all oysters, regardless of ancestry, had patent infections (Figure 1). At this time, nearly all parasites were still located extracellularly in gill and palp epithelia. Myhre concluded that all oysters in each group were equally susceptible to the establishment of infections, and that the true prevalence was probably 100% in each group.

In May and June 1968, important differences emerged between selected and unselected strains: in

the unselected groups, prevalence rose to 90%, and the intensity of infections greatly increased as parasites spread from the localized lesions into all tissues (Figure 1). At the same time, prevalence decreased to 20% in the selected stocks. Differences became more pronounced by July when prevalence in selected oysters dropped to 7%, but prevalence remained close to 90% in unselected spat, and more than half of the infections were systemic.

Between initial exposure in September 1967 and November 1968, nearly 90% of the unselected oysters and somewhat more than half of the selected stocks died (Figure 1). Most of the differential mortality occurred after March and coincided with the period of greatest infection differences between strains. The selected strains not only had better survival, they also grew faster. Through April, there were no differences in size between groups. From early May through July, coincident with intensification of infection in unselected strains and reduction of infection in selected oysters, the latter grew more than twice as fast as unselected individuals (Figure 2).

Myhre (1973) concluded that the survival mechanism in selected eastern oysters was innate, because it was expressed on first exposure to the parasite; that it involved restriction of parasites to small, nonlethal lesions, usually in the gill or palp epithelium; and that it was probably not effective at low temperature. Subsequent field observations and lab-

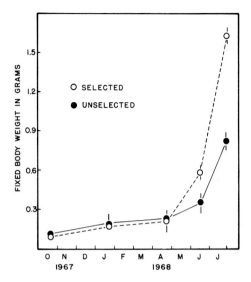

FIGURE 2.—Shucked meat weight of selected and unselected eastern oyster strains exposed to *Haplosporidium nelsoni* (see Figure 1). Oysters were fixed in Davidson's fixative before they were weighed. Each point is the mean of three strains, 20 oysters/strain. Vertical lines through points are standard deviations. (From Myhre 1973.)

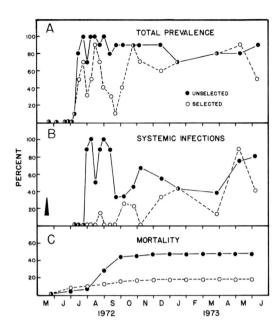

FIGURE 3.—Infection and mortality patterns in one selected strain and one unselected wild stock of eastern oyster during initial exposures to *Haplosporidium nelsoni*. Percent prevalence is based on a 10-oyster sample from each group. The percent systemic infections is based on the total number of infected oysters in the sample. Arrow indicates time of initial exposure. (From Douglass 1977.)

oratory experiments (Haskin and Douglass 1971; Ford and Haskin 1982) led to the conclusion that the survival mechanism is not as effective at temperatures below 18–20°C as it is at higher temperatures.

Douglass (1977) continued to investigate survival mechanisms, but, in contrast to Myhre, he examined early infection patterns in selected and unselected strains exposed to *H. nelsoni* infections in early summer. At this time, response to initial infection was not inhibited by low temperature. The selected eastern oysters were a mixture of 2nd- and 3rd-generation laboratory-reared strains. They were 2–3 years old and had been maintained in nonenzootic water (Cape May harbor, New Jersey) before the study. Unselected oysters came from a natural population in New Haven harbor, Connecticut, and were at least 2 years old. Oysters were sampled approximately weekly during the early infection period to provide a detailed histological picture of host–parasite interactions.

Both groups were placed in Delaware Bay on 21 May 1972. Parasites were detected in the gill epithelium of individuals in both groups by the first week of July (Figure 3A). Two weeks later, 100% of the unselected and 70% of the selected group had patent infections. From that time until the end of the study in early June 1973, prevalence remained mostly between 80 and 100% in the unselected oysters. In contrast, the initial prevalence peak of 70% in the selected group was followed by several fluctuations with peaks of 90% in mid August 1972, late October 1972, and early May 1973 (Figure 3A). Each peak was followed by a decline of variable extent and duration, the most dramatic being a drop from 90 to 10% between mid-August and mid-September 1972.

One month after the first patent infection appeared, all unselected eastern oysters had systemic infections (Figure 3B). In comparison, almost no systemically infected oysters appeared in the selected group until October 1972, and then the proportion remained less than 50% until early May 1973, when it reached 90%. At that time, both prevalence and infection intensity declined substantially in the selected group but remained high in unselected oysters (Figure 3A, B), a pattern similar to that found by Myhre (1973). Most mortality occurred in August and September 1972, after infections had become systemic in the unselected oysters (Figure 3C). The mean monthly death rate during that 2-month period was 3 and 22% for selected and unselected groups, respectively.

The Douglass (1977) experiment demonstrated that even at the elevated temperatures of early summer, when the defense mechanism was presumed to be fully expressed, nearly all the selected eastern oysters initially became infected. The small sample size, 10 oysters per group at each collection, raises the possibility that subsequent fluctuations in parasite levels resulted from sampling error. Whereas some error is unquestionably present in the data, two factors support the argument that the major fluctuations were real: (1) in the summer and early fall, large increases and decreases were shown by several sample points and (2) patterns of fluctuations were similar for two or more variables. For instance, the pattern for total prevalence in the selected strain (Figure 3A) was similar with slight delay, to that for the proportion of systemic infections in the unselected oysters (Figure 3B). The loss of infections in the selected oysters could not be attributed to mortality of parasitized individuals, and even the mean monthly death rate of 22% in August and September 1972 of unselected oysters was not enough to completely explain the decrease in proportion of systemically infected individuals between early and late September (Figure 3B, C).

In both groups, then, a dynamic interaction appeared to be occurring which involved two factors: the acquisition of new infections, the proliferation of already-established parasites, or both; and a condition in the oysters or a response by the oysters that suppressed or eliminated them. Prevalence and intensity peaks in August and October could well have been due to reinfection since the infective period extends through October. Relapse of subpatent infections in early fall has also been documented (Ford 1985), and the spring peak is due to both relapse and proliferation of infections acquired during the previous year's infection period (Haskin et al. 1965; Andrews 1966). In selected oysters, the response or condition that eliminated parasites occurred while the latter were still in the gill epithelium, as was the case in Myhre's (1973) study. As Myhre pointed out, when *H. nelsoni* plasmodia initially invade the gill, they are found between epithelial cells; therefore, to be precise, they are epizoic or outside the oyster when they are being suppressed or eliminated at this stage.

Tolerance.—The selected eastern oysters in Douglass' (1977) experiment clearly had fewer and lighter infections than did unselected animals during the summer and early fall. However, infection levels in the selected oysters increased during the late fall, so that over the winter, parasite burdens were similar in both groups (Figure 3A, B). Early histological investigations also found infections in eastern oysters that had survived several years of epizootic disease, although no seasonal patterns were described (Haskin et al. 1965). Ford (1985) reported that chronic infections of *H. nelsoni* of several years' duration alternated between local and systemic distributions in the same host. More recently, a variety of unselected and selected (F_3–F_6) groups were examined histologically in early August and again in late November, during 2 years of heavy infection activity (1981 and 1982) and 1 year of relatively light pressure (1983) (Table 2; see also Ford and Haskin 1987). In August, three of five selected groups had few or no patent infections, whereas all but one group of unselected oysters had high prevalences, often 80–100%, and intense infections. In late November, most of the selected and unselected groups had high prevalences, although, in any given year, infections never became as heavy in selected oysters as they did in unselected individuals. Despite the sharply increased parasite burdens in selected groups, annual cumulative mortality, with one exception, was less than 30%, whereas that in the unselected oysters was generally above 70%. This difference occurred even though November infection levels in some selected strains (e.g., in 1982) approached or exceeded those in certain unselected groups.

The results suggest that improved survival is associated with tolerance of systemic infections during periods of low temperature and with restriction of parasites during periods of high temperature. Tolerance as a survival mechanism implies the ability to live and function while carrying parasite levels that kill "susceptible" individuals. In the case of *H. nelsoni* infections, the ability of eastern oysters to survive infections that become systemic in late fall may be enhanced by low temperature, which depresses the metabolic rate of both parasite and host. When the temperature rises in spring, restriction and elimination again become important defense mechanisms.

Ford and Haskin (1987) reported that when highly selected strains of eastern oyster were exposed to repeated infection by *H. nelsoni*, most hosts eventually died from parasite-caused disease. In unselected groups, annual mortalities were initially high and decreased in succeeding years, whereas death rates in more highly selected oys-

TABLE 2.—Comparison of summer and winter prevalence and annual mortality in eastern oysters selected and unselected for resistance to mortality caused by the parasite *Haplosporidium nelsoni*.

Year and oyster stock[a]	Prevalence				Annual mortality (%)[d]
	Percent[b]		Weighted[c]		
	August	November	August	November	
Selected groups					
1981					
Strain NA, F_4	0	90	0	3.3	15
Strain OF, F_4	0	30	0	1.7	23
1982					
Strain OJ, F_3	20	60	0.4	4.1	30
Strain A, F_6	70	80	1.3	4.8	37
1983					
Strain LA, F_6	0	10	0	1.5	20
Unselected groups					
1981					
Virginia progeny	90	80	8.3	9.0	81
Virginia import	90	90	7.3	6.6	83
Virginia import	90	90	7.5	11.0	84
1982					
Virginia progeny	70	60	4.5	1.3	71
Long Island progeny	60	60	4.1	9.4	79
Maryland import	100	95	11.1	9.0	90
Virginia import	80	90	5.4	8.2	90
1983					
Virginia progeny		50		5.0	51
Virginia import	10	100	0.2	10.4	62

[a] Stocks were reared in the laboratory from parents that had (selected) or had not (unselected) experienced *Haplosporidium nelsoni*-associated mortality. Native imports were brought from areas with little or no *H. nelsoni*-associated mortality.
[b] Samples of 10 eastern oysters exposed to natural infections in lower Delaware Bay.
[c] Mean infection intensity for the entire sample, including uninfected oysters. Based on a scale of 0 (uninfected) to 15 (heavy, systemic infection).
[d] From July 1 of listed year to June 30 of following year.

ters increased with each year of exposure. These data suggested that although the capacity to tolerate chronic and repeated infection increased with selective breeding, it was limited, and eventually all oysters were debilitated beyond the point of recovery. Thus, selection simply delayed death, but many selected strains reached market size (75–100 mm in 2–3 years) without significant mortality.

Specificity and Dose Dependency

Valiulis (1973) examined several eastern oyster strains selected for resistance to *H. nelsoni*, to determine whether they would survive when challenged by another oyster pathogen, *Perkinsus marinus*, formerly called *Dermocystidium marinum* or *Labyrinthomyxa marina* (Mackin et al. 1950; Levine 1978). His results were not conclusive in all respects, but they did show distinct differences in resistance to *P. marinus*-caused mortality among the various stocks tested, and two experiments suggested that improved survival might be correlated with previous selection by *H. nelsoni*.

In the first experiment, eastern oysters known to be infected with *P. marinus*, which is transmitted directly from oyster to oyster, were placed in trays containing stocks with different histories of selection by *H. nelsoni*. The trays were kept in a location where *H. nelsoni* was not known to occur. Duplicate trays of the same stocks, but without the *P. marinus*-infected oysters, were exposed to natural *H. nelsoni* infections in lower Delaware Bay. The stock showing the highest *H. nelsoni* infection and mortality levels in Delaware Bay originated in the Potomac River, Virginia (Table 3). The same group also had the highest infection levels when exposed to *P. marinus*. Stocks originating in Great South Bay, Long Island, and in Delaware Bay had considerably lower *H. nelsoni* mortality and infection levels and fewer *P. marinus* infections than those of Potomac River origin. However, differences in mortality among the experimental *P. marinus*-exposed stocks were not correlated with infection level or with resistance to *H. nelsoni* mortality.

TABLE 3.—Mortality and infection statistics for five stocks of eastern oysters, replicates of which were exposed separately to the parasites *Haplosporidium nelsoni* (June 1969–September 1970) and *Perkinsus marinus* (July 1969–January 1972) to determine whether stocks resistant to one parasite were resistant to another. (From Valiulis 1973.)

| | Oysters exposed to *Haplosporidium nelsoni* | | | | Oysters exposed to *Perkinsus marinus* | | | |
| | | | % prevalence | | | | | |
Oyster stock[a]	Percent mortality	Number examined[b]	Total	Systemic	Percent mortality	Number examined[b]	Percent prevalence	Weighted incidence[c]
Potomac River, Virginia	92	56	70	48	51	180	49	1.12
Great South Bay, Long Island								
Stock A	60	60	32	7	78	179	28	0.65
Stock B	60	60	25	17	66	179	16	0.24
Delaware Bay								
Stock A	57	59	42	20	47	181	32	0.50
Stock B	51	60	38	10	42	179	13	0.02

[a] One replicate of each stock was exposed to natural infection of *Haplosporidium nelsoni* in lower Delaware Bay, the other to *Perkinsus marinus* by proximity to *P. marinus*-infected oysters at a site nonenzootic for *H. nelsoni*.
[b] Live oysters were examined histologically for *H. nelsoni* or by thioglycollate culture (Ray 1954) for *P. marinus*.
[c] Based on a scale of 0 (uninfected) to 5 (heavy infection) and averaged for entire sample.

For such a highly contagious pathogen as *P. marinus*, prevalences were unexpectedly low, suggesting that the doses received were low and perhaps variable. The results of this experiment, while hinting at a link between resistance to mortality caused by *H. nelsoni* and by *P. marinus*, were not wholly convincing.

The field test was followed by a better controlled experiment involving injection of known amounts of *P. marinus* into individual eastern oysters. Two strains of oysters selected for resistance to *H. nelsoni* and two unselected control groups, all laboratory-reared, were each split into two groups as in the previous trial. Oysters from each strain were exposed to natural *H. nelsoni* infections in lower Delaware Bay. The exposure to *P. marinus*, however, was provided by direct inoculation of parasites (obtained from heavily infected oysters) in doses ranging from 10 to 100,000 cells per oyster. Inoculated groups were maintained in the laboratory where they were not exposed to *H. nelsoni*.

Death rates of the groups inoculated with 10 to 10,000 *P. marinus* cells were in the same order as their replicates exposed to *H. nelsoni*: the two unselected strains had the highest losses and the two selected strains, the lowest (Table 4). However, the pattern broke down when oysters were injected with 100,000 *P. marinum* cells. One of the selected strains that had exhibited relatively good survival at lower doses suffered very high mortality at this very high dose. Samples diagnosed for the presence of parasites showed that dead oysters in the replicate groups were infected with the pathogen to which they had been exposed.

Valiulis (1973) interpreted his results to indicate that resistance to *H. nelsoni* mortality may be nonspecific, may also operate against at least one other serious oyster pathogen, and may be dose dependent. Although the data do suggest such an interpretation, they should be treated cautiously. During the 1960s, *P. marinus* was present in experimental eastern oysters undergoing selection by *H. nelsoni*, although it was controlled by placing trays at a distance from each other (Andrews 1967) and lethal infections were rare during the first 3 years of exposure. The question must be asked whether the strains that Valiulis used had actually been selected for resistance to both *P. marinus* and *H. nelsoni*. The record shows that both of his selected groups did experience some *P. marinus*-associated mortality in the parental generation. This occurred, however, only after 40 and 90% losses caused by *H. nelsoni* had already taken place, and in each case, involved only about 8–10% of the remaining oysters.

A more troublesome problem arose because all of the stocks used by Valiulis had experienced light *H. nelsoni* exposure in the preceeding fall, and 20% of one unselected group was patently infected at the start of the experiment, which probably caused the relatively high loss in the control groups (Table 4). When control results were subtracted from experimental figures, differences between selected and unselected groups were diminished considerably; however, the strain with highest *H. nelsoni*-related mortality remained the one with the greatest *P. marinus* losses. The strain most resistant to *P. marinus*-

TABLE 4.—Mortality in four strains of eastern oysters selected and unselected for resistance to mortality caused by *Haplosporidium nelsoni,* when replicated groups were inoculated with *Perkinsus marinus* in the laboratory or exposed to *H. nelsoni* in the field. (From Valiulis 1973.)

	Percent mortality after infection with parasites							
	Injection[a]							Exposure to *Haplosporidium nelsoni*[c]
	None	Saline only	*Perkinsus marinus*[b] cells per oyster					
Oyster stock			10	10^2	10^3	10^4	10^5	
Selected groups								
Navesink River, New Jersey								
Total	0	2	0	4	18	29	27	43
Corrected[d]			0	3	17	28	26	
Delaware Bay								
Total	6	2	0	18	35	44	57	48
Corrected[d]			0	14	31	40	55	
Unselected groups								
James River, Virginia								
Total	14	16	17	21	41	60	34	73
Corrected[d]			2	6	26	45	25	
Navesink River, New Jersey								
Total	28	20	41	73	77	78	67	94
Corrected[d]			17	49	53	54	50	

[a] Mortality calculated from day 0–95 postinjection, except day 0–75 for the 10^5 dose.
[b] Obtained from heavily infected oysters.
[c] Mortality cumulated during field exposure in Delaware Bay from November 1970 through June 1972.
[d] Mean control mortality subtracted from total mortality.

caused mortality also had the fewest deaths when exposed to *H. nelsoni.*

Valiulis (1973) concluded that correlation of survival rates after exposure to the two pathogens existed only at the "extreme ends of the resistance-susceptibility spectrum." Although his results do not provide entirely compelling evidence for a common survival mechanism against the two pathogens, they are intriguing enough to indicate that additional research in this area would be fruitful.

Potential Defense Mechanisms against *Haplosporidium nelsoni*

Cellular and humoral defense mechanisms in bivalve molluscs are reviewed elsewhere in this volume (Chu 1988; Feng 1988). Much of the experimental work, particularly that describing cellular responses, has involved injection of particles or substances that the animals would never encounter in nature. There has been far less work with natural pathogens, largely because of the difficulty in transmitting many of the known parasites and because considerable effort is required to develop and maintain resistant strains in which to measure protective responses. The availability of eastern oyster strains resistant to mortality from *H. nelsoni,* strains that exhibit clearly different infection patterns, presents an opportunity to determine which, if any, of the

potential defense mechanisms actually plays a protective role against this parasite.

Cellular Mechanisms

Hemocytosis.—Among the cellular responses against nonself most commonly described are phagocytosis–pinocytosis, encapsulation, and hemocytosis. Hemocytosis, or infiltration of tissues by hemocytes, resembles an inflammatory response in vertebrates and is a clear reaction to *H. nelsoni* infection (Farley 1965b, 1968; Myhre and Haskin 1968; Newman 1971; Kern 1976; Ford 1985). Localized infections can often be found by scanning tissue sections for aggregations of hemocytes (Figure 4). Farley (1968) indicated that hemocytosis did not occur when parasites were in epithelial lesions, but became pronounced, with a large increase in the proportion of hyaline hemocytes, when the parasites became subepithelial.

Myhre (1973), in his comparison of selected and unselected strains, found no evidence of hemocytosis in infected eastern oysters until June and July, at which time "hemocytic infiltration into infected areas was a consistent response in *all oysters*" (emphasis added). He concluded that the lack of hemocytosis during the winter and early spring was caused by a low heart rate and small

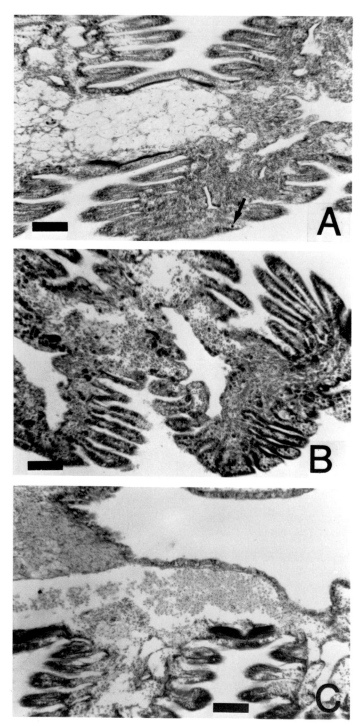

FIGURE 4.—**A.** Rare, epithelial gill infection of eastern oyster with *Haplosporidium nelsoni* showing intense hyaline hemocyte infiltration. Arrow points to parasite. **B.** Large numbers of parasites, mostly epithelial, eliciting intense hyaline hemocyte response. **C.** Granulocyte accumulation in gill blood vessel and tissue; contrast with hyaline hemocyte infiltration in A and B. (Bars = 100 μm.)

numbers of blood cells in circulation, which in turn were associated with low temperature (Feng 1965). Another explanation is that, before June, most infections were epithelial and might not have elicited a response. Although he did not report numbers, Myhre apparently found no differences between strains that could account for the loss of infections that he reported in selected oysters in late spring and early summer.

Farley (1975) and Kern (1976) reported that native eastern oysters in Pocomoke Sound (lower Chesapeake Bay), which had survived several years of exposure to *H. nelsoni*, had lighter infections and less cellular response than did introduced populations experiencing infections for the first time. Ford (1985), on the other hand, found no reduction in hemocytosis, or in infection intensity, in chronically infected eastern oysters over time. Indeed, intense hemocyte infiltration was associated with outbreaks of infection in the digestive gland.

In his comparison of selected and unselected eastern oyster strains exposed to *H. nelsoni*, Douglass (1977) counted hemocytes in tissue sections. Counts were made randomly and included infection sites as well as other areas. He found no consistent difference between groups in either infected or uninfected oysters (Figure 5A, B), and 95% confidence limits overlapped on most sampling dates. Unfortunately, the pooling of data for infected oysters, regardless of infection intensity, masked the fact that selected oysters had considerably lighter infections than did unselected ones (Figure 3B). It follows that, for a given infection level, there was probably more hemocytosis in the selected oysters, although it is not possible to quantify this argument with the available data.

It is puzzling, given the reports cited earlier, that Douglass measured only small differences between infected and uninfected animals during the summer and fall (Figure 5A, B). This result is probably because counts were averaged over the entire tissue section, thus minimizing the effect of localized lesions in which large numbers of hemocytes may be present in a relatively small area.

Douglass (1977) also determined the percent hyaline hemocytes in the tissue sections and reported it according to infection type. There were no differences between strains of eastern oyster without patent infections, in which the proportions of hyaline hemocytes were generally less than 10% (data not shown). In animals with epithelial infections, the proportion of hyaline hemocytes increased gradually throughout the study from less than 5% in early July to nearly 40% by the following May (Figure 5C). Douglass interpreted the increase as an indication that apparent epithelial infections were actually proliferating outside the plane of section and thus elicited a greater response. An alternative interpretation is that hemocytes were responding to cumulative damage by relatively small numbers of parasites. The proportion of hyalinocytes associated with epithelial infections was generally higher in unselected animals, particularly over the winter.

The spread of infections to underlying tissues resulted in a considerable increase in the proportions of hyaline hemocytes, to 20–50%, in both selected and unselected groups (Figure 5D). In these oysters, also, there was a trend toward a larger proportion of hyaline hemocytes over time. Caution should be used in interpreting the apparent extreme fluctuations in hyalinocyte proportions because relatively few oysters may have been involved (i.e., selected oysters with systemic infections and unselected oysters with epithelial infections).

Douglass (1977) concluded that hemocytic "responses to [*H. nelsoni*]-disease was similar in both mortality-resistant and mortality-susceptible oysters." His work should be expanded to include an analysis of numbers and types of hemocytes at the sites of infection. Also, numbers and types of circulating cells should be compared with the same parameters in tissue sections to help determine whether hemocytosis results from an increase in total cell counts or simply from concentration of existing cells.

Phagocytosis.—Phagocytosis has been described as the primary line of defense against foreign material in molluscs and has been demonstrated to rid oysters rapidly and effectively of a variety of injected substances (Stauber 1950; Tripp 1960; Feng 1966; Foley and Cheng 1975). Oyster phagocytes are also known to engulf at least two parasites, *Nematopsis* sp. and *Perkinsus marinus*. The first is nonpathogenic, and a dynamic equilibrium exists between infection and loss of parasites, apparently resulting from phagocytosis and disposal by diapedesis through epithelial walls (Feng 1958). In the case of *P. marinus*, Mackin (1951) reported that it was "common to find parasites in well over 50% of the leucocytic cells which invade infected tissues," but parasites continued to develop inside blood cells, eventually destroying them.

Myhre (1973) reported that in spring and summer, "when many [*H. nelsoni*] infections became subpatent or were aborted . . . phagocytosis was

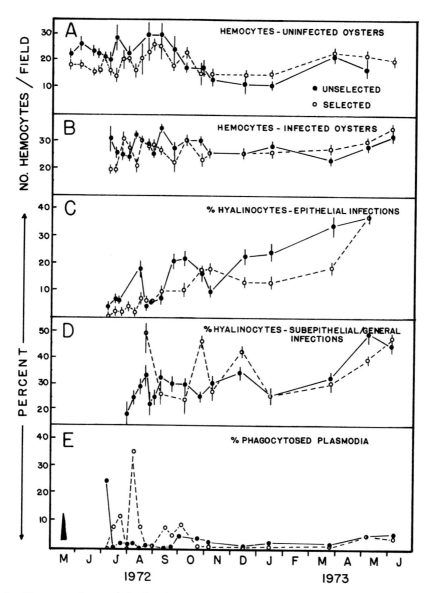

FIGURE 5.—Hemocyte characteristics in selected and unselected eastern oysters exposed to *Haplosporidium nelsoni* (see Figure 3). Hemocytes were counted in 12 random fields per oyster, six in the gill and six in the visceral mass (one field = 24.4 × 10³ μm²). Hyaline hemocytes were identified on the basis of a large nucleus-to-cytoplasm ratio. Vertical lines through points represent 95% confidence limits. Percent phagocytosed parasites was determined from counts of as many as 200 plasmodia per oyster. Note that there was complete overlap of sample ranges for percent phagocytosis for all collections (ranges not shown). Arrow indicates time of initial exposure. (From Douglass 1977.)

consistently observed in low intensity infections with the parasites, apparently in poor condition (dense and coarsely granular, sometimes with the cytoplasm withdrawn from the plasma membrane)." He saw little evidence of phagocytosis in heavy infections and reported no differences in phagocytosis rates between selected and unse-

lected strains. He also reported that phagocytosed *H. nelsoni* were exclusively in granulocytes and observed "parasite-laden phagocytes migrating through gill, palp and/or mantle epithelia in some infected spat." Myhre concluded that "this cellular response [presumably including hemocytosis, phagocytosis, and diapedesis] is probably one of the oyster's principal defense mechanisms against *H. nelsoni*''; however, he did not present quantitative evidence of strain differences. Had there been clearly significant differences, he presumably would have measured and reported them.

Douglass (1977) compared the number of phagocytosed *H. nelsoni* plasmodia in tissue sections from his selected and unselected strains (Figure 5E). With the exception of the initial sample, in which there was a single infected oyster, the proportion of phagocytosed parasites in unselected oysters was less than 5% at all times. There was considerably more fluctuation in the selected oysters, with the percentage reaching 10% on two occasions and 36% in one sample during the summer. The ranges about these means were considerable, however, and the phagocytic peaks did not coincide with infection peaks. Douglass (1977) was of the opinion that the role of hemocytes was primarily to remove parasites already damaged "by some other mechanism(s)."

Most investigators have found that phagocytosis of *H. nelsoni* is rare overall, although occasional individuals are found with a high proportion of parasites in phagocytes. Clear areas are frequently present around parasites situated in the gill epithelia (Farley 1965b, 1968; Myhre 1973; my personal observation), but because of the compact nature of this tissue, it is not always possible to determine whether these parasites are in phagocytes, have lysed surrounding tissues, or are showing cytoplasm contracted from the plasma membrane (Figure 6A, B). It is often difficult to determine whether a parasite is in a phagocyte or simply in close juxtaposition to one (Figure 6C–E).

The infiltration of hemocytes, especially hyalinocytes, into tissues where parasites are located is unquestionably the most consistent measured response to infection by *H. nelsoni*. Whether the reaction is to the recognition of nonself or to the considerable tissue damage associated with *H. nelsoni* infections, however, remains an important question. The involvement of hemocytes in wound repair has been well documented (see reviews by Cheng 1984; Sparks 1985; Sparks and Morado 1988, this volume), although it is not always clear what type of cells are involved. The evidence that blood cells recognize the parasite is much weaker. In vitro observations on fresh hemocyte–*H. nelsoni* preparations indicate that eastern oyster blood cells do not tend to aggregate around parasites (Figure 6F), and, in fact, do not appear to react at all (my unpublished observations), despite a clear ability to recognize and engulf foreign particles. The recogni-

tion and phagocytic capability of hemocytes toward other nonself material does not seem to be compromised in eastern oysters infected by *H. nelsoni* because spores of *Nematopsis* sp., in concurrent infections, are always in phagocytes (Figure 6G).

The low phagocytic rate of *H. nelsoni* reported by Douglass (1977) and others, added to our in vitro observations, suggests that hemocytes rarely recognize live parasites as foreign. Their infiltration into sites of infections is more likely to be in response to tissue damage.

The large number of hyalinocytes at sites of *H. nelsoni* infection is intriguing, especially since no specific function has been ascribed to this type of blood cell. They have very few granules (lysosomes), which argues against an enzyme-based cytotoxic function. The relative increase in numbers of hyalinocytes associated with infection occurs without comparable change in total hemocyte numbers (Figure 5B, C). Perhaps these cells represent a developmental stage of a single cell line that includes both hyaline and granular cells (Mix 1976; Auffret 1988, this volume). Are they, for instance, granulocyte precursors or, perhaps, spent granulocytes (see Cheng 1984)?

Humoral Responses

In contrast to the abundance of work on molluscan hemocytes, there is relatively little information on the kinds and functions of soluble substances in molluscan hemolymph. Further, it has been much more difficult to demonstrate a protective role against pathogenic agents. Many changes in humoral components measured in molluscs infected by parasites may be symptoms of disease rather than defenses against the pathogen.

Disease symptoms.—The decrease in hemolymph protein, including certain enzymes, and in free amino acids in eastern oysters parasitized by *H. nelsoni* (Mengebier and Wood 1969; Feng et al. 1970; Douglass 1977; Ford 1986) illustrates an altered humoral state that is most likely symptomatic of disease. Such depletion is probably the result of competition for nutrients or interference by the parasite in physiological or metabolic processes. Ford (1986) argued that such competition, rather than an "agglutination response" against *H. nelsoni*, was the cause for the large loss of total hemolymph protein, correlated with increasing infection intensity, in parasitized eastern oysters from an unselected strain. Total protein remained high in a selected strain exposed at the same time, which showed relatively few and low-intensity infections. Even though a large fraction of hemolymph protein

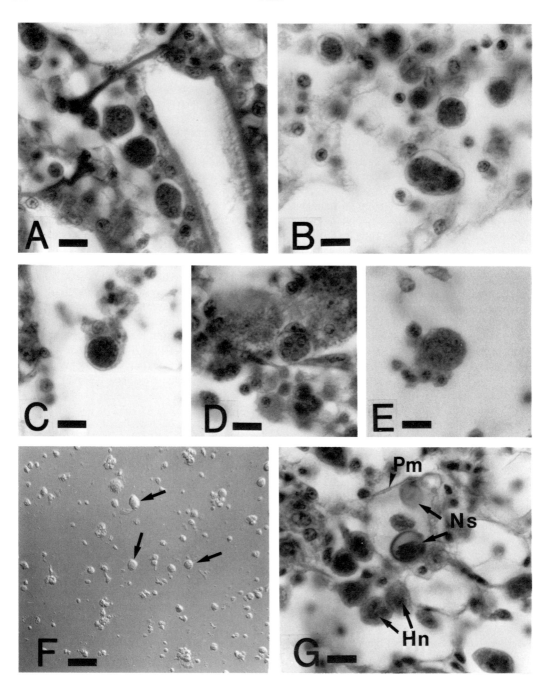

FIGURE 6.—**A.** Plasmodial stages of *Haplosporidium nelsoni* parasites in "clear areas" of gill epithelium of eastern oyster. **B.** Plasmodia, with cytoplasm withdrawn from plasma membrane, give the appearance of being in phagocytes, but no hemocyte nucleus or cytoplasm is visible. **C.** Phagocytosed plasmodium. **D.** Phagocytosed plasmodium. Note apparently degenerating, nonphagocytosed plasmodia in background. **E.** Hyaline hemocytes associated with plasmodium in blood vessel. **F.** Fresh blood preparation (Nomarski differential interference contrast illumination) of plasmodia (arrows) and hemocytes. Note lack of aggregation of hemocytes around parasites. **G.** Phagocytosed *Nematopsis* sp. spores (Ns) of the same size as nearby nonphagocytosed *Haplosporidium nelsoni* (Hn) plasmodia. Pm = plasma membrane of phagocytic cell containing the *Nematopsis* spores. Bars in A–E, G = 8 μm; bar in F = 60 μm.

in oysters has agglutinating capacity (Tripp 1966; Acton et al. 1969), there is no histological evidence that agglutination of parasites occurs in vivo, nor have we found any clear indication that it occurs in vitro (my unpublished data). Ford (1986) also pointed out that if the loss of protein were due to agglutination, it afforded the oyster no protection since the greatest depletion was in the most severely diseased individuals.

Abnormally high concentrations of some serum enzymes may also signal disease when they have leaked from damaged tissues. Douglass and Haskin (1976) reported increases in hemolymph concentrations of phosphohexose isomerase and alanine and aspartate amino transferases in oysters infected with *H. nelsoni*, especially those with gill infections. They proposed that these increases were attempts by the oyster to replace metabolites utilized by the parasite as well as those lost from damaged host cells. The evidence for such an interpretation is inconsistent, however. Hammen (1968) reported that aminotransferase activity was closely correlated with rates of amino acid excretion, and Feng et al. (1970) found reduced alanine concentration in the hemolymph of *H. nelsoni*-infected eastern oysters, both of which would support Douglass and Haskin's (1976) contention. Feng et al. (1970), however, found that aspartate was elevated in infection in one comparison but reduced in the other. Also, in a later experiment, Douglass (1977) found no consistent association of gill parasitism with enzyme concentrations and generally depressed levels in systemically infected individuals. Also, Mengebier and Wood (1969) documented a loss of phosphohexose isomerase activity in *H. nelsoni*-parasitized oysters, at all levels of infection.

It is important to determine infection levels when changes in biochemical constitutents of hemolymph are interpreted, because a low-level infection may produce quite different results than a heavy infection. For instance, concentration increases may be more likely in light infections where tissue damage is the major pathological change. Reduced concentrations, on the other hand, would be expected with heavy parasite burdens, which deplete circulating substrates through competition and interference with normal metabolic processes.

Anti-pathogen activity.—Agglutinating activity is often cited as a potential defense mechanism, but it does not appear to operate effectively, if at all, against *H. nelsoni*, as detailed above. Agglutinins act also as opsonins, facilitating phagocytosis by oyster (and other molluscan) hemocytes in vitro (Tripp 1966; Arimoto and Tripp 1977). Because there is little evidence that phagocytosis is an effective defense against *H. nelsoni*, this potential protective function is also doubtful.

Cheng (1983) considered that lytic enzymes may play an important role in defense. Lysozyme, proteases, lipases, phosphatases, and β-glucuronidase are found in oyster hemolymph and hemocytes (Tripp 1966; Feng and Canzonier 1970; Cheng and Rodrick 1975), and there is some information on alterations in their activity in the presence of pathogens. Feng and Canzonier (1970) measured lysozyme activities in eastern oysters parasitized by *H. nelsoni* and the trematode *Bucephalus* sp. Activities varied with season and differed according to the type of parasite. They were generally lower in summer than in winter, elevated in animals with *Bucephalus*, and depressed (in summer) in those with *H. nelsoni*. The authors suggested that the latter difference might be due to the "benign [*Bucephalus*] vs lethal [*H. nelsoni*] nature of the two types of infection," but they did not speculate as to whether lysozyme, which is bacteriolytic, was playing a protective role. Their results may illustrate a contrast between leakage of cellular contents due to tissue damage by (benign) *Bucephalus* and reduction of substrates resulting from competition and metabolic disruption by (lethal) *H. nelsoni*.

Yoshino and Cheng (1976) reported that leucine aminopeptidase (LAP) activity in eastern oyster hemocytes incubated for 15 min with heat-killed *Bacillus megaterium* was significantly higher than untreated controls, but they also found that incubation with sterile seawater elevated enzyme activity to nearly the same degree. They found no differences in the serum fraction, however, and in preliminary investigations, I have found no association of serum LAP activity with *H. nelsoni* parasitism (unpublished data). Finally, Mengebier and Wood (1969) found that eastern oysters with *H. nelsoni* infections had lower levels of hemolymph alkaline phosphatase activity than did patently uninfected individuals, although there was considerable variability among the latter, and differences were not statistically significant. Their results, however, are supported by histochemical evidence (Eble 1966) that tissue alkaline phosphatase levels decrease in heavily infected oysters.

With electrophoresis, Feng and Canzonier (1970) reported a shift in the ratios of two protein components in the hemolymph of infected compared to uninfected eastern oysters and thought that this might represent a humoral response. Ford (1986),

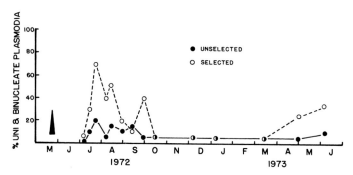

FIGURE 7.—Percent uni- and binucleated *Haplosporidium nelsoni* plasmodia in selected and unselected eastern oysters during initial exposure. Percentages represent counts made on as many as 200 plasmodia per oyster. Arrow indicates time of initial exposure. (From Douglass 1977.)

however, analyzed hemolymph from selected and unselected oysters collected at several times during the year and was unable to detect consistent differences in the electrophoretic patterns that correlated with either infection or ancestry. Both studies documented a large amount of individual, seasonal, and locality variation in hemolymph protein constituents, suggesting that it may be difficult to find, among this variability, specific serum proteins that are diagnostic for *H. nelsoni* disease or associated with resistance to the parasite.

Epithelial Barrier

Haplosporidium nelsoni usually exists in oysters as a multinucleated plasmodium (Haskin et al. 1966; Farley 1967). Occasionally, however, plasmodia are found with only one or two nuclei. Douglass' (1977) comparison of selected and unselected strains included a description of parasite morphology during early infection development. The stage seen in the earliest gill lesions was multinucleated, confirming the reports of Farley (1967), Myhre and Haskin (1968), and Myhre (1973). In unselected groups, Douglass found that most plasmodia remained multinucleated throughout the year (Figure 7). In contrast, up to 70% of all parasites in the selected strain were uni- or binucleated just at the time when early infections were being contained in epithelial lesions, suppressed, or eliminated by these oysters (compare Figures 5A, B, 7).

Because our knowledge of the life cycle of *H. nelsoni* is incomplete (Farley 1967), the exact meaning of uninucleated and binucleated forms relative to multinucleated ones is not clear. However, it is meaningful that when selected eastern oysters were most effectively suppressing *H. nelsoni*, most parasites had a much different appearance than they did

in unselected strains, in which plasmodia multiplied rapidly and spread throughout the tissues.

Synthesis, Conclusions, and Hypotheses

As we currently understand them, survival mechanisms operating against *Haplosporidium nelsoni* involve restriction of parasite development and tolerance of infection. Eastern oysters selected for improved survival become infected, but parasites remain epizoic, rarely developing beyond the gill epithelium during periods of high water temperature (above approximately 20°C), and some infections may be completely eliminated at this time. Infections frequently become systemic at lower temperatures either from relapse or reinfection, but still remain less intense and cause far lower mortalities than in unselected groups. Recovery through remission or elimination occurs when temperatures are again elevated. The ability to restrict and tolerate parasitism is limited, however, and eventually, most oysters in enzootic areas succumb to the stress of chronic and repeated reinfection with *H. nelsoni* (see Ford and Haskin 1987). In practical terms, selected strains are considered successful if they reach market size before significant losses have occurred.

These conclusions are based on results of experiments conducted in an area that experiences extremely heavy infection pressure, lower Delaware Bay. In areas that have lower pressure (i.e., fewer infective particles), such as the York River, Virginia, at the Virginia Institute of Marine Science (see Haskin and Andrews 1988, this volume), selected eastern oysters may suppress or eliminate parasites so effectively that infection levels are always low or even undetectable. This effect might explain Andrew's (1968) finding that eastern oysters selected against *H. nelsoni* at the Virginia Institute of Marine Science had consistently low infection and mortality levels.

The most consistent and obvious host response to infection is an infiltration of tissues with large numbers of hemocytes, mostly hyalinocytes. The earliest infections, localized in the gill epithelia and involving only a few parasites, show little hemocytosis. Hemocytosis becomes more pronounced as infections develop, and eventually, even localized infections are characterized by intense hemocytosis. There is no good evidence to date that response is correlated with selection for survival. Phagocytosis of *H. nelsoni* plasmodia, both in vivo and in vitro, is low compared to that reported for many injected inert particles as well as for certain natural parasites. It appears that oyster hemocytes may not recognize parasites as foreign until the parasites are already moribund, at which time phagocytosis occurs. No humoral responses have been associated with recovery from infection.

A most intriguing correlate with restriction and elimination of early infections is the morphological appearance of parasites themselves. The large percentage of uni- and binucleated plasmodia found in selected oysters when most parasites are confined to the gill epithelium suggests that the basal lamina in such hosts provides a physical barrier to the parasites, which are unable to survive in the epithelium, or that the parasites are faced with conditions in the epithelium of selected oysters that inhibit their ability to survive, proliferate, or both.

Read (1958) discussed resistance and susceptibility in terms that are helpful in understanding what may be happening in eastern oyster–*H. nelsoni* interactions. Whereas these terms are generally thought of as opposites of the same phenomenon, Read considered them to be separate and distinct conditions, both of which can exist in the same host. According to his definition, a susceptible host is one that provides a parasite with its life needs, and in which the parasite grows and develops. An insusceptible host, on the other hand, fails to provide the proper environment for parasite development. Resistance is an *altered* condition representing "a *response* of the host to *present or previous experience* with the parasite." Thus, a host may be highly susceptible and demonstrate high, low, or intermediate resistance. Presumably, a totally insusceptible individual would have no opportunity to "respond" to the parasite, but it is possible that hosts providing a marginal environment for the parasite (i.e., slight

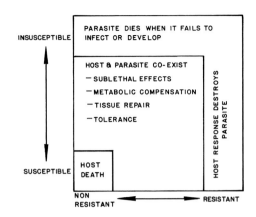

FIGURE 8.—Graphic representation of hypothetical relationship between resistance and susceptibility to *Haplosporidium nelsoni* in eastern oysters.

susceptibility) might show varying degrees of resistance (Stauber 1961).

The idea that resistance, as defined above, can be improved by selection is well accepted, but less attention is given to the possibility that selection may operate equally well to lessen susceptibility. Both susceptibility and resistance may exist, to different degrees, within individuals belonging to the same population (Figure 8). This hypothesis can be illustrated by the *H. nelsoni–* oyster relationship. Whereas the vast majority of individuals in unselected stocks provide the parasite with the proper requirements for proliferation of the plasmodial form, very few are found with spore stages. The rarity of spore stages has been noted by all investigators and has led to speculation that oysters are abnormal hosts that do not provide the requirements that would allow the parasite to complete its life cycle (Farley 1965a; Andrews 1979a; Ford and Haskin 1982). An extension of this idea is that individual oysters provide environments with different suitability for growth of the plasmodial forms. A range of parasite burdens can be found in unselected oysters after exposure to presumably similar infection doses in nature (see Ford 1986). Oysters are variable in almost every biochemical and physiological parameter yet measured. It would thus be easy to conceive of tremendous individual variation in the production of a substance or substances that would enhance or inhibit growth of *H. nelsoni*. Because the substance would be present before encounter with the parasite, it could not be considered a true response to infection. This idea could be extended to argue that seasonal, and perhaps even site-specific, variability in the "internal milieu" of oysters would change their susceptibility to *H. nelsoni*. Selection would then favor the individuals that provided the least favorable environment (Ford and Haskin 1987).

Parasites, too, should exhibit individual variability so that some could exist in "marginal host environments." As discussed above, there is no reason why an oyster that limits parasite development by some degree of insusceptibility could not also actively respond (i.e., show resistance as defined by Read 1958) against the few parasites that it does harbor. In the case of *H. nelsoni*, the functions performed by the masses of hemocytes that infiltrate infected areas surely deserves more attention. In particular, the role of the cells designated as hyalinocytes must be clarified.

Tolerance as a survival mechanism also deserves further study. Are selected oysters physiologically and metabolically better able to compensate for the burden placed on them by parasites, and if so, how? Are they better at gathering and processing food, or are they metabolically more efficient? Can they repair damage or dispose of potentially toxic metabolites of the parasites more rapidly? Would they be able to tolerate other types of stress better?

Tolerance becomes an extremely important attribute in individuals with intermediate degrees of susceptibility, resistance, or both (Figure 8). Highly susceptible, nonresistant individuals will die rapidly, and highly insusceptible or highly resistant ones will suffer little or no debilitation when challenged. Between these extremes fall the animals in which parasites develop to some degree. It is in these individuals that external influences, acting on both host and parasite, may play a critical role in determining the outcome of the interaction. Additional stress on the host may overwhelm tolerance and hasten death, whereas favorable ambient conditions may enhance this ability and prolong survival.

In summary, resistance to *Haplosporidium nelsoni*-caused mortality in oysters, and perhaps to other pathogens of molluscs, appears to be a combination of attributes. It must be considered not only in terms of the classical cellular and humoral "responses," but also as a set of physiological and metabolic conditions in the host that do or do not provide a suitable environment for parasite development. Finally, any analysis of host–parasite relationships in marine and estuarine molluscs must recognize that they are continually altered by external environmental influences.

Acknowledgments

I am indebted to H. H. Haskin and his students, J. Myhre, R. Douglass, and G. Valiulis, for the data on which much of this chapter relies. J. D. Andrews, F. Kern, W. S. Fisher, and H. H. Haskin provided helpful criticism of the manuscript. The studies described were supported by the National Marine Fisheries Service under Public Law 88-309 and the New Jersey Department of Environmental Protection. This paper is New Jersey Agricultural Experiment Station publication F 32504-2-88 supported by state funds.

References

Acton, R. T., J. C. Bennet, E. E. Evans, and R. W. Schrohenloher. 1969. Physical and chemical characterization of an oyster hemagglutinin. Journal of Biological Chemistry 244:4128–4135.

Andrews, J. D. 1966. Oyster mortality studies in Virginia. V. Epizootiology of MSX, a protistan parasite of oysters. Ecology 47:19–31.

Andrews, J. D. 1967. Interaction of two diseases of oysters in natural waters. Proceedings National Shellfisheries Association 57:38–48.

Andrews, J. D. 1968. Oyster mortality studies in Virginia. VII. Review of epizootiology and origin of *Minchinia nelsoni*. Proceedings National Shellfisheries Association 58:23–36.

Andrews, J. D. 1979a. Oyster diseases in Chesapeake Bay. U.S. National Marine Fisheries Service Marine Fisheries Review 41(1–2):45–53.

Andrews, J. D. 1979b. Pelecypoda: Ostreidae. Pages 293–342 *in* A. C. Giese and J. S. Pearse, editors. Reproduction of marine invertebrates, volume 5. Academic Press, London.

Andrews, J. D., and M. Castagna. 1978. Epizootiology of *Minchinia costalis* in susceptible oysters in seaside bays of Virginia's Eastern Shore, 1959–1976. Journal of Invertebrate Pathology 32:124–138.

Andrews, J. D., and W. G. Hewatt. 1957. Oyster mortality studies in Virginia. II. The fungus disease caused by *Dermocystidium marinum* in oysters of Chesapeake Bay. Ecological Monographs 27:1–26.

Arimoto, R., and M. R. Tripp. 1977. Characterization of a bacterial agglutinin in the hemolymph of the hard clam, *Mercenaria mercenaria*. Journal of Invertebrate Pathology 30:406–413.

Auffret, M. 1988. Bivalve hemocyte morphology. American Fisheries Society Special Publication 18:169–177.

Canzonier, W. J. 1968. Present status of attempts to transmit *Minchinia nelsoni* under controlled conditions. Proceedings National Shellfisheries Association 58:1.

Canzonier, W. J. 1973. Tissue grafts in the American oyster, *Crassostrea virginica*. Proceedings National Shellfisheries Association 64:92–101.

Cheng, T. C. 1983. The role of lysosomes in molluscan inflammation. American Zoologist 23:129–144.

Cheng, T. C. 1984. A classification of molluscan hemocytes based on functional evidences. Pages 111–146 in T. C. Cheng, editor. Comparative pathobiology, volume 6. Plenum, New York.

Cheng, T. C., and G. E. Rodrick. 1975. Lysosomal and other enzymes in the hemolymph of Crassostrea virginica and Mercenaria mercenaria. Comparative Biochemistry and Physiology B, Comparative Biochemistry 52:443–447.

Chu, F.-L. E. 1988. Humoral defense factors in marine bivalves. American Fisheries Society Special Publication 18:178–188.

Douglass, W. R. 1977. Minchinia nelsoni disease development, host defense reactions, and hemolymph enzyme alterations in stocks of oysters (Crassostrea virginica) resistant and susceptible to Minchinia nelsoni-caused mortality. Doctoral dissertation. Rutgers University, New Brunswick, New Jersey.

Douglass, W. R., and H. H. Haskin. 1976. Oyster–MSX interactions: Alterations in hemolymph enzyme activities in Crassostrea virginica during the course of Minchinia nelsoni disease development. Journal of Invertebrate Pathology 27:317–323.

Eble, A. F. 1966. Some observations on the seasonal distribution of selected enzymes in the American oyster as revealed by enzyme histochemistry. Proceedings National Shellfisheries Association 56:37–43.

Elston, R. A. 1986. An intranuclear pathogen [Nuclear Inclusion X(NIX)] associated with massive mortalities of the Pacific razor clam, Siliqua patula. Journal of Invertebrate Pathology 47:93–104.

Farley, C. A. 1965a. Acid-fast staining of haplosporidian spores in relation to oyster pathology. Journal of Invertebrate Pathology 7:144–147.

Farley, C. A. 1965b. Pathologic responses of the oyster, Crassostrea virginica (Gmelin), to infection by the protistan parasite, MSX. Bulletin of the American Malacological Union Incorporated 32:23–24.

Farley, C. A. 1967. A proposed life cycle of Minchinia nelsoni (Haplosporida, Haplosporidiidae) in the American oyster Crassostrea virginica. Journal of Protozoology 14:615–625.

Farley, C. A. 1968. Minchinia nelsoni (Haplosporida) disease syndrome in the American oyster Crassostrea virginica. Journal of Protozoology 15:585–599.

Farley, C. A. 1975. Epizootic and enzootic aspects of Minchinia nelsoni (Haplosporida) disease in Maryland oysters. Journal of Protozoology 22:418–427.

Feng, S. Y. 1958. Observations on distribution and elimination of spores of Nematopsis ostrearum in oysters. Proceedings National Shellfisheries Association 48:162–173.

Feng, S. Y. 1965. Heart rate and leucocyte circulation in Crassostrea virginica (Gmelin). Biological Bulletin (Woods Hole) 128:198–210.

Feng, S. Y. 1966. Experimental bacterial infections in the oysters Crassostrea virginica. Journal of Invertebrate Pathology 8:505–515.

Feng, S. Y. 1988. Cellular defense mechanisms of oysters and mussels. American Fisheries Society Special Publication 18:153–168.

Feng, S. Y., and W. J. Canzonier. 1970. Humoral responses in the American oyster (Crassostrea virginica) infected with Bucephalus sp. and Minchinia nelsoni. American Fisheries Society Special Publication 5:497–510.

Feng, S. Y., E. A. Khairallah, and W. J. Canzonier. 1970. Hemolymph-free amino acids and related nitrogenous compounds of Crassostrea virginica infected with Bucephalus sp. and Minchinia nelsoni. Comparative Biochemistry and Physiology 34:547–556.

Foley, D. A., and T. C. Cheng. 1975. A quantitative study of phagocytosis by hemolymph cells of the pelecypods Crassostrea virginica and Mercenaria mercenaria. Journal of Invertebrate Pathology 25:189–197.

Ford, S. E. 1985. Chronic infections of Haplosporidium nelsoni (MSX) in the oyster Crassostrea virginica. Journal of Invertebrate Pathology 45:94–107.

Ford, S. E. 1986. Comparison of hemolymph proteins between resistant and susceptible oysters, Crassostrea virginica, exposed to the parasite Haplosporidium nelsoni (MSX). Journal of Invertebrate Pathology 47:283–294.

Ford, S. E., and Haskin, H. H. 1982. History and epizootiology of Haplosporidium nelsoni (MSX), an oyster pathogen, in Delaware Bay, 1957–1980. Journal of Invertebrate Pathology 40:118–141.

Ford, S. E., and Haskin, H. H. 1987. Infection and mortality patterns in strains of oysters Crassostrea virginica selected for resistance to the parasite Haplosporidium nelsoni (MSX). Journal of Parasitology 73:368–376.

Grizel, H. 1983. Impact of Marteilia refringens and Bonamia ostreae on oyster culture of Brittany. International Council for the Exploration of the Sea, C.M. 1983/Gen:9, Copenhagen.

Hammen, C. S. 1968. Aminotransferase activities and amino acid excretion of bivalve molluscs and brachiopods. Comparative Biochemistry and Physiology 26:697–705.

Haskin, H. H., and J. D. Andrews. 1988. Uncertainties and speculations about the life cycle of the eastern oyster pathogen Haplosporidium nelsoni (MSX). American Fisheries Society Special Publication 18:5–22.

Haskin, H. H., W. J. Canzonier, and J. L. Myhre. 1965. The history of MSX on Delaware Bay oyster grounds, 1957–65. Bulletin of the American Malacological Union Incorporated 32:20–21.

Haskin, H. H., and W. R. Douglass. 1971. Experimental approaches to oyster–MSX interactions. Proceedings National Shellfisheries Association 61:4.

Haskin, H. H., and S. E. Ford. 1979. Development of resistance to Minchinia nelsoni (MSX) mortality in laboratory-reared and native oyster stocks in Delaware Bay. U.S. National Marine Fisheries Service Marine Fisheries Review 41(1–2):54–63.

Haskin, H. H., and S. E. Ford. 1986. Breeding for disease resistance in molluscs. Pages 431–441 in K.

Tiews, editor. Proceedings of a world symposium on selection, hybridization, and genetic engineering in aquaculture, volume 18–19. Bundesforschungsanstalt für Fisherei, Berlin.

Haskin, H. H., L. A. Stauber, and J. A. Mackin. 1966. *Minchinia nelsoni* n. sp. (Haplosporida, Haplosporidiidae): causative agent of the Delaware Bay oyster epizootic. Science (Washington, D.C.) 153: 1414–1416.

Kern, F. G. 1976. *Minchinia nelsoni* (MSX) disease of the American oyster. U.S. National Marine Fisheries Service Marine Fisheries Review 38(10):22–24.

Levine, N. D. 1978. *Perkinsus* gen. n. and other new taxa in the protozoan phylum Apicomplexa. Journal of Parasitology 64:549.

Longwell, A. C., and S. S. Stiles. 1973. Gamete cross incompatibility and inbreeding in the commercial American oyster, *Crassostrea virginica* Gmelin. Cytologia (Tokyo) 38:521–533.

Mackin, J. G. 1951. Histopathology of infection of *Crassostrea virginica* (Gmelin) by *Dermocystidium marinum* Mackin, Owen, and Collier. Bulletin of Marine Science of the Gulf and Caribbean 1:72–87.

Mackin, J. G., H. M. Owen, and A. Collier. 1950. Preliminary note on the occurrence of a new protistan parasite, *Dermocystidium marinum* n. sp. in *Crassostrea virginica* (Gmelin). Science (Washington, D.C.) 111:328–329.

Mengebier, W. L., and L. Wood. 1969. The effects of *Minchinia nelsoni* infection on enzyme levels in *Crassostrea virginica*—II. Serum phosphohexose isomerase. Comparative Biochemistry and Physiology 29:265–270.

Mix, M. C. 1976. A general model for leucocyte renewal in bivalve mollusks. U.S. National Marine Fisheries Service Marine Fisheries Review 38(10):37–41.

Myhre, J. L. 1973. Levels of infection in spat of *Crassostrea virginica* and mechanisms of resistance to the haplosporidan parasite *Minchinia nelsoni*. Master's thesis. Rutgers University, New Brunswick, New Jersey.

Myhre, J. L., and H. H. Haskin. 1968. Some observations on the development of early *Minchinia nelsoni* infections in *Crassostrea virginica* and some aspects of the host–parasite relationship. Proceedings National Shellfisheries Association 58:7.

Myhre, J. L., and H. H. Haskin. 1970. MSX infections in resistant and susceptible oyster stocks. Proceedings National Shellfisheries Association 60:9.

Needler, A. W. H., and R. R. Logie. 1947. Serious mortalities in Prince Edward Island oysters caused by a contagious disease. Transactions of the Royal Society of Canada 41:73–89.

Newman, M. W. 1971. A parasite and disease survey of Connecticut oysters. Proceedings National Shellfisheries Association 61:59–63.

Ray, S. M. 1954. Biological studies of *Dermocystidium marinum*, a fungus parasite of oysters. Rice Institute Pamphlet (special issue), The Rice Institute, Houston, Texas.

Read, C. P. 1958. Status of behavioral and physiological "resistance." Rice Institute Pamphlet 45:36–58. (The Rice Institute, Houston, Texas).

Sparks, A. K. 1985. Synopsis of invertebrate pathology exclusive of insects. Elsevier, Amsterdam.

Sparks, A. K., and J. F. Morado. 1988. Inflammation and wound repair in bivalve molluscs. American Fisheries Society Special Publication 18:139–152.

Stauber, L. A. 1950. The fate of india ink injected intracardially into the oyster, *Ostrea virginica* Gmelin. Biological Bulletin (Woods Hole) 98:273–282.

Stauber, L. A. 1961. Immunity in invertebrates, with special reference to the oyster. Proceedings National Shellfisheries Association 50:7–20.

Tripp, M. R. 1960. Mechanisms of removal of injected microorganisms from the American oyster, *Crassostrea virginica* (Gmelin). Biological Bulletin (Woods Hole) 119:273–282.

Tripp, M. R. 1966. Hemagglutinin in the blood of the oyster *Crassostrea virginica*. Journal of Invertebrate Pathology 8:478–484.

Valiulis, G. A. 1973. Comparison of the resistance to *Labyrinthomyxa marina* with resistance to *Minchinia nelsoni* in *Crassostrea virginica*. Doctoral dissertation. Rutgers University, New Brunswick, New Jersey.

Vrijenhoek, R. C., and S. E. Ford. 1987. Maintenance of heterozygosity in oysters during selective breeding for tolerance to MSX infections. Journal of Shellfish Research 7:179

Yoshino, T. P., and T. C. Cheng. 1976. Experimentally induced elevation of aminopeptidase activity in hemolymph cells of the American oyster, *Crassostrea virginica*. Journal of Invertebrate Pathology 27:367–370.

American Fisheries Society Special Publication 18:225–237, 1988

ENVIRONMENTAL INFLUENCE ON HOST RESPONSE

Environmental Influence on Bivalve Hemocyte Function

WILLIAM S. FISHER[1]

Horn Point Environmental Laboratories, Center for Environmental and Estuarine Studies
University of Maryland, Post Office Box 775, Cambridge, Maryland 21613, USA

Abstract.—It is well known that the environment influences an animal's defense capabilities, but the means are rarely understood. Marine bivalves are poikilothermic and osmoconforming, that is, their blood cells, or hemocytes, are exposed to the same thermal and salinity changes that occur in the environment. The hemocytes, active in phagocytosis, encapsulation, inflammation, and would healing, form the primary line of internal defense for bivalves. Some of their defense-related cellular activities are altered by environmental factors. Hemocyte aggregation, spreading, locomotion, and foreign-particle binding are influenced by oyster habitat and season, as well as acute and short-term changes in salinity and temperature. Increased salinity retards hemocyte spreading and locomotion, whereas salinity reductions retard hemocyte spreading only at the lowest salinities. Increased temperature can enhance cell activities, but summertime temperatures in estuarine oyster habitats appear to stress the hemocytes so that they are less responsive. Effects of stress and acclimation on specific defense-related cellular mechanisms and potential enviromental influences over nonself recognition, chemotaxis, and lysosomal destabilization are discussed.

Hemocytes (or leucocytes) are primarily responsible for the defense of marine bivalves, functioning in inflammation, wound repair, encapsulation, and phagocytosis. These cells exist individually within the hemolymph and interstitial spaces of the animal, where they are exposed to variations of salinity and temperature and to nutrients and other substances in the hemolymph. Natural environmental variation is known to play an important role in the defense response for many animals, but it may be particularly true for the marine and estuarine bivalve molluscs, which are poikilothermic osmoconformers (Galtsoff 1964; Shumway 1977). As such, their internal milieu mimics the salinity and temperature of the environment, which may change with seasonal, diurnal, and tidal cycles or with the movement of animals from one locality to another.

The activities of hemocytes that accompany their known functions in the defense response are varied and complex. For example, phagocytosis relies on nonself recognition, locomotion, binding and ingestion of foreign matter, and intracellular digestion. These processes require simultaneous participation by the cytoplasm, cytoskeleton, membrane, surface receptors, and possibly the nucleus. Impairment of the organelles or restriction of any biochemical reaction related to these activities could diminish the capacity of the cell in defense.

Because the effects of the environment on cellular functions are not well understood for marine bivalve molluscs, there is a great opportunity for productive study. Hemocytes can be easily obtained from hemolymph sinuses without unduly harming the animal (Feng et al. 1971; Ford 1986), and they are easy to manipulate in vitro (Eble and Tripp 1969). However, a major disadvantage is the inability to distinguish the types of hemocytes morphologically and functionally. Cheng (1981) has noted that there is "a variety of interpretations as to how many types of hemocytes occur in bivalves . . . [and] little agreement as to the designations of these cells." Because cells are not easily distinguished, assigning specific functions to them is especially difficult. Even within the groups of cells that can be distinguished, i.e., granular and agranular hemocytes, there is no consistency between species. In some species, phagocytosis is primarily accomplished by granular hemocytes (Foley and Cheng 1975; Feng et al. 1977; Poder et al. 1982; Auffret 1986); in others, it is accomplished by agranular hemocytes (Nakahara and Bevelander 1969; Ruddell 1971; Reade and Reade 1976). Even within a single species, there are several types of cells and several different functions (Cheng et al. 1980), and the number of cells of each type may fluctuate. Hence, there is much we do not understand, and great potential for confusion arises

---

[1]Present address: Marine Biomedical Institute, University of Texas Medical Branch, 200 University Boulevard, Galveston, Texas 77550, USA.

if hemocytes are perceived as a functionally homogenous population. Yet they must often be treated as such in present-day studies. As cell separation techniques become more refined, studies combining morphological data with cell activity measurements will begin to shed light on the functional distinctions between hemocyte types.

In addition to their better-known defensive responsibilities, i.e., phagocytosis, inflammation, encapsulation, and wound repair, hemocytes also function in pinocytosis, elimination by diapedesis, intracellular digestion, secretion of humoral components, and, possibly, heavy metal detoxification. Many of the underlying cell activities, the cellular machinery for these defense-related functions, are instrumental in more than one of the functions. Locomotion, for example, takes part in phagocytosis, inflammation, wound repair, and diapedesis. Thus, research should focus not only on defensive mechanisms but also on the cell activities that underlie them.

Recent studies which have examined the effects of the environment on defense-related cell activities will be summarized in this chapter. Continued research will eventually allow us to pinpoint the cell processes involved in a host–environment interaction. It will no longer be sufficient to state that a disease was caused by an "environmental stress"; rather, we will need to demonstrate how a stressful environment affects specific cell activities and consequently impairs vital defense mechanisms.

Defensive Functions of the Hemocytes

Our present understanding of bivalve defense mechanisms has been formed by many studies which have already been reviewed (Cheng 1975; Bayne 1983; Fisher 1986). Only a sketch of the basic mechanisms will be given here, and emphasis will be placed on measurement of the cell activities involved in each mechanism and their response to the environment.

Endocytosis

The uptake of soluble (pinocytosis) or particulate (phagocytosis) material by cells is collectively known as endocytosis. Phagocytosis by hemocytes is considered to be the most important defense response of bivalves (Foley and Cheng 1975). It relies on a sequence of cellular events that can include nonself recognition, cell spreading, locomotion, binding of a foreign particle, ingestion, and intracellular digestion (Fisher 1986). The most common in vivo method to assay phagocytosis is to inject traceable foreign materials such as india

ink (Stauber 1950), vertebrate erythrocytes (Tripp 1958; Feng and Feng 1974), bacteria (Tripp 1960; S. Y. Feng 1966), or viruses (J. S. Feng 1966; Acton and Evans 1968) into a bivalve and monitor their elimination by serial withdrawals of hemolymph. Some of these studies (J. S. Feng 1966; S. Y. Feng 1966; Acton and Evans 1968; Feng and Feng 1974) showed that cooler temperatures decreased the rate of particle clearance. These results were confirmed in vitro by Foley and Cheng (1975), who demonstrated that binding and ingestion of various bacteria by hemocytes of northern quahog *Mercenaria mercenaria* were greatly reduced at 4°C relative to those at 22 and 37°C.

Inflammation

In bivalves, the essential component of inflammation is leucocytosis, or the movement of hemocytes to an infection site. Presumably, hemocytes clear the area of cellular debris, disease agents, or both by phagocytosis and release lysosomal enzymes for extracellular killing of biotic agents (Cheng 1983a). Indirect evidence of restricted leucocytosis at lower temperatures was provided by Feng and Feng (1974) who found fewer hemocytes in the epithelial layers of oysters at 4°C injected with avian erythrocytes.

Wound Repair

Repair of wounds, which has been reviewed by Sparks (1972) and Sparks and Morado (1988, this volume), depends on the hemocytes' ability to move to a break in the epithelial lining and form a seal. Initially, the procedure appears to be similar to the aggregation (clumping) of hemocytes in vitro (Foley and Cheng 1972) whereby cells connect with one another by filipodia and form a tight aggregate. Cells in vitro eventually spread and migrate away from the aggregate, whereas in vivo they become "fibrocytic" in appearance and maintain contact until they are replaced by the inward migration of adjacent epithelial cells (Cheng 1981). Certain in vivo emboli, however, have been observed to disaggregate (Narain 1973; Feng and Feng 1974). Thus, cell activities in wound repair may include nonself recognition, locomotion, aggregation, cell spreading (to a fibrocyte-like shape), and enzyme release.

Encapsulation

When a particle or parasite is too large to be phagocytosed by a single cell, a group of hemocytes surrounds the particle to separate it from host tissue (Cheng and Rifkin 1970; Feng 1988, this volume).

This encapsulation process appears to combine certain properties of phagocytosis and wound repair. Possibly, each hemocyte attempts to phagocytose the particle but can only spread thinly over a large surface area. Other hemocytes may follow suit until the particle's foreignness is masked by the growing layer of cells. Presumably, biotic agents are incapacitated by this process and can invade no further. In insect studies, quantitative assays of encapsulation have been developed as measures for nonself recognition factors (Lackie 1979; 1983).

Elimination by Diapedesis

Foreign matter may also be eliminated by evacuation, i.e., when particle-laden phagocytes move through epithelial barriers (diapedesis) into digestive and excretory lumina to be passed out of the body. Elimination of foreign matter by diapedetic migration was first observed by Stauber (1950) with ink particles but was later found to exist with biotic and soluble material as well (Tripp 1960; Feng 1965). Possibly, the first priority for a phagocyte is to digest the material intracellularly, but if this is not possible, it may resort to evacuation. This function requires the same cell activities as those involved in phagocytosis, particularly locomotion.

Enzyme Secretion

Lysosomal enzymes may be released by the hemocytes for (1) extracellular killing of parasites or other disease agents in the hemolymph and interstitial spaces, (2) to "unmask" certain foreign particles, and possibly (3) to attract more hemocytes to the site (Cheng 1983a). Enzyme secretion can require the formation of lysosomes, enzyme synthesis, release of enzymes into the cytoplasm, and extracellular secretion (exocytosis or degranulation) of enzymes. The stimulation may be from foreign material (Foley and Cheng 1977) and possibly from stressful environmental conditions, as has been found with lysosomes of bivalve digestive gland cells (Moore 1976; IMW 1980).

Defense-Related Hemocyte Activities

Each of the cellular defense functions outlined above relies on several underlying cell activities. Some of these cell activities have been isolated for study by a variety of techniques, and their responses to changing environmental conditions have been measured.

Aggregation

Bang (1961) and Feng and Feng (1974) have observed "clots" of hemocytes in oyster tissue, an apparent response to wounding. These cell-to-cell aggregations are facilitated by protoplasmic extensions, known as "spikes" (Davies and Partridge 1972), "bristle-like processes" (Takatsuki 1934) or "filopodia" (Narain 1973), which form interconnecting bridges that shorten and thicken to draw the cells closer together. Davies and Partridge (1972) found that spike formation in hemocytes from the limpet *Patella* sp. began within 30 s of cell withdrawal and was temperature dependent, requiring 3 and 12 min at 20 and 0°C, respectively. Cell aggregation was also faster and more complete at 20°C, an occurrence attributed to greater adhesion between cells. In the limpets, divalent cations were required for cell aggregation but not for spike formation. Aggregation for many marine invertebrates can be inhibited in vitro by divalent cation chelators such as EDTA, EGTA, or N-ethylmaleimide (Noble 1970; Kenney et al. 1972; Edds 1977; Kanungo 1982). Aggregation has been quantified by counting single cells on a hemacytometer before and after disaggregation with a chelating substance (Kanungo 1982). Among oyster hemocytes, filopodia and aggregrates are clearly present in EDTA solutions (Fisher 1986), and this technique is not applicable.

Unlike vertebrate blood clots, marine invertebrate hemocyte aggregates are reversible, both in vivo (Narain 1973; Feng and Feng 1974) and in vitro (Eble and Tripp 1969; Edds 1977). Cell aggregation may be the initial response to an "inducer" such as a foreign particle or physical perturbation. Inducers may cause displacement of surface divalent cations, resulting in greater permeability of the plasmalemma (Prusch and Hannafin 1979). This response could initiate spike formation and enhanced adhesion. If no foreign particles are present to be encapsulated or phagocytosed, then the cells disengage and migrate away from the aggregate, possibly to seek out foreign material or to return to an uninduced state. Such a scenario would require a multiple-step response by the hemocytes that depends on their capacity to be induced, to recognize self (other hemocytes), to recognize nonself, and to recognize the absence of nonself material while in an induced state. These activities have not been investigated, and it is not known how the environment might affect each step. It appears that there is some variability in hemocyte aggregation in eastern oysters *Crassostrea virginica*; during a relatively hot, low-salinity season (summer, 1984) in the Chesapeake Bay, Maryland, aggregate formation in hemolymph samples was practically nonexistent (Fisher, unpublished data).

Cell Spreading

Hemocytes must change their shape to be functional in defensive roles. Uninduced hemocytes in the hemolymph are spherical, yet they are found as flattened and elongated cells in wound plugs and capsules. In addition, an ameboid shape is necessary for locomotion (Fisher and Newell 1986). The ability to change shape then, is of primary importance and undoubtedly must integrate complex changes in membrane fluidity and the cytoskeleton. Little is known of hemocyte spreading activities of bivalves, but research on sea urchin coelomocytes (Edds 1984) has shown actin and microfilaments to be essential in cell spreading. It has been noted that sea urchin coelomocytes spread spontaneously when the external salinity is reduced ("hypotonic shock") (Edds 1979).

One method to assay cell spreading in bivalves (Fisher and Newell 1986) is to measure the time required for hemocytes to transform from a spherical shape to an ameboid morphology in vitro (Figure 1). Time to hemocyte spreading (TTS) was first used to demonstrate the influence of salinity change on eastern oyster hemocytes. This measurement was made by noting the time required for hemocytes to become ameboid and begin to migrate away from aggregates while being observed with an inverted microscope. Acute in vitro increases in salinity increased TTS for hemocytes from oysters acclimated to low salinities at a rate proportional to the salinity increase (Figure 2). This finding implied that an iso- or hyperosmotic cell was prerequisite for cell spreading or that volume regulation took precedent over spreading for cell mechanisms or energy sources used by both activities.

Yet, during a year of high estuarine salinity, oyster hemocyte spreading was retarded at low salinity (Fisher and Tamplin 1988). Differences between this and previous results are believed to be due to different ambient conditions. Further research (Fisher, unpublished data) has supported these conjectures: (1) regardless of ambient or acclimation salinity, an acute increase in salinity requires a proportionally longer TTS; (2) an acute salinity decrease does not affect TTS except when hemocytes are placed in very low salinities, near 10‰. The 10‰ threshold nears the lower limit of the salinity range for eastern oysters (Galtsoff 1964).

It appears that volume regulation is the factor that retards cell spreading when the ambient salinity is increased. Hemocytes may need to form osmotically active molecules to become isoos-

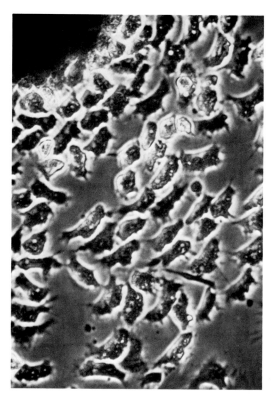

FIGURE 1.—Hemocytes from the eastern oyster form aggregates when withdrawn from the hemolymph sinuses. After settling to the surface of a slide or culture dish, the hemocytes along the periphery of the aggregate (upper left) spread to an ameboid shape and begin to locomote. Eventually, all healthy cells become locomotory. (From Fisher and Newell 1986.)

motic before they can spread. The factor responsible for retarding cell spreading at very low salinity is unknown. Recent studies (Fisher, unpublished) have indicated that hemocytes have a residual "imprint" of their ambient salinity that may take several weeks to erase. Figure 3 shows the different responses of hemocytes from oceanic (32‰) and estuarine (12‰) eastern oysters to a range of in vitro salinities. An acclimation period of 1–2 weeks in the alternative salinity is sufficient to reverse the patterns.

Temperature also plays a role in the ability of the hemocytes to spread. Recent experiments have shown that warmer temperatures enhance cell activity (shorter TTS) of nonstressed eastern oyster hemocytes but can also create stress that results in slower activity (lower TTS). Hemocytes from estuarine oysters sampled during high-temperature summer conditions showed longer TTS than during lower-temperature spring conditions.

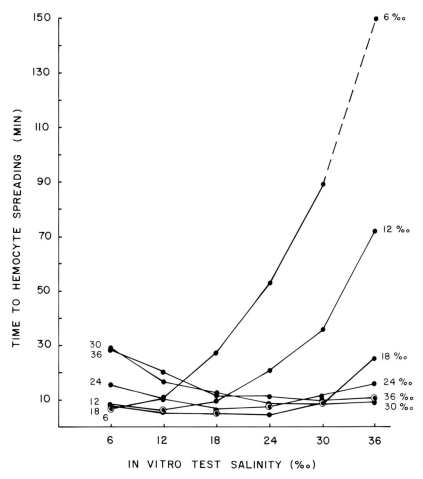

FIGURE 2.—The time required for hemocytes to spread to an ameboid shape is influenced by the acclimation salinity and the test salinity. Eastern oysters were acclimated for 1 month to a range of salinities noted along the right side of the graph, then hemolymph was withdrawn and the hemocytes were immersed in one of six test salinities (abscissa) in vitro. Hemocytes from oysters acclimated to low salinities were strongly affected at high test salinities and required much longer to spread. Time to hemocyte spreading was consistently faster when test salinity and salinity of acclimation were equal (circled data points). (From Fisher and Newell 1986.)

When some of these oysters was held for a week at cooler temperatures (15°C), their hemocyte activity increased over the activity in animals kept at 25°C (Figure 4). Experiments are presently being conducted to determine whether this is the result of stress on spreading activity or on acute temperature acclimation.

Locomotion

In most defense-related functions, the hemocytes must be capable of locomotion. Cell movement is especially important in endocytosis, probably the most important defense response of bivalves. Moreover, if bivalve hemocytes are similar to mammalian macrophages, endocytosis and locomotion are linked to the same cell activity, i.e., the recycling of the cell membrane (Bretscher 1984). Membrane recycling occurs during endocytosis as material is enclosed and internalized. Locomotion may result when internalized portions of membrane are then replaced to the leading edge of the hemocyte. As a corollary, stimulation of the hemocyte surface by a foreign particle or foreign substrate may initiate locomotion. This could be a mechanism for chemotaxis, to be discussed later.

Locomotion can be measured as a rate or a direction (Wilkinson et al. 1982). Rates of locomotion (ROL) for eastern oyster hemocytes have been measured by the use of a video monitor connected

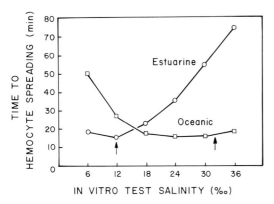

FIGURE 3.—Hemocytes of eastern oysters from an estuarine (Tred Avon River, Maryland) and an oceanic (Wachapreague, Virginia) habitat, collected in April 1987, exhibited different patterns of response to acute in vitro changes in salinity. The differences were due to the influence of the ambient salinities, denoted by an arrow for each habitat. When the oceanic oysters were placed in 12‰ salinity and the estuarine oysters were placed in 32‰ in the laboratory, 1–2 weeks were sufficient for the salinity-response patterns to reverse (acclimation). As shown in Figure 2, time to hemocyte spreading was fastest when measured at the ambient salinities. Sample size was 12 for each data point.

to an inverted microscope and have demonstrated a salinity effect (Figure 5) similar to that for cell spreading: ROL for hemocytes from oysters acclimated to low salinities was retarded by higher in vitro salinities (Fisher and Newell 1986). Moreover, ROL was enhanced (with certain exceptions) for

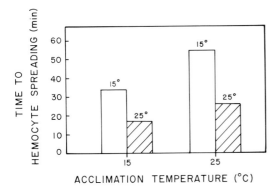

FIGURE 4.—Spreading of eastern oyster hemocytes can be retarded by high temperatures. Hemocytes from estuarine oysters acclimated in the laboratory for 2 weeks at 25°C took longer to spread than those of oysters acclimated at 15°C. This was found whether hemocytes were measured at 15 or 25°C (testing temperatures are shown above histogram bars). Acclimation and testing were conducted at the ambient (12‰) salinity; $N = 12$ for each histogram bar.

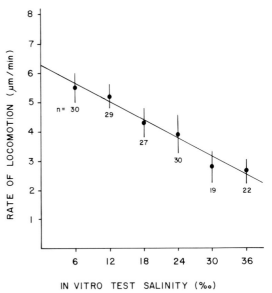

FIGURE 5.—Rates of hemocyte locomotion for eastern oysters from both estuarine and oceanic habitats showed a linear response to acute changes in salinity. Oysters from both habitats were acclimated to a range of salinities (6–24‰), and the withdrawn hemocytes were exposed to six in vitro test salinities (abscissa). The rate of locomotion was enhanced at low salinities and depressed at high salinities regardless of the acclimation condition. Sample sizes (n) are given for each test salinity. Vertical lines are 95% confidence intervals. (Adapted from Fisher and Newell 1986.)

hemocytes tested at salinities below their acclimation salinity. These studies also implied that hemocytes have an imprint of previous salinity exposure; hemocytes from low-salinity oysters had slower ROL at high salinities even after a month in high-salinity laboratory conditions.

It might seem that, due to many similar responses, TTS and ROL measure the same cell process. However, observations on individual eastern oysters and averaged responses of subpopulations have shown that ROL is independent of TTS. Acclimation to a higher salinity by hemocytes from low-salinity oysters, for example, took less time for TTS than for ROL (Fisher and Newell 1986). Analysis of 132 samples at a variety of temperatures and salinities showed no significant correlation between TTS and ROL (Fisher and Tamplin 1988); there was a retardation of cell spreading for low-salinity oysters but no effect on locomotion. During temperature–EDTA studies, TTS was sharply increased at cold and warm temperatures, but ROL was not affected. These results indicate that once the cells have reached

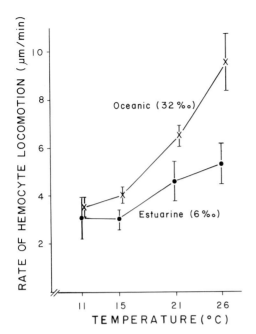

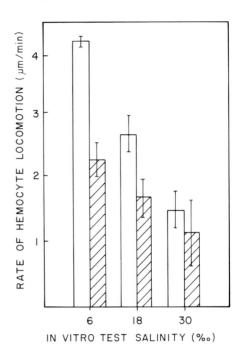

FIGURE 6.—Eastern oysters from an oceanic site (Wachapreague, Virginia) and an estuarine site (Tred Avon River, Maryland) were maintained for 2 weeks at ambient salinities but at four different temperatures (abscissa). Rates of hemocyte locomotion measured at the acclimation temperatures showed that the oceanic oysters had significantly faster hemocytes at the higher temperatures. Sample size was 12 for each data point, and 95% confidence intervals are shown as vertical lines.

FIGURE 7.—Rate of hemocyte locomotion for estuarine eastern oysters was slower after 2 weeks at 25°C (hatched bars) than after 2 weeks at 15°C (open bars). Measurements were made at 15°C, so the differences may be due to temperature acclimation, high-temperature stress, or both. This graph also shows the effect of different in vitro test salinities (abscissa) on the rate of hemocyte locomotion. Standard errors are shown as vertical lines, and $N = 6$ for each bar.

an osmotic balance sufficient to allow spreading, they are able to locomote without further hindrance. In preliminary studies, the antifouling agent tributyltin retarded ROL yet had no effect on TTS. Locomotion, then, depends on the ameboid shape but is a separate cell process that may or may not be affected by the same factors that influence cell spreading.

Warm water temperatures have enhanced locomotion of hemocytes from nonstressed eastern oysters. It has also been observed that hemocytes of eastern oysters from Wachapreague, Virginia, an oceanic habitat, increased their activity more at higher temperatures than those of oysters from Tred Avon River, Maryland, an estuarine habitat (Figure 6). This may be of importance to future selective breeding programs if the ability to respond faster at high temperatures is a major factor in defense.

Rate of locomotion can also demonstrate high-temperature stress; eastern oysters from long-term high-temperature conditions in nature had slower ROL than subsamples of the same oysters

maintained in the laboratory for 1 week at 15°C (Figure 7). Figure 7 shows the combined effects of high-temperature stress and acute in vitro salinity stress; as salinity increased, ROL decreased. In all cases, ROL of hemocytes from oysters kept under natural conditions (23–27°C) were slower than those of hemocytes from oysters maintained at 15°C. The greatest difference between lots was at the lowest salinity (6‰), implying a high-temperature–low-salinity stress interaction.

Nonself Recognition

Fundamental to the defense response is the ability to distinguish between self and nonself material. This perception is accomplished by receptors on the external surface of the hemocyte membrane (Cheng 1983b). At least one of the hemocyte membrane receptors in oysters is the same as a lectin found in the humoral component of the hemolymph (Vasta et al. 1982). Differences in receptor specificity could explain differences in the ability of bivalve hemocytes to bind various foreign particles (Bang 1961; Tripp and Kent 1967; Anderson and Good 1976; Hardy et al. 1977; Bayne et al. 1979). Recently, investigators have suggested that cell subpopulations with distinct sets of surface receptors exist within a single organism (Cheng et al. 1980; Yoshino and Granath 1985). Of special interest are the differences found between granular and agranular hemocytes (Cheng et al. 1980), which may explain differences in phagocytic ability.

Because recognition sites are found on the surface of the membrane, any environmental factors that affect the membrane could also affect the ability of the hemocyte to recognize foreign material. Temperature is known to have dramatic effects on biological membranes (Hochachka and Somero 1973), and this influence might alter receptor site configurations. However, no research of this type has been conducted.

Nonself recognition should remain, at least conceptually, separate from foreign-particle binding and phagocytosis. Binding is perhaps the most direct means to measure recognition, but there may be separate cell activities involved. Chemotactic hemocytes, for example, exhibit nonself recognition without binding the foreign particle. Presumably, they bind molecules released by the particle, but these molecules may not be the ones involved in subsequent binding processes. Studies summarized below (on chemotaxis and foreign-particle binding) are equally applicable to nonself recognition because this capacity is integral to those measurements.

Chemotaxis

The movement of hemocytes along a chemical gradient facilitates their contact with foreign particles for phagocytosis or encapsulation. Chemotaxis by bivalve hemocytes has been demonstrated by Cheng and Howland (1982a) and includes the cellular activities of nonself recognition and locomotion. Cheng and Howland (1982b) showed that random movement of eastern oyster hemocytes was inhibited by cytochalasin B and colchicine; locomotion thus depended on the synthesis of microfilaments and microtubules. But colchicine inhibited directed movement more than random movement, implying a role for microtubules in chemotaxis. The surface receptors at work in chemotaxis are believed to be the same as those that bind the foreign particles (Cheng 1985). There have not been any studies on bivalves to examine the effects of the environment on the chemotactic ability of hemocytes, but fish mac-

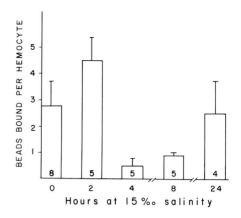

FIGURE 8.—Hemocytes from edible oysters showed an initial increase and then a decrease in their ability to bind fluorescent latex beads after the oysters were immersed in low-salinity (15‰) water. Ambient salinity was 30‰ for this oceanic species. Bead binding did, however, return to normal levels within 24 h. Positive values of standard errors (vertical lines) and sample sizes are shown on the histogram bars. (From Fisher et al. 1987.)

rophage chemotaxis is disrupted by certain anthropogenic pollutants (Weeks et al. 1986).

Foreign-Particle Binding

In phagocytosis, the culmination of many cell activities is the actual contact and binding of a foreign particle to the external surface of the hemocyte. Foreign-particle binding is sometimes considered synonymous with ingestion, because the former nearly always leads to the latter (Pearsall and Weiser 1970). In some cases, however, binding may be differentiated from ingestion (Rabinovitch 1967; Smith and Rommel 1977), distinguishing these as separate steps in the phagocytic process. Latex beads have been employed to measure the binding capabilities of insect hemocytes both in vivo (Brehelin and Hoffman 1980) and in vitro (Rowley and Ratcliffe 1979; Huxham and Lackie 1988). The effect of salinity and temperature on binding of eastern oyster hemocytes after several weeks of acclimation has been examined with use of fluorescent latex beads (Fisher and Tamplin 1988). Salinity and temperature factors altered both the number of adhesive hemocytes and the number of beads bound per adhesive hemocyte. Binding was reduced at the lowest (6‰) salinity and enhanced at the highest (28°C) temperature. Short-term acclimation studies on the edible oyster *Ostrea edulis* have shown (Figure 8) that salinity reduction from 30 to 15‰

retarded binding by hemocytes between 2–8 h after exposure, but control levels returned by 24 h for these normally oceanic bivalves (Fisher et al. 1987). A strikingly similar response was found for the short-term effect of a 5°C temperature increase, although potential connections between the two responses are unknown.

In vitro studies by Foley and Cheng (1975) showed that binding of the bacterium *Bacillus megaterium* by northern quahog hemocytes was less at 4 than at 22 and 37°C. Based on the assumption that contact between the cells and the bacteria was identical at all temperatures, Foley and Cheng concluded that the binding process was influenced by temperature. This conclusion may be true, but the assumption is questionable. Measurement of latex bead binding by hemocytes of eastern oysters (Fisher and Tamplin 1988) showed a significant positive correlation between the rate of hemocyte locomotion and foreign particle binding under a variety of environmental conditions. This relationship implies that faster cell movement (at higher temperatures) may increase the encounter rate between hemocytes and foreign particles and result in greater binding rates.

According to the above notion, foreign particle binding should be closely related to the rate of locomotion as long as the foreign particles bind identically to the hemocytes upon contact. But particles, particularly biotic particles, do not always bind identically. Parasites and other pathogens can develop means to disguise or mask their nonself nature or to block the hemocyte receptors (Damian 1964; Salt 1968; Cheng 1988, this volume). Bang (1961) was the first to report differences in the binding of various bacteria by eastern oyster hemocytes. The measurement of binding capacity, in addition to the information gained from rate of locomotion, can indicate the relative ability of hemocytes to recognize and bind a foreign particle. Fisher (1988) used both latex beads and purified parasite preparations to isolate these factors. Two species of oysters from the Baie de Quiberon in Bretagne, France, were compared in their abilities to bind the beads and the parasites. The parasites, *Bonamia ostreae*, have caused unprecedented mortalities in edible oysters but are not harmful to Pacific oysters, *Crassostrea gigas* (Balouet et al. 1983). Results from a variety of environmental conditions showed no differences in the ability of the two species to bind latex beads (Figure 9), but Pacific oyster hemocytes were always more successful in binding the parasites. So, it appears that edible oyster hemo-

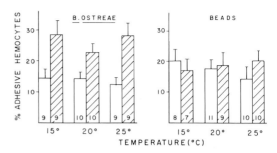

FIGURE 9.—Hemocytes from edible oysters (open bars) were equal to hemocytes from Pacific oysters (hatched bars) in their ability to bind latex beads (right), but they were less able to bind the protistan pathogen *Bonamia ostreae* (left). This implies that they were equally active but less able to recognize and bind the pathogen. No differences were found among tests at 15, 20 and 25°C. Positive values of standard errors (vertical lines) and sample sizes are shown on the histogram bars. (From Fisher 1988.)

cytes, although equally active with latex particles, were less able to recognize the pathogen, to bind it, or both. This reduced ability may be a factor in their lack of resistance against bonamiasis.

Lysosomal Destabilization

The release of lysosomal enzymes, either during intracellular digestion of foreign matter (Feng et al. 1977) or release into the serum (Foley and Cheng 1977), requires that lysosomes respond to a stimulus. During intracellular digestion, they fuse with a phagosome to mix the foreign matter with degradatory enzymes, often leading to glycogen synthesis (Cheng 1975; Auffret 1986). During release into the serum, lysosomes migrate to the ectoplasm from the endoplasm and then bud off (degranulate, exocytose) from the plasma membrane (Mohandas 1985). Cheng and Yoshino (1976) found that injection of bacteria increased lipase in both the serum and the hemocytes and suggested that synthesis and release had been induced. This finding prompted Cheng (1983b) to suggest that three recognition sites were required for lysosomal synthesis, the plasmalemma, the nuclear envelope, and the lysosomal granule.

Although lysosomal latency of mussel digestive cells has been studied as a biological monitoring tool (IMW 1980; Newell and Barber 1988, this volume), environmental influences on hemocyte lysosomes has not been studied. Environmental conditions may influence not only the processes involved in the release of lysosomal enzymes but also the potential killing power or degradative properties of the enzymes after their release (Roderick and Cheng 1974). This is a promising area for research into environmental effects on the bivalve defense response.

Discussion

The environmental factors that most influence marine and estuarine bivalve molluscs are salinity and temperature. As can be seen from the preceding examples, these natural environmental variables can affect the defense-related activities of blood cells in many ways. The most important effect of salinity appears to be the need for upward volume regulation at the expense of other cell functions, such as cell spreading and locomotion, or as a prerequisite to them. Temperature, on the other hand, probably influences the hemocyte membrane, affecting its structure and permeability. Temperature also profoundly affects the metabolism of the cells. A confounding factor in any attempt to understand environmental effects on hemocyte activities are the cellular imprints of past environmental conditions. Experiments are undoubtedly influenced by environmental history, seasonal variation, or even the availability of certain nutrients and energy sources. For example, van der Knapp et al. (1983) showed that, in addition to pH and temperature, phagocytosis by snail hemocytes depends on the feeding condition of the snails.

Of all the cell activities examined here, perhaps the most important to bivalve defense is that of locomotion. It is required for endocytosis, wound healing, inflammation, and encapsulation. Concomitantly, locomotion is probably the most vulnerable to environmental variation because of its dependence on cell membrane functions and its reliance on the continuous recycling of the cell membrane, a process which requires the energy and machinery for breakdown and synthesis of the plasmalemma.

It is likely that different environmental conditions, even within a single habitat, create the high variation often found in measured responses of bivalves. When animals are maintained under identical circumstances in the laboratory, individual variation decreases. Conversely, environmental change greatly increases variation among animals, an obvious consequence of individual resilience.

None of the measured hemocyte activities nor any specific defensive function will give a complete picture of the influence of the environment on the bivalve defense system. We know, for example, that anaphylaxis in humans is a strong defense response, but it is not necessarily beneficial to the whole organism. Continued measurement of defense activities and defense-related functions of hemocytes may, however, begin to give us insight to the seasonality, periodicity, and incidence of diseases. For example, it is known that *Haplosporidium nelsoni* (MSX) occurs in high-salinity regions (Haskin and Ford 1982), that *Perkinsus marinus* (Andrews 1984) and the bacteria associated with "summer mortalities" (Lipovsky and Chew 1972) are most severe in warm waters, and that *Bonamia ostreae* is more likely to infect oysters in crowded conditions (Grizel 1985). Although these relationships may be due largely to environmental influence on parasite activity, probably the host defense system is, at least in a relative sense, compromised. It is particularly relevant that eastern oyster hemocyte activities are depressed by the same high water temperatures that are believed to increase parasite activity.

Continued study of defense processes in bivalve pathology is directly applicable to disease prevention. There are only a few strategies available to combat diseases: (1) prevent the spread of disease agents by limiting the movement of stock and monitoring for the agents, (2) manage densities and habitat to avoid disease situations, or (3) purposefully select resistant stock. The hatchery technology now available makes the last alternative a realizable goal, and several programs have been implemented to select bivalve stocks based on their survival under natural disease pressures. These programs are very valuable and can be improved by our ability to know the mechanism of resistance for survivors and whether the same stock would be as successful in a different habitat and against a new or different pathogen. Comparisons of defense processes and defense-related hemocyte activities will undoubtedly play a vital role in the further development of stock selection for disease resistance.

Acknowledgments

I thank the University of Maryland Sea Grant and the Maryland Department of Natural Resources for partial funding of this research, and M. Chintala, D. Feeney, L. Gale, and M. Moline for their technical assistance. This is contribution 1899 from the Center for Environmental and Estuarine Studies, University of Maryland.

References

Acton, R. T., and E. E. Evans. 1968. Bacteriophage clearance in the oyster (*Crassostrea virginica*). Journal of Bacteriology 95:1260–1266.
Anderson, R. S., and R. A. Good. 1976. Opsonic in-

volvement in phagocytosis by mollusk hemocytes. Journal of Invertebrate Pathology 27:57–64.

Andrews, J. D. 1984. Epizootiology of diseases of oysters (*Crassostrea virginica*) and parasites of associated organisms in eastern North America. Helgoländer Meeresuntersuchungen 37:149–166.

Auffret, M. 1986. Internal defense in bivalve molluscs: ultrastructural observations on the fate of experimentally injected bacteria in *Ostrea edulis* granular hemocytes. Pages 351–356 in C. P. Vivares, J.-R. Bonami, and E. Jaspers, editors. Pathology in marine aquaculture. European Aquaculture Society, Special Publication 9, Bredene, Belgium.

Balouet, G., M. Poder, and A. Cahour. 1983. Haemocytic parasitosis: morphology and pathology of lesions in the French flat oyster, *Ostrea edulis* L. Aquaculture 34:1–14.

Bang, F. B. 1961. Reaction to injury in the oyster (*Crassostrea virginica*). Biological Bulletin (Woods Hole) 121:57–68.

Bayne, C. J. 1983. Molluscan immunobiology. Pages 407–486 in A. S. M. Saleuddin and K. M. Wilbur, editors. The Mollusca, volume 5. Academic Press, New York.

Bayne, C. J., M. N. Moore, T. H. Carefoot, and R. J. Thompson. 1979. Hemolymph functions in *Mytilus californianus*: the cytochemistry of hemocytes and their responses to foreign implants and hemolymph factors in phagocytosis. Journal of Invertebrate Pathology 34:1–20.

Brehelin, M., and J. A. Hoffman. 1980. Phagocytosis of inert particles in *Locusta migratoria* and *Galleria mellonella*: study of ultrastructure and clearance. Journal of Insect Physiology 26:103–111.

Bretscher, M. S. 1984. Endocytosis: relation to capping and cell locomotion. Science (Washington, D.C.) 224:681–686.

Cheng, T. C. 1975. Functional morphology and biochemistry of molluscan phagocytes. Annals of the New York Academy of Sciences 266:343–379.

Cheng, T. C. 1981. Bivalves. Pages 233–300 in N. A. Radcliffe and A. F. Rowley, editors. Invertebrate blood cells. Academic Press, New York.

Cheng, T. C. 1983a. The role of lysosomes in molluscan inflammation. American Zoologist 23:129–144.

Cheng, T. C. 1983b. Triggering of imunologic defense mechanisms of molluscan shellfish by biotic and abiotic challenge and its applications. Marine Technology Society Journal 14:18–25.

Cheng, T. C. 1985. Evidences for molecular specificities involved in molluscan inflammation. Pages 129–142 in T. C. Cheng, editor. Comparative pathobiology, volume 8. Plenum, New York.

Cheng, T. C. 1988. Strategies employed by parasites of marine bivalves to effect successful establishment in hosts. American Fisheries Society Special Publication 18:112–129.

Cheng, T. C., and K. H. Howland. 1982a. Chemotactic attraction between hemocytes of the oyster, *Crassostrea virginica*, and bacteria. Journal of Invertebrate Pathology 33:204–210.

Cheng, T. C., and K. H. Howland. 1982b. Effects of colchicine and cytochalasin B on chemotaxis of oyster (*Crassostrea virginica*) hemocytes. Journal of Invertebrate Pathology 40:150–152.

Cheng, T. C., J. W. Huang, H. Karadogan, L. R. Renwrantz, and T. P. Yoshino. 1980. Separation of oyster hemocytes by density gradient centrifugation and identification of their surface receptors. Journal of Invertebrate Pathology 36:35–40.

Cheng, T. C., and E. Rifkin. 1970. Cellular reactions in marine molluscs in response to helminth parasitism. American Fisheries Society Special Publication 5: 443–496.

Cheng, T. C., and T. P. Yoshino. 1976. Lipase activity in the serum and hemolymph cells of the soft-shelled clam, *Mya arenaria*, during phagocytosis. Journal of Invertebrate Pathology 27:243–245.

Damian, R. T. 1964. Molecular mimicry: antigen sharing by parasite and host and its consequences. American Naturalist 98:129–149.

Davies, P. S., and T. Partridge. 1972. Limpet haemocytes: studies on aggregation and spike formation. Journal of Cell Science 11:757–769.

Eble, A. F., and M. R. Tripp. 1969. Oyster leucocytes in tissue culture: a functional study. Proceedings National Shellfisheries Association 59:5.

Edds, K. T. 1977. Dynamic aspects of filopodial formation by reorganization of microfilaments. Journal of Cell Biology 73:479–491.

Edds, K. T. 1979. Isolation and characterization of two forms of a cytoskeleton. Journal of Cell Biology 83: 109–115.

Edds, K. T. 1984. Differential distribution and function of microtubules and microfilaments in sea urchin coelomocytes. Cell Motility 4:269–281.

Feng, J. S. 1966. The fate of a virus, *Staphylococcus aureus* Phage 80, injected into the oyster, *Crassostrea virginica*. Journal of Invertebrate Pathology 8: 496–504.

Feng, S. Y. 1965. Pinocytosis of proteins by oyster leucocytes. Biological Bulletin (Woods Hole) 128: 95–105.

Feng, S. Y. 1966. Experimental bacterial infections in the oyster, *Crassostrea virginica*. Journal of Invertebrate Pathology 8:505–511.

Feng, S. Y. 1988. Cellular defense mechanisms of oysters and mussels. American Fisheries Society Special Publication 18:153–168.

Feng, S. Y., and J. S. Feng. 1974. The effect of temperature on cellular reactions of *Crassostrea virginica* to the injection of avian erythrocytes. Journal of Invertebrate Pathology 23:22–37.

Feng, S. Y., J. S. Feng, C. N. Burke, and L. H. Khairallah. 1971. Light and electron microscopy of the leucocytes of *Crassostrea virginica* (Mollusca: Pelecypoda). Zeitschrift für Zellforschung und mikroskopische Anatomie 120:222–245.

Feng, S. Y., J. S. Feng, and T. Yamasu. 1977. Roles of *Mytilus coruscus* and *Crassostrea gigas* blood cells in defense and nutrition. Pages 31–67 in L. A. Bulla, Jr., and T. C. Cheng, editors. Comparative pathobiology, volume 3. Plenum, New York.

Fisher, W. S. 1986. Structure and functions of oyster

hemocytes. Pages 25–35 *in* M. Brehelin, editor. Immunity in invertebrates. Springer-Verlag, Berlin.

Fisher, W. S. 1988. In vitro binding of the parasite *Bonamia ostreae* and latex particles by hemocytes of susceptible and insusceptible oysters. Developmental and Comparative Immunology 12:43–53.

Fisher, W. S., M. Auffret, and G. Balouet. 1987. Acclimation of European flat oyster (*Ostrea edulis*) hemocytes to acute salinity and temperature changes. Aquaculture 67:179–190.

Fisher, W. S., and R. I. E. Newell. 1986. Salinity effects on the activity of granular hemocytes of American oysters, *Crassostrea virginica*. Biological Bulletin (Woods Hole) 170:122–134.

Fisher, W. S., and M. Tamplin. 1988. Environmental influence on activity and foreign-particle binding by hemocytes of American oysters, *Crassostrea virginica*. Canadian Journal of Fisheries and Aquatic Sciences 45:1309–1315.

Foley, D. A., and T. C. Cheng. 1972. Interaction of molluscs and foreign substances: the morphology and behavior of hemolymph cells of the American oyster, *Crassostrea virginica, in vitro*. Journal of Invertebrate Pathology 19:383–394.

Foley, D. A., and T. C. Cheng. 1975. A quantitative study of phagocytosis by hemolymph cells of the pelecypods *Crassostrea virginica* and *Mercenaria mercenaria*. Journal of Invertebrate Pathology 25:189–197.

Foley, D. A., and T. C. Cheng. 1977. Degranulation and other changes of molluscan granulocytes associated with phagocytosis. Journal of Invertebrate Pathology 29:321–325.

Ford, S. E. 1986. Effect of repeated hemolymph sampling on growth, mortality, hemolymph protein, and parasitism of oysters, *Crassostrea virginica*. Comparative Biochemistry and Physiology A, Comparative Physiology 85:465–470.

Galtsoff, P. S. 1964. The American oyster. U.S. Fish and Wildlife Service Fishery Bulletin 64:1–480.

Grizel, H. 1985. Etude des récentes épizooties de l'huître plate *Ostrea edulis* Linné et de leur impact sur l'ostréiculture Bretonne. Doctoral dissertation. Université des Sciences et Techniques du Languedoc, Montpellier, France.

Hardy, S. W., T. C. Fletcher, and J. A. Olafsen. 1977. Aspects of cellular and humoral defense mechanisms in the Pacific oyster, *Crassostrea gigas*. Pages 59–66 *in* J. B. Solomon and J. D. Horton, editors. Developmental immunobiology. Elsevier, Amsterdam.

Haskin, H. H., and S. E. Ford. 1982. *Haplosporidium nelsoni* (MSX) on Delaware Bay seed beds: a host-parasite relationship along a salinity gradient. Journal of Invertebrate Pathology 40:388–405.

Hochachka, P. W., and G. N. Somero. 1973. Strategies of biochemical adaptation. Saunders, Philadelphia.

Huxham, I. M., and A. M. Lackie. 1988. Behaviour in vitro of separated fractions of haemocytes of the locust *Schistocerca gregaria*. Cell and Tissue Research 251:677–684.

IMW (International Mussel Watch). 1980. Environmental studies board commission on natural resources.

National Research Council, National Academy of Sciences, Washington, D.C.

Kanungo, K. 1982. *In vitro* studies on the effects of cell-free coelomic fluid, calcium, and/or magnesium on clumping of coelomocytes of the sea star *Asterias forbesi* (Echinodermata: Asteroida). Biological Bulletin (Woods Hole) 163:438–452.

Kenney, D. M., F. A. Belamarich, and D. Shepro. 1972. Aggregation of horseshoe crab (*Limulus polyphemus*) amebocytes and reversible inhibition of aggregation by EDTA. Biological Bulletin (Woods Hole) 163:438–452.

Lackie, A. M. 1979. Cellular recognition of foreign-ness in two insect species, the American cockroach and the desert locust. Immunology 36:909–914.

Lackie, A. M. 1983. Effect of substratum wettability and charge on adhesion in vitro and encapsulation in vivo by insect haemocytes. Journal of Cell Science 63:181–190.

Lipovsky, V. P., and K. K. Chew. 1972. Mortality of Pacific oysters (*Crassostrea gigas*): the influence of temperature and enriched seawater on oyster survival. Proceedings National Shellfisheries Association 62:72–82.

Mohandas, A. 1985. An electron microscope study of endocytosis mechanisms and subsequent events in *Mercenaria mercenaria* granulocytes. Pages 143–161 *in* T. C. Cheng, editor. Comparative pathobiology, volume 8. Plenum, New York.

Moore, M. N. 1976. Cytochemical demonstration of latency of lysosomal hydrolases in digestive cells of the common mussel, *Mytilus edulis*, and changes induced by thermal stress. Cell and Tissue Research 175:279–285.

Nakahara, H., and G. Bevelander. 1969. An electron microscope study of ingestion of thorotrast by amoebocytes of *Pinctada radiata*. Texas Reports on Biology and Medicine 27:101–109.

Narain, A. S. 1973. The amoebocytes of lamellibranch molluscs, with special reference to the circulating amoebocytes. Malacological Review 6:1–12.

Newell, R. I. E., and B. J. Barber. 1988. A physiological approach to the study of bivalve molluscan diseases. American Fisheries Society Special Publication 18:269–280.

Noble, P. B. 1970. Coelomocyte aggregation in *Cucumaria frondosa*: effect of ethylenediamine tetraacetate, adenosine and adenosine nucleotides. Biological Bulletin (Woods Hole) 139:549–556.

Pearsall, N. N, and R. S. Weiser. 1970. The macrophage. Lea and Febiger, Philadelphia.

Poder, M., A. Cahour, and G. Balouet. 1982. Réactions hémocytaires à l'injection de corps bactériens ou des substances inertes chez *Ostrea edulis* L. Malacologia 22:9–14.

Prusch, R. D., and J. A. Hannafin. 1979. Calcium distribution in *Amoeba proteus*. Journal of General Physiology 74:511–521.

Rabinovitch, M. 1967. The dissociation of the attachment and ingestion phases of phagocytosis by macrophages. Experimental Cell Research 46:19–28.

Reade, P., and E. Reade. 1976. Phagocytosis in invertebrates: studies on the hemocytes of the clam *Tridacna maxima*. Journal of Invertebrate Pathology 28:281–290.

Roderick, G. E., and T. C. Cheng. 1974. Biochemistry of molluscan phagocytosis. American Zoologist 14:1263.

Rowley, A. F., and N. A. Ratcliffe. 1979. An ultrastructural and cytochemical study of the interactions between latex particles and the hemocytes of the wax moth *Galleria mellonella* in vitro. Cell and Tissue Research 199:127–137.

Ruddell, C. L. 1971. The fine structure of oyster agranular amebocytes from regenerating mantle wounds in the Pacific oyster, *Crassostrea gigas*. Journal of Invertebrate Pathology 18:260–268.

Salt, G. 1968. The resistance of insect parasitoids to the defense reactions of their host. Biological Reviews of the Cambridge Philosophical Society 43:200–210.

Shumway, S. E. 1977. Effect of salinity fluctuation on the osmotic pressure and Na^+, Ca^{2+}, and Mg^+ ion concentrations in changing salinities. Marine Biology (Berlin) 41:153–177.

Smith, D. L., and F. Rommel. 1977. A rapid micromethod for the simultaneous determination of phagocytic–microbiocidal activity of human peripheral blood leukocytes in vitro. Journal of Immunological Methods 17:241–247.

Sparks, A. K. 1972. Invertebrate pathology: noncommunicable diseases. Academic Press, New York.

Sparks, A. K., and J. F. Morado. 1988. Inflammation and wound repair in bivalve molluscs. American Fisheries Society Special Publication 18:139–152.

Stauber, L. A. 1950. The fate of india ink injected intracardially into the oyster, *Ostrea virginica* Gmelin. Biological Bulletin (Woods Hole) 98:227–241.

Takatsuki, S. 1934. On the nature and functions of the amoebocytes of *Ostrea edulis*. Quarterly Journal of Microscopical Science 76:379–431.

Tripp, M. R. 1958. Disposal by the oyster of intracardially injected red blood cells of vertebrates. Proceedings National Shellfisheries Association 48:143–147.

Tripp, M. R. 1960. Mechanisms of removal of injected microorganisms from the American oyster, *Crassostrea virginica* (Gmelin). Biological Bulletin (Woods Hole) 119:210–223.

Tripp, M. R., and V. E. Kent. 1967. Studies on oyster cellular immunity. In Vitro (Rockville) 3:129–135.

van der Knaap, W. P. W., T. Sminia, R. Schutte, and L. H. Boerrigter-Barendsen. 1983. Cytophilic receptors for foreignness and some factors which influence phagocytosis by invertebrate leucocytes: in vitro phagocytosis by amoebocytes of the snail *Lymnaea stagnalis*. Immunology 48:377–383.

Vasta, G. R., J. T. Sullivan, T. C. Cheng, J. J. Marchalonis, and G. W. Warr. 1982. A cell membrane-associated lectin of the oyster hemocyte. Journal of Invertebrate Pathology 40:367–377.

Weeks, B. A., J. E. Warinner, P. L. Mason, and D. S. McGinnis. 1986. Influence of toxic chemicals on the chemotactic response of fish macrophages. Journal of Fish Biology 28:653–658.

Wilkinson, P. C., J. M. Lackie, and R. B. Allan. 1982. Methods for measuring leucocyte locomotion. Pages 145–193 *in* N. Catsimpoolas, editor. Cell analysis, volume 1. Plenum, New York.

Yoshino, T. P., and W. O. Granath, Jr. 1985. Surface antigens of *Biomphalaria glabrata* (Gastropoda) hemocytes: functional heterogeneity in cell subpopulations recognized by a monoclonal antibody. Journal of Invertebrate Pathology 45:174–186.

American Fisheries Society Special Publication 18:238–242, 1988

Effects of Anthropogenic Agents on Bivalve Cellular and Humoral Defense Mechanisms

ROBERT S. ANDERSON

Chesapeake Biological Laboratory, University of Maryland
Post Office Box 38, Solomons, Maryland 20688, USA

Abstract.—The field of immunotoxicology of bivalves and other marine invertebrates is in its infancy. The influence of environmental pollutants on immunocompetency is well documented for vertebrates, and probably similar phenomena take place in bivalves. Preliminary evidence is now available to support this idea. Comparatively low levels of polychloroaromatic hydrocarbon and chlorinated phenol pollutants have little immunosuppressive activity and may instead cause slight hemocytic activation. However, at higher body burdens, the ability of bivalves to eliminate injected bacteria seems to be substantially impaired. These laboratory studies require confirmation and extension, and the extent to which they can be extrapolated to field situations of pollutant exposure needs to be determined.

An organism's ability to resist infectious agents that it normally encounters is determined in part by the functional capacity of its internal defense mechanisms. Animals are usually protected by an enveloping layer of skin, exoskeleton, etc., and these external defenses may be augmented by the presence of mucus, other antimicrobial secretions, or both. Once microorganisms or multicellular parasites penetrate the external barriers, they encounter the potentially protective hemocytes and humoral factors described elsewhere in this volume (Chu 1988; Feng 1988). If these defense mechanisms are compromised, the outcome of the infection will probably be unfavorable to the host, whereas if the normal defense mechanisms are functional, or in some way enhanced, the infection will be more satisfactorily resolved. Xenobiotics are known to impair immune mechanisms in mammals (Vos 1977; Koller 1979; Dean et al. 1982); increased susceptibility to infection is one of the most common consequences of such impairment. Similar synergism between pollutant stress and alterations in internal defense capabilities is found in aquatic invertebrates, and some examples in bivalve molluscs will be discussed in this chapter. Throughout this paper, I will use the terms "internal defense" and "immune" mechanisms interchangeably, although I recognize that the responses of bivalves lack the specificity and memory characteristic of mammalian adaptive immunity. The study of invertebrate immunotoxicology is early in its development but promises to extend our understanding of the physiological basis of disease in natural populations. As we gain insight into the importance of immunocompetency to disease resistance in bi-

valves and other invertebrates, the picture doubtless will become more complicated. For example, the chemical microenvironment of host tissues may play a major role in the response of eastern oysters *Crassostrea virginica* to *Haplosporidium nelsoni* (MSX), and defense mechanisms could be a minor or irrelevant component in this response (Ford and Haskin 1987). Indeed, a major research task for the future is to determine the "real-world" importance of alterations in cellular activity and humoral factor titers observed following various experimental maneuvers. These alterations are most commonly recorded by in vitro assays, but such phenomena must be related to the more complex in vivo situation. Such considerations are practically and academically important. We already have a good understanding of the direct toxicity of many environmental contaminants, as measured by acute toxicity data. It is now necessary to determine if chronic exposure to pollutants at sublethal levels has deleterious effects on disease resistance, reproductive capacity, and other vital factions. Exposure to xenobiotics may make bivalves less resistant to infection by both pathogenic and opportunistic nonpathogenic microorganisms.

Immunity in bivalves is based on nonspecific defense mechanisms mediated by phagocytic hemocytes and probably involves several classes of nonimmunoglobulin serum proteins. Although adaptive immune responses typical of mammals have not been shown in bivalves, prior stimulation by foreign material can produce, in some cases, at least transient hemocyte activation or increments in the titers of humoral factors. The limited studies of xenobiotic effects on bivalve defense mechanisms have concentrated on quantifying immediate changes in hemocyte activity or serum factor levels in otherwise untreated animals. It also would be interesting to determine the effects of prior exposure to pollutants on the inducibility of responses.

Background Studies in Bivalves and Other Marine Invertebrates

Although they only indirectly mimicked typical environmental exposures, some efforts were made to describe the cellular responses of bivalves to foreign particulates and toxic organic liquids injected into the hemocoel or muscles. In the case of particulates, usually a local infiltration of hemocytes was followed by phagocytosis or encapsulation of the foreign material. Enzymatically digestible particulates usually underwent intracellular degradation (Tripp 1958), whereas undigestible material was cleared from the animals after diapedesis of particle-laden phagocytes across epithelial linings of the digestive tract (Stauber 1950). Inert materials, such as talc, produced a local inflammatory response (hemocyte infiltration and edema) and localization of the particles by granuloma formation (Pauley and Sparks 1965, 1966). These authors reported that the response to injection of an irritant liquid (turpentine) was more dramatic, producing loss of tissue function and systemic necrosis. Heavy leukocytic infiltration was seen in the necrotic area.

Exposure of northern quahogs *Mercenaria mercenaria* to soluble irritants, such as phenol at concentrations exceeding 1 mg/L, can produce damage to gill and digestive tract epithelia, which may render the clams more susceptible to microbial infection and disease (Fries and Tripp 1976). Subsequently, Fries and Tripp (1980) reported that phenol exposure depressed phagocytic activity, and caused cytoplasmic disorganization and selective lysis of hemocytes. At concentrations as low as 10 μg/L, hemocyte lysis and inhibition of phagocytosis in intact hemocytes were observed. These responses reached their maxima at about 100 μg/L phenol (about 25% lysis and about 10% inhibition) and showed no further increase up to 50 mg/L, the highest concentration tested. The hemocytes showed differential sensitivity to phenol; no fibrocytes were observed after treatment, granulocytes had fewer cytoplasmic components, and hyalinocytes were not affected. Differential cell counts indicated a marked increase in the hyalinocyte: granulocyte ratio at 1–100 μg/L phenol. These results, coupled with the damaging effects on protective epithelia, imply that phenol may be considered as immunosuppressive in bivalves.

A limited number of in vivo observations indicate that exposure to xenobiotics may increase disease susceptibility in marine invertebrates. To identify a stress syndrome in northern quahogs from polluted areas of Narragansett Bay, Rhode Island, Jeffries (1972) studied morphological and histological changes induced by pollutants as well as rates of infection, carbohydrate levels, and patterns of occurrence and concentration of amino acids and fatty acids. In addition to other responses to the predominantly hydrocarbon-polluted environment, the clams showed an abnormally high (5–10%) incidence of *Polydora* sp. infestation of their shells. This parasite is fairly common in eastern oysters, but is rare in molluscs that burrow into sediments. In this situation, the clams may have been forced to emerge from the polluted sediments, thereby exposing themselves to the *Polydora* larvae.

Studies more relevant to our central thesis have been carried out with a system that includes a penaeid shrimp, *Baculovirus* sp., and a pollutant. Preliminary studies with shrimp having a low prevalence of *Baculovirus* indicated that exposure to a 1–3 μg/L solution of polychlorinated biphenyl (Aroclor 1254) produced a higher prevalence and increased intensity of infection (Couch 1976). Further studies exposed the shrimp to a nominal Aroclor 1254 concentration of 0.708 μg/L continuously for 35 d in a flow-through system. The chemically stressed shrimp had progressively higher prevalences of *Baculovirus* infection than the controls throughout the experiment (Couch and Courtney 1977). Although the underlying mechanism(s) of this response was not known, results indicated that low concentrations of pollutants could influence natural pathogen–host interactions, possibly by immunosuppression.

Nimmo et al. (1977) reported that pink shrimp *Penaeus duorarum* exposed to cadmium developed *Fusarium* sp. infections in the gills, but unexposed control shrimp did not. Cadmium exposure resulted in the development of black gill lesions and eventual gill necrosis. Presumably, disruption of the structural integrity of the gills by Cd allowed *Fusarium* to become established. Other more subtle effects of Cd on host defense mechanisms have not been demonstrated.

Some Effects of Pollutants on Bivalve Immune Functions

Cellular and humoral defense mechanisms were studied with northern quahogs in an attempt to identify aspects of immune capability that were affected by a suite of model marine pollutants. These assays included total and differential hemocyte counts, presence and extent of phagocytosis, and titers of hemolysin, hemagglutinin, lysozyme, and bacterial agglutinin (Anderson 1981a). The pollutants tested were pentachlorophenol (PCP), hexachlorobenzene (HCB), or benzo[a]pyrene (BP) administered either at high doses in recirculated seawater systems or at low doses in flow-through systems. The short-term tests in recirculated seawater lasted no more than 3 weeks; body burdens of about 1,000–2,000 µg PCP/L or about 1,500–2,000 µg HCB/L were typically recorded. The experimental animals had no higher rates of mortality or morbidity than the controls during the experiment. Gaping or nonresponsive clams were considered moribund and immediately removed from the aquaria. Six of these short-term, high-exposure assays were run with PCP and HCB. Three long-term, flow-through exposure studies were also carried out; two of these experiments lasted 8 weeks and one lasted 18 weeks. Tissue concentrations in these exposed northern quahogs were 200–500 µg PCP/L, 40–130 µg HCB/L, and 2–4 µg BP/L; in controls, the tissue burdens were about 8 µg PCP/L, 0.4 µg HCB/L, and 0.015 µg BP/L.

Consistent with vertebrate studies, not all the internal defense mechanisms assayed were sensitive to the test pollutants. Several of the immune variables showed no change regardless of the experimentally imposed body burdens. For example, no significant changes in total or differential hemocyte counts or the titers of serum hemolysin, hemagglutinin, and bacterial agglutinin were detected. In control clams, the total hemocyte count was 2,148 ± 790/mL (N = 110); 76.5 ± 2.4% of these were granulocytes, and the rest were mainly hyalinocytes. Perhaps pollutant-induced changes in total hemocyte counts were difficult to show because of the high inherent variability in such counts. However, the percent of the total hemocytes that were granulocytes was quite constant among controls and was unaffected by the experimental protocols. Northern quahog hemolymph contains a naturally occurring hemolysin which may play a protective role in bacterial lysis via complementlike activity (Anderson 1981b). Its activity was quantified against rabbit erythrocytes by both microtiter analysis (titer range = 16–64) and by spectrophotometric analysis of hemoglobin released from lysed erythrocytes. The serum titers of northern quahog hemagglutinin (tested against aldehyde-treated rabbit erythrocytes) and bacterial agglutinin (tested against aldehyde-

treated marine Flavobacterium strain 807098) were determined. Hemagglutinins have been postulated to act as opsonic factors in bivalves (Tripp and Kent 1967) and thus may function in immune recognition. Bacterial agglutinins in the clam (Arimoto and Tripp 1977) may also act as humoral recognition factors to facilitate phagocytosis after reacting with particular saccharide moieties on bacterial cell surfaces.

Although the above-mentioned variables seemed insensitive to the chemical exposure protocols, the model pollutants produced significantly altered rates of in vitro phagocytosis and levels of intrahemocytic lysozyme. These effects were more pronounced in the long-term studies than in the shorter experiments, even though actual tissue burdens were lower. Both in vitro phagocytosis and intrahemocytic lysozyme levels were slightly elevated in some of the high-exposure clams, but the increments were rarely of statistical significance. In the chronic studies, phagocytic indexes and intrahemocytic lysozyme levels were significantly increased for several experimental groups. Nonspecific increases in phagocytic activity, with accompanying rises in hydrolytic enzyme content, are characteristic of macrophage activation. Crustaceans also respond to injection of bacterial vaccines or endotoxin by hemocytic activation (McKay and Jenkin 1970; Paterson and Stewart 1979), indicating that this capability is not restricted to vertebrates. Such a nonspecific response to stress might be expected to play an important role in animals whose capacity for adaptive immune responses are limited. In another study, the activity of a hemocyte-derived bactericidal glycoprotein in a marine polychaete was consistently induced by exposure of the host to PCP or HCB (Anderson et al. 1984). Similar reactions were seen in mouse macrophages, where lead or cadmium exposure produced increased intracellular hydrolase activity and stimulated phagocytic activity (Koller and Roan 1977).

A novel method was used to quantify bacterial clearance by northern quahogs (Anderson et al. 1981b). The hemolymph of normal clams, when obtained under aseptic conditions, is virtually sterile. A marine Flavobacterium suspension (standardized turbidometrically) was injected into the hemolymph sinus of the anterior adductor muscle. Hemolymph samples were subsequently withdrawn from the posterior adductor muscle after dilution equilibration of the injected bacteria. This bacterial species was apparently nonpathogenic, and the dose was carefully selected such that all untreated, healthy clams would clear more than 90% in 4 h. Hemolymph samples were diluted serially and plated to determine viable bacteria present. Exposure of the clams to high doses of PCP or HCB produced a highly reproducible inability to clear the injected bacteria. Most of the exposed clams were still capable of a subnormal rate of bacterial clearance, but some showed no reduction, or even an increase, in bacterial colony-forming units present in the hemolymph 4 h after injection. The pathological implications of such impairment to resistance to bacterial infection are obvious. The reduction in clearance did not seem to depend on the duration of pollutant exposure but was determined solely by the tissue burden.

These data imply that very low-level chronic exposure to PCP or HCB does not compromise bivalve defensive capabilities. Such exposure may cause hemocytic activation, as indicated by increased intracellular enzyme levels and increased phagocytic potential. Enzyme levels may also rise because of the destabilizing effects of the pollutants on lysosomal membranes (Moore et al. 1978). The state of increased hemocytic activation and enzyme levels may not be entirely beneficial to the host; the release of cytotoxic enzymes and antimicrobial factors (such as activated oxygen intermediates) has been shown to mediate tissue damage in higher animals. The circumstances seem to be different at more elevated PCP or HCB body burdens; instead of activation, the clams' defense mechanisms were depressed. The ability of the clams to eliminate bacteria from their hemolymph was partially or completely impaired.

Immunotoxicity and Other Indexes of Stress in Bivalves

Alterations in bivalve immune status that result from environmental stress may have implications for the survival of individuals in a population. Immunocompetency may play an important role in determining the outcome of microbe–host or parasite–host interactions but it is probably not even indirectly reflected in such widely accepted indexes of physiological condition as the scope for growth (Bayne et al. 1979) or the oxygen:nitrogen ratio. However, situations that reduce the potential for somatic growth and the production of gametes also may be presumed to reduce the energy stores available for mounting a vigorous immune response. When the scope for growth is strongly negative, hydrolases are released into the cytoplasm via lysosomal membrane destabilization,

producing various cytotoxic effects. Changes in the labilization period of lysosomal enzymes are most commonly measured in bivalve digestive gland cells (Moore 1977) but likely also occur in other cells such as hemocytes. As in mammals, molluscan lysosomes have been shown to accumulate various toxic chemicals (George et al. 1976, 1978; Moore 1977); it is thought that the lysosomal membranes are rendered unstable in the process and release membrane-associated hydrolases. The relationship of increments in lysosomal enzyme activity measured in the hemolymph after pollutant exposure to lysosomal destabilization or to stimulatory events associated with macrophage activation needs to be established.

Much work remains to determine the utility of measuring putative immune variables of bivalves in assaying the biological responses of individuals for the assessment of pollution effects. Immune variables may have some unique potential in predicting the individual's responses to subsequent challenges such as parasitism or other infection. However, there is no direct evidence that the observed effects of pollutants on immune status are linked with the ecological fitness of individuals or populations of bivalves. These assays do show encouraging signs of sensitivity, because atypical activities seem to be routinely measured at low, sublethal pollutant levels. Specificity has not been shown yet, and adequate data on the responses to various classes of pollutants are not available. Whether or not a seasonal cycle of immunocompetency exists in bivalves must be established to better recognize anomalies induced by pollutants.

Bivalves, particularly blue mussel *Mytilus edulis*, seem to be the animals of choice in many contaminant monitoring programs (Goldberg et al. 1978). Information on pollutant levels in these animals is important but must be combined with some assessment of toxicity more subtle than acute lethal tests. The activities of various components of a critical survival mechanism, such as the immune system, may prove to be useful in future monitoring programs. These responses, as well as cytochemical and biochemical indexes, probably will show greater sensitivity and less natural variability than indexes of general physiological condition.

References

Anderson, R. S. 1981a. Effects of carcinogenic and noncarcinogenic environmental pollutants on immunological functions in a marine invertebrate. Pages 319–331 *in* C. J. Dawe, J. C. Harshbarger, S.

Kondo, T. Sugimura, and S. Takayama, editors. Phyletic approaches to cancer. Japan Scientific Societies Press, Tokyo.

Anderson, R. S. 1981b. Inducible hemolytic activity in *Mercenaria mercenaria* hemolymph. Developmental and Comparative Immunology 5:575–585.

Anderson, R. S., C. S. Giam, and L. E. Ray. 1984. Effects of hexachlorobenzene and pentachlorophenol on cellular and humoral immune parameters in *Glycera dibranchiata*. Marine Environmental Research 14:317–326.

Anderson, R. S., C. S. Giam, L. E. Ray, and M. R. Tripp. 1981. Effects of environmental pollutants on immunological competency of the clam *Mercenaria mercenaria*: impaired bacterial clearance. Aquatic Toxicology (New York) 1:187–195.

Arimoto, R., and M. R. Tripp. 1977. Characterization of a bacterial agglutinin in the hemolymph of the hard clam, *Mercenaria mercenaria*. Journal of Invertebrate Pathology 30:406–413.

Bayne, B. L., M. N. Moore, J. Widdows, and D. R. Livingstone. 1979. Measurements of the responses of individuals to environmental stress and pollution: studies with bivalve molluscs. Philosophical Transactions of the Royal Society of London B, Biological Sciences 286:563–581.

Chu, F.-L. E. 1988. Humoral defense factors in marine bivalves. American Fisheries Society Special Publication 18:178–188.

Couch, J. A. 1976. Attempts to increase *Baculovirus* prevalence in shrimp by chemical exposure. Progress in Experimental Tumor Research 20:304–314.

Couch, J. A., and L. Courtney. 1977. Interaction of chemical pollutants and virus in a crustacean: a novel bioassay system. Annals of the New York Academy of Sciences 298:497–504.

Dean, J. H., M. L. Luster, and G. A. Boorman. 1982. Immunotoxicology. Research Monographs in Immunology 4:349–398.

Feng, S. Y. 1988. Cellular defense mechanisms of oysters and mussels. American Fisheries Society Special Publication 18:153–168.

Ford, S. E., and H. H. Haskin. 1987. Infection and mortality patterns in strains of oyster *Crassostrea virginica* selected for resistance to the parasite *Haplosporidium nelsoni* (MSX). Journal of Parasitology 73:368–376.

Fries, C., and M. R. Tripp. 1976. Effects of phenol on clams. U.S. National Marine Fisheries Service Marine Fisheries Review 38(10):10–11.

Fries, C. R., and M. R. Tripp. 1980. Depression of phagocytosis in *Mercenaria* following chemical stress. Developmental and Comparative Immunology 4:233–244.

George, S. G., B. J. S. Pirie, A. R. Cheyne, T. L. Coombs, and P. T. Grant. 1978. Detoxification of metals by marine bivalves: an ultrastructural study of the compartmentation of copper and zinc in the oyster *Ostrea edulis*. Marine Biology (Berlin) 45:147–156.

George, S. G., B. J. S. Pirie, and T. L. Coombs. 1976.

The kinetics of accumulation and excretion of ferric hydroxide in *Mytilus edulis* (L.) and its distribution in the tissues. Journal of Experimental Marine Biology and Ecology 23:71–84.

Goldberg, E. D., and nine coauthors. 1978. The mussel watch. Environmental Conservation 5(2):101–125.

Jeffries, H. P. 1972. A stress syndrome in the hard clam, *Mercenaria mercenaria*. Journal of Invertebrate Pathology 20:242–251.

Koller, L. D. 1979. Effects of environmental chemicals on the immune system. Advances in Veterinary Science and Comparative Medicine 23:267–295.

Koller, L. D., and J. G. Roan. 1977. Effects of lead and cadmium on mouse peritoneal macrophages. Journal of the Reticuloendothelial Society 21:7–12.

McKay, D., and C. R. Jenkin. 1970. Immunity in the invertebrates. Correlation of the phagocytic activity of haemocytes with resistance to infection in the crayfish (*Parachaeraps bicarinatus*). Australian Journal of Experimental Biology and Medical Science 48:609–617.

Moore, M. N. 1977. Lysosomal responses to environmental chemicals in marine invertebrates. Pages 143–154 *in* C. S. Giam, editor. Pollutant effects on marine organisms. Heath, Lexington, Massachusetts.

Moore, M. N., D. M. Lowe, and P. E. M. Fieth. 1978. Lysosomal responses to experimentally injected anthracene in the digestive cells of *Mytilus edulis*. Marine Biology (Berlin) 48:297–302.

Nimmo, D. R., D. V. Lightner, and L. H. Bahner. 1977. Effects of cadmium on the shrimps, *Penaeus duorarum*, *Palaemonetes pugio* and *Palaemonetes vulgaris*. Pages 131–183 *in* F. J. Vernberg, A. Calabrese, F. P. Thurberg, and W. B. Vernberg, editors. Physiological responses of marine biota to pollutants. Academic Press, New York.

Paterson, W. D., and J. E. Stewart. 1979. Rate and duration of phagocytic increase in lobsters induced by *Pseudomonas perolens* endotoxin. Developmental and Comparative Immunology 3:353–357.

Pauley, G. B., and A. K. Sparks. 1965. Preliminary observations on the acute inflammatory reaction in the Pacific oysters, *Crassostrea gigas* (Thunberg). Journal of Invertebrate Pathology 7:248–256.

Pauley, G. B., and A. K. Sparks. 1966. The acute inflammatory reaction in two different tissues of the Pacific oyster, *Crassostrea virginica*. Journal of the Fisheries Research Board of Canada 23:1913–1921.

Stauber, L. A. 1950. The fate of India ink injected intracardially into the oyster, *Ostrea virginica* Gmelin. Biological Bulletin (Woods Hole) 98:227–241.

Tripp, M. R. 1958. Disposal by the oyster of intracardially injected red blood cells of vertebrates. Proceedings National Shellfisheries Association 48:143–147.

Tripp, M. R., and V. E. Kent. 1967. Studies on oyster cellular immunity. In Vitro (Rockville) 3:129–135.

Vos, J. G. 1977. Immune suppression as related to toxicology. CRC Critical Reviews in Toxicology 5:67–101.

American Fisheries Society Special Publication 18:243–245, 1988

MANAGEMENT PRACTICES

Management Strategies to Control Diseases in the Dutch Culture of Edible Oysters

P. VAN BANNING

Netherlands Institute for Fishery Investigations
Post Office Box 68, 1970 AB IJmuiden, The Netherlands

Abstract.—The infection characteristics of the protistan parasites *Marteilia refringens, Haplosporidium armoricanum,* and *Bonamia ostreae* in the Dutch cultures of edible oysters *Ostrea edulis* are discussed. Infection characteristics have led to disease-control experiments based on intensified cleaning of infected oyster beds followed by new plantings of healthy edible oysters. As a result, no *M. refringens* and *H. armoricanum* have been observed in the Dutch oyster areas since 1978. The incidence of *B. ostreae,* however, had a different history: a sharp decrease of prevalence during the first two experimental years was followed by a slower decline during the following years, and the parasite was not wholly eliminated. This asymptotic trend suggests that short-term control of this oyster pathogen may be difficult.

Dutch oyster culture and fisheries concentrate primarily on the edible oyster *Ostrea edulis*. Culture of this species has been carried out as an organized bottom culture since 1870 in Yerseke Bank, a rather small area about 16 km long and 8 km wide. This area is a part of the sea arm Oosterschelde, a tidal water of the estuary system in the southern Netherlands. It has a constant and relatively high salinity (annual average of about 28‰) and average minimum–maximum temperatures of 3°C in winter to 19°C in summer. The oyster culture can be classified as a productive, high-density culture and, therefore, highly vulnerable to outbreaks of parasites and diseases.

Until the 1970s, shell disease caused by the fungus *Ostracoblabe implexa* and pit disease caused by the flagellate *Hexamita* sp. were the major parasite problems in the oyster fishery (Korringa 1976). In most cases, management strategies to control these diseases included intensified cleaning of infected areas, better selection of seed and culture areas, and improved storage systems (Korringa 1976).

Between 1968 and 1979, several new diseases caused by (probably introduced) protistans were discovered in European edible oyster cultures, each appearing first in Brittany, France. They included *Marteilia refringens* in 1968 (Grizel et al. 1974), *Haplosporidium armoricanum (formerly Minchinia armoricana)* in 1974 (van Banning 1977), and *Bonamia ostreae* in 1979 (Pichot et al. 1979). Because of existing commercial importa-

tion of French oysters into the Netherlands, all three oyster pathogens were introduced into the Dutch oyster culture area of the Yerseke Bank (van Banning 1979).

Starting in 1974, continuous monitoring in the Netherlands showed that *M. refringens* infections were restricted to recently imported oysters from France and never extended to the native oysters present in the same area of Yerseke Bank. It was concluded that the Yerseke Bank ecosystem was probably missing an important transmission factor, perhaps specific intermediate hosts (van Banning 1979) or specific environmental conditions (Alderman 1979; Balouet 1979). Balouet et al. (1979) observed a low prevalence in the Pacific oyster *Crassostrea gigas* but felt this resulted from the capture of food and was not a real infection. The general data indicated that *M. refringens* was very specific for edible oysters (Grizel and Tigé 1973). This specificity, together with the restricted distribution in the Dutch oyster area, opened the possibility for successful control of the introduced *M. refringens* pathogen. Measures were taken to check all oyster importations; recently planted oysters and all infected lots were harvested or destroyed. As a result, no *M. refringens* has been observed in edible oysters of the Yerseke Bank area since 1978.

The second protistan oyster pathogen introduced into the Dutch oyster culture, *Haplosporidium armoricanum,* showed characteristics comparable to *M. refringens* and offered some chance

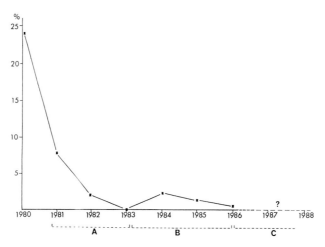

FIGURE 1.—Mean prevalences of bonamiasis in edible oysters in the Dutch disease-control experiments since 1980. A = period of small-scale test monitoring after beds had been cleaned and new plantings had ceased. B = period of large-scale test monitoring. C = period when no new experimental oyster plantings were made (1986–1987), to minimize the chance that *Bonamia ostreae* would survive. New experimental challenge experiments are planned for 1988 or 1989.

for successful control measures. *Haplosporidium armoricanum* seemed very specific for edible oysters and the disease never extended to the native oysters of the Yerseke Bank or increased in prevalence. It was concluded that this pathogen also needed an intermediate host or specific environmental conditions for its life cycle or mode of infection, that such factors were missing in the Yerseke Bank area, and that the control measures taken for *M. refringens* could also control *H. armoricanum*. *Haplosporidium armoricanum* has not been observed in the Yerseke Bank oysters since 1978, so cleaning of infected oyster beds seems to have eradicated this parasite.

The third pathogenic oyster protistan, *Bonamia ostreae*, was introduced in 1980 to the Dutch oyster fishery in imported lots from France (van Banning 1982). The disease spread quickly in the autumn of that year to the native Dutch oysters and successfully established itself. Because this new oyster pathogen was quite different from *M. refringens* and *H. armoricanum*, its control was expected to be more difficult.

Industry-Wide Efforts to Control
Bonamia ostreae

In the year *B. ostreae* was introduced to The Netherlands, samples were taken of Pacific oysters, blue mussels *Mytilus edulis*, and the cockle *Cerastoderma edulis*; these bivalve species were considered potential carriers or transmitters of the pathogen. However, no signs of *B. ostreae* could

be detected in the samples, probably because this pathogen was specific for edible oysters. Because of the pathogen's quick spread to the native oysters, direct infection was considered possible; this was proven shortly after by experiments in France (Poder et al. 1982). The specificity of bonamiasis for edible oysters suggested that the pathogen could be controlled by cleaning infected beds and halting the planting of new oysters for some period, but the potential effects and success of such control measures were unknown and had to be assessed. A strict program of cleaning edible oysters out of the infected Yerseke Bank was started at the end of 1980, and, with the full cooperation of Dutch oyster growers, all commercial plantings ceased. The effects of cleaning and cessation of oyster culture were studied with yearly small-scale experimental plantings of disease-free indicator oysters (van Banning 1986). A remarkable decrease in the presence and prevalence of bonamiasis were observed in the first 2 years of the indicator experiments, and they declined apparently to zero by 1983 (Figure 1). This underscored the basic concept that *B. ostreae* depends for its survival and spread on edible oysters, and that other aquatic organisms (including other bivalves) are not involved as important carriers or transmitters. Pacific oysters, blue mussels, and cockles were continuously abundant in the same area during the years of the oyster experiments. The rapid decrease of bonamiasis in the indicator oysters could not have happened had these bivalves been major carriers of the parasite.

Monitoring of Large-Scale Plantings

Having reached the zero point in the small-scale indicator experiments, the next step was taken in 1984 with larger, commercial-scale plantings. The initial results of these challenge experiments, however, were disappointing because some foci of infection were still present and, consequently, the control measures had to be extended. Yearly challenge tests were also necessary to estimate the real disease-free status of the Yerseke Bank area for the next period. The experiments of 1985 and 1986 showed only a slowly decreasing, asymptotic trend (van Banning 1987; Figure 1). These characteristics make it difficult to estimate the moment of absolute zero level infection, despite the previous low level of less than 1% prevalence in 1986.

The unexpected ability of B. ostreae to maintain itself for a long period in low density and repeatedly changed oyster stocks raises some questions concerning the pathogen. Perhaps heretofore unrecognized chronic or resting stages are present in the oyster, or perhaps the pathogen has spore stages outside the oyster.

Proposed Management Strategy against B. ostreae

The asymptotic decrease of bonamiasis in the Dutch oyster fishery has indicated that control measures for this pathogen (cessation of planting and yearly cleaning) must be continued for several years. In the case that bonamiasis persists by repeatedly forming new foci of infection, a proposed culture strategy is to rotate oyster beds and to employ only a short-term culture period (one or two growing seasons) with medium-sized oysters originating from a disease-free region. Such a strategy can keep the prevalence of bonamiasis at levels sufficiently low that losses of edible oysters stay within commercially acceptable ranges.

References

Alderman, D. J. 1979. Epizootiology of *Marteilia refringens* in Europe. U.S. National Marine Fisheries Service Marine Fisheries Review 41(1–2):67–69.

Balouet, G. 1979. *Marteilia refringens*—considerations of the life cycle and development of Abers disease in *Ostrea edulis*. U.S. National Marine Fisheries Service Marine Fisheries Review 41(1–2):64–66.

Balouet, G., C. Chastel, A. Cahour, A. Quillard, and M. Poder. 1979. Etude épidémiologique et pathologique de la maladie de l'huître plate en Bretagne. Science et Pêche 289:13–24.

Grizel, H., M. Comps, J. R. Bonami, F. Cousserans, J. L. Duthoit, and M. A. Le Pennec. 1974. Recherche sur l'agent de la maladie de la glande digestive de *Ostrea edulis* Linné. Science et Pêche 240:7–30.

Grizel, H., and G. Tigé. 1973. La maladie de la glande digestive d'*Ostrea edulis* Linné. International Council for the Exploration of the Sea, C.M. 1973/K:13, Copenhagen.

Korringa, P. 1976. Farming the flat oysters of the genus *Ostrea*. Elsevier, Amsterdam.

Pichot, Y., M. Comps, G. Tigé, H. Grizel, and M. A. Rabouin. 1979. Recherches sur *Bonamia ostrea* gen.n.,sp.n., parasite nouveau de l'huître plate *Ostrea edulis* L. Revue des Travaux de l'Institute des Pêches Maritimes 43(1):131–140.

Poder, M., A. Cahour, and G. Balouet. 1982. Hemocytic parasitosis in European oyster *Ostrea edulis* L.: pathology and contamination. Proceedings of the International Colloquium on Invertebrate Pathology 3:254–257. (University of Sussex, Brighton, England.)

van Banning, P. 1977. *Minchinia armoricana* sp.nov. (Haplosporida), a parasite of the European flat oyster, *Ostrea edulis*. Journal of Invertebrate Pathology 30:199–206.

van Banning, P. 1979. Haplosporidian diseases of imported oysters, *Ostrea edulis*, in Dutch estuaries. U.S. National Marine Fisheries Service Marine Fisheries Review 41(1–2):8–18.

van Banning, P. 1982. Some aspects of the occurrence, importance and control of the oyster pathogen *Bonamia ostreae* in the Dutch oyster culture. Proceedings of the International Colloquium on Invertebrate Pathology 3:261–265. (University of Sussex, Brighton, England.)

van Banning, P. 1986. Case-history of the *Bonamia ostreae* control in the Dutch oyster culture. Pages 23–27 *in* C. P. Vivarès, J. R. Bonami, and E. Jaspers, editors. Pathology in marine aquaculture. European Aquaculture Society, Special Publication 9, Bredene, Belgium.

van Banning, P. 1987. Further results of the *Bonamia ostreae* challenge tests in the Dutch oyster culture. Aquaculture 67(1–2):191–194.

American Fisheries Society Special Publication 18:246–248, 1988

Circumvention of Mortalities Caused by Denman Island Oyster Disease during Mariculture of Pacific Oysters

SUSAN M. BOWER

Department of Fisheries and Oceans, Biological Sciences Branch
Pacific Biological Station, Nanaimo, British Columbia, V9R 5K6, Canada

Abstract.—Denman Island disease was first observed in British Columbia in 1960, when it was associated with about 34% mortality of Pacific oyster *Crassostrea gigas*. Gross symptoms of this disease include a beige digestive gland and green pustules on surfaces of the body and mantle. The disease has been found only in the spring and early summer and usually does not kill Pacific oysters under 2 years of age. The causative agent is an intracellular organism called a microcell. However, recent descriptions of actinomycete bacteria causing similar pustules on the oysters have placed early epizootiological data, based on gross signs, in question. Well-timed plantings and harvests of Pacific oysters in relation to tidal levels can help circumvent mortalities from this disease.

History of Disease

The first occurrence of Denman Island disease in Pacific oysters *Crassostrea gigas* was reported by Quayle (1961). It was observed in April 1960 in Henry Bay, at the northwest end of Denman Island in Baynes Sound near Comox, British Columbia (Figure 1). About 1 month earlier, there had been no indication of abnormal conditions. However, by mid-April, a Pacific oyster mortality of about 10% was observed at the lowest tide level. The digestive gland of moribund oysters ("gapers") was beige in color instead of green, which would have been normal at this time of year. In addition, moribund Pacific oysters had green pustules on the surface of the body and mantle, pus-filled sinuses within the body wall, or both. Identical lesions were also present in 33% of the live animals examined. Affected Pacific oysters were between 5 and 7 years old and large (50–60 animals per 4.5 L shucked, or per 30 L intact). No deaths were observed among 2-year-old oysters that were attached to many of the diseased older oysters. By mid-August 1960, the disease appeared to have run its course. The percentage of Pacific oysters that died varied from 17% at the 1.2-m tide level to 53% at the 0.3-m tide level and averaged 34% within the 0–5-m tide range. About 25% of the survivors had residual scars on the shell that were being covered with nacre, and the position of former pustules was indicated by a slight depression on the body surface. Also, gametogenesis had not proceeded normally. In surviving Pacific oysters with scars, the gonads had a grey, mottled appearance quite unlike that of normal animals at the same time of year.

From early April to mid-May of most years since 1960, the prevalence of green pustules in samples of about 100 larger Pacific oysters from the low-tide level has been determined. Over this time, the prevalence has varied between a high of 39% in 1972 and a low of 11% in 1967 (Quayle 1988). Surface seawater temperature and salinity data collected at the nearest recording station on Chrome Island (approximately 18 km south of Henry Bay) showed no anomalies that might account for the variation in annual prevalence of infection (Quayle 1982).

By 1969, J. G. Mackin had recognized the causative agent as an intracellular organism he called a microcell (Farley et al. 1988). Histological examinations of about 25 Pacific oysters from Denman Island collected in various months of 1967–1969 and in 1980 revealed that lesions and microcells were consistently present in the April and May samples, sometimes in the March and June samples, and never at other times of the year (Farley et al. 1988). This seasonality was confirmed during recent studies in our laboratory at the Pacific Biological Station (Bower, unpublished). However, microcells were observed in some barely visible greenish pustules in Pacific oysters collected on 26 January 1987. Also, typical lesions were present in 35 and 29% of the animals collected on 9 June and 5 August 1986, respectively, but no microcells were observed histologically. Although an etiological agent was not observed in the lesions of Pacific oysters collected in June (the majority of the microcells may have been destroyed by the host response by this time), about 25% of the lesions excised in August contained actinomycete bacteria as described by Friedman et al. (1987) and Elston et al. (1987).

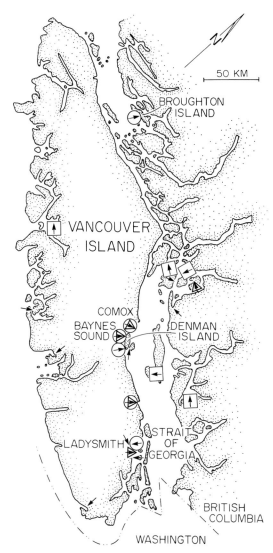

FIGURE 1.—Map of Vancouver Island and adjacent coast in British Columbia, Canada, indicating 21 localities from which Pacific oysters were examined (arrows) in 2 years since fall 1985. In five of these localities, a few animals had idiopathic green lesions (arrows in squares). In two localities, Pacific oysters had green lesions apparently caused by actinomycete bacteria (arrows in triangles). In three localities, only microcells were associated with the lesions (arrows in circles). In three localities, either actinomycetes or microcells were associated with the lesions (arrows in circled triangles).

It is now recognized that actinomycete infections in Pacific oysters can result in lesions that may be grossly identical with those caused by the microcell (Bower, unpublished). From fall 1985 to fall 1987, samples of 40 Pacific oysters from each of 21 localities in southwestern British Columbia have been examined for parasites and symbionts (Figure 1). In five of these localities, a few (less than 5%) Pacific oysters had idiopathic greenish lesions or discoloration of tissues. In five other localities, about 10% of the animals had green lesions or pustules and these contained actinomycete bacteria.

To date, microcells have been observed in Pacific oysters from 6 of these 21 localities. However, Pacific oysters from only 12 of the 21 localities were examined during February–June, when microcells are known to be present in tissues of infected animals. In addition to infections in Pacific oysters from Henry Bay on Denman Island, microcell-infected animals were observed in two nearby localities of Metcalf Bay on the south end of Denman Island and in Fanny Bay adjacent to Denman Island on Vancouver Island; in Nanoose Bay and in Shell Bay near Ladysmith, approximately 60 and 110 km south of Henry Bay, respectively; and in Booker Lagoon on Broughton Island, about 210 km north of Henry Bay (Figure 1). Infections with actinomycete bacteria were also observed in Pacific oysters from three of these localities: Henry Bay, Fanny Bay, and Nanoose Bay. Although both pathogens can occur concurrently, a dual infection in a single Pacific oyster has yet to be detected. Past reports on the distribution of Denman Island disease in British Columbia (Quayle 1982) were based on the presence of green lesions and not histological observations, and thus may not reflect the true distribution of microcells among Pacific oysters in British Columbia.

Management Practices to Circumvent Mortalities

Without knowing the etiological agent of Denman Island disease, Quayle (1961, 1969, 1982) drew the following conclusions from field and laboratory observations.

(1) The greatest mortality and highest prevalence of disease occurs in Pacific oysters at the lowest tide level.

(2) Pacific oysters up to 2 years of age are less affected than older animals.

(3) The disease is seasonal, occurring in spring when water temperatures are about 10–15°C.

(4) About 10% of the apparently infected animals recover from the disease.

(5) Aquarium tests indicate that the disease can be acquired by direct transmission.

From these observations, Quayle (1969, 1982) indicated that judicious timing of planting and harvesting in relation to tide levels would circumvent the impact of Pacific oyster mortalities caused by this disease. Because the greatest infection occurs at or near the lowest tide level, Pacific oysters should be moved to higher tide levels prior to March, and lower tide levels should not be planted before June. In addition, it is possible to hold Pacific oysters less than 2 years old at the lowest tide levels because they apparently resist the disease. Implementation of these management procedures has allowed the successful and profitable culture of Pacific oysters in Henry Bay on Denman Island despite the continued presence of the microcell disease.

Acknowledgments

I thank Dan Quayle for his valuable comments on earlier drafts and Johanne Laliberté and Gary Meyer for their technical assistance in conducting the oyster parasite survey work.

References

Elston, R. A., J. H. Beattie, C. Friedman, R. Hedrick, and M. L. Kent. 1987. Pathology and significance of fatal inflammatory bacteraemia in the Pacific oyster, *Crassostrea gigas* Thünberg. Journal of Fish Diseases 10:121–132.

Farley, C. A., P. H. Wolf, and R. A. Elston. 1988. A long term study of "microcell" disease in oysters with a description of a new genus, *Mikrocytos* (g.n.), and two new species, *Mikrocytos mackini* (sp.n.) and *Mikrocytos roughleyi* (sp.n.). U.S. National Marine Fisheries Service Fishery Bulletin 86:581–593.

Friedman, C. S., H. Beattie, R. Elston, and R. P. Hedrick. 1987. Isolation of the bacterium causing focal necrosis (fatal inflammatory bacteremia) in Pacific oysters (*Crassostrea gigas*). American Fisheries Society, Fish Health Section Newsletter 15(1):3. (Bethesda, Maryland.)

Quayle, D. B. 1961. Denman Island oyster disease and mortality, 1960. Fisheries Research Board of Canada Manuscript Report 713, Ottawa.

Quayle, D. B. 1969. Pacific oyster culture in British Columbia. Fisheries Research Board of Canada Bulletin 169.

Quayle, D. B. 1982. Denman Island oyster disease 1960–1980. British Columbia Shellfish Mariculture Newsletter 2(2):1–5. (Victoria, Canada.)

Quayle, D. B. 1988. Pacific oyster culture in British Columbia. Canadian Bulletin of Fisheries and Aquatic Sciences 218.

American Fisheries Society Special Publication 18:249–256, 1988
© Copyright by the American Fisheries Society 1988

Management Strategies for MSX (*Haplosporidium nelsoni*) Disease in Eastern Oysters

SUSAN E. FORD AND HAROLD H. HASKIN

Shellfish Research Laboratory, New Jersey Agricultural Experiment Station
Cook College, Rutgers University, Port Norris, New Jersey 08349, USA

Abstract.—MSX disease in oysters is caused by a protozoan parasite, *Haplosporidium nelsoni*. Since discovered in 1957, it has caused heavy mortalities of the eastern oyster *Crassostrea virginica* in the middle Atlantic section of the USA. Management strategies vary according to region but rely, in large measure, on the inhibition of MSX by low salinity. If possible, eastern oysters should be grown where salinity is low enough to prevent serious MSX activity, even if growth is reduced. If susceptible seed oysters must be moved from low to high salinity for final growth and conditioning before they are marketed, their exposure to high salinity should be limited to a single growing season. Planters can further minimize mortality by moving eastern oysters after the early summer infection period (June and July) and harvesting by the end of the year; however, loss of time in the best growing areas must be balanced against better survival, and results may vary according to location and time. Epizootiological studies are essential to define infection and mortality patterns for each region. Continuous monitoring and early diagnosis of infections also are important because they allow mortality to be predicted so growers and managers can make informed decisions on when or whether to plant and harvest. After an epizootic, surviving eastern oysters should be preserved as brood stock that may help to establish a population with increased resistance. Selective breeding has produced strains of eastern oysters that are highly resistant to MSX-caused mortality, and use of these strains is recommended for intensive culture operations that deploy hatchery-produced seed.

MSX disease in oysters is caused by a protozoan (Ascetospora: Stellatosporea) parasite, *Haplosporidium* (formerly *Minchinia*) *nelsoni* (Haskin et al. 1966). When it was first seen, the name "MSX" (multinucleate sphere unknown) was given to the parasite because of its microscopic appearance and because its exact classification was then unknown. Since it was discovered in 1957, it has caused heavy mortalities of the eastern oyster *Crassostrea virginica* in the middle Atlantic states of the USA, primarily in Delaware and Chesapeake bays (Haskin et al. 1965; Andrews 1968; Farley 1975; Haskin and Andrews 1988, this volume). Eastern oyster stocks suffered heavy losses (up to 90–95%) in the late 1950s and early 1960s, and chronic disease pressure with occasional epizootic outbreaks continues to suppress eastern oyster production in both estuaries. Nevertheless, the industry in each locality has persisted by altering traditional culture and harvest methods. This paper describes management strategies currently in practice for large-scale planting and harvesting of natural seed in areas where MSX is present. It also describes additional options available in more tightly controlled situations, such as those involving hatchery-produced seed and protected growout.

Practical methods for dealing with the MSX parasite face several problems. Infective stages of the parasite are water-borne for an unknown period before they contact an oyster. Proximity to diseased oysters does not affect infection rates in newly exposed animals (Andrews 1966; Ford and Haskin 1982). Most oyster fisheries are in large estuaries having free exchange of water with the ocean, important tidal and wind-driven currents, and variable flushing from river inputs. The possibility of controlling the parasite is further hampered by uncertainty about its life cycle and method of transmission (from oyster to oyster or through a reservoir host: see Haskin and Andrews 1988). Nevertheless, enough is known about the seasonality of infections, the disease process, and environmental constraints on the parasite to suggest several management strategies.

Protection by Low Salinity

Early studies of MSX showed that the parasite occurred primarily in estuarine regions where salinities were above 15‰ and was almost entirely absent in areas below 10‰ (Andrews 1964, 1983; Haskin et al. 1965; Haskin and Ford 1982). A major reason is that the parasite is intolerant of low salinity (Sprague et al. 1969; Andrews 1983; Ford

1985; Ford and Haskin, in press). It is also probable that infective stages are produced in the lower estuary and diluted in an upstream direction (Ford and Haskin 1982; Andrews 1983), although the "source" may penetrate upbay during droughts. Much of the eastern oyster fishery in areas affected by MSX traditionally involved transplantation of seed oysters out of low-salinity, upper-estuary seed beds—where the parasite is restricted or absent, but where oyster growth is often poor—to higher-salinity water—where the oysters achieve the size and meat quality demanded by the market, but where the parasite is also active.

After MSX appeared, oyster planters in Virginia largely abandoned high-salinity planting grounds in favor of plantings in low-salinity regions (Andrews and Frierman 1974). In most of the Maryland portion of Chesapeake Bay, salinity is low enough, except during droughts, to prevent incursion of MSX. In Delaware Bay, planters have moved away from the most saline, downbay areas toward the upbay edge of the planting grounds, but MSX is prevalent even there in most years. In 1981, a new section of planting grounds was created in the lower seed bed area, but MSX activity has been high there as well. Thus, planters in Delaware Bay must use grounds where MSX is rarely inhibited by low salinity, and they must employ tactics that are different from those used in Virginia or Maryland.

Long-term planning that involves the use of low- or intermediate-salinity planting beds should consider the potential effects of drought, and also the predicted rise in sea level (Hull and Titus 1986), which will (temporarily and perhaps permanently) increase salinity in formerly "protected" areas. Some areas that were previously free of MSX-caused mortalities are no longer so because of drought during the 1980s (Haskin and Andrews 1988).

Shortened Planting Cycle

In pre-MSX days, eastern seed oysters were often transplanted to the growing grounds in Delaware Bay at a very small size and left for several years until they were large enough to market. This practice is no longer possible because mortality due to MSX increases with length of exposure and a 2–3-year planting cycle brings cumulative losses to economically prohibitive levels (Ford and Haskin 1982).

Oyster planters in Delaware Bay now attempt to plant only seed large enough to reach market size in a single growing season. Eastern oysters planted in May and harvested by the end of December suffer an average of only 12–15% mortality due to MSX, although mortality can rise to nearly 40% in years of exceptionally heavy disease activity (Haskin and Ford 1983). This practice, of course, places a premium on large seed oysters, which are found primarily on the lower-bay beds, and it makes little use of the small oysters found on the upbay beds. Thus, a complementary strategy is to institute "two-stage" moves: small eastern oysters are transplanted from low- to intermediate-salinity areas where they can achieve moderate growth with modest loss to MSX (and to predators such as the Atlantic oyster drill *Urosalpinx cinerea*); then they are transferred on to high-salinity water once they are large enough that a single growing season in the lower bay will bring them to market size.

Altered Planting Times

In the two major estuaries most intensively studied, Chesapeake and Delaware bays, the timing of infections has been fairly well established: between mid May (Chesapeake Bay) or early June (Delaware Bay) and late October (Andrews 1982; Ford and Haskin 1982). Eastern oysters exposed in spring and early summer become infected immediately and parasites are detectable by mid to late July; mortalities begin in August and peak in September and October. If eastern oysters are not exposed until late in the season, however, infections often remain subpatent until the following spring and mortalities do not occur until June and July, 8–9 months after exposure (Haskin et al. 1965; Andrews 1966, 1982). Also, fall infections are much more variable in timing and intensity than early-summer ones, and more likely to fail altogether (Andrews 1982; Ford and Haskin 1982).

In some areas, including Delaware Bay, eastern oyster seed are transplanted in late spring, just in time for the early summer infection period. If planting times were delayed to avoid the major early summer infection period, would yields improve, even though the growing season would be shortened? Conversely, could the growing season be safely extended if it included a fall infection period? We examined these questions in a set of spring and fall "plantings" made in 1975 and 1976. In each year, eastern oysters were brought from an MSX-free seed bed in early June and placed in a tray on a planting ground in lower Delaware Bay. Another tray, containing eastern oysters from the same source, was established at the same site in late September. Mortality, shell growth, and MSX

TABLE 1.—Mortality, growth, and MSX infections of eastern oysters first exposed to MSX in either spring or fall, lower Delaware Bay.

First exposure	Sampling period					
	1975		1976			1977
	Sep	Dec	May	Sep	Dec	Dec
Percent mortality[a]						
1975						
Spring		17		47		
Fall		<1		38		
1976						
Spring					20	67
Fall					1	39
Shell height (mm)[b]						
1975						
Spring	65 (3)	72 (3)			78 (3)	
Fall	59 (2)	66 (2)			72 (2)	
1976						
Spring				87 (3)	83 (3)	84 (2)
Fall				74 (2)	79 (3)	86 (2)
MSX infection (% total prevalence)[c]						
1975						
Spring	80 (30)	100 (75)			75 (20)	
Fall	10 (10)	25 (15)			70 (10)	
1976						
Spring					80 (35)	83 (67)
Fall					55 (20)	55 (30)

[a] Nonpredation mortality accumulated from initial exposure to sample date.

[b] One-half 95% confidence intervals are in parentheses.

[c] Infection rates are based on 20 animals per sample. Systemic infection percentages are in parentheses.

infection rates were followed in all trays from the time they were set up until December of the following year, 15–19 months later. In both 1975 and 1976, essentially no mortality occurred among fall plants by the first December (Table 1), and losses were still 20–40% lower among fall plants than among spring plants by the end of the following year. Fall plants, in both years, had not reached the same size as spring plants by the first December, but the difference was not great (Table 1). In the second years, fall 1975 plants were still somewhat smaller than, but fall 1976 plants were the same size as, the respective spring plants. In December 1976, however, fall 1975 plants had accumulated more mortality, and were still smaller, than spring 1976 plants (Table 1). Thus, a fall planting does not necessarily add enough extra growing time (that fall and the following spring) to balance higher mortality and provide an economic advantage over a subsequent spring planting.

These results indicate that delayed planting to avoid the major early-summer infection period clearly reduces mortality. Whether better survival compensates for the loss of growing time in the lower bay, however, depends on growing conditions during the interval between fall planting and harvesting. Attempts to extend the growing season by planting in the fall would be successful only if the fall infection period failed. J. Andrews (Virginia Institute of Marine Science, personal communication) found no differences, after 2 years, in mortality of spring and fall plants when late-summer infections did occur. When infections failed or were very light, fall plants grew well and survived better than spring imports.

Another consideration is location. For instance, Andrews (1982) reported that plantings made as early as August 1 near southern Chesapeake Bay would not suffer mortalities until the following early summer. Our experience to the north in Delaware Bay suggests that introductions should be delayed until September to avoid fall and winter losses. On balance, the safest strategy is to keep susceptible seed as long as possible in waters that are reasonably free of MSX. Seed should be exposed to as little infection pressure as possible, even if it shortens time in the best growing areas.

Andrews (1968) reported that planting eastern oyster seed a month before the start of the June infection period delayed mortalities in Virginia, presumably because the animals had time to acclimate before they were exposed to infection. We have failed to find a similar situation in lower Delaware Bay, possibly because of much heavier infection pressure there. We have also been unable to demonstrate that dredging or experimentally induced shell damage makes newly exposed eastern oysters more susceptible to infection and subsequent mortality. Once oysters become infected, however, dredging for harvest will accelerate losses (our unpublished data). Therefore, when infected eastern oysters are to be harvested, dredging should be done over as short a time as possible.

Early Diagnosis

When eastern oyster seed must be planted in areas where MSX is present, accurate and speedy diagnosis of infections is an essential management tool. This is because the timing and intensity of the infections vary from year to year (Andrews and Frierman 1974; Ford and Haskin 1982). Particularly when eastern oysters have some resistance to the disease, there is a period of several weeks between the earliest (histologically) detectable infections and the onset of mortality. If infection prevalence is determined early enough, and if the relationship of infection to mortality pattern is known well enough for the locality and the oyster stock in question, planters will have

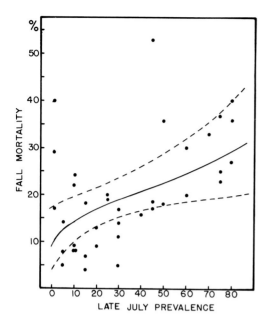

FIGURE 1.—Regression of MSX prevalence (%) in histological samples from eastern oysters collected in late July against nonpredation mortality occurring between early August and late December (mostly in September and October). Samples were collected from eastern oysters planted in lower Delaware Bay in late May and early June between 1966 and 1985. Prevalence and mortality data were arcsine-transformed before regression analyses were performed. The regression (solid line) and 95% confidence interval (dashed lines) were then back-transformed before they were plotted ($N = 37$; $r^2 = 0.22$; $P < 0.01$).

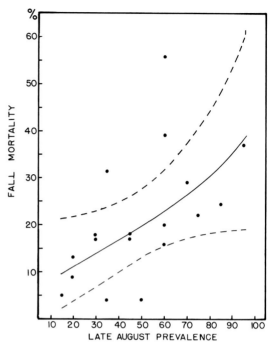

FIGURE 2.—Regression of MSX prevalence (%) in histological samples from eastern oysters collected in late August against nonpredation mortality occurring between early August and late December (mostly in September and October). Samples were collected from eastern oysters planted in lower Delaware Bay in late May and early June between 1966 and 1985. Prevalence and mortality data were arcsine-transformed before regression analyses were performed. The regression (solid line) and 95% confidence interval (dashed lines) were then back-transformed before they were plotted ($N = 15$; $r^2 = 0.30$; $P < 0.02$).

time to decide when or whether to harvest (i.e., how best to balance additional growth against potential mortality).

In lower Delaware Bay, late July is the earliest time that current-year infections can be diagnosed with sufficient accuracy to predict fall mortality, most of which occurs in September and October (Ford and Haskin 1982). High July prevalences are a clear warning of impending losses (Figure 1); however, low prevalences do not necessarily correlate with low mortalities because timing of the infection period varies. If infections are late, they will not be patent by late July. Thus, a low July prevalence is not a reliable predictor of low mortality, and a second sampling in late August is advisable. By then, failure to find infections is a reliable indication that serious mortalities are not likely to occur that fall (Figure 2).

If infections are acquired late in the season, they may remain sublethal during the fall but cause mortalities during the late winter or early

summer (Andrews 1966; Ford and Haskin 1982). Diagnosis in late fall will signal late-winter losses, and early May would be an appropriate time to sample for potential mortalities in June and July.

Monitoring for the parasite is also important when seed oysters are transplanted from low to high salinities. In some cases, eastern oyster seed have had low-level infections that proliferated under the influence of high salinity, causing rapid and sometimes serious losses (Andrews 1964; Haskin and Andrews 1988).

At present, microscopic examination of slides of stained tissue is the only method for accurately detecting the earliest infections, which occur in the gill epithelium. Blood smears are more rapid and less expensive, but are accurate only after parasites reach the circulatory system (Burreson et al. 1988; Ford and Kanaley 1988). Immunological or nucleic acid detection methods currently

TABLE 2.—Comparison of MSX infection statistics and cumulative mortality in eastern oysters placed at the top and bottom of the water column in Delaware Bay, 1968.

| Water column location | 1968 date | Number of eastern oysters | | | | % Cumulative mortality[a] |
		Examined	Infected	Infection systemic	Infection advanced	
Top	Aug 7	20	2	0	0	
	Oct 9	20	13	4	4	16.5
	Nov 11	20	12	4	3	No record
	Total	60	27	8	7	
Bottom	Aug 7	20	4	0	0	
	Oct 9	20	9	4	1	15.5
	Nov 11	20	13	8	7	16.5
	Total	60	26	12	8	

[a] Calculated from June 1, 1968.

being developed should enable rapid and inexpensive diagnosis of early gill infections.

Low-Salinity Purging of MSX

Eastern oysters tolerate considerably lower salinity than does MSX, and several studies have demonstrated that parasites are eliminated from infected oysters placed in low-salinity water (Sprague et al. 1969; Andrews 1983; Ford 1985). At salinities of 10‰ or less, this process can occur in as little as 2 weeks in the spring and summer (Andrews 1983; Ford 1985); however, it is considerably slower, and may not occur, at temperatures of 10°C or less (our unpublished data). Our data indicate that the reduced salinities act directly on the parasite, not by enhancing the host's ability to eliminate the parasite (as suggested by Andrews 1983). In vitro viability of MSX at different salinities closely parallels that reported in vivo (Ford and Haskin, in press). Thus, the effect of low salinity should not depend on the relative resistance or susceptibility of the oysters, except that resistant individuals tolerate parasitism better than susceptible ones, and presumably have a better chance of recovering from an infection (Ford 1988, this volume).

Considerable experimental work remains to be done to define time–temperature–salinity interactions, as well as to determine at what point an infection becomes so debilitating that an eastern oyster will not recover even if appropriately "treated" in low salinity. Also, eastern oysters can become reinfected when returned to growing areas (Andrews, personal communication), perhaps necessitating repeated immersions. Low-salinity "treatment" of large numbers of eastern oysters planted on the bottom would probably not be economically feasible, but it might be a reasonable method for animals grown under intensive culture

(i.e., in containers). A system for purging infected eastern oysters of MSX would be practical only if there were rapid and accurate diagnoses to inform the planter of the proper time for moving a stock, economical means of moving it, low-salinity areas where the stock could be placed, and reasonable assurance that harvesting could take place before reinfection becomes a problem, or that repeated immersions are practical.

Placement in the Water Column

In areas of intensive aquaculture, oysters are often protected from predators by holding them above the bottom in bags or cages, or by placing them in intertidal locations while they are still small. Would such practices also reduce the risks of MSX infections? To test this possibility, we placed eastern oyster seed in a wire basket suspended just below the low-tide mark, approximately 5 m off the bottom. We then compared the onset and development of MSX infections and associated mortality between these animals and an identical group on the bottom. Both groups were placed in lower Delaware Bay in late May. Infections were acquired and intensified at approximately the same rate, regardless of location in the water column, and mortality was the same, at least through mid October, in both groups (Table 2). Eastern oysters in intertidal locations of lower Delaware Bay are also subjected to infection pressure equal to that in subtidal sites (Haskin and Ford 1979), as are intertidal animals on pilings in lower Chesapeake Bay (Andrews, personal communication). Thus, infective stages appear to be equally distributed from top to bottom of the water column, and protection from MSX infections cannot be provided in the same manner as protection from predators. Eastern oysters kept off the bottom usually grow faster and have lower predation losses than those on the bottom, however, so overall production might be improved by using such a system, even if MSX is not deterred.

Resistant Oysters

The development of resistant stocks, either in nature or through artificial propagation, offers another opportunity for controlling losses due to the MSX parasite. The native population in Delaware Bay is now three to four times more resistant (as measured by the number of survivors after a 3-year exposure period) as it was before the 1957–1959 epizootic (Haskin and Ford 1979). There are indications that local stocks in lower Chesapeake Bay have also become resistant (Farley 1975; Andrews 1966, 1968). When an epizootic such as that caused by MSX occurs in a self-replenishing population, planters should be encouraged not to harvest the survivors, but to leave them as brood stock and perhaps even to concentrate them in spawning sanctuaries. If mortalities have been severe enough and if selected oysters produce a significant fraction of new recruits into the seed supply, offspring should be considerably more resistant than the original population.

Controlled selection and breeding programs at Rutgers University and the Virginia Institute of Marine Science have produced strains of eastern oysters that are considerably more resistant to MSX-caused mortality than are natives (Andrews 1968; Haskin and Ford 1979). Resistance does not necessarily prevent infection but enables the hosts to restrict and tolerate parasites. Because these oysters must be produced in a hatchery, their higher cost must be outweighed by improved survival relative to natural seed. Between setting and market size, the best Rutgers strains have a survival of about 70%, compared to 30% for Delaware Bay natives and 7% for unselected imports (Ford and Haskin 1987). This ratio means that, in Delaware Bay at least, hatchery-produced seed can be no more than about 2.5 times as costly as wild seed of the same size in order to make economical use of resistant strains. The cost ratio would increase if growth and meat yields were also improved by selection for resistance, but differences in these traits appear minor compared to survival differences (our unpublished data). In regions where native set is sparse and MSX is also a problem, and particularly if hatchery seed has already proved effective, MSX-resistant eastern oysters should definitely be used as brood stock.

Restocking natural habitats with hatchery-produced MSX-resistant eastern oysters may be possible in small, enclosed areas where native oysters are extremely sparse. In larger estuaries where residual populations exist, it is likely that the cost of producing enough resistant individuals to outcompete the native population would be entirely prohibitive. Resistance can be overwhelmed by extremely heavy infection pressure; a majority of selected eastern oysters eventually die with MSX infections after 5 or 6 years of constant exposure (Ford and Haskin 1987). This risk should be considered in any restocking plan involving hatchery seed.

Restriction on Oyster Shipments

Although we lack evidence that transmission of MSX occurs directly from infected to uninfected oysters, there is concern that transplantation of diseased eastern oysters into Wellfleet and Cotuit, Massachusetts, caused epizootics in areas that were previously free of the parasite (Krantz et al. 1972; Haskin and Andrews 1988; G. Matthiessen, Ocean Pond Corporation, and F. Kern, National Marine Fisheries Service, personal communications). The possibility that direct transmission occurs, or that an infected reservoir host may be included in shipments, makes it extremely unwise for planters to bring eastern oysters from enzootic areas into waters where the parasite is absent. Certification that eastern oysters are free of MSX is impossible because infections may be subpatent, particularly at certain times of the year (Andrews 1964; Ford 1985). Elimination of MSX from eastern oysters by immersion in low-salinity water might be used to treat seed, or even selected brood stock, before shipment; however, more needs to be known about the threshold requirements for complete elimination of the parasite before this can be used with confidence.

Conclusions

From a fisheries management perspective, the most important aspect of MSX disease is that the parasite is most prevalent in high-salinity areas of estuaries and that mortality increases with length of exposure, even in oysters that have been selected for resistance. To reduce the risks of infection, susceptible stocks should be kept in nonenzootic areas (in most cases, this means in low-salinity water) for as long as possible, even if growth is reduced. If eastern oysters must be moved to high salinity for final growth and conditioning before market, this period should be limited to a single growing season. Planters can further minimize mortality by moving oysters after the early-summer infection period (June and July) and harvesting by the end of the year. Loss of time in the best growing areas must be balanced against better survival, however, and results may vary according to location and time. Epizootiological studies are essential to define infection and mortality patterns that will suggest management options for specific localities. Continuous monitoring and early diagnosis of infections is also important because they allow mortality to be predicted and growers and managers to make informed decisions on when or whether to plant or harvest.

After an epizootic, surviving eastern oysters should not be harvested, but should be preserved as brood stock that may help to establish a population with increased resistance. Selective breeding has produced strains of eastern oyster that are highly resistant to MSX-caused mortality. Use of these strains is recommended for intensive culture operations that employ hatchery-produced seed.

Infected eastern oysters can be cleansed of MSX by immersing them in low-salinity water (<10‰) in spring and summer, but exact time–temperature–salinity thresholds for this procedure need further definition before they are adopted by planters.

Acknowledgments

We are indebted to the many people who, over the past 30 years, have contributed to our understanding of MSX and eastern oysters in Delaware Bay. D. Kunkle, J. Myhre, D. O'Connor, and W. Canzonier deserve special thanks. Support for this effort has come from the National Marine Fisheries Service under Public Law 88-309 and the New Jersey Department of Environmental Protection. J. D. Andrews kindly reviewed the manuscript and provided guidance on matters dealing with oyster management in Virginia. This is New Jersey Agricultural Experiment Station publication F 32504-1-87, supported by state funds.

References

Andrews, J. D. 1964. Oyster mortality studies in Virginia. IV. MSX in James River public seed beds. Proceedings National Shellfisheries Association 53: 65–84.

Andrews, J. D. 1966. Oyster mortality studies in Virginia. V. Epizootiology of MSX, a protistan parasite of oysters. Ecology 47:19–31.

Andrews, J. D. 1968. Oyster mortality studies in Virginia. VII. Review of epizootiology and origin of *Minchinia nelsoni*. Proceedings National Shellfisheries Association 58:23–36.

Andrews, J. D. 1982. Epizootiology of late summer and fall infections of oysters by *Haplosporidium nelsoni*, and comparison to annual life cycle of *Haplosporidium costalis*, a typical haplosporidan. Journal of Shellfish Research 1:15–23.

Andrews, J. D. 1983. *Minchinia nelsoni* (MSX) infections in the James River seed-oyster area and their expulsion in spring. Estuarine, Coastal and Shelf Science 16:255–269.

Andrews, J. D., and M. Frierman. 1974. Epizootiology of *Minchinia nelsoni* in susceptible wild oysters in Virginia, 1959–1971. Journal of Invertebrate Pathology 24:127–140.

Burreson, E. M., M. E. Robinson, and A. Villalba. 1988. A comparison of paraffin histology and hemolymph analysis for the diagnosis of *Haplosporidium nelsoni* (MSX) in *Crassostrea virginica* (Gmelin). Journal of Shellfish Research 7:19–24.

Farley, C. A. 1975. Epizootic and enzootic aspect of *Minchinia nelsoni* (Haplosporida) disease in Maryland oysters. Journal of Protozoology 22:418–427.

Ford, S. E. 1985. Effects of salinity of survival of the MSX parasite *Haplosporidium nelsoni* (Haskin, Stauber, and Mackin) in oysters. Journal of Shellfish Research 5:85–90.

Ford, S. E. 1988. Host–parasite interactions in eastern oysters selected for resistance to *Haplosporidium nelsoni* (MSX) disease: survival mechanisms against a natural pathogen. American Fisheries Society Special Publication 18:206–224.

Ford, S. E., and H. H. Haskin. 1982. History and epizootiology of *Haplosporidium nelsoni* (MSX), an oyster pathogen, in Delaware Bay, 1957–1980. Journal of Invertebrate Pathology 40:118–141.

Ford, S. E., and H. H. Haskin. 1987. Infection and mortality patterns in strains of oysters *Crassostrea virginica* selected for resistance to the parasite *Haplosporidium nelsoni* (MSX). Journal of Parasitology 73:368–376.

Ford, S. E., and H. H. Haskin. In press. Comparison of in vitro salinity tolerance of the oyster parasite *Haplosporidium nelsoni* (MSX) and hemocytes from the host, *Crassostrea virginica*. Comparative Biochemistry and Physiology.

Ford, S. E., and S. A. Kanaley. 1988. An evaluation of hemolymph diagnosis for detection of the oyster parasite *Haplosporidium nelsoni* (MSX). Journal of Shellfish Research 7:11–18.

Haskin, H. H., and J. D. Andrews. 1988. Uncertainties and speculations about the life cycle of the eastern oyster pathogen *Haplosporidium nelsoni* (MSX). American Fisheries Society Special Publication 18: 5–22.

Haskin, H. H., W. J. Canzonier, and J. L. Myhre. 1965. The history of MSX on Delaware Bay oyster grounds, 1957–65. American Malacological Union Bulletin 32:20–21.

Haskin, H. H., and S. E. Ford. 1979. Development of resistance to *Minchinia nelsoni* (MSX) mortality in laboratory-reared and native oyster stocks in Delaware Bay. U.S. National Marine Fisheries Service Marine Fisheries Review 41(1–2):54–63.

Haskin, H. H., and S. E. Ford. 1982. *Haplosporidium*

nelsoni (MSX) on Delaware Bay seed oyster beds: a host–parasite relationship along a salinity gradient. Journal of Invertebrate Pathology 40:388–405.

Haskin, H. H., and S. E. Ford. 1983. Quantitative effects of MSX disease (*Haplosporidium nelsoni*) on production of the New Jersey oyster beds in Delaware Bay, USA. International Council for the Exploration of the Sea, C.M. 1983/Gen:7/Mini-Symp., Copenhagen.

Haskin, H. H., L. A. Stauber, and J. A. Mackin. 1966. *Minchinia nelsoni* n.sp. (Haplosporida, Haplosporidiidae): causative agent of the Delaware Bay oyster epizootic. Science (Washington, D.C.) 153:1414–1416.

Hull, C. H. J., and J. G. Titus. 1986. Greenhouse effect, sea level rise, and salinity in the Delaware estuary. Environmental Protection Agency, EPA 230-05-86-010, Washington, D.C.

Krantz, E. L., L. R. Buchanan, C. A. Farley, and A. H. Carr. 1972. *Minchinia nelsoni* in oysters from Massachusetts waters. Proceedings National Shellfisheries Association 62:83–88.

Sprague V., E. A. Dunnington, and E. Drobeck. 1969. Decrease in incidence of *Minchinia nelsoni* in oysters accompanying reduction of salinity in the laboratory. Proceedings National Shellfisheries Association 59:23–26.

American Fisheries Society Special Publication 18:257–264, 1988

Management Strategies to Control the Disease Caused by *Perkinsus marinus*

JAY D. ANDREWS

School of Marine Science, Virginia Institute of Marine Science
Gloucester Point, Virginia 23062, USA

SAMMY M. RAY

Marine Biology Department, Texas A&M University of Galveston
Galveston, Texas 77553, USA

Abstract.—*Perkinsus marinus* is a warm-season protozoan pathogen (Phylum Apicomplexa) that parasitizes eastern oysters *Crassostrea virginica* from Chesapeake Bay to the Gulf of Mexico. In years of normal rainfall in Chesapeake Bay, *P. marinus* is limited by low salinities in the estuaries and is dormant during winter and spring. A few subpatent infections persist to initiate new generations of infections from June to October. During 6 years of drought in the 1980s, the disease spread throughout Chesapeake Bay, killing eastern oysters at record rates. In 1987, it was more widely distributed and more destructive than the disease caused by *Haplosporidium nelsoni* (MSX). The invasion by *P. marinus* of normally low-salinity seed areas in Chesapeake Bay, which brought the pathogen to an abundance of hosts and accelerated dispersal of the disease, was a critical turning point. Control depends upon return of normal rainfall and low estuarine salinities to suppress the disease, which remains active during summer at levels of 12–15‰. *Perkinsus marinus* is widely distributed in the Gulf of Mexico with very high prevalences. There, *P. marinus* maintains infections throughout the year, though intensities decline during the winter. Warm year-round water temperatures and few low-salinity refuges make management of the disease difficult in the gulf. Eastern oysters planted in late fall have been harvested by the following summer to avoid mortalities. This practice has been aided by reduction of legal size limits in Texas. Removal of diseased oysters and planting of only disease-free oysters will also limit destruction by *P. marinus*.

The occurrence of *Perkinsus marinus* in eastern oysters *Crassostrea virginica* in Chesapeake Bay was discovered in 1950 (Andrews and Hewatt 1957) soon after the pathogen was first described in the Gulf of Mexico (Mackin et al. 1950). Extensive reviews of the biology of this warm-water pathogen were made by Ray (1954) and Mackin (1962). The activity of the disease in Chesapeake Bay during years of normal salinity regimes was described by Andrews and Hewatt (1957). As a result of many dry years during the 1980s, salinities increased throughout the bay, and *P. marinus* is now widely distributed in nearly all estuarine systems of Chesapeake Bay. Only the uppermost oyster-growing sectors of the large rivers escaped the rapid spread of the disease during the dry period of 1985–1987. The enzootic area during periods of normal lower salinities extended from the Choptank River (Figure 1) downbay into every small estuary and large portions of the large rivers in lower Maryland and upper Virginia waters where eastern oysters are grown.

For a current account of the history, epizootiology, and life cycle of *P. marinus*, see Andrews (1988, this volume). The pathogen is persistent at summer salinities below 12–15‰ which are minimal levels for eastern oyster growth and reproduction; therefore, during wet years it maintains low-level infections at marginal sites, and it causes extensive mortalities during dry periods. The pathogen thrives on planted private beds where eastern oyster densities are high, and it survives on low-density public beds, isolated piers, and other structures. These hard-bottom public sites are persistent foci of infection if some recruitment of eastern oysters occurs. Based on 37 years of monitoring data, the disease probably will never be eradicated from its enzootic areas, but return to normal salinities will suppress its activity in upper Virginia and most Maryland estuaries.

Most infections of *P. marinus* are produced by close proximity of prospective host eastern oysters to disintegrating gapers with high-intensity infections (Andrews 1965, 1967). The disease is usually slow to spread to new beds and new areas except by introduction of infected seed oysters. This transplantation of infected animals occurs commonly during dry periods such as the mid-1960s and 1985–

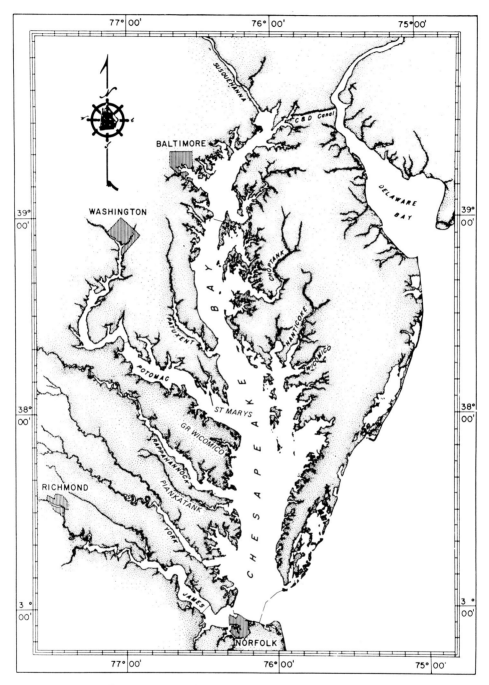

FIGURE 1.—Map of Chesapeake Bay and its river systems.

1987, because then the disease spreads into seed areas and these are not monitored adequately to warn oyster planters. Infested areas must be closed to additional transplantations. Once the pathogen becomes established in growing areas, even in low-salinity waters, years are required to eradicate it from specific beds, which must be fallowed until all oysters are dead. On public beds, where even light recruitment occurs sporadically, the disease persists interminably. Control of *P. marinus* is more difficult on public beds than on private grounds for two major reasons: (1) eastern oysters tend to smother on soft muddy or shifting sandy bottoms of private grounds, but die slowly of diseases and old age on hard shelly public rocks; and (2) harvesting of private beds by dredges is regular and as complete as possible, whereas sufficient cultch remains on public beds to sustain wild populations.

Nearly all high-salinity areas in lower Chesapeake Bay below New Point Comfort at the mouth of Mobjack Bay (immediately north of the mouth of the York River: Figure 1) have not been planted with eastern oysters since 1960, when a new disease caused by *Haplosporidium nelsoni* (MSX) became enzootic (Andrews and Wood 1967). Some eastern oysters are planted in the upper York River, where both diseases are enzootic, but losses are high because James River seed animals planted there are highly susceptible to MSX. Some cultural success has been achieved in upper Mobjack Bay and its four river tributaries by use of native animals resistant to MSX. However, *P. marinus* is always present because regular setting occurs in these areas, and care must be exercised to harvest early (as soon as eastern oysters are marketable) and to fallow beds between crops. The duration of culture varies with the size of seed animals and growth rate in the area. During 1985–1987, both diseases took a heavy toll in this area when exceptional MSX virulence inexplicably overwhelmed even native eastern oysters that had previously resisted infection. In Chesapeake Bay, MSX-resistant eastern oysters are available only in native populations of the lower York River and Mobjack Bay, where all populations have been selected intensively and continuously over 27 years by high mortalities (Andrews and Frierman 1974).

Planting of MSX-susceptible James River seed eastern oysters has occurred only in relatively low-salinity waters since the disease struck in 1959. No significant planting has occurred since 1960 below the mouth of the Rappahannock River, where summer salinities rarely exceed 20‰. James River seed oysters are usually free of both diseases (Andrews 1983), except during drought periods such as 1985–1987, but are routinely planted only in low-salinity areas of upper Virginia and Maryland. During those droughty years, infected seed oysters that were moved from the lower James River seed area to Virginia tributaries of the Potomac River had disastrous mortalities from *P. marinus*, and the grounds there are now contaminated by the disease. Because transplanting practices have been careless, wider distribution of *P. marinus* occurs during each period of successive dry years.

Management practices to avoid the effects of *P. marinus* (and of MSX) are different in the two types of Chesapeake estuaries, that is, large rivers and small coastal plain estuaries. Rivers with large drainage areas (James, Rappahannock, and Potomac rivers) have large winter–spring discharges of fresh water that depress salinities below the level required for *P. marinus* and MSX to infect and multiply in eastern oysters (<15‰). Salinities below 10‰ eradicate MSX and delay development of *P. marinus* until late summer, thereby reducing summer multiplication of the latter pathogen that otherwise would produce additional deaths and new generations of infected hosts. Because, normally, few eastern oysters carry overwintering *P. marinus* infections, foci of infection are few and late developing (gapers are required). Seed areas should be free of diseases; therefore, the James River seed area with its protective, annual, freshwater discharges each winter and spring has always been the major source of seed animals in Chesapeake Bay.

The small coastal plain estuaries (Great Wicomico, Piankatank, Choptank, and St. Marys rivers are examples: Figure 1) have little freshwater runoff, and bay salinity regimes tend to dominate the whole eastern oyster-growing sectors. When the bay is salty, overwintering infections there may persist and develop early and thereby initiate two or three generations of infections by November, when temperatures below 20°C halt development and mortalities. These are called trap-type estuaries because setting is more intensive and larval distribution patterns are different than those in large, flushing-type rivers and the bay. During normal years, in upper Virginia and lower Maryland, summer salinities are near or below 15‰, and spring salinities usually drop below 10‰, which is low enough to eradicate MSX but not *P. marinus* (Stroup and Lynn 1963). In autumn, salinities in most of these coastal plains seed areas usually exceed 15‰, which allows both *P. marinus* and MSX to infect eastern oysters and multiply.

These small coastal plain estuaries are important seed areas in Virginia and Maryland, but, with low salinities, their eastern oysters tend to have slow growth, poor meat condition, and thickened shells. Because eastern oysters are slow to reach market size (75 mm), *P. marinus* has ample time to infect them on public and private beds. The rivers are small with shallow flats that expose eastern oysters to foci of infection along the shore as well as to those in public beds with immature animals. This exposure inhibits the use of these waters as seed areas because transplanting infected animals at any prevalence level causes early epizootic mortalities on growing beds. But *P. marinus* does not usually attack young oysters in Chesapeake Bay (in contrast to MSX), and natural spatfall on planted shells can be grown 1 or 2 years before the animals acquire the disease. These small estuaries should be used, therefore, only for production of seed oysters that are regularly transplanted early, before the second summer of exposure, to avoid infection. Each estuary must be operated entirely for seed production by the state or by private growers to avoid holding infected populations of eastern oysters to market size.

In Virginia, planters could buy few seed eastern oysters from the James River during the 1986–1987 and 1987–1988 planting seasons. Instead, seed animals were tonged for direct marketing at US$12–22/bushel (1 bushel = 35.2 L) despite their small size and poor condition. During the previous season, the seed eastern oysters of the same size were sold to planters for $3/bushel. One fall or spring season in a good growing area, such as the Rappahannock River, would more than double the yield of meats from such animals. During 1987, private planters deliberately moved seed oysters infected with both diseases from the Great Wicomico River to very-low-salinity areas in the Rappahannock River because this was their only alternative if they were to remain in business. They also bought private grounds in good setting areas of the James River and Mobjack Bay with the intent of transplanting seed oysters at early ages from disease-enzootic areas. After 3 years of drought, *P. marinus* has moved up the James River seed area above Wreck Shoal, and most eastern oysters below this bed were killed by MSX and *P. marinus*. Apparently, when the latter disease reached the comparatively dense eastern oyster populations in the seed area, the abundance of infective stages released from dying gapers increased so rapidly that dosage needed for infection was achieved from bed to bed in 1987, in contrast to the oyster–to–oyster transmission that usually occurred. Furthermore, overwintering infections occurred at 80–90% levels and began multiplying in June, whereas the overwintering prevalence is usually 10–15% and development is later in the summer. Intensive harvesting of seed oysters for market from the few remaining upriver beds may deplete the brood stocks needed for reliable setting in the James River.

During years of normal rainfall, the disease caused by *P. marinus* can be controlled in most low-salinity areas of Chesapeake Bay by proper management. Because isolation of beds is the most important method of controlling the disease, the contiguous location of public and private beds complicates management. The vast area of public eastern oyster beds in Chesapeake Bay is far greater than state efforts can rehabilitate. Overfishing and disease mortalities have reduced brood stocks to record low levels and spat setting has declined accordingly. Private culture could not readily restore these areas either, because seed oysters and capital are not available. Persistent losses from extreme weather conditions have discouraged private oystermen from planting. Importation and repacking of eastern oysters from the Gulf of Mexico and the Pacific coast are presently more-economical and less-risky practices.

The first management requirement is to restore the supply of seed eastern oysters for private planters. The present Virginia management plan proposes to plant more seed oysters on public growing beds at state expense. Further, recruitment in the James River seed area is uncertain due to 2 years' intensive fishing of brood stocks for market sale. The state, which controls most seed areas, must provide sources of seed oysters for private planters. This provision could be accomplished by releasing all public beds to private planters in two of the trap-type estuaries, e.g., the Great Wicomico and the Piankatank rivers. These estuaries could be declared seed areas subject to strict limitations on duration of culture to avoid diseases. This is a control system for *P. marinus* disease. Planters must have access to shells for planting from Maryland or, if fossil shells are dredged, from Virginia.

Monitoring of diseases is inadequate in both Maryland and Virginia, and managers must become familiar with the methods to avoid infections. Correct timing for monitoring diseases and transplantation of oysters is essential to limit *P. marinus* distribution. Testing must be done in late summer and fall because it is not technically feasible during most of the fall–winter planting season. *Perkinsus marinus* is highly seasonal in activity because it depends upon warm temperatures to multiply. It kills eastern oysters mostly during late summer and early fall, which are the best monitoring periods for disease prevalence. Overwintering infections cannot be easily determined in the laboratory. Therefore, each growing bed at risk must be sampled in the fall to assay the level of infection; that will determine

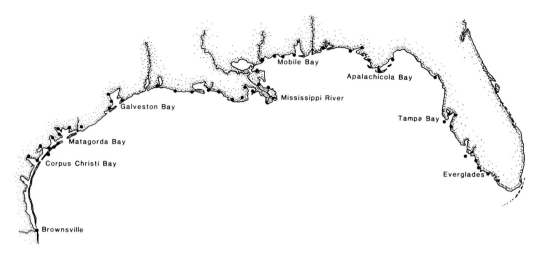

FIGURE 2.—Map of the coastline of the Gulf of Mexico with major landmarks noted. Forty-eight of the 50 oyster sampling sites used by Brooks et al. (1987, 1988) are marked with solid circles.

whether animals can be safely held through another summer. Not much can be done if an epizootic kill occurs during the warm season. Meat yields are drastically reduced by high levels of infection, which may reach 100% during the second summer of *P. marinus* activity, and many deaths may occur.

Because MSX restricts culture of eastern oysters to low-salinity areas, *P. marinus*, which maintains low-intensity infections in nearly all these areas, becomes destructive if summer temperatures are prolonged in fall. A small increase in salinities (above 12–15‰) provides conditions for epizootic mortalities caused by *P. marinus*. To avoid losses, isolation of disease-free stocks is essential. The following management practices are recommended for periods of normal weather conditions. Control is nearly impossible when two successive 3-year periods of drought occur during one decade.

(1) Transplant only disease-free stocks. Even low prevalence of *P. marinus* will accelerate mortality.

(2) Select growing beds isolated 0.4 km from any other bed with infected oysters. Dispersion of infective stages limits chances of dosages large enough to produce infection. This procedure has not been effective during the 1985–1987 drought.

(3) Early harvest followed by fallowing of beds limits mortality and distribution of the disease. Fall and spring periods of fattening occur in eastern oysters, and meat yields depend more on condition than size of animals.

(4) To monitor beds of growing eastern oysters for *P. marinus*, test 25 animals at 2 years of age or older by the thioglycollate method in late summer

or early fall. All planted beds should be examined by managers each late summer or fall for dead animals to determine if harvesting or another year of culture are desirable or necessary.

Some drastic changes must occur in Chesapeake Bay if the eastern oyster industry is to survive. The rapid decline of oyster production from private beds indicates that planters are fearful of natural risks and state management decisions. Politics and economics have always dominated decisions on oyster culture practices, but proper management to avoid diseases is essential whether oyster production is on private or public grounds. Recovery from the drought-induced expansion of disease ranges may be slow, but it will certainly occur when salinities return to normal.

Perkinsus marinus in the Gulf of Mexico

Perkinsus marinus was first associated with widespread epizootic mortality of eastern oysters during warm seasons in high-salinity (>15–20‰) waters of Louisiana during the late 1940s. Since 1950, *P. marinus* has been widely distributed in high-salinity waters of the Gulf of Mexico, and studies by many investigators have shown this organism to be the most destructive oyster parasite along the gulf coast and in some areas along the south Atlantic coast (Andrews 1988).

Recent surveys show that *P. marinus* continues to be widely distributed along the gulf coast. Brooks et al. (1987) sampled eastern oysters at 50 sites from Brownsville, Texas, (South Bay) to the Florida Everglades (Figure 2). Although the survey was made during the cool part of the year

(February and March 1986), none of the samples were free of this parasite. The prevalence (percent infection) of *P. marinus* was at least 50% at all but one site—Bay Garderne, Louisiana. Furthermore, the prevalence was less than 75% at only 10 sites. The lowest prevalences were encountered at sites in Louisiana near the Mississippi River, where freshwater input limited the disease. Many of the infections were light, which is to be expected during the winter. The same sites were sampled again in January and February 1987, and preliminary results appear to be similar to 1986 results except that intensities of infection were somewhat lower, due probably to cooler winter temperatures (Brooks et al. 1988).

Soniat and Gauthier (in press) reported the occurrence of *P. marinus* in 23 samples of eastern oysters from 16 sites along the midnorthern Gulf of Mexico (from Galveston Bay, Texas, to Mississippi Sound) in 1985 and 1986. The prevalence of infection exceeded 50% in 16 of 23 samples. This parasite was detected at all but 2 of the 16 sites—Hackberry and Vermillion bays in Louisiana.

Thus, the results of recent surveys suggest that few areas on the gulf coast with significant eastern oyster populations are free of *P. marinus*. Even the usually highly productive Louisiana seed grounds east of the Mississippi River currently have high prevalences of the parasite as a result of 3 years of drought (M. Chatry, Louisiana Department of Wildlife and Fisheries, personal communication). The degree and extent of infections appear to be directly related to salinity regimes of the systems in question. During prolonged droughts, the parasite may invade areas where salinity usually is barely sufficient to support eastern oyster populations.

Since 1983, Ray (unpublished data) has monitored the occurrence of *P. marinus* from monthly samples of 25 market-sized (7.6-cm-long) eastern oysters from Confederate Reef at the east end of West Bay, which is a high-salinity area of the Galveston Bay system, Texas. Although data through December 1987 show that weighted incidence (intensity of infection) may decrease during the cool months, prevalence of infection remains above 90% throughout the year. Such highly enzootic areas as West Bay provide constant foci of infection for nearby eastern oyster populations as well as for newly recruited spat and juvenile oysters.

The problem of managing the eastern oyster industry to control *P. marinus* is much more difficult in warm waters of the Gulf of Mexico than for the Chesapeake Bay area. Because the water temperatures of the Gulf of Mexico usually exceed 20°C for at least 6 months of the year, the relatively long cool season that retards *P. marinus* does not prevail in the gulf.

For many years, the Louisiana oystermen avoided losses of eastern oysters from summer mortalities (probably caused by disease and predation in high-salinity areas such as lower Barataria Bay, Louisiana) by limiting oyster planting to the cool season. From the 1930s through the mid-1970s, oystermen commonly transplanted seed oysters from state-owned seed grounds, located primarily east of the Mississippi River, to private leases west of the river. Plantings usually took place in late summer and early fall, and the eastern oysters were usually harvested before the next warm season to avoid predictable summer mortalities. Since the animals remained on the grounds for only 3–6 months, relatively few were large enough for the shucked and half-shell trade, and most were used for canning. The profitable Louisiana canning trade died out about 1975. According to Chatry (personal communication), this industry declined because it could not compete with imported canned oysters.

The apparent resistance of young eastern oysters to infection with *P. marinus* during their first year has been noted by several investigators. Some have suggested that this attribute may be used to time the harvest of eastern oysters, especially from private leases, prior to the development of lethal infections. This strategy would reduce losses of oysters and also the number of infective elements released from dying oysters (see Andrews 1988). More recent studies (Hofstetter 1977; Ray 1987) show that spat less than 2.5 cm long may become infected with *P. marinus* in Texas bays. Present data (Ray, unpublished) indicate that the prevalence of this parasite in spat is related to the prevalence in adult populations. As mentioned earlier, most infections with *P. marinus* are transmitted by disintergrating gapers with heavy infections (see Andrews 1988).

Thus, some of the possible strategies available for controlling *P. marinus* in the Chesapeake Bay seem to be negated by the long warm season in the gulf. Also, early infection of small eastern oysters reduces the potential of exploiting age-related resistance in disease-control strategies.

The early harvest of eastern oysters, especially on private leases, can minimize the effects of disease caused by *P. marinus*. Texas management authorities, in 1963, lowered the minimum legal market length for eastern oysters harvested from public reefs from 8.9 cm to 7.6 cm. This step greatly increased Texas eastern oyster production. It is likely that the biomass gained by the increased growing period to reach 8.9 cm did not compensate for the biomass lost during this period to natural mortality. In establishing an 8.9-cm minimum length, resource managers probably intended to provide more adult oysters as brood stock. With better knowledge regarding the ravages and spread of *P. marinus*, it may be a good trade-off to allow harvesting that reduces the breeding populations but thereby reduces the foci of parasite infection. Moreover, the requirement of a minimum legal market length for eastern oysters harvested from private leases may be counterproductive.

The primary source of recruitment of young eastern oysters in the gulf is market- and submarket-sized oysters that spawn during late spring and early summer. Yet, juvenile oysters become sexually mature during the first summer at ages of 1–2 months. Thus, the view that large numbers of market-sized oysters, even if diseased, are necessary to maintain eastern oyster populations may be incorrect.

Although the occurrence and intensity of *P. marinus* is easily assayed by the thioglycollate technique (Ray 1966), routine monitoring of eastern oyster beds for diseases is conducted only sporadically by most management entities in the gulf states. Because the level of infection depends on salinity regimes, all major oyster growing areas should be monitored on a seasonal basis. The monitoring data would be useful in permitting a flexible management approach. For example, if late-winter or early-spring data indicate a high level of infection, it may be prudent to extend the harvest season to encourage the removal of oysters. Besides salvaging many eastern oysters for consumption that may not survive to the next oyster season, the removal of infected individuals would reduce the number of infective elements released by dying animals.

A deliberate effort should be made to remove as many eastern oysters as possible from areas of consistently high prevalence for *P. marinus* prior to the end of the oyster season in late spring. Currently, there are such areas in the east end of West Bay in the Galveston Bay system.

From a practical point of view, sufficient freshwater inflow to maintain an average salinity of 15‰ or less, especially during the warm season, is the primary means for controlling disease caused by *P. marinus* in the Gulf of Mexico. Thus, diligence must be exercised in protecting eastern oyster growing areas from salinity increases due to reduc-

tions in freshwater inflow reduction, or to saltwater intrusions resulting from human activities. Substrate (reefshell) for setting of spat must be protected or enhanced where there is a clear-cut demonstration that its absence in an area has led to a depletion of eastern oysters. No amount of reef enhancement, however, will restore a reef if high-salinity conditions exist that favor diseases, predators, and fouling organisms. Oysters will flourish only in areas where sufficient freshwater inflows occur on a regular basis.

Management measures that support the diversion of fresh water into high-salinity bay systems, such as is currently being proposed for Matagorda Bay, Texas, and Breton Sound, Louisiana, offer great potential for reviving or enhancing areas that are only marginally productive for eastern oysters except during wet periods. Limited freshwater diversion from the Mississippi River to public seed grounds east of the river has been effective (Chatry, personal communication). Feasibility and design studies for structures to divert fresh water from the Mississippi River to control salinity at Barataria and Breton Sound basins, Louisiana, have been conducted by the New Orleans District of the U.S. Army Corps of Engineers (1984, 1986).

Acknowledgment

This is contribution 1496 of the Virginia Institute of Marine Science.

References

Andrews, J. D. 1965. Infection experiments in nature with *Dermocystidium marinum* in Chesapeake Bay. Chesapeake Science 6:60–67.

Andrews, J. D. 1967. Interaction of two diseases of oysters in natural waters. Proceedings National Shellfisheries Association 57:38–49.

Andrews, J. D. 1983. *Minchinia nelsoni* (MSX) infections in the James River seed oyster area and their expulsion in spring. Estuarine, Coastal and Shelf Science 16:255–269.

Andrews, J. D. 1988. Epizootiology of the disease caused by the oyster pathogen *Perkinsus marinus* and its effects on the oyster industry. American Fisheries Society Special Publication 18:47–63.

Andrews, J. D., and M. Frierman. 1974. Epizootiology of *Minchinia nelsoni* in susceptible wild oysters in Virginia, 1959–1970. Journal of Invertebrate Pathology 24:127–140.

Andrews, J. D., and W. G. Hewatt. 1957. Oyster mortality studies in Virginia. II. The fungus disease caused by *Dermocystidium marinum* in oysters of Chesapeake Bay. Ecological Monographs 27:1–25.

Andrews, J. D., and J. L. Wood. 1967. Oyster mortality studies in Virginia VI. History and distribution of

Minchinia nelsoni, a pathogen of oysters in Virginia. Chesapeake Science 8:1–13.

Brooks, J. M., and seven coauthors. 1987. Analysis of bivalves and sediments for organic chemicals and trace elements from Gulf of Mexico estuaries. Pages 8.1–8.29 *in* First annual report to the U.S. National Oceanic and Atmospheric Association, Rockville, Maryland.

Brooks, J. M., and seven coauthors. 1988. Analysis of bivalves and sediments for organic chemicals and trace elements from Gulf of Mexico estuaries. Pages 8.1–8.17 *in* Second annual report to the U.S. National Oceanic and Atmospheric Association, Rockville, Maryland.

Hofstetter, R. P. 1977. Trends in population levels of the American oyster *Crassostrea virginica* Gmelin on public reefs in Galveston Bay, Texas. Texas Parks and Wildlife Department, Technical Series 24, Austin.

Mackin, J. G. 1962. Oyster diseases caused by *Dermocystidium marinum* and other microorganisms in Louisiana. Publications of the Institute of Marine Science, University of Texas 7:132–229.

Mackin, J. G., H. M. Owen, and A. Collier. 1950. Preliminary note on the occurrence of a new protistan parasite, *Dermocystidium marinum* n. sp. in *Crassostrea virginica* (Gmelin). Science (Washington, D.C.) 111:328–329.

Ray, S. M. 1954. Biological studies of *Dermocystidium marinum*. Rice Institute Pamphlet 41 (special issue), The Rice Institute, Houston, Texas.

Ray, S. M. 1966. A review of the culture method for detecting *Dermocystidium marinum*, with suggested modifications and precautions. Proceedings National Shellfisheries Association 54:55–69.

Ray, S. M. 1987. Salinity requirements of the American oyster, *Crassostrea virginica*. NOAA (National Oceanic and Atmospheric Administration) Technical Report NMFS (National Marine Fisheries Service) SEFC-189:E.1–E.28.

Soniat, T. M., and J. D. Gauthier. In press. The prevalence and intensity of *Perkinsus marinus* from the midnorthern Gulf of Mexico, with comments on the relationship of the oyster parasite to temperature and salinity. Tulane Studies in Zoology and Botany 27.

Stroup, E. O., and R. J. Lynn. 1963. Atlas of salinity and temperature distributions in Chesapeake Bay 1952–1961 and seasonal average 1959–1961. Johns Hopkins University, Chesapeake Bay Institute, Graphical Summary Report 2, Baltimore, Maryland.

U.S. Army Corps of Engineers. 1984. Louisiana coastal area, Louisiana, freshwater diversion to Barataria and Breton Sound basins, feasibility study, volume 1, main report, environmental impact statement, September 1984. New Orleans District, New Orleans, Louisiana.

U.S. Army Corps of Engineers. 1986. Caernarvon freshwater diversion structure. Biological, hydrologic, water and sediment quality monitoring program, design memorandum no. 1, supplement no. 1, December 1986. New Orleans District, New Orleans, Louisiana.

American Fisheries Society Special Publication 18:265–268, 1988

Summer Mortality of Pacific Oysters

J. Harold Beattie

Washington State Department of Fisheries, Shellfish Laboratory
1000 Point Whitney Road, Brinnon, Washington 98320, USA

J. P. Davis, S. L. Downing, and K. K. Chew

School of Fisheries, University of Washington
Seattle, Washington 98195, USA

Abstract.—During the past 20 years, researchers in Japan and the United States have sought to determine the causes of summer mortality among Pacific oysters *Crassostrea gigas*. Two hypotheses are now preferred: one revolves about physiological stress associated with spawning, the other focuses on disease caused by a pathogenic agent. The development of large hatcheries on the west coast of the USA has allowed oyster growers to reduce summer mortalities through improved husbandry practices. Breeding studies at the University of Washington have produced oyster stocks resistant to summer mortality.

Over the last 50 years, oyster culture in Japan and the Pacific coasts of Canada and the USA has focused almost exclusively on cultivation of the Pacific oyster *Crassostrea gigas*. Mass summer mortalities of Pacific oysters have become a recurring obstacle for oyster growers in these countries since the 1950s. The term "summer mortality" (regardless of etiology) has been used to describe any mortality of 30% or greater among 2-year-old or older Pacific oysters anytime between June and November.

Summer mortality is an intermittent occurrence. Mortalities of over 30% occurred on the west coast of the USA in the late 1950s and intensified in the early 1960s (Glude 1975). During the early 1970s, there was little oyster mortality in Washington state, but mortalities began to increase in 1976 and peaked in 1979–1981. There were up to 60% losses in some bays (Perdue et al. 1981). Since 1982, mortalities have been at a low but notable annual level (10–20%) in most oyster bays in south Puget Sound.

Several hydrographic and biological factors coincide with summer mortality. Mortality areas are characterized as poorly circulating upper reaches of estuaries in areas with turbid, highly productive waters and temperatures exceeding 18°C. Mortalities coincide with maximal development of reproductive tissues. In Japan, results of physiological studies indicated that dying Pacific oysters had reduced glycogen content and lower metabolic activity in association with gonad overmaturation (Imai et al. 1965; Mori et al. 1965; Mori 1979). Other studies showed that affected Pacific oysters exhibited associated inflammation of the digestive

diverticulae due to normal gonad resorption that occurs during and after spawning (Tamate et al. 1965). Studies in several inlets of south Puget Sound, Washington, also showed that Pacific oyster mortalities were associated with high reproductive activity (Glude 1975). Annual dinoflagellate blooms often coincide with summer mortality in Washington (Glude 1975), and have also been associated with mortalities of Pacific oyster larvae (Cardwell et al. 1979). Some observers have suggested that mortality occurs predominantly among faster-growing oysters; Koganezawa (1975) reported that the inception of summer mortality in Japan coincided with the introduction of hanging culture, a system that results in rapid growth of Pacific oysters.

The coincidence of summer mortality with these various factors led researchers to suggest several hypotheses concerning the causes of mass death. Two of these seem to have the most merit at the present time. The first is stress-induced mortality related to excessive gonad maturation. Up to 65% of the dry meat weight of sexually mature Pacific oysters is germinal tissue; the effort to produce and maintain this level of reproductivity is stressful and could contribute to death. This hypothesis was explored by Perdue (1983), who based his work on some speculation by Mann (1979) concerning the relationship of glycogen level and mortality for juvenile Pacific oysters. Perdue's work is discussed later in the text. The other hypothesis of promise centers on the possibility of death as the result of pathogenic bacteria. Previous bacteriological studies showed no causative relationship between various species of *Vibrio* and Pacific oyster mortality (Lipovsky and Chew 1972; Grischkowsky and Liston 1974). Recently, Friedman (University of California–Davis, personal communication) and Elston et al. (1987) examined the role of an actinomycete in the death of Pacific oysters. This condition, denoted Pacific oyster nocardiosis by Friedman, was first called "focal necrosis" by earlier researchers (Katkanski and Warner 1974). It remains to be seen whether bacteria or gonadogenic stress might independently or synergistically cause mortality among Pacific oysters.

Selective Breeding For Improved Survival

Selection and breeding for higher survival has been established for other oyster species and specific diseases (Ford and Haskin 1986; Haskin and Ford 1987). The course of our experiments in selective breeding of Pacific oysters at the University of Washington has its foundations in laboratory experiments on the nature of summer mortality conducted by Lipovsky and Chew (1972), who created mass mortality by duplicating the summer conditions of elevated temperature and nutrient concentrations. This method was adapted as a technique for selecting brood stock and for challenging progeny (Beattie et al. 1978, 1980) in work designed to explore the possibility of producing mortality-resistant stocks through selection and breeding (Hershberger et al. 1984). Our breeding work coincided with a severe summer mortality period that occurred in south Puget Sound from 1976 to 1982. During this time, progeny from our experiments were planted in bays with a known history of summer Pacific oyster mortality.

The results of these plantings proved promising. Some F_1 progeny exhibited superior survival, and these were selected as brood stock for later breeding. The first field challenge of the F_2 progeny occurred in 1982. Survival of Pacific oysters in full-sib experimental families ranged from 69 to 92%. Survival of most families was over 80%; some survived significantly better ($P \leq 0.05$) than the control stock (less than 60% survival). After 1982, the occurrence of severe mortalities subsided among all Pacific oysters throughout our study areas, making further field experiments on survival impossible.

Mortality, Carbohydrate Level, and Growth

Using data from several breeding seasons, we demonstrated conclusively that inbreeding among first-generation siblings had an adverse effect on growth (Beattie et al. 1987). Slow growth also may be related to a genetic interplay among the heritable factors of summer survival, growth, and carbohydrate level. Perdue (1983) found a positive correlation between survival and glycogen level among families of full-sib Pacific oysters. Pongthana (1987) showed an inverse relationship between growth and carbohydrate level among progeny of stocks selected for carbohydrate level. By deduction, we see that growth would be inversely related to survival; thus, stocks selected for high summer survival would, on the whole, be expected to exhibit slower growth. Of course, use of slower-growing stocks resistant to summer mortality might still be of use to oyster growers in the face of continuing mortality.

Strategies for Dealing with Mortality

During previous summer mortality periods, several schemes were proposed to deal with the incidence of summer mortality. The Japanese hardened seed by holding them high in the intertidal area through the winter with the expectation that the seed would produce hardier oysters that survive better. We do not believe hardening could be successful as a means to counteract any severe summer mortality. Before the use of hatchery seed became widespread, the Japanese traditionally hardened seed for shipment to the Pacific coast of the USA. Our experiments with selective breeding used hardened Japanese seed as the control, and Pacific oysters grown from this seed consistently exhibited the poorest survival.

The advent of hatcheries has changed Pacific oyster culture considerably in the USA, and at least one of these changes may reduce oyster mortality. In the years prior to the 1970s, when the industry essentially depended upon importation of Japanese seed, all planting was timed to the arrival of seed from Japan in mid-April to mid-May. Today, hatchery seed is available from early spring to early fall. By varying planting times, Pacific oyster growers have been able to maximize use of their growing areas and to actually increase productivity. Those growers who work in areas where summer mortality has occurred feel a manipulation of planting times will enable them to circumvent summer mortality. When Pacific oysters were planted in the spring, heavy mortality usually struck in the second summer, a few months prior to harvest. If seed is planted in the late summer, however, Pacific oysters survive the second summer and are harvested during the following winter or spring, having experienced no excessive mortality.

Knowledge of the etiology of the disease or diseases associated with summer mortality will greatly facilitate the search for a solution to this problem. C. S. Friedman and R. P. Hedrick (University of California–Davis) have successfully isolated and cultured a bacterium associated with field mortalities in Washington and British Columbia. This bacterium is gram-positive, acid-fast, branched, and beaded as typical for actinomycetes; analyses indicate that it is a species of *Nocardia*. Their current work focuses on infecting healthy Pacific oysters with the isolated pathogen and on cohabitation experiments. Success in this laboratory work will enable more succinct and reliable breeding studies to produce stocks of Pacific oysters resistant to summer mortality.

Summary

Summer mortality of Pacific oysters occurs in the warm enriched waters at heads of bays. The two current hypotheses on the cause of this mortality are stress related to gonadogenesis and bacteria-induced disease.

Experiments in selective breeding were successful in producing mortality-resistant stocks. There has been, in general, an inverse relationship between survival and growth.

Seed hardening does not effectively reduce summer mortality. Due to the extended period of hatchery seed availability, summer mortality may be circumvented by manipulation of planting time.

Work in defining the etiology of the associated pathogen will facilitate combatting summer mortality through improved breeding strategies and husbandry techniques.

References

Beattie, J. H., K. K. Chew, and W. K. Hershberger. 1980. Differential survival of selected strains of Pacific oysters (*Crassostrea gigas*) during summer mortality. Proceedings National Shellfisheries Association 70:184–189.

Beattie, J. H., W. K. Hershberger, K. K. Chew, C. Mahnken, E. F. Prentice, and C. Jones. 1978. Breeding for resistance to summertime mortality in the Pacific oyster (*Crassostrea gigas*). University of Washington, Washington Sea Grant Progress Report, WSG 78-3, Seattle.

Beattie, J. H., J. A. Perdue, W. K. Hershberger, and K. K. Chew. 1987. Effects of inbreeding on growth in the Pacific oyster (*Crassostrea gigas*). Journal of Shellfish Research 6:25–28.

Cardwell, R. D., S. Olsen, M. I. Carr, and E. W. Sanborn. 1979. Causes of oyster larvae mortality in south Puget Sound. NOAA (National Oceanic and Atmospheric Administration) Technical Memorandum ERL (Environmental Research Laboratories) MESA–39, Boulder, Colorado.

Elston, R. A., J. H. Beattie, C. Friedman, R. P. Hedrick, and M. L. Kent. 1987. Pathology and significance of fatal inflammatory bacteraemia in the Pacific oyster, *Crassostrea gigas* Thünberg. Journal of Fish Diseases 10:121–132.

Ford, S. E., and H. H. Haskin. 1986. Infection and mortality patterns in strains of oysters, *Crassostrea virginica*, selected for resistance to the parasite *Haplosporidian nelsoni* (MSX). Journal of Parasitology 73:368–376.

Glude, J. B. 1975. A summary report of the Pacific Coast oyster mortality investigations 1965–1972. Pages 1–28 *in* the Proceedings of the Third U.S.–Japan Meeting on Aquaculture. Japanese Government Fishery Agency and Japan Sea Regional Fisheries Research Laboratory, Special Publication, Tokyo.

Grischkowsky, R. S., and J. Liston. 1974. Bacterial pathogenicity in laboratory induced mortality of the Pacific oyster (*Crassostrea gigas*, Thunberg). Proceedings National Shellfisheries Association 64:82–91.

Haskin, H. H., and S. E. Ford. 1987. Breeding for disease resistance in molluscs. Pages 27–30 *in* Proceedings of the EIFAC/FAO symposium on selection, hybridization and genetic engineering in aquaculture of fish and shellfish for consumption and stocking, volume 2. Schriften der Bundesforschungsanstadlt für Fischerei Hamburg, Band 18/19.

Hershberger, W. K., J. A. Perdue, and J. H. Beattie. 1984. Genetic selection and systematic breeding in Pacific oyster culture. Aquaculture 39:237–245.

Imai, T., K. Numachi, J. Oizumi, and S. Sato. 1965. Search for the cause of mass mortality and the possibility to prevent it by transplantation experiments. Bulletin of the Tohoku Region Fisheries Research Laboratory 25:27–38.

Katkansky, S. C., and R. W. Warner. 1974. Pacific oyster disease and mortality studies in California. California Department of Fish and Game, Marine Resources Technical Report 25:22–23.

Koganezawa, A. 1975. Present status of studies on the mass mortality of cultural oysters in Japan and its prevention. Pages 29–34 *in* Proceedings of the Third U.S.–Japan meeting on aquaculture. Japanese Government Fishery Agency, and Japan Sea Regional Fisheries Research Laboratory, Special Publication, Tokyo.

Lipovsky, V. P., and K. K. Chew. 1972. Mortality of Pacific oysters (*Crassostrea gigas*): the influence of temperature and enriched seawater on oyster survival. Proceedings National Shellfisheries Association 62:72–82.

Mann, R. 1979. Some biochemical and physiological aspects of growth and gametogenesis in *Crassostrea gigas* and *Ostrea edulis* grown at sustained elevated temperatures. Journal of the Marine Biological Association of the United Kingdom 58:95–110.

Mori, K. 1979. Effect of artificial eutrification on the metabolism of the Japanese oyster, *Crassostrea gigas*. Marine Biology (Berlin) 53:361–369.

Mori, K., T. Imai, K. Toyoshima, and I. Usuki. 1965. Changes in the physiological activity and the glycogen content of the oyster during the stages of sexual maturation and spawning. Bulletin of the Tohoku Region Fisheries Research Laboratory 25:49–63.

Perdue, J. A. 1983. The relationship between the gametogenic cycle of the Pacific oyster, *Crassostrea gigas*, and the summer mortality phenomenon in strains of selectively bred oysters. Doctoral dissertation. University of Washington, Seattle.

Perdue, J. A., J. H. Beattie, and K. K. Chew. 1981. Some relationships between gametogenic cycle and the summer mortality phenomenon in the Pacific oyster, *Crassostrea gigas*, in Washington State. Journal of Shellfisheries Research 1:9–16.

Pongthana, N. 1987. Genetics of growth rate and carbohydrate content in the Pacific oyster *Crassostrea gigas* (Thunberg). Doctoral dissertation. University of Washington, Seattle.

Tamate, H., K. Numachi, K. Mori, O. Itikawa, and T. Imai. 1965. Studies on the mass mortality of the oyster in Matsushima Bay: pathological studies. Bulletin of the Tohoku Region Fisheries Research Laboratory 25:89–104.

American Fisheries Society Special Publication 18:269–280, 1988
© Copyright by the American Fisheries Society 1988

POTENTIAL RESEARCH TOOLS AND TECHNOLOGY

A Physiological Approach to the Study of Bivalve Molluscan Diseases

ROGER I. E. NEWELL

Horn Point Environmental Laboratories, Center for Environmental and Estuarine Studies
University of Maryland, Box 775, Cambridge, Maryland 21613, USA

BRUCE J. BARBER

Rutgers University, Shellfish Research Laboratory
Post Office Box 687, Port Norris, New Jersey 08349, USA

Abstract.—The life history, distribution, and pathological effects of the diseases and parasites of many species of bivalve molluscs have been well documented. In few of these studies, however, have the physiological or biochemical consequences to the host been elucidated. Moreover, little consideration has been paid to the relationship between environmental stress and disease susceptibility. A physiological approach can yield unique insights into the consequences of disease and parasitism in bivalve molluscs. This approach is especially relevant to studies of nonfatal diseases and the initial stages of fatal diseases, when the host is sublethally stressed. In such situations, the parasite may compromise the host's ecological fitness, even when standard histopathological techniques cannot identify large-scale deleterious effects. The techniques that we describe are designed to quantify effects on host physiological rate functions and enable predictions concerning effects on ecological fitness. We also review the literature for information in support of the hypothesis that bivalves that experience environmental stress suffer increased susceptibility to disease and parasitic invasion. Although we found little direct evidence to support this hypothesis, sufficient information is available to indicate that this area is an important one for future research.

Many species of bivalve molluscs are potential hosts for several genera of invertebrate pathogens. During the last century, the many investigations of these symbioses have elucidated various aspects of the geographical distribution and cellular occurrence of these parasites as well as some aspects of their life cycles (see the first section of this volume). Only few investigations, however, were related to the physiological basis of the interaction in terms of the consequences to the host in general and to the susceptibility of the host to infection in particular. In this paper, we identify physiological approaches that have been successful in previous studies of molluscan diseases. The most complete understanding of the causes and consequences of molluscan diseases can only be obtained when physiological techniques are employed in conjunction with more routine pathological and histological methods. This paper is, therefore, not intended to be a complete literature review of the diseases of bivalve molluscs, a subject which has already been reviewed by Lauckner (1983) and by others in this volume. Instead, we selected those studies that

exemplify the value of physiology in the study of bivalve–disease interactions.

To determine the effect of a host–parasite relationship, the degree to which the health of the host is impaired must be quantified. In those studies that have considered the health of the host, many have concentrated on the later stages when the host is obviously traumatized. This approach ignores the period of time before the infection becomes patent, a period which may be important because of its effect on the ecological fitness of the individual and, under epizootic conditions, on the population.

The approach of measuring physiological functions to quantify the relative health of marine bivalve individuals and populations was used successfully to study anthropogenic pollutants (Bayne et al. 1985). These existing techniques can readily be adapted to the study of the consequences of disease and parasitism on the ecological fitness of bivalve molluscs. The first part of this paper reviews some of the methods suitable for quantifying the degree to which a parasite stresses the host organism. Examples from the literature are given to demonstrate how such an approach can lead to a greater understanding of the host–parasite relationship.

The importance of environmental factors in regulating molluscan physiological processes is well documented (Newell 1979; Bayne and Newell 1983). However, the relationship between environmental stress and host–pathogen interaction is virtually unknown. The second part of this paper is concerned with how conditions of environmental stress, which lead to physiological dysfunction, might also lead to increased disease susceptibility. This could occur as the result of a general stress reaction, in which "scope for growth" is reduced and the energetic balance between host and pathogen favors the pathogen. Increased susceptibility might also result from stress-related debilitation of specific cellular and humoral defense mechanisms. The relationships between molluscan disease prevalence, mortality, and potential environmental stressors are thus examined as a basis for this concept.

Effects of Disease on Ecological Fitness

The most direct way to estimate the deleterious effect of a parasite on the ecological fitness[1] of a bivalve host is to measure the reduction in the rate of somatic and germinal growth. Several direct methods estimate growth, but the simplest involves weighing a living animal suspended in water (Andrews 1961). Because the density of bivalve tissue approximates that of water, effectively only the shell is weighed. If the same individual is weighed at intervals, its rate of shell growth can be related to the degree of parasitic infection, as subsequently determined histologically. Andrews (1961) successfully used this method to determine when specimens of the eastern oyster *Crassostrea virginica* had ceased to grow due to heavy infection of *Perkinsus marinus* ("dermo") (Figure 1). A disadvantage of the wet-weighing technique is that it only measures a change in the weight of the shell, which does not generally decrease in weight in response to poor physiological condition.

Condition index (CI) methods relate the tissue weight of the animal to some estimate of shell size, usually the internal shell volume, by the formula (Galtsoff 1964; Lawrence and Scott 1982),

---

[1]Ecological fitness may be defined and measured in several ways. Here, it means the organism's survival, growth, reproductive output, activity patterns, and resistance to predation, other diseases, and stressful conditions.

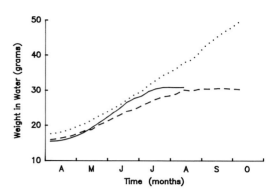

FIGURE 1.—Progressive weekly weights of three live eastern oysters weighed in water. One oyster (dotted line) exhibited normal growth throughout the growing season, whereas the other two ceased to gain weight due to infection with *Perkinsus marinus*. One oyster (solid line) died in mid-August and the other (broken line) survived until the following March. (Redrawn from Andrews 1961.)

$$CI = \frac{\text{dry tissue weight}}{\text{internal shell volume}} \times 100.$$

The advantage of this method over wet weighing is that actual losses in weight also can be measured, although increased mass of the parasite at the expense of the host tissue will mask actual changes in the weight of the host tissue. A disadvantage compared to the wet weighing method is that the animal has to be killed to obtain its dry tissue weight, and so the same individual cannot be monitored over time. However, in future studies, both methods can be used to determine the relationship between shell growth, tissue growth, and degree of parasitism.

Perhaps the most extensive use of condition indexes to quantify the effects of parasitism is the work of Kent (1979). He reported that the polychaete *Polydora ciliata*, which bores into the shell of the blue mussel *Mytilus edulis*, significantly reduced the condition index of the mussel. Over an annual cycle, mussels with a heavy infestation of *P. ciliata* generally had a significantly lower condition index (Figure 2).

Although both wet weighing and condition index methods have been used to study the effects of parasitism, they offer little insight into the actual mechanisms whereby parasites interfere with normal growth of the host. Energy-budgeting techniques, in which the animal's physiological rate functions are quantified in energy units, can be used to construct an instantaneous estimate of the energy the animal has available for growth (Crisp 1971;

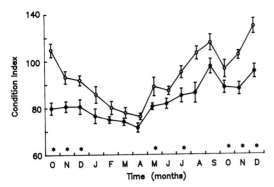

FIGURE 2.—Monthly changes in the mean (±SE) condition index (100 × dry tissue weight/internal shell volume) of blue mussels lightly (open circles) or heavily (solid circles) infested with the shell parasite *Polydora ciliata*. In months marked with an asterisk (*), the means were significantly different at $P < 0.05$. (Redrawn from Kent 1979.)

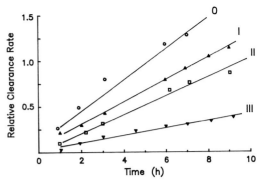

FIGURE 3.—Relative clearance rates of methylene blue by healthy Portuguese oysters (0) and individuals with infections of an iridovirus causing the "gill disease" syndrome. Portuguese oysters with stage-I infections had gills with slight damage on a single filament. Those with stage-II infections had gills with profound alterations on one or more filaments. Stage-III infections were associated with almost total degeneration of the entire gills. (Redrawn from Lauckner 1983 after His 1969.)

Bayne et al. 1985). The magnitude of a positive energy balance indicates the quantity of energy available for assimilation into somatic and germinal tissue or to be sequestered in nutrient reserves. Conversely, if there is a negative energy balance, the organism can survive only while it has nutrient reserves to utilize. Thus, the index of energy balance, or "scope for growth" (Bayne et al. 1976a, 1979b), is a sensitive method for rapidly quantifying parasite effects on bivalve growth potential, even before there is a measurable decline in condition index. This approach also provides information concerning the degree to which specific physiological functions are perturbed by a parasite. The methods for quantifying the physiological rate functions required to construct a balanced energy budget for molluscs in both field and laboratory have been fully described by Bayne et al. (1977, 1985).

The rate of bivalve food consumption can be one of the first physiological rate functions disrupted by a parasitic infection. The gills are frequently the initial site of parasite infection due to their intimate contact with the milieu. The integrated functioning of various types of gill cilia is required to provide a ventilatory water current and to capture food particles (Galtsoff 1964; Jorgensen 1966, 1981). In addition, lesions produced by the parasitic infections may stimulate mucus production which will interfere with normal gill function (Jorgensen 1966). Bang and Bang (1980) demonstrated that pathogenic bacteria stimulate mucus production in *Sipunculus nudus*, and physical disturbance of the gill stimulates mucus production in many bivalve molluscs (Jorgensen 1966).

Several bivalve species are host organisms for the parasitic pea crab *Pinnotheres* sp., which uses its pereiopods to attach to the gill filaments. Pearce (1966) reported substantial breakdown of the gill filaments in the northern horsemussel *Modiolus modiolus* as a result of the physical abrasion caused by the pea crab *Fabia subquadrata*. In blue mussels, there is a reduced clearance rate in pea crab-infested mussels, as measured by the clearance of neutral red dye particles (Progenzer 1979).

The effect of damage to the gill surface on the feeding activity of a bivalve is well illustrated by studies of an iridovirid virus infection, which causes "gill disease" in the Portuguese oyster *Crassostrea angulata*. Symptoms of the disease include deep indentations and pustules on the gill, which enlarge and eventually destroy entire filaments. His (1969) demonstrated that the severity of the gill infection is correlated with loss in tissue weight. This correlation was attributed to a depression of feeding activity (Figure 3), although the clearance rate of natural food particles was not measured.

The influence of the parasite *Haplosporidium nelsoni* (MSX) on the physiology of the eastern oyster was studied by Newell (1985), who found that even lightly MSX-infected eastern oysters had significantly depressed rates of feeding compared to noninfected individuals (Figure 4). The rates of oxygen consumption of infected and

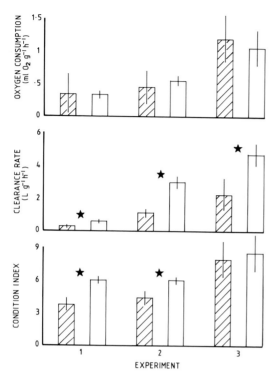

FIGURE 4.—Metabolic rates (oxygen consumption), feeding activities (clearance rate), and meat weights expressed as a function of shell size (condition index) of eastern oysters infected with the parasite *Haplosporidium nelsoni* (hatched bars), and of nonparasitized oysters (open bars). For each of the three experiments, mean values (±SE) are shown. Stars indicate significant differences between parasitized and nonparasitized oysters in paired groups ($P < 0.05$). (Data from Newell 1985.)

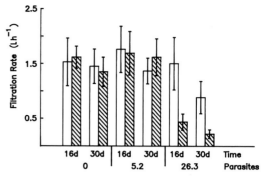

FIGURE 5.—Filtration rates (mean ± SD) of blue mussels, measured 16 and 30 d after the mussels were infected with an average number of 5.2 or 26.3 gut parasites, *Mytilicola intestinalis*, compared to uninfected control blue mussels (0 parasites). The mussels were maintained at either 10°C (open bars) or 22°C (hatched bars). (Redrawn from Bayne et al. 1979a.)

noninfected individuals did not differ significantly. Such a decline in feeding activity may account for the decreased condition indexes in MSX-infected eastern oysters observed by Andrews (1961), Newell (1985), and Barber et al. (1988). Several studies have reported that MSX plasmodia initially attack the animal via the gill epithelia, where they form lesions (for review see Farley 1968). From there, the parasites penetrate the basement membrane, cause a sloughing of gill and palp epithelia, and enter the hemolymph. Thus, the infection and disruption of the epithelia and ciliary function of the gills is the most likely explanation for the reduction in clearance rate in MSX-infected eastern oysters (Newell 1985).

Ward and Langdon (1986) found a significant reduction in tissue and shell growth of eastern oysters infested by the impressed odostome *Boonea impressa*, a gastropod ectoparasite. Previous researchers (reviewed by Ward and Langdon 1986)

attributed such a reduction of growth to the direct consumption of oyster tissue and nutrients by the parasite. However, Ward and Langdon (1986) calculated that less than 40% of the reduction in growth was due to direct consumption. They attributed most of the reduction in growth to the disruption of the oyster's normal feeding activity associated with the mechanical irritation of the parasite. Such important effects of parasitism can only be demonstrated by extensive physiological studies.

The intestinal copepod parasite *Mytilicola intestinalis* has frequently been implicated in the mass mortalities of the blue mussel (Lauckner 1983). Bayne et al. (1979a) found that the clearance rates of blue mussels heavily infested with *M. intestinalis* and maintained at high temperatures were significantly depressed (Figure 5). However, this effect was not permanent because normal levels of feeding were restored once the number of parasites was reduced. Although there was no measurable influence on the mussels' rate of oxygen consumption, the scope for growth of infected blue mussels at 22°C was significantly reduced as a consequence of the reduction in feeding activity. Thus, it was only under the highest parasite burdens in combination with another stress, such as elevated temperature or low food availability, that a deleterious influence on the mussels' physiology was recorded (Bayne et al. 1979a). This emphasizes the importance of combined stress factors in the deleterious effects of parasitism.

Disruption of the digestive process may be an important effect of *M. intestinalis* and some species of trematodes (Bucephalidae) that infect the digestive gland. The parasites may actually reduce the efficiency of the digestive process and compete with the host for nutrients. The evidence on the effects of parasites on digestion in molluscs is not clear (Lauckner 1983); in some studies no difference in carbon and nitrogen absorption was found (Davis and Farley 1973); in others, severe cellular (Moore and Halton 1977) and enzyme disruptions (Eble 1966) were discovered. Future emphasis should be placed on quantifying changes in food absorption efficiency, such as with the accurate dual radioisotope labeling technique (Calow and Fletcher 1972; Bricelj et al. 1984). This technique involves providing a ^{14}carbon-labeled food source (eg., cultured phytoplankton) mixed with ^{51}chromium. The chromium is not absorbed by the bivalve, and so the difference in the ratio ^{51}Cr:^{14}C between the food and biodeposits (i.e., combined feces and pseudofeces) is a measure of digested carbon. When the tissues of animals fed such a diet are prepared histologically for thin-layer autoradiography (Rogers 1966), the cellular location of the ^{14}C label in the host and parasite can be determined. If, in an appropriate time-series experiment, thin-layer autoradiography is performed on tissues from animals at various stages of parasitic infection, the competition between host and parasite for nutrients can be followed.

As stated above, a negative scope for growth implies that the animal must use its stored reserves to satisfy its metabolic energy requirements. The catabolism of protein can be quantified by calculating the ratio between oxygen consumed and nitrogen excreted (O:N, calculated in atomic equivalents). A high rate of protein catabolism relative to carbohydrate and lipid catabolism results in a low O:N ratio (Bayne et al. 1976b). Values below 30 for this ratio generally indicate that the animals are severely stressed (Bayne et al. 1976a, 1976b, 1979b, 1985). In some animals such as the blue mussel, however, seasonal changes in the catabolism of protein are associated with the reproductive and nutrient storage cycles. Therefore, data on the O:N ratio of parasitized bivalves must be compared to those of nonparasitized individuals in the same reproductive condition.

Energy-budgeting techniques can provide a sensitive measure of the degree to which a parasite disrupts the scope for growth of its host and of which specific functions are affected, but they yield no information as to whether somatic or germinal growth is reduced. Also, some parasites may not disrupt any of the measurable physiological rate functions but may only absorb nutrients or tissue

necessary for the host's gamete production. Reduced fecundity has obvious ecological implications. Trematode parasites of the family Bucephalidae are perhaps the best known examples of parasites that reduce the reproductive potential of bivalve molluscs (Lauckner 1983). Numerous species have been identified that infect the gonad, although some species primarily infect the digestive gland. Few studies have actually quantified the reduction in gamete production in bivalves infected with Bucephalus sp.; however, Cheng and Burton (1965) found that infected females had significantly smaller ova than noninfected individuals. A reduction in the size of bivalve eggs means that the quantity of nutrients in an egg, primarily lipids, is reduced (Sastry 1979; Bayne et al. 1978). Bayne et al. (1975, 1978) showed that decreased egg nutrient reserves can significantly reduce the viability of blue mussel larvae. Thus, there can be a deleterious effect on recruitment even if the total number of gametes is not reduced. The effects of the parasite may not be readily discernible in an individual but the parasite may have profound effects at the population level by reducing recruitment success. This subtle effect of parasitism may be more common than is thought.

Perhaps the most accurate technique to assess the influence of parasitism on the reproductive output of bivalves is the "gamete volume fraction" stereological method, in which the volume of germinal material is quantified from analysis of histological sections (Lowe et al. 1982). In some bivalves, such as oysters, this method is more difficult to use because the gonad does not form a discrete entity but surrounds and invaginates the visceral mass. For these bivalves, a different histological method is available in which the area of the germinal layer is expressed as a proportion of the visceral area (Mori 1979; Perdue et al. 1981). Barber et al. (1988) used this method to demonstrate that eastern oysters systemically infected with Haplosporidium nelsoni had significantly reduced germinal production. These quantification methods are superior to the "index" method of Chipperfield (1953), in which the reproductive condition of the animal was classed according to a subjective assessment of its reproductive stage.

Parasitism may interfere with the ability of the host organism to store nutrients. Studies by Cheng (1965) and Cheng and Burton (1966) documented a decrease in glycogen and lipid reserves of the eastern oyster parasitised by Bucephalus sp., and Barber (unpublished data) found that glycogen content in eastern oysters was reduced by *H. nelsoni* infection. Normally, these reserves supply the animal with nutrients for catabolism during winter dormancy and to initiate gametogenesis. Any decrease in these reserves may compromise both the survival of the individual during the winter and its reproductive capacity. Nutrient reserves may also be measured by the histochemical method outlined by Lowe et al. (1982); the size and number of cells in which glycogen is stored in the mantle connective tissues are quantified by a stereological technique similar to that mentioned above for assessing germinal output. The advantage of this method is that the specific type of cell that stores nutrients and that is affected by the parasite can be identified. Recently, Bayne et al. (1982) and Gabbott (1983) found that, at least in the blue mussel, nutrients sequestered for maintenance metabolism are stored in tissues and cells different from those used for production of gametes.

We discuss below the possible role of environmental stressors in increasing the susceptibility of a bivalve to parasitism, but it is equally important to consider how parasitism renders an animal more vulnerable to stress. Parasitized animals may have lower nutrient reserves, which can reduce their ability to survive winter dormancy when the animal is generally not feeding (Bayne and Newell 1983). Lunetta and Vernberg (1971) postulated that parasites alter the host's fatty acid composition, especially the longer-chain C_{20} fatty acids that are important in maintaining membrane integrity. Changes in the lipid composition of the membrane must take place in order to compensate for temperature-induced changes in the fluidity of the membrane. If a parasitized individual cannot alter the composition of its membranes, it will be less able to survive high- and low-temperature stress.

Cockles parasitized in the foot by a trematode were significantly less able to rebury themselves (Lauckner 1972). This crippling may interfere with the cockle's normal feeding process and render it more susceptible to predation by birds and crabs. In addition, parasitized bivalves frequently cannot close their valves completely (Lauckner 1983), and this inability makes them more susceptible to predation and prevents their isolation from the environment. Shell closure is the primary mechanism by which bivalves avoid desiccation from exposure to air and stress from abrupt changes in salinity.

In all physiological studies of the effects of parasitism on ecological fitness, there must be a coordinated histological investigation to quantify the intensity and site of the parasitic infection. For any group of animals, however, it is usually impossible to determine the exact degree and stage of parasitism before physiological measurements are made. This apparent problem can actually be a benefit because it leads to a double-blind experimental protocol in which neither the physiologist nor the histologist knows the outcome of the other's work until the results are combined and analyzed statistically. To quantify the true effects of a parasite, individuals with parasitic infections that range from none to severe must be sampled from within the same population. If uninfected animals cannot be obtained due to epizootic conditions, specimens from an external population must serve as controls. This procedure complicates the interpretation of the physiological data considerably because the time required to acclimatize to environmental conditions is largely unknown. Research has shown that it takes about 14 d for bivalve molluscs to acclimate to a temperature change (for review see Newell 1979; Bayne and Newell 1983). However, individuals collected from different habitats may have very different physiological and reproductive cycles (Newell et al. 1982), and may take over a year to acclimatize fully to uniform conditions (Newell, unpublished data). Therefore, data must be interpreted carefully from experiments in which the control (uninfected) animals are brought to the test site from a different location.

Effects of Stress on Disease Susceptibility

The concept that stress lowers resistance and makes the individual more susceptible to invading parasites and pathogens is intuitively attractive. Stress has been defined as "a measurable alteration of a physiological steady-state which is induced by an environmental change and which renders the individual more vulnerable to further environmental change" (Bayne et al. 1976a). An organism that experiences stress has less energy available for normal metabolic functions (including the processes associated with disease resistance) and would become more susceptible to parasitism and disease and subsequent deleterious effects. Some mass mortalities might arise in this way (Laird 1961).

Temperature is a major environmental determinant of metabolism and level of activity in invertebrates. Bivalves acclimate to normal seasonal changes in temperature by altering feeding and respiration rates so that their scope for growth is relatively stable (Newell 1979). However, high water temperatures can still impose a stress on bivalves and may be important in controlling their susceptibility to the effects of pathogens. "Summer disease" of Pacific oyster *Crassostrea gigas* in Willapa and Rocky bays, Washington, may be caused by a bacterial infection (Sparks 1981). Mortalities in late July and August are associated with elevated water temperature (Lipp et al. 1976; Beattie et al. 1988, this volume). Lipovsky and Chew (1972) found that Pacific oysters infected with suspected pathogenic bacteria (vibrios) die at a rate that correlates with temperature; all animals kept at 21°C died, but those kept at 9°C did not. Similarly, Feng (1966) found that experimental infections of bacteria caused mortality of eastern oysters at 23°C but not at 9°C. The softshell *Mya arenaria* is also susceptible to *Vibrio* sp. at elevated temperatures (20–22°C) (Tubiash 1971). Bacterial (vibrio) counts in Pacific oysters from Puget Sound, Washington, were directly correlated with water temperature (Baross and Liston 1968, 1970). In addition, healthy Pacific oysters exposed to vibrios isolated from dying Pacific oysters died at high temperatures (Lipp et al. 1976). High temperature was also responsible for increased levels of parasitism and mortality of oysters by *Perkinsus marinus* (Mackin 1962).

Levels of dissolved oxygen below saturation occur in lagoons, enclosed areas, muddy areas, and salt marshes, and in association with abnormally high levels of biological oxygen demand during so called "anoxic events." Hypoxia is experienced by intertidal organisms during periods of shell closure at low tide. Adaptive responses to environmental hypoxia include increased volumes of water passed over the gills, increased extraction efficiency of oxygen, and a general reduction in metabolic activity; those responses tend to maintain a constant scope for growth over a wide range of oxygen tensions. However, when the adaptive capacity is exceeded, scope for growth decreases (Bayne et al. 1985). Environmentally induced anoxia has been cited by Laird (1961) as responsible for increased disease susceptibility and mortality of eastern oysters and edible oysters *Ostrea edulis*.

Nutritive stress caused by low quantity or poor quality of food occurs seasonally in coastal areas (Widdows et al. 1979; Berg and Newell 1986). The amount of energy absorbed by a bivalve from ingested food has a direct influence on scope for growth. When adjustments in filtration rate and respiration rate cannot compensate for a decrease in the food supply, scope for growth decreases (Bayne et al. 1973, 1985). Bayne et al. (1979a)

reported that only blue mussels held at high temperatures or under low-ration conditions exhibit a significant depression of feeding activity associated with high infections with the copepod, *Mytilicola intestinalis* (Figure 5). These results imply that, under conditions of normal temperatures or high food availability, blue mussels can compensate for the effect of the parasite. When subjected to an additional stress, the capacity of the animals to compensate is exceeded.

Production of gametes in the majority of bivalve molluscs depends on a combination of stored nutrient reserves and incoming food. Spawning can cause a reduction in filtration rate (Newell and Thompson 1984), and postspawning animals are frequently depleted of energy reserves (Bayne et al. 1976a, 1982; Barber and Blake 1981). Thus, reproduction can be considered a stress because energy is diverted from somatic growth, maintenance, and defense-related functions. Mass mortalities of Pacific oysters in Matsushima Bay, Japan, characterized by "focal necrosis" (= fatal infectious bacteremia: Elston et al. 1987) were attributed to physiological stress associated with spawning and high water temperature (Mori et al. 1965a, 1965b; Imai et al. 1968; Beattie et al. 1988). Conversely, Sparks (1981) suggested that summer mortalities of Pacific oysters are probably of bacterial etiology and that physiological stress from gonad maturation and spawning plays a secondary role. However, Beattie et al. (1988) suggest that bacterial and physiological stresses may operate either synergistically or independently as causative agents of the summer mortality.

As discussed earlier, marine bivalves are parasitized by a variety of pathogens that stress the host to varying degrees. Although cause and effect are difficult to determine in hosts infected by more than one species of parasite, infection by *Cercaria cerastodermae* may increase the susceptibility of *Cardium edule* to other digenean species (Lauckner 1983).

Overcrowding (space competition) is mostly a concern in aquaculture, but density-dependent growth has also been found in natural populations (Broom 1982). Potential problems that arise from overcrowding include reduced food availability, increased levels of toxic waste products, and an increased likelihood of disease transmission, all of which increase stress levels and reduce scope for growth. Problems occur in hatcheries when overcrowding results in the proliferation of bacterial pathogens (Tubiash et al. 1965; Elston and Leibovitz 1980).

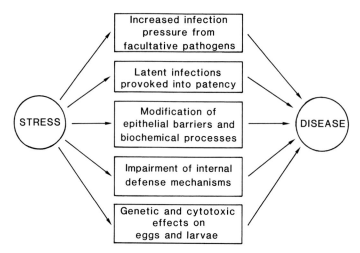

FIGURE 6.—Schematic of potential interactions between stress and disease in bivalve molluscs. (After Sinderman 1983.)

Anthropogenic stressors (pollutants) are becoming increasingly prevalent as human population pressures change estuarine and coastal waters where commercially important shellfish are grown. Sublethal responses to pollution include reduced filtration (feeding efficiency), elevated respiration rate, elevated rates of protein catabolism, increased glucose catabolism, and reduced reproductive output. Scope for growth decreases in proportion to levels of various pollutants in both laboratory and field studies (Bayne et al. 1979b, 1985; Widdows et al. 1981). Oil pollution has been associated with an increased prevalence of neoplasia in softshells (Barry and Yevich 1975; Yevich and Barszcz 1977; Reinisch et al. 1984). Hydrocarbon pollution has been implicated in increased infections of northern quahog *Mercenaria mercenaria* by *Polydora* sp. (Jeffries 1972). In addition, Barszcz et al. (1978) found that the number and types of parasites in oysters increased with exposure to crude oil. Infection of eastern oysters by *Polydora* sp. (Tinsman and Maurer 1986) and *Perkinsus marinus* (Quick 1971) increases as a consequence of thermal pollution.

Apparently, a variety of environmental stressors may reduce host resistance to disease-causing agents and mortality (Laird 1961; Sindermann 1983). Although the evidence is broadly based on energetic considerations, specific mechanisms are most likely involved (Figure 6). Environmental factors such as temperature and salinity potentially affect cellular defense mechanisms by altering hemocyte behavior. The rate of locomotion, ability to spread, and ability to adhere of hemocytes are retarded in eastern and edible oysters by increased salinity, in both acute and acclimated conditions (Fisher and Newell 1986; Fisher et al. 1987). Increased temperature generally increases hemocytic activity in eastern oysters up to the point at which high temperature stress occurs and causes decreased activity (Fisher 1988, this volume). Thus, environmental conditions that are considered to be within the normal tolerance range can affect hemocyte activity.

The relationship between immune response and chemical (pollution) stress in molluscs is also being investigated (see Anderson 1988, this volume). Northern quahogs exposed to several pollutants (hexachlorobenzene, pentachlorophenol, and benzo[a]pyrene) exhibited alterations in intracellular and serum lysozyme (a bactericidal enzyme) and impaired bacterial clearance (Anderson 1981). Similarly, chemical stress results in a depression of phagocytic response in northern quahogs (Fries and Tripp 1980). Thus, at least for this species, pollution stress can result in an altered host resistance through both cellular and humoral responses.

In summary, enough evidence exists to suggest that, although estuarine bivalves have a well-developed ability to compensate for changing environmental conditions, their adaptive capabilities can be exceeded with a consequent reduction in scope for growth. This may compromise their ability to successfully resist the effects of other stressors, including pathogens. Decreased resistance could be the result of debilitated cellular or humoral defense mechanisms. This concept is most applicable in cases (probably the majority) in which pathogen and host organisms have had time to coevolve. By weakening the host, environmental stress favors the pathogen. At the level of the individual, this relationship would account for the residual (low-level) population mortalities generally attributed to disease. If enough individuals were affected over a short time period, mass mortalities might result (Laird 1961).

In contrast, there are cases in which sudden mass mortalities occur in the absence of discernible environmental stress. Such mass mortalities might arise when a pathogen is first introduced to a susceptible host population or when a more virulent form of an existing pathogen appears. An example is the *Haplosporidium nelsoni* epizootic in Delaware and Chesapeake bays, in which populations of eastern oysters suffered high mortalities between 1957 and 1959 (Haskin et al. 1966; Andrews and Wood 1967). Prior to 1957, there was no evidence to suggest that oysters in these regions were stressed, and the extensive nature of the outbreaks makes it unlikely that localized environmental stressors were at work.

Conclusions

The ecological fitness of bivalve molluscs can be compromised by parasitic infections and diseases. Few studies have been designed to quantify the degree of stress to which affected individuals or populations have been subjected. Therefore, future physiological studies should be designed to quantify the effects of sublethal parasite levels on bivalve molluscs. Long-term and sublethal effects of parasitism on host reproduction, for example, may exert a more detrimental effect on a species' ecological fitness and recruitment than is generally recognized.

Care must be taken to select the experimental approach and methods appropriate for each parasite (e.g., for a gill parasite and a gonad parasite, appropriate assays would be changes in feeding rate and germinal output, respectively). In addition, physiological studies must be fully integrated with histological and pathological studies so that a comprehensive picture of parasite effects may be obtained.

Is a stressed organism more susceptible to parasitic infection or disease? Little direct evidence in the literature supports or refutes such a hypothesis, and more research is required. At the environmental level, relationships between potential stressors and the prevalence and intensity of disease need investigation. This research requires monitoring physical, chemical, and biological aspects of water quality as well as disease prevalence. Over time, it may be possible to detect whether particular environmental factors regulate rates of infection and mortality. At the organismal level, links between environmental stress levels (quantified by scope for growth) and resistance to parasitism and disease (measured by incidence and mortality) are needed. This examination could be made along a stress gradient in a natural setting or in laboratory experiments in which stress levels can be more closely controlled. More work is needed to establish the relationships between stress, host physiological response, and specific defense mechanisms. Once we better understand some of these relationships, improved shellfish production may be realized through careful management of stocks and, when practical, reduction of environmental stress.

Acknowledgments

This work was supported, in part, by the National Sea Grant College Program under grant NA84-AAD-00014 to R. I. E. Newell and T. W. Jones. This paper is Center for Estuarine and Environmental Studies publication 1905 and New Jersey Agricultural Experiment Station publication F-32504-1-88, supported by state funds. B. J. Barber received support from the New Jersey Fisheries and Aquacultural Technology Extension Center and the Bureau of Shellfisheries, New Jersey Department of Environmental Protection.

References

Anderson, R. S. 1981. Effects of carcinogenic and noncarcinogenic environmental pollutants on immunological functions in a marine invertebrate. Pages 319–331 *in* C. J. Dawe, J. C. Harshbarger, S. Kondo, T. Sugimura, and S. Takayama, editors. Phyletic approaches to cancer. Japan Scientific Societies Press, Tokyo.

Anderson, R. S. 1988. Effects of anthropogenic agents on bivalve cellular and humoral defense mechanisms. American Fisheries Society Special Publication 18:238–242.

Andrews, J. D. 1961. Measurement of shell growth in oysters by weighing in water. Proceedings National Shellfisheries Association 52:1–11.

Andrews, J. D., and J. L. Wood. 1967. Oyster mortality studies in Virginia. VI. History and distribution of *Minchinia nelsoni*, pathogen of oysters, in Virginia. Chesapeake Science 8:1–13.

Bang, B. G., and F. B. Bang. 1980. The urn cell complex of *Sipunculus nudus*: a model for study of mucus-stimulating substances. Biological Bulletin (Woods Hole) 159:571–581.

Barber, B. J., and N. J. Blake. 1981. Energy storage and

utilization in relation to gametogenesis in *Argopecten irradians concentricus* (Say). Journal of Experimental Marine Biology and Ecology 52:121–134.

Barber, B. J., S. E. Ford, and H. H. Haskin. 1988. Effects of the parasite MSX (*Haplosporidium nelsoni*) on oyster (*Crassostrea virginica*) energy metabolism. I. Condition index and relative fecundity. Journal of Shellfish Research 7:25–31.

Baross, J., and J. Liston. 1968. Isolation of *Vibrio parahaemolyticus* from the northwest Pacific. Nature (London) 217:1263–1264.

Baross, J., and J. Liston. 1970. Occurrence of *Vibrio parahaemolyticus* and related hemolytic vibrios in marine environments of Washington State. Applied Microbiology 20:179–186.

Barry, M. M., and P. P. Yevich. 1975. The ecological, chemical and histopathological evaluation of an oil spill site. Part III. Histopathological studies. Marine Pollution Bulletin 6(11):171–173.

Barszcz, C., P. P. Yevich, L. R. Brown, J. D. Yarbrough, and C. D. Minchew. 1978. Chronic effects of three crude oils on oysters suspended in estuarine ponds. Journal of Environmental Pathology and Toxicology 1:879–896.

Bayne, B. L., and nine coauthors. 1985. The effects of stress and pollution on marine animals. Praeger, Westport, Connecticut.

Bayne, B. L., A. Bubel, P. A. Gabbott, D. R. Livingstone, D. M. Lowe, and M. N. Moore. 1982. Glycogen utilization and gametogenesis in *Mytilus edulis* L. Marine Biology Letters 3:89–105.

Bayne, B. L., P. Gabbott, and J. Widdows. 1975. Some effects of stress in the adult on the eggs and larvae of *Mytilus edulis* L. Journal of the Marine Biological Association of the United Kingdom 55:675–689.

Bayne, B. L., J. M. Gee, J. T. Davey, and C. J. Scullard. 1979a. Physiological responses of *Mytilus edulis* L. to parasitic infestation by *Myticola intestinalis*. Journal du Conseil, Conseil International pour l'Exploration de la Mer 38:12–17.

Bayne, B. L., D. L. Holland, M. N. Moore, D. M. Lowe, and J. Widdows. 1978. Further studies on the effects of stress in the adult on the eggs of *Mytilus edulis*. Journal of the Marine Biological Association of the United Kingdom 58:825–841.

Bayne, B. L., M. N. Moore, J. Widdows, D. R. Livingstone, and P. Salkeld. 1979b. Measurement of the responses of individuals to environmental stress and pollution: studies with bivalve molluscs. Philosophical Transactions of the Royal Society of London B, Biological Sciences 286:563–581.

Bayne, B. L., and R. C. Newell. 1983. Physiological energetics of marine molluscs. Pages 407–515 *in* A. S. M. Saleuddin and K. M. Wilbur, editors. The Mollusca, volume 4. Academic Press, New York.

Bayne, B. L., R. J. Thompson, and J. Widdows. 1973. Some effects of temperature and food on the rate of oxygen consumption by *Mytilus edulis* L. Pages 181–193 *in* W. Wieser, editor. Effects of temperature on ectothermic organisms. Springer-Verlag, Berlin.

Bayne, B. L., R. J. Thompson, and J. Widdows. 1976a. Physiology: I. Pages 121–206 *in* B. L. Bayne, editor. Marine mussels: their ecology and physiology. Cambridge University Press, Cambridge, England.

Bayne, B. L., J. Widdows, and R. I. E. Newell. 1977. Physiological measurements on estuarine bivalve molluscs in the field. Pages 57–68 *in* B. K. Keegan, P. O'Ceidigh, and P. J. S. Boaden, editors. Biology of benthic organisms. Pergamon Press, New York.

Bayne, B. L., J. Widdows, and R. J. Thompson. 1976b. Physiological integrations. Pages 261–291 *in* B. L. Bayne, editor. Marine mussels: their physiology and ecology. Cambridge University Press, Cambridge, England.

Beattie, J. H., J. P. Davis, S. L. Downing, and K. K. Chew. 1988. Summer mortality of Pacific oysters. American Fisheries Society Special Publication 18:265–268.

Berg, J., and R. I. E. Newell. 1986. Temporal and spatial variations in the composition of seston available to the suspension feeder *Crassostrea virginica*. Estuarine, Coastal and Shelf Science 23:375–386.

Bricelj, V. M., A. E. Bass, and G. R. Lopez. 1984. Absorption and gut passage time of microalgae in a suspension feeder: an evaluation of the ^{51}Cr:^{14}C twin tracer technique. Marine Ecology Progress Series 17:57–63.

Broom, M. J. 1982. Analysis of the growth of *Anadara granosa* (Bivalvia: Arcidae) in natural, artificially seeded and experimental populations. Marine Ecology Progress Series 9:69–79.

Calow, P., and C. R. Fletcher. 1972. A new radiotracer technique involving ^{14}C and ^{51}Cr, for estimating the assimilation efficiencies of aquatic primary consumers. Oecologia (Berlin) 9:155–170.

Cheng, T. C. 1965. Histochemical observations on changes in the lipid composition of the American oyster *Crassostrea virginica* (Gmelin), parasitized by the trematode *Bucephalus* sp. Journal of Invertebrate Pathology 7:398–407.

Cheng, T. C., and R. W. Burton. 1965. Relationships between *Bucephalus* sp. and *Crassostrea virginica*: histopathology and sites of infection. Chesapeake Science 6:3–16.

Cheng, T. C., and R. W. Burton. 1966. Relationships between *Bucephalus* sp. and *Crassostrea virginica*: a histochemical study of some carbohydrates and carbohydrate complexes occurring in the host and parasite. Parasitology 56:111–122.

Chipperfield, P. N. J. 1953. Observations on the breeding and settlement of *Mytilus edulis* (L.) in British waters. Journal of the Marine Biological Association of the United Kingdom 34:449–476.

Crisp, D. J. 1971. Energy flow measurements. Methods for study of marine benthos. IBP (International Biological Programme) Handbook 16:197–279.

Davis, D. S., and J. Farley. 1973. The effect of parasitism by the trematode *Cryptocotyle lingua* (Creplin) on digestive efficiency in the snail host, *Littorina saxatilis* (Olivi). Parasitology 66:191–197.

Eble, A. F. 1966. Some observations on the seasonal distribution of selected enzymes in the American oyster as revealed by enzyme histochemistry. Proceedings National Shellfisheries Association 56:37–42.

Elston, R. A., and L. Leibovitz. 1980. Detection of vibriosis in hatchery reared larval oysters: correlation between clinical, histological and ultrastructural observations in experimentally induced disease. Proceedings National Shellfisheries Association 70:122–123.

Elston, R. A., J. H. Beattie, C. Friedman, R. Hedrick, and M. L. Kent. 1987. Pathology and significance of fatal inflammatory bacteraemia in the Pacific oyster, *Crassostrea gigas* Thunberg. Journal of Fish Diseases 10:121–132.

Farley, C. A. 1968. *Minchina nelsonia* (Haplosporidia) disease syndrome in the American oyster, *Crassostrea virginica*. Journal of Protozoology 15:585–599.

Feng, S. Y. 1966. Experimental bacterial infections in the oyster *Crassostrea virginica*. Journal of Invertebrate Pathology 8:505–511.

Fisher, W. S. 1988. Environmental influence on bivalve hemocyte function. American Fisheries Society Special Publication 18:225–237.

Fisher, W. S., N. Auffret, and G. Balouet. 1987. Response of European flat oyster (*Ostrea edulis*) hemocytes to acute salinity and temperature changes. Aquaculture 67:179–190.

Fisher, W. S., and R. I. E. Newell. 1986. Salinity effects on the activity of granular hemocytes of American oysters, *Crassostrea virginica*. Biological Bulletin (Woods Hole) 170:122–134.

Fries, C. R., and M. R. Tripp. 1980. Depression of phagocytosis in *Mercenaria* following chemical stress. Developmental and Comparative Immunology 4:233–244.

Gabbott, P. A. 1983. Developmental and seasonal metabolic activities in marine molluscs. Pages 165–217 *in* P. W. Hochachka, editor. The Mollusca, volume 2. Academic Press, New York.

Galtsoff, P. S. 1964. The American oyster *Crassostrea virginica* Gmelin. U.S. Fish and Wildlife Service Fishery Bulletin 64, Washington, D.C.

Haskin, H. H., L. A. Stauber, and J. A. Mackin. 1966. *Minchinia nelsoni* n. sp. (Haplosporida, Haplosporidiidae): causative agent of the Delaware Bay oyster epizootic. Science (Washington, D.C.) 153:1414–1416.

His, E. 1969. Recherche d'un test permetant de comparer l'activité respiratoire des huîtres au cours de l'évolution de la maladie des branchies. Revue des Travaux de l'Institut des Pêches Maritimes 33:171–175.

Imai, T., L. Mori, Y. Sugawara, H. Tamate, J. Oizumi, and O. Itikawa. 1968. Studies on the mass mortality of oysters in Matsushima Bay. VII. Pathogenetic investigation. Tohoku Journal of Agriculture Research 19:250–265.

Jeffries, H. P. 1972. A stress syndrome in the hard clam, *Mercenaria mercenaria*. Journal of Invertebrate Pathology 20:242–251.

Jorgensen, C. B. 1966. The biology of suspension feeding. Pergamon Press, Oxford, England.

Jorgensen, C. B. 1981. A hydromechanical principle for particle retention in *Mytilus edulis* and other ciliary suspension feeders. Marine Biology 61:277–282.

Kent, R. M. L. 1979. The influence of heavy infestations of *Polydora ciliata* on the flesh content of *Mytilus edulis*. Journal of the Marine Biological Association of the United Kingdom 59:289–297.

Laird, M. 1961. Microecological factors in oyster epizootics. Canadian Journal of Zoology 39:449–485.

Lauckner, G. 1972. Zur Taxonomie, Okologie und Physiologie von *Cardium edule* L. und *Cardium lamarcki* Reeve. Doctoral dissertation. University of Kiel, West Germany. (Not seen; cited in Lauckner 1983.)

Lauckner, G. 1983. Diseases of Mollusca: Bivalvia. Pages 477–962 *in* O. Kinne, editor. Diseases of marine animals, volume 2. Biologische Anstalt Helgoland, Hamburg, West Germany.

Lawrence, D. R., and G. I. Scott. 1982. The determination and use of condition index of oysters. Estuaries 5:23–27.

Lipovsky, V. P., and K. K. Chew. 1972. Mortality of Pacific oysters (*Crassostrea gigas*): the influence of temperature and enriched seawater on oyster survival. Proceedings National Shellfisheries Association 62:72–82.

Lipp, P. R., B. Brown, J. Liston, and K. K. Chew. 1976. Recent findings on the summer diseases of Pacific oysters. Proceedings National Shellfisheries Association 65:9–10.

Lowe, D. M., M. N. Moore, and B. L. Bayne. 1982. Aspects of gametogenesis in the marine mussel *Mytilus edulis*. Journal of the Marine Biological Association of the United Kingdom 62:133–140.

Lunetta, J. E., and W. B. Vernberg. 1971. Fatty acid composition of parasitized and non-parasitized tissue of the mud-flat snail, *Nassarius obsoleta* (Say). Experimental Parasitology 30:244–248.

Mackin, J. G. 1962. Oyster disease caused by *Dermocystidium marinum* and other microorganisms in Louisiana. Publications of the Institute of Marine Science, University of Texas 7:132–229.

Moore, M. N., and D. W. Halton. 1977. The cytochemical localization of lysosomal hydrolases in the digestive cells of littorinids and changes induced by larval trematode infection. Zeitschrift für Parasitenkunde 53:115–122.

Mori, K. 1979. Effects of artificial eutrophication on the metabolism of the Japanese oyster *Crassostrea gigas*. Marine Biology 53:361–369.

Mori, K., T. Imai, K. Toyoshima, and I. Usuki. 1965a. Studies on the mass mortality of the oyster in Matsushima Bay. IV. Changes in the physiological activity and the glycogen content of the oyster during the stages of sexual maturation and spawning. Bulletin of Tohoku Regional Fisheries Research laboratory 25:49–63.

Mori, K., H. Tamate, T. Imai, and O. Itikawa. 1965b. Studies on the mass mortality of the oyster in Matsushima Bay. V. Changes in the metabolism of lipids and glycogen of the oyster during the stages of sexual maturation and spawning. Bulletin of

Tohoku Regional Fisheries Research Laboratory 25:65–88.

Newell, R. C. 1979. Biology of intertidal animals. Marine Ecological Surveys, Faversham, England.

Newell, R. I. E. 1985. Sublethal physiological effects of the parasite MSX (*Haplosporidium nelsoni*) on the oyster *Crassostrea virginica*. Journal of Shellfish Research 5:91–95.

Newell, R. I. E., T. J. Hillish, R. K. Koehn, and C. J. Newell. 1982. Temporal variation in the reproductive cycle of *Mytilus edulis* L. (Bivalvia, Mytilidae) from localities on the east coast of the United States. Biological Bulletin (Woods Hole) 162:299–310.

Newell, R. I. E., and R. J. Thompson. 1984. Reduced clearance rates associated with spawning in the mussel, *Mytilus edulis* (L.) (Bivalvia, Mytilidae). Marine Biology Letters 5:21–33.

Pearce, J. B. 1966. The biology of the mussel crab *Fabia subquadrata*, from the waters of the San Juan Archipelago, Washington. Pacific Science 20:3–35.

Perdue, J. A., J. H. Beattie, and K. K. Chew. 1981. Some relationships between gametogenic cycle and summer mortality phenomenon in the Pacific oyster (*Crassostrea gigas*) in Washington State. Journal of Shellfish Research 1:9–16.

Progenzer, C. 1979. Effect of *Pinnotheres hickmani* on neutral red clearance by *Mytilus edulis*. Australian Journal of Marine and Freshwater Research 30: 547–550.

Quick, J. A. 1971. Pathological and parasitological effects of elevated temperatures on the oyster *Crassostrea virginica* with emphasis on the pathogen *Labyrinthomyxa marina*. Florida Department of Natural Resources, Marine Research Laboratory Professional Papers Series 15:105–171.

Reinisch, C. L., A. M. Charles, and A. M. Stone. 1984. Epizootic neoplasia in soft shell clams collected from New Bedford Harbor. Hazardous Waste 1:73–81. (Mary Ann Liebert, Incorporated, New York.)

Rogers, A. W. 1966. Techniques of autoradiography. Elsevier, London.

Sastry, A. N. 1979. Pelecypoda (excluding Ostreidae). Pages 113–292 *in* A. C. Giese and J. S. Pearse, editors. Reproduction of marine invertebrates. Academic Press, New York.

Sindermann, C. J. 1983. An examination of the relationships between pollution and disease. Rapports et Procès-Verbaux des Réunions, Conseil International pour l'Exploration de la Mer 182:37–43.

Sparks, A. K. 1981. Bacterial diseases of invertebrates other than insects. Pages 323–363 *in* E. W. Davidson, editor. Pathogenesis of invertebrate microbial diseases. Allanheld and Osmun, Totowa, New Jersey.

Tinsman, J. C., and D. Maurer. 1986. The relationship between disease in *Crassostrea virginica* (Gmelin) and thermal effluents in the Chesapeake–Delaware Bay area. Internationale Revue der gesamten Hydrobiologie 71:495–509.

Tubiash, H. S. 1971. Soft-shell clam, *Mya arenaria*, a convenient laboratory animal for screening pathogens of bivalve mollusks. Applied Microbiology 22: 321–324.

Tubiash, H. S., P. E. Chanley, and E. Leifson. 1965. Bacillary necrosis, a disease of larval and juvenile bivalve mollusks. I. Etiology and epizootiology. Journal of Bacteriology 90:1036–1044.

Ward, J. E., and C. J. Langdon. 1986. Effects of the ectoparasite *Boonea* (= *Odostomia*) *impressa* (Say) (Gastropoda: Pyramidellidae) on the growth rate, filtration rate and valve movements of the host *Crassostrea virginica* (Gmelin). Journal of Experimental Marine Biology and Ecology 99:163–180.

Widdows, J., P. Fieth, and C. M. Worrall. 1979. Relationships between seston, available food and feeding activity in the common mussel, *Mytilus edulis*. Marine Biology 50:295–207.

Widdows, J., D. K. Phelps, and W. Galloway. 1981. Measurement of physiological condition of mussels transplanted along a pollution gradient in Narragansett Bay. Marine Environmental Research 4:181–194.

Yevich, P. P., and C. A. Barszcz. 1977. Neoplasia in soft-shell clams (*Mya arenaria*) collected from oil-impacted sites. Annals of the New York Academy of Sciences 298:409–426.

American Fisheries Society Special Publication 18:281–285, 1988

Cell Separation by Centrifugal Elutriation

Evelyne Bachère, Dominique Chagot, and Henri Grizel

Institut Français de Recherche pour l'Exploitation de la Mer
Laboratoire de Pathologie et de Génétique des Invertébrés Marins
17390 La Tremblade, France

Abstract.—There is an ongoing need to isolate and purify cells in bivalve mollusc disease research. Many separation techniques differentiate cells solely by a single factor and usually require additives in the medium that may damage live cells. Centrifugal elutriation is a technique that separates particles by size and density simultaneously and can be conducted in physiological media without additives. Living cells can be separated without chemical damage and, because the procedure can be conducted aseptically, cells can be maintained in vitro after elutriation. A description of the technique, methods to determine flow rates and rotor speeds, and a preliminary separation of oyster hyalinocytes is presented.

In molluscan pathology and immunology, methods to isolate and purify cells are needed, be they hemocytes involved in the immunity processes, cells susceptible to infection by parasites, or the parasites themselves. These methods must be simultaneously quantitative and qualitative and must preserve the structural and functional integrity of the cells for in vitro studies.

Classical methods of cell isolation are generally based on differential and isopycnic centrifugations in a density (pressure) gradient. The techniques are efficient but present some disadvantages related to the chemical nature of the products used (sucrose, cesium chloride, Percoll, meglumine diatrizoate, metrizamide) and the physicochemical changes they induce (e.g., in ionic force, osmotic pressure, or viscosity) at the concentrations used (Rickwood 1984). Moreover, these methods generally require several stages of centrifugation and collection.

An alternative method, counterstreaming centrifugation, was originated by Lindahl (1948) and later developed by Beckman Instruments under the name of centrifugal elutriation (McEwen et al. 1968). This procedure, which can eliminate the disadvantages previously cited, has been used to prepare different vertebrate cells, e.g., whole blood and hemopoietic cells, cells from brain and liver, transformed and tumor cells, cells from testis and ovary (Pretlow and Pretlow 1979), and also to purify protozoan parasites such as *Plasmodium* sp. (Russman et al. 1982) or *Eimeria* sp. (Stotish et al. 1977). This technique is efficient because it is rapid and because a large quantity of cells can be purified at once.

Principles of Centrifugal Elutriation

Centrifugal elutriation is a form of velocity sedimentation in which cells are separated according to their rates of sedimentation. Separation is proportional to the size of the cells and to the difference between the densities of the cells and the medium. Thus, velocity sedimentation is particularly useful in separating cells of the same density (Pretlow and Pretlow 1982). In centrifugal elutriation, cells are exposed to a centrifugal force while suspended in a selected medium which flows continuously in a centripetal direction (Pretlow and Pretlow 1979). The Beckman Instruments system designed for this technique is shown schematically in Figure 1.

In the elutriation chamber, each cell tends to migrate to a zone where its sedimentation rate is exactly balanced by the flow rate of the fluid. Because the geometry of the chamber produces a gradient of flow rates from one end to the other, cells with a wide range of sedimentation rates can be held in suspension. By increasing the flow rate of the elutriating fluid in steps, or by decreasing the rotor speed, successive populations of homogeneous cell sizes can be washed from the chamber (Figure 2). The major theoretical aspects of elutriation were developed by Lindahl (1948), McEwen et al. (1968), Pretlow et al. (1975) and Sanderson et al. (1976).

Elutriation Methodology

Meistrich (1983) defined and studied the various methodological factors of centrifugal elutriation.

Preparation of Single-Cell Suspension

Some cell systems, such as those of peripheral blood, are already in single-cell suspension and may require only treatment to avoid aggregation (Mackler et al. 1977). However, cells in organs and tissues have to be dissociated, either by

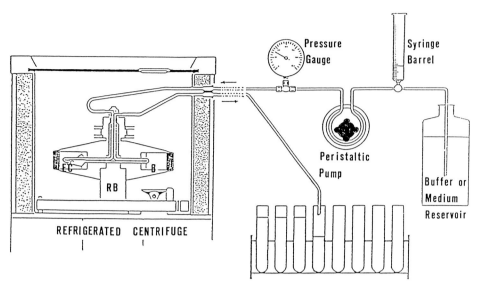

FIGURE 1.—The elutriator system with the loading setup. EC = elutriation chamber; RB = rotor body; B = bypass.

gentle mechanical methods or by different enzymes (trypsin, hyaluronidase, etc.) sometimes together with EDTA treatment (Meistrich 1972).

Temperature Choice and Regulation

The choice of temperature for elutriation should be compatible with cell physiology. Generally, 4°C is recommended to lessen the phenomenon of aggregation. However, Sanderson et al. (1977) claimed that aggregation of mouse spleen cells was greater at 4°C than at 30°C and that temperature consistency was a more important factor.

Choice of Medium

Any buffer and culture medium can be used for elutriation. The essential criterion in the choice of buffer or medium is its capacity to maintain the physiological integrity of the cells and to prevent cell aggregation.

Loading System

The Beckman Instruments system for loading cells into the rotor is designed to avoid passing cells through the peristaltic pump. However, this system is cumbersome because the loading time is long and cell loss is not negligible (Meistrich 1983). In a simpler loading system (Figure 1), such as one by Meistrich (1983), cells pass through the peristaltic pump but without apparent cell damage. A three-way valve with a syringe barrel is inserted between the buffer reservoir and the

pump. A small volume of cells in suspension is poured into the barrel and then pumped into the rotor. This loading system reduces the loading time and the formation of aggregates. A wide range of cells, from less than 10^7 to more than 10^9 mL^{-1}, can be used, but the number will vary according to the nature and volume of the cells.

Flow Rates and Rotor Speeds

Cell volume and distribution are the only factors that need to be known before new cell types are separated by elutriation. Cell volume can be used to compute the sedimentation rate (S, mm/h) by a generalized form of Stokes' law (Wyrobek et al. 1976),

$$S = (\rho c - \rho o)V^{2/3}a/\eta f;$$

ρc and ρo are the densities of the cell and the suspending medium, respectively; V is the cell volume; a is the acceleration due to centrifugation; η is the viscosity of the fluid; and f is a coefficient dependent only on the particle shape.

For the standard Beckman chamber, and given $\rho c - \rho o = 0.05$ g/cm^3, $\eta = 1.43 \times 10^{-2}$, and $f = 11.7$, the preceding equation can then be simplified to

$$S/a = 0.105\ V^{2/3}.$$

The sedimentation time S/a can thus be predicted solely on the basis of cell volume (V), which can be estimated either by Coulter counter

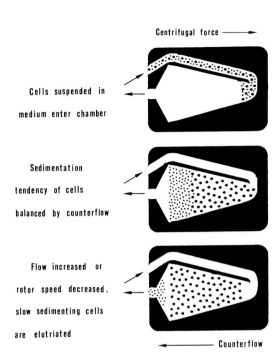

Centrifugal force ⟶

Cells suspended in
medium enter chamber

Sedimentation
tendency of cells
balanced by counterflow

Flow increased or
rotor speed decreased,
slow sedimenting cells
are elutriated

⟵ Counterflow

FIGURE 2.—Elutriation process within the separation chamber. The elutriation actually occurs at the wide point of the chamber. The cell population collected will contain cells that are larger or more dense than those of the previous fraction. (From Beckman Instruments applications data.)

or by microscopic measurements. Rotor speed k (1,000 revolutions/min) and flow rate r (mL/min), can be chosen to determine elutriation conditions corresponding to the point of elutriation, i.e., $S/a = 1.93r/k^2 = 0.105 \, V^{2/3}$. Thus, when separation of a new cell mixture is desired, values of r and k are chosen for the smallest and largest cells in the population. If there are wide ranges of cell size and sedimentation rate in a given population, two runs will be needed to obtain cell separation.

Collection of Cells

Cells can be eluted by changing either the flow rate or the rotor speed. In some cases, alternately decreasing rotor speed and increasing flow rate during the same run may be best for collecting the different fractions.

Separation of Oyster Hyaline Cells

Ultrastructural studies of hyaline cells in molluscs have shown considerable polymorphism, the ontogenic and physiological significance of

which is not well known. To determine the specificities of different morphological types, we conducted separation experiments of hemocytes from the Pacific oyster *Crassostrea gigas* using the technique of elutriation.

In the methodology developed for these experiments, the hemolymph is withdrawn by intrapericardiac puncture and diluted 2:3 in Alsever's solution (glucose, 20.8 g/L; sodium citrate, 8 g/L) to which is added EDTA (3.36 g/L) as a chelating agent and Tween 80 (0.1%) as an emulsifier. Sodium chloride is used to make the solution isotonic for the hemocytes; osmotic pressure is adjusted to 1,000 mosmol by means of a cryoosmometer. Under these conditions, cell aggregation is substantially avoided.

The hemocyte suspension is submitted to isopycnic centrifugation in a Percoll discontinuous gradient to separate hyalinocytes and granulocytes by their density differences. A homogeneous population of granulocytes is distributed in high-density fractions (60–70%); the less-dense fractions (10%, 20%, 30%) contain a heterogeneous population of hyaline cells. We hope that this potentially harmful step with Percoll can be eliminated in the future.

Centrifugal elutriation is performed to separate hyalinocyte subpopulations on the basis of their size differences. The elutriation medium is Alsever's solution (12°C), and the rotor is equipped with a Sanderson chamber. Hyalinocytes are eluted in eight fractions according to the nomogram presented in Figure 3. This protocol is effective and reproducible; cell recovery is about 60%. Nearly 90% of the cells fractioned in this manner were viable when tested by the trypan blue exclusion method. Cell integrity was verified by examination with an electron microscope.

The ultrastructural features of elutriated cells led to the identification by size and morphology of an evolutive series. This series was characterized in the first fractions by cells measured 5–6 μm and having a high nucleocytoplasmic ratio and rough endoplasmic reticulum, often located near the nucleus. In the last fractions collected, the hemocytes measured 11–14 μm and had a low nucleocytoplasmic ratio, numerous small vesicles within the cytoplasm, a considerable system of lamellae and ergastoplasmic cisternae, and some glycogen granules.

In vitro maintenance of these elutriated cells revealed that their capacity for adhesion and displacement remained intact. Moreover, long-term in vitro culture can be attempted because the

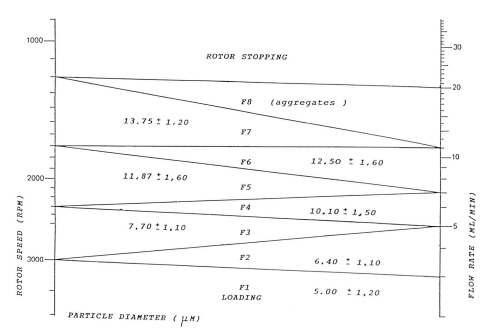

FIGURE 3.—Nomogram of the elutriation of Pacific oyster hyalinocytes. Each successive fraction (F) yields a larger particle diameter as the rotor speed is decreased or the flow rate is increased. RPM = revolutions/min.

system can be sterilized with 70% ethanol prior to an aseptic separation run.

Discussion

The method of elutriation to separate viable hyaline cells of Pacific oysters is new to malacology; it makes possible certain investigations in molluscan immunology and pathology. The ability to obtain consistently homogenous populations of hyalinocytes that can be maintained in vitro should enable us to investigate their biochemical characteristics and their functions in immune mechanisms. Furthermore, this method is now available for the study of cells of other species of bivalves.

Bivalve hemocytes are associated with diseases caused by viruses, e.g., iridovirus in *Crassostrea angulata* (Comps and Duthoit 1976) and retrovirus in softshells *Mya arenaria* (Oprandy and Chang 1983), and by protozoa, e.g., *Bonamia ostreae* in edible oysters *Ostrea edulis* (Pichot et al. 1980). Because these diseases are economically important, we need to understand the relationships of host and parasite at the cellular level. We presently are studying the relationships between oyster hemocytes and the pathogen *B. ostreae*.

References

Comps, M., and J. L. Duthoit. 1976. Infection virale associé à la maladie des branchies de l'huître por-

tugaise *Crassostrea angulata* Lmk. Comptes Rendus Hebdomadaires des Séances de l'Académie des Sciences (Paris) 283:1595–1597.

Lindahl, P. E. 1948. Principle of a counterstreaming centrifuge for the separation of particles of different sizes. Nature (London) 161:648–649.

Mackler, B. F., P. A. O'Neill, and M. L. Meistrich. 1977. T lymphocyte induction of non T cell-mediated non-specific cytotoxicity. I. Induction mechanisms. European Journal of Immunology 7:55–61.

McEwen, C. R., R. W. Stallard, and E. T. Juhos. 1968. Separation of biological particles by centrifugal elutriation. Analytical Biochemistry 23:369.

Meistrich, M. L. 1972. Separation of mouse spermatogenic cells by velocity sedimentation. Journal of Cellular Physiology 80:299–312.

Meistrich, M. L. 1983. Experimental factors involved in separation by centrifugal elutriation. Pages 33–61 *in* T. G. Pretlow and T. P. Pretlow, editors. Cell separation: methods and selected applications. Academic Press, New York.

Oprandy, J. J., and P. W. Chang. 1983. 5-bromodeoxyuridine induction of hematopoietic neoplasia and retrovirus activation in the soft-shell clam, *Mya arenaria*. Journal of Invertebrate Pathology 42:196–206.

Pichot, Y., M. Comps, G. Tigé, H. Grizel, and M. A. Rabouin. 1980. Recherches sur *Bonamia ostreae* gen.n., sp.n., parasite nouveau de l'huître plate *Ostrea edulis* L. Revue des Travaux de l'Institut des Pêches Maritimes 43(1):131–140.

Pretlow, T. G., and T. P. Pretlow. 1979. Centrifugal

elutriation (counterstreaming centrifugation) of cells. Cell Biophysics 1:195–210.

Pretlow, T. G., and T. P. Pretlow. 1982. Sedimentation of cells: an overview and discussion of artifacts. Pages 41–57 in T. G. Pretlow and T. P. Pretlow, editors. Cell separation. Academic Press, New York.

Pretlow, T. G., E. E. Wier, and J. G. Zettergren. 1975. Problems connected with the separation of different kinds of cells. International Review of Experimental Pathology 14:91–204.

Rickwood, D. 1984. The theory and practice of centrifugation. Pages 1–43 in D. Rickwood, editor. Centrifugation, a practical approach. IRL Press, Oxford, England.

Russmann, L., A. Jung, and H. G. Heidrich. 1982. The use of Percoll gradients, elutriation rotor elution and mithramycin staining for the isolation and identification of intraerythrocytic stages of *Plasmodium berghei*. Zeitschrift für Parasitenkunde 66:273–280.

Sanderson, R. J., K. E. Bird, N. F. Palmer, and J. Brenman. 1976. Design principles for a counterflow centrifutation cell separation chamber. Analytical Biochemistry 71:615–622.

Sanderson, R. J., F. T. Shepperdson, A. E. Vatter, and D. W. Talmage. 1977. Isolation and enumerations of peripheral blood monocytes. Journal of Immunology 118:1409–1414.

Stotish, R. L., P. M. Simashkevich, and C. C. Wang. 1977. Separation of sporozoites, sporocysts and oocysts of *Eimeria tenella* by centrifugal elutriation. Journal of Parasitology 63:1124–1126.

Wyrobeck, A. J., M. L. Meistrich, R. Furrer, and W. R. Bruce. 1976. Physical characteristics of mouse sperm nuclei. Biophysical Journal 16:811–825.

American Fisheries Society Special Publication 18:286–291, 1988

Flow Cytometry: A Tool for Cell Research in Bivalve Pathology

WILLIAM S. FISHER[1]

University of Maryland, Center for Environmental and Estuarine Studies
Horn Point Environmental Laboratories
Cambridge, Maryland 21613, USA

SUSAN E. FORD

Rutgers University Shellfish Research Laboratory
Port Norris, New Jersey 08349, USA

Abstract.—Flow cytometry employs a fluid system to pass individual cells through a focused beam of light, and it measures the reflected or emitted light with optical detectors. This technique can be used to simultaneously measure size, shape, index of refraction, and auto- or labelled fluorescence of the individual cells. This method provides an integrated picture of the characteristics of a single cell rather than an estimate derived from independently measured population averages. Because a flow cytometer can also physically sort cells on the basis of measured characteristics without harming their viability, in vitro assays of sorted cell subpopulations are possible. Flow cytometry may eventually be applied to disease-related research to quantify sarcomatous lesions, to determine morphological differences between hemocyte types pertaining to their functional roles in defense, to separate parasites from host tissue for in vitro study, and to quantify specific cell characteristics with fluorescent probes.

Flow cytometry has been used in mammalian biomedical research to analyze heterogenous populations of cells (Herzenberg et al. 1976; Horan and Wheeless 1977). The following overview of the technical concepts of flow cytometry was condensed from Phinney et al. (1987).

Several types of flow cytometers are available for specific research needs (Melamed et al. 1979; Shapiro 1985), but all of them share some basic concepts. The principle of flow cytometry is the passage of individual cells in a fluid through focused beams of illumination and the measurement of the reflected or emitted light from each cell with optical detectors. A major advantage of flow cytometry is the ability to rapidly and simultaneously measure size, shape, index of refraction, and fluorescence of individual cells. This capability provides an integrated picture of the characteristics of each cell rather than an estimate derived from population averages. Consequently, absolute correlation of single-cell characteristics is realizable.

Each cell and three separate technical systems (fluid, illumination, and detection) converge at a single site, the interrogation point (Figure 1A). The fluid system transports the cells in suspension to the interrogation point in a single file. This arrangement is accomplished by injecting the suspended cells into the center of a flow chamber with faster-moving, laminar flow sheath fluid. There is no mixing between the sheath fluid and the sample fluid, which passes directly through the interrogation point. If the concentration of cells in the sample fluid is maintained at 10^5–10^7 cells/mL and the velocity of the cells (dependent on the pressure differential between sample and sheath fluid reservoirs) is 1–10 m/s, samples may be assayed at a rate of 1,000 cells/s; each cell remains only 2–3 μs at the interrogation point.

The illumination system is the source of energy for (1) light scattering by the cells and (2) excitation of natural pigments or introduced fluorescent stains within the cells as they pass through the interrogation point. Cellular measurements vary according to the stability, wavelength, intensity, beam resolution (spot size), and power distribution (spot shape) of an illumination source. Two types of illumination are used in flow cytometry, arc lamps and lasers, each with qualities that make them preferable under different circumstances (Peters 1979). Arc lamps give less intense light and are less stable but are inexpensive to purchase and operate. Arc lamps may be sufficient for some procedures, such as those that use stains to enhance cell characteristics. Lasers are used for small particles or low fluorescence. Single-laser systems are usually capable of simulta-

[1]Present address: Marine Biomedical Institute, University of Texas Medical Branch, 200 University Boulevard, Galveston, Texas 77550, USA.

A   SINGLE CELL ANALYSIS

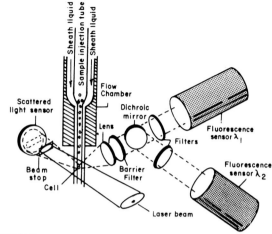

B   CELL SORTING

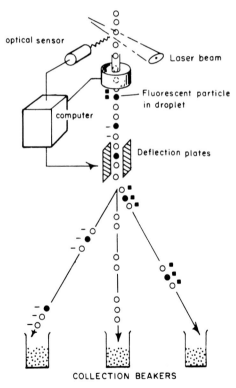

FIGURE 1.—**A.** Diagram of the fluid, illumination, and detection systems of a flow cytometer converging at the point of interrogation at a cell. Optical sensors detect light emitted or reflected from sample particles, which are transported to the point of interrogation by a stream of fluid in an injection tube. Illumination is provided by lasers or arc lamps. **B.** Diagram of the sorting apparatus in a flow cytometer. Drops of fluid are formed as the sample exits from the injection tube, and flow rates are adjusted so that each drop has a maximum of one particle in it. Just as the drop exits the tube, it is electrically charged according to the particular variable to be measured. Charged deflection plates pull the droplets from the line of flow into separate receptacles. (From Yentsch et al. 1983, courtesy of the Bigelow Laboratory for Ocean Sciences.)

neously measuring cell size (forward light scatter), orthagonal light scatter, and two bands of fluorescence. Multiple-laser beam instruments permit simultaneous measurements of light scatter at various wavelengths and more than two bands of fluorescence. Several optical filters and devices to pass or block bands of specific wavelengths are available to manipulate the illuminating beam to desired parameters. Cell volume may also be estimated without illumination, by means of impedance (Coulter) volume sensing.

The detection system is tightly integrated with the wavelength and intensity of illumination. Photodiodes may be used to detect high-intensity emissions such as forward-angle light scatter, but more sensitive photomultiplier tubes must be used for low levels of light emitted from fluorescence or orthagonal light scatter. Optical filters are necessary to discriminate and eliminate background and unwanted light from the detectors. Most commercial flow cytometers can simultaneously measure two bands of fluorescence by the appropriate placement of blocking and interference filters (Figure 1A) to direct bands of different wavelength to two separate photomultiplier tubes.

Each of the three technical systems contribute to measurement error, but coefficients of variation (CV = standard deviation/mean intensity) of 2% can be obtained. Photomultiplier tubes that receive light emitted or reflected from a cell at the interrogation point create an electrical pulse with a Gaussian shape, a base width of 2–3 μs, and a peak height that is determined by the wavelength and intensity of the excitation and emission beams, the optical configuration of the flow cytometer, and the amplification of the detectors. These variables make it imperative to standardize each run on the flow cytometer, even for similar samples measured at different times. For each event detected, the pulse height is measured, digitized, and stored in one of the (256–1,024) modal channels. Each channel represents a unit of increasing intensity, and counts are kept of the number of events stored in each channel. Measurement of a single variable (e.g., cell fluorescence) results in a simple graph of the number of events versus intensity (Figure 2). When simultaneous measurements are made, scatter diagrams and three-dimensional plots are generated (Figures 3A, 4).

Most commercial flow cytometers have the ability to sort cells based on the intensity of a measured characteristic. Windows for sorting, or gates, are established by defining discretionary levels. The flow cytometer processes data from

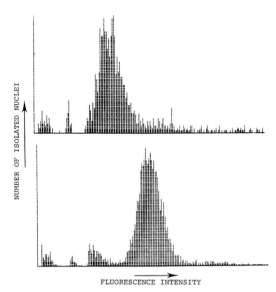

FIGURE 2.—Fluorescence intensity for hemocyte nuclei from normal softshells (above) and sarcomatous softshells (below); nearly all the nuclei from the sarcomatous clams exhibited increased DNA content. Hemocyte nuclei from the clams were isolated from the cytoplasm and stained for DNA with propidium iodide (Fisher, unpublished data).

the interrogation point and makes a "decision" about each cell based on the measured characteristic. Once a decision is made, the flow stream is charged positively or negatively or left uncharged as the cell exits the capillary tube (Figure 1B). To achieve this, the time delay required for the decision-making circuitry must exactly equal the time taken for the cell to pass from the interrogation point to its release from the tip of the capillary. Cells are isolated from each other by the continuous vibration of the capillary tube, which forms drops as the stream exits from the tube. Sample concentration and sample flow rate must be low enough that each drop contains no more than one cell. High-voltage deflection plates immediately below the capillary tube force the drop right, left, or center as determined by the electrical charge on the drop. The ability to collect sorted cells from these fluid streams is one of the outstanding features of flow cytometry.

Application of Flow Cytometry to Bivalve Pathology

Flow cytometry is potentially very useful in bivalve pathology. Several cell characteristics can be measured simultaneously, and specific components can be labeled with fluorescent probes for quantifi-

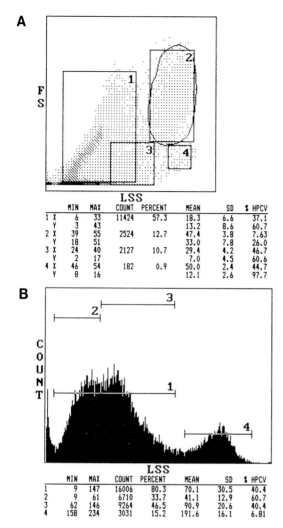

		MIN	MAX	COUNT	PERCENT	MEAN	SD	% HPCV
1	X	6	33	11424	57.3	18.3	6.6	37.1
	Y	3	43			13.2	8.6	60.7
2	X	39	55	2524	12.7	47.4	3.8	7.63
	Y	18	51			33.0	7.8	26.0
3	X	24	40	2127	10.7	29.4	4.2	46.7
	Y	2	17			7.0	4.5	60.6
4	X	46	54	182	0.9	50.0	2.4	44.7
	Y	8	16			12.1	2.6	97.7

	MIN	MAX	COUNT	PERCENT	MEAN	SD	% HPCV
1	9	147	16006	80.3	70.1	30.5	40.4
2	9	61	6710	33.7	41.1	12.9	60.7
3	62	146	9264	46.5	90.9	20.6	40.4
4	158	234	3031	15.2	191.6	16.1	6.81

FIGURE 3.—Analysis of eastern oyster hemocytes for size (FS = forward light scatter) and granularity (LSS = log$_{10}$side-light scatter). Hemocytes were diluted with Alsever's solution and analyzed with an EPICS Profile Analyzer (Coulter Corporation). **A.** Boxes 1–4 delineate possible hemocyte subpopulations: large, granular hemocytes are circled. Data for each box are summarized below the diagram. **B.** Granularity of hemocytes may fall into two (bars 1 and 4) or three (bars 2, 3, and 4) categories. % HPCV = coefficient of variation at half peak height. Boxes in **A** do not correspond to bars in **B**.

cation. The ability to sort live cells without damage provides a means for comparative physiological research on sorted cells. Preliminary use of flow cytometry in bivalve pathology has included investigation of hemolymph sarcomas (neoplasias), host hemocyte morphology and surface structure, and parasite isolation from host hemolymph.

Sarcomas

Sarcomatous cells that circulate in the hemolymph have a high rate of mitotic activity and an increased nuclear DNA content. Sarcomatous cells also have enlarged nuclei and are readily diagnosed in hemolymph smears (Brown et al. 1977). The percentage of cells that are affected by neoplasia can be estimated from the smears. With certain fluorescent dyes (e.g., propidium iodide) that bind quantitatively to DNA, flow cytometry can be used to detect increases in nuclear DNA and to rapidly evaluate the number of sarcomatous cells in a population. Preliminary attempts to identify sarcomatous cells in bivalves (W. S. Fisher, unpublished data) were confounded by interference from cytoplasmic DNA that resulted in overlapping peaks and high CVs. Techniques to strip the nuclei of cytoplasm (Jakobsen 1983; Vindelov et al. 1983) reduced CVs to less than 5% in the bivalve samples, but results were inconsistent and varied with different bivalve species. In another study (Fisher, unpublished data), isolated nuclei of neoplastic hemocytes of softshells *Mya arenaria* had higher DNA content than normal hemocytes, as measured by fluorescence intensities (Figure 2), but the increase was not an exact multiple of the diploid DNA content. Nuclear DNA content in bivalves generally varies greatly among individuals, whereas isolated nuclei from fish red blood cells have exhibited a consistent DNA content among individuals with low CVs (about 2%: Fisher, unpublished data). Because the technique and the instrument were the same in both cases, difficulties with bivalve cells probably are due to cell heterogeneity or sample preparation rather than to limitations of the instrument.

Hemocyte Morphology

An understanding of the functions of different hemocyte types has been hindered by the inability to easily separate each type by distinctive morphological characteristics. Separation by differential centrifugation, when possible, is time consuming and requires several manipulations and immersion of the cells in a separation medium (e.g., Percoll or Ficoll). These processes stress the cell and limit the ability to test the functions of the cells after separation. Flow cytometry can sort cells suspended in physiological medium rapidly and with minimal handling. Separations are restricted, however, by the characteristics that can be measured by the flow cytometer without specific staining.

Preliminary analysis of hemocytes from eastern oysters *Crassostrea virginica* (Fisher, unpublished

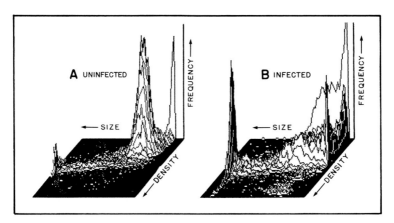

FIGURE 4.—Three-dimensional plots of hemocytes from eastern oysters that were uninfected (**A**) or infected (**B**) with the parasite *Haplosporidium nelsoni*. The plots compare forward-angle light scatter (cell diameter or size) and $\log_{10}$orthagonal light scatter ($\log_{10}$ density). The shift in the peaks presumably represents the parasites. Hemocytes were withdrawn into cold Alsever's solution and analyzed on a Coulter model 752 Dye-laser Flow Cytometer (Ford, unpublished data).

data) demonstrated a broad band of hemocytes (containing more than 50% of the cells: Figure 3A) that increased in granularity (orthogonal, or side, scatter) as the cells increased in size (forward scatter). There was also a distinct subpopulation of hemocytes (12.7%) that were large and granular (circled in Figure 3A). Possibly, there were three hemocyte subpopulations based on granularity, because the large, broadbased peak seen in Figure 3B may be bimodal. The hemocytes depicted in Figure 3 were diluted in Alsever's solution to inhibit aggregation; however, Alsever's solution is not a physiological medium and may have altered the results. Defining a physiological medium for oyster hemocytes that prevents cell-to-cell aggregation is still a major problem for in vitro studies of hemocyte subpopulations.

Flow cytometry has also been used to examine surface characteristics of oyster hemocytes. A fluorescein-conjugated antiserum made against eastern oyster lectin was applied to hemocytes, which were then analyzed simultaneously for light scatter (size) and fluorescence (Vasta et al. 1984). However, no subpopulations of hemocytes were identified by size or fluorescence criteria.

Parasite Isolation

Flow cytometry is also being explored as a method for separating the oyster parasite *Haplosporidium nelsoni* (MSX) from host hemocytes; the parasites occur free in the hemolymph of heavily infected animals. To establish sorting windows, hemolymph from eastern oysters that had been diagnosed by fresh smears (Ford and Kanaley 1988)

as lacking systemic infections was compared with hemolymph from oysters with advanced infections. Hemolymph was immediately diluted (2:3) in cold Alsever's solution to minimize hemocyte aggregation. Particles in the hemolymph, which were either host hemocytes or parasite agents, were classified by size and "density" (surface texture and cell inclusions) by means of forward and orthogonal angle light scatter, respectively.

The results demonstrated both the potential and the limitations of flow cytometry in analyzing cell morphology. Forward-angle light scatter showed clear changes in the size distribution of particles (hemocytes and parasites) with increasing infection intensity (Figure 4); however, the changes appeared as broadened distributions of cells rather than discrete modes. These results were different from direct microscopic measurement of cell diameters, which revealed distinct modes for different cell types. A possible explanation for this discrepancy is that forward-angle light scatter may be altered by surface texture and cell inclusions (Shapiro 1985). Both of these cell characteristics tend to absorb light and reduce the intensity of scatter signals. Parasites have a textured surface and numerous internal structures, and granulocytes contain many granules. Thus, both MSX parasites and granulocytes might be "seen" by the flow cytometer as smaller than they really are.

The profiles described by orthogonal light scatter (density) resembled the actual diameters more closely. It may be that laser-based flow cytometry can better distinguish between oyster hemocytes and the MSX parasite by using cell content rather than cell size as a variable. Preliminary experiments indicated that host hemocytes may be segregated from MSX parasites by applying a fluorescein-conjugated antihemocyte antiserum and then sorting out the fluorescent particles (S. E. Ford, unpublished data).

The above examples clearly demonstrate a valuable potential for flow cytometry in bivalve pathology. Continued research is necessary to refine the techniques and to develop systematic procedures that will reduce the variability found thus far in bivalve hemolymph samples. The potential for labeling host defensive cells or parasites with fluorescent dyes could produce extremely valuable results. The potential to physically sort viable subpopulations of cells (e.g., host hemocytes) for in vitro analysis is highly important.

Acknowledgments

We thank D. A. Phinney, T. L. Cucci, and C. M. Yentsch, Bigelow Laboratory for Ocean Sciences, for use of their training document, and D. Laskin and S. O'Connell, New Jersey Center for Advanced Biotechnology and Medicine at Rutgers. This paper is New Jersey Agricultural Experiment publication F-32504-4-88, supported by state funds, and is contribution 1900 from the Center for Environmental and Estuarine Studies, University of Maryland.

References

Brown, R. S., R. E. Wolke, S. B. Saila, and C. W. Brown. 1977. Prevalence of neoplasia in 10 New England populations of the soft-shell clam (*Mya arenaria*). Annals of the New York Academy of Sciences 298:522–534.

Ford, S. E., and S. A. Kanaley. 1988. An evaluation of hemolymph diagnosis for detection of the oyster parasite *Haplosporidium nelsoni* (MSX). Journal of Shellfish Research 7:11–18.

Herzenberg, L. A., R. G. Sweet, and L. A. Herzenberg. 1976. Fluorescence activated cell sorting. Scientific American 234(3):108–117.

Horan, P. K., and L. L. Wheeless. 1977. Quantitative single cell analysis and sorting. Science (Washington, D.C.) 198:149–157.

Jakobsen, A. 1983. The use of trout erythrocytes and human lymphocytes for standardization in flow cytometry. Cytometry 4:161–165.

Melamed, M. R., P. F. Mullaney, and M. L. Mendelsohn, editors. 1979. Flow cytometry and sorting. Wiley, New York.

Peters, D. C. 1979. A comparison of mercury arc lamp and laser illumination for flow cytometers. Journal of Histochemistry and Cytochemistry 27:241–245.

Phinney, D. A., T. L. Cucci, and C. M. Yentsch. 1987. Perspectives on aquatic flow cytometry, I: instrumentation and analysis. Bigelow Laboratory for Ocean Sciences, West Boothbay Harbor, Maine.

Shapiro, H. M. 1985. Practical flow cytometry. Alan R. Liss, New York.

Vasta, G. R., T. C. Cheng, and J. J. Marchalonis. 1984. A lectin on the hemocyte membrane of the oyster (*Crassostrea virginica*). Cellular Immunology 88:475–488.

Vindelov, L. L., I. J. Christensen, and N. I. Nissen. 1983. A detergent-trypsin method for the preparation of nuclei for flow cytometric DNA analysis. Cytometry 3:323–327.

Yentsch, C. M., F. C. Mague, P. K. Horan, and K. Muirhead. 1983. Flow cytometric DNA determinations on individual cells of the dinoflagellate *Gonyaulax tamarensis* var. *excavata*. Journal of Experimental Marine Biology and Ecology 67:175–183.

American Fisheries Society Special Publication 18:292–297, 1988

Chemiluminescence: An Advanced Tool for Measuring Phagocytosis

Arieh Wishkovsky

Aquaculture and Fisheries Program, Department of Medicine, School of Veterinary Medicine
University of California, Davis, California 95616, USA

Abstract.—Hemocytes are a major component in the bivalve defensive system; they function in phagocytosis and encapsulation of foreign material such as potential pathogens. Phagocytosis is initiated by stimulation of the cell membrane. This elicits a "respiratory burst" in which the cells consume increased amounts of oxygen and produce hydrogen peroxide, superoxide anions, hydroxyl radicals, and singlet oxygen. Energy release in the form of light emission (chemiluminescence) accompanies this metabolic process. Chemiluminescence from phagocytic cells can be measured with liquid scintillation counters and luminometers. Luminol, a cyclic hydrazide, amplifies the light and greatly increases the sensitivity of the test. Several substances elicit chemiluminescence from mammalian phagocytes, and many of these are now being used to study fish and mollusc cells. This technique offers great potential for monitoring the effects of natural or anthropogenic stressors on bivalve molluscs.

Stimulation of phagocyte membranes activates a "respiratory burst" along metabolic pathways. This provides oxygen radicals that kill microorganisms. The membrane is activated when its receptors are stimulated or when it is penetrated by ionophores that cause sodium ions to enter the cells and change the resting cell membrane potential and alter the movement of calcium ions into the cell. Calcium ion movement mediates the respiratory burst activation (Van Dyke et al. 1981; Fischer et al. 1987; Van Dyke 1987a). Activation of calmodulin (a calcium-binding protein) provokes cellular functions such as superoxide formation and stimulation of NADPH oxidase (Takeshige and Minakami 1981, 1987; Jones et al. 1982). Protein kinase C, which functions in the phosphorylation of proteins, may be involved in the oxidase activation (Fischer et al. 1987; O'Flaherty and Nishihira 1987).

The respiratory burst is characterized by an increase in the consumption of oxgyen, an activation of the hexose monophosphate shunt (HMPS), and the production of large quantities of superoxide anion (O_2^-), hydrogen peroxide (H_2O_2), hydroxyl radical ($\cdot OH$) and singlet oxygen ($^1O_2$). With NADPH as an electron donor, a membrane-associated NADPH oxidase catalyzes the reduction of O_2 to O_2^-. In the presence of the enzyme superoxide dismutase, two molecules of O_2^- combine to form H_2O_2. A variety of reactions involving O_2^- and H_2O_2 lead to the formation of $\cdot OH$ and $^1O_2$. During the respiratory burst, the oxygen uptake increases 10–15-fold within a few seconds. The $NADP^+$ is regenerated to NADPH by a step in the hexose monophosphate pathway (Babior 1980; Drutz and Mills 1982; Klebanoff 1982; Rossi et al. 1982; Werb and Goldstein 1982; Bellavite et al. 1987; Takeshinge and Minakami 1987; Van Dyke 1987a). The mechanism mediated by the enzyme myeloperoxidase (MPO) in conjunction with H_2O_2 and halide is most widely implicated for the bactericidal activity of mammalian phagocytes (Klebanoff 1975; Klebanoff et al. 1982).

Chemiluminescence

The phenomenon of light emission or chemiluminescence (CL) during phagocytosis was originally described by Allen et al. (1972). Bacteria and latex bead particles that were opsonized (coated with serum components such as immunoglobulins or complement proteins) elicited CL when they were phagocytosed by polymorphonuclear leukocytes (PMNs). Chemiluminescence, measured in a liquid scintillation counter operated in an out-of-coincidence mode, was correlated with the number of cells, the concentration of the stimulus, and the HMPS activity. Allen (1975) found CL in crude preparations of MPO in the presence of H_2O_2 and Cl^-. He also emphasized that O_2 was not necessary for the engulfment of bacteria by phagocytes, but it was necessary for microbiocidal activity and the CL response (Allen 1980).

Kearns (1971) showed that singlet oxygen is in an energetic state and that there is a release of energy that can be measured in terms of CL during its conversion to the triplet ground state. Klebanoff (1982) suggested two general mechanisms, one involving peroxidase and the other involving the superoxide anion, for the CL response of phagocytes. Experimental evidence for the mechanism involving peroxidase includes (1) decreased CL response from MPO-deficient PMNs, (2) inhibition of light emission by azide (a $^1O_2$ scavenger) in the MPO–H_2O_2–halide system of normal PMNs with an inability of azide to decrease the CL in MPO-deficient PMNs, and (3) spectral similarity of light emitted from the MPO–H_2O_2–halide system to that from phagocytosing PMNs (Rosen and Klebanoff 1976; Klebanoff 1982). Inhibition of the CL response by superoxide dismutase in MPO-deficient cells and inhibition of light emission from a xanthine oxidase system (a system that generates O_2^-) by superoxide dismutase, but not by azide, suggests that O_2^- is involved in the emission of light (Rosen and Klebanoff 1976; Klebanoff 1982). Catalase and benzoate (a scavenger of hydroxyl radicals $\cdot OH$) minimize the CL response, implying that H_2O_2 and $\cdot OH$ are involved in the emission of light by phagocytes (Webb et al. 1974). Cheson et al. (1976) reported that the spectrum of the light emitted by PMNs was not the same as that of $^1O_2$. Brestel (1987) concluded that the CL of stimulated PMNs was a product of multiple reactions with absolute dependence on O_2^- production and partial dependence on the MPO–H_2O_2–Cl^- system and was not dependent on the production of $^1O_2$.

Modification of the CL assay by addition of luminol, a cyclic hydrazide that can be oxidized to emit light (Allen and Loose 1976), greatly amplified the sensitivity of the assay. Chemiluminescence obtained in the luminol system is approximately 1,000 times more intense than the "native" CL, thus permitting detection with fewer cells. Luminol was added to cells of low-intensity CL, including monocytes, macrophages, basophils, and mast cells, to enhance their CL (Allen 1980; Klebanoff 1982). DeChatelet et al. (1982), Dahlgren and Stendahl (1983), and Brestel (1985) showed that the MPO-mediated reaction had a major role in the luminol-enhanced CL and hypothesized that emission of light was due to hypochlorite (HOCl) generation. Cheung et al. (1985) and Schopf et al. (1987) postulated that the source of luminol-dependent CL of neutrophils phagocytosing zymosan was secondary to the metabolism of arachidonic acid via the lipoxygenase and cyclooxygenase pathways. Brestel (1987) stated that CL of luminol was likely a cooxidative phenomenon in the presence of both HOCl and H_2O_2. Singlet oxygen and superoxide anion do not participate in the emission of light via luminol.

Many substances elicit a CL response from phagocytes (Takeshinge and Minakami 1987; Van Dyke 1987b). The most common stimuli are zymosan, a yeast extract (Webb et al. 1974; DeChatelet et al. 1982); latex beads (Allen et al. 1972); phorbol 12-myristate 13-acetate (PMA), a soluble tumor-promoting compound (DeChatelet et al. 1976); n-formyl-methionyl-leucyl-phenylalanine (Hatch et al. 1978); many bacteria and bacterial particles; and other soluble and insoluble activators (see Van Dyke 1987b). In most studies, opsonized particles elicited a higher CL response than nonopsonized particles (Allen et al. 1972; Stevens et al. 1978).

Scintillation counters are most frequently used to measure light emission because they are readily available. However, the use of scintillation counters presents two major problems: the inability to measure fast CL reactions and the inability to control the experimental temperature (Van Dyke 1987a). Although these problems have been solved with luminometers, the appropriate adaptation of scintillation counters for CL studies is most convenient. More information concerning instrumentation can be found in Van Dyke and Castranova (1987).

Because the CL response represents the bactericidal mechanisms during the respiratory burst, it is widely used for studying phagocytosis. A correlation between the bactericidal activity and the CL response was shown by Welch (1980) and Horan et al. (1982). Chemiluminescence has been demonstrated from phagocytes of different animals including humans (Allen et al. 1972; Rosen and Klebanoff 1976), bovines (Roth et al. 1981), dogs (Andersen and Amirault 1979), mice (Tomita et al. 1981), rabbits (Allen and Loose 1976), rats, guinea pigs (Trush et al. 1978), fish, snails, and oysters (see below).

Chemiluminescent Response of Fish Phagocytes

Scott and Klesius (1981) were the first to use the CL assay in determining phagocytic responses of fish cells. Other studies have characterized the CL response of leukocytes from different fish species: channel catfish *Ictalurus punctatus* (Scott and Klesius 1981; Scott et al. 1985), striped bass *Morone saxatilis* (Stave et al. 1983, 1984, 1985, 1987; Stave and Roberson 1985), rainbow trout *Oncorhynchus mykiss* (formerly *Salmo gairdneri*) (Sohnle and Chusid 1983; Elsasser et al. 1986; Wishkovsky et al. 1987), hogchoker *Trinectes maculatus* (Warinner et al. 1988; Wishkovsky et al., in press), spot *Leiostomus xanthurus* (Warinner et al. 1988), oyster toadfish *Opsanus tau*, and Atlantic croaker *Micropogonias undulatus* (Wishkovsky et al., in press).

Scott and Klesius (1981), Sohnle and Chusid (1983), and Scott et al. (1985) used peripheral blood cells for their assays, but the other researchers used cells from the fish kidney. Wishkovsky and associates (unpublished data) found that kidney (mainly pronephros) and spleen cells had high CL potentials, whereas blood, gill, and liver cells had little or no response. The separation of the leukocytes from the red blood cells (Scott and Klesius 1981; Stave et al. 1983, 1984, 1985, 1987; Stave and Roberson 1985; Scott et al. 1985) may not be justified, however, because unseparated oyster toadfish cells exhibited a higher CL response than separated phagocytes (Wishkovsky, unpublished data).

Although Scott and Klesius (1981) and Scott et al. (1985) measured CL in a luminometer, liquid scintillation counters in out-of-coincidence mode (or in a special mode for measuring light for newer models) were used successfully in other studies. Luminol has been used to amplify the response in all fish cell assays. Scott and Klesius (1981) and Wishkovsky (unpublished data) found better CL responses after dissolving the luminol in potassium hydroxide and boric acid than in dimethyl sulfoxide.

Dose-dependent CL responses of fish cells were elicited by zymosan, PMA, latex beads, and various living or killed bacterial (Scott and Klesius 1981; Stave et al. 1983, 1984; Elsasser et al. 1986; Wishkovsky, unpublished data). Zymosan is the prefered stimulus because it is inexpensive, easy to use, and elicits a high response. Although opsonization of stimulants with autologous serum generated high CL responses in some studies (Scott and Klesius 1981; Stave et al. 1984; Scott et al. 1985), other studies have shown high CL responses without opsonization (Wishkovsky et al. 1987; Wishkovsky, unpublished data). Opsonization with nonautologous serum generated lower responses than those elicited by nonopsonized particles (Scott and Klesius 1981; Wishkovsky, unpublished data).

Stave et al. (1985) showed that *Vibrio anguillarum*, a fish pathogen, consistently elicited CL responses that increased more rapidly with higher peaks and were sustained for longer periods of time than those elicited by other nonvirulent *Vibrio* species. However, studies with *Yersinia ruckeri* strains showed low CL responses elicited by the virulent strains in comparison to the nonpathogenic strains (Stave et al. 1987).

Experiments to study the effect of environmental factors on fish phagocytes have been carried out by applying the CL response. Cells of channel catfish maintained at 18–20°C emitted optimal CL responses at 35°C (Scott and Klesius 1981) whereas cells of rainbow trout generated a peak response at their maintenance temperature (15°C) (Sohnle and Chusid 1983). In vitro exposure of fish leukocytes to hydrocortisone acetate to stimulate the primary physiological consequences of stress reduced the CL potential of the leukocytes (Stave and Roberson 1985).

Wishkovsky et al. (1987) showed suppression of CL responses of phagocytes exposed to tetracyclines in vitro. The oxytetracycline dosage recommended for aquaculture suppressed the CL by approximately 50%. Elsasser et al. (1986) exposed rainbow trout leukocytes to heavy metals in vitro. Copper caused a decrease in the emission of the light to the baseline level. Aluminum also caused a partial suppression, whereas cadmium enhanced the CL response when added within 1 h prior to the assay or caused variable results after a 24-h exposure. The CL response was suppressed in vitro after phagocytes of three estuarine fish were exposed to tributyltin (TBT), a known aquatic toxicant (Wishkovsky et al., in press). Exposure to TBT concentrations of 40–400 µg/L immediately prior to the assay caused a great decrease of the response, and lower doses suppressed the response of cells exposed to the TBT for 24 h. Preliminary studies to compare the CL response of cells obtained from normal and pollutant-exposed fish exhibited a decrease in the response of the exposed fish (Warinner et al. 1988).

Chemiluminescent Response of Molluscan Hemocytes

Dikkeboom et al. (1987) used the luminol-enhanced CL assay to study oxygen radical production by hemocytes of the swamp lymnaea *Lymnaea stagnalis*. Among the different stimuli tested (zymosan, latex beads, PMA, and killed bacteria), zymosan was most effective. These results agreed with those obtained from mammalian and fish cells (Tomita et al. 1981; Stave et al. 1984; Wishkovsky et al. 1987). Total hemolymph showed higher CL responses than isolated hemocytes, which was probably due to opsonizing properties of the snail plasma (Sminia et al. 1979). Azide and superoxide dismutase decreased the CL response, an indication that the response of the snail was comparable to mammalian (Allen 1980; Klebanoff 1982) and fish responses (Scott et al. 1985).

The effects of environmental pollutants on the CL responses of hemocytes of eastern oysters *Crassostrea virginica* have been studied by Larson et al. (in press), and similar studies have been conducted on eastern and Pacific oysters *C. gigas* by W. S. Fisher and associates (University of Maryland, unpublished data). Luminol-enhanced CL assays and liquid scintillation counters were used to measure light emission; PMA and zymosan were used by Larson et al. and Fisher et al., respectively, to elicit the CL responses.

The CL response of eastern oyster hemocytes was depressed after the cells were exposed in vitro to low concentrations of copper (0.4 mg/L) and dieldrin (1 mg/L), and to high concentrations of aluminum, cadmium, and zinc at 320 mg/L, and chlordane, naphthalene, and 2,4-dinitrophenol at 100 mg/L (Larson et al., in press). Acclimation of cells after preincubation with some of these pollutants was similar to that obtained in experiments with fish cells (Elsasser et al. 1986).

Fisher et al. (unpublished data) observed the suppression of the CL responses emitted by hemocytes of three populations of oysters (a Pacific oyster population and eastern oysters from two different locations) upon exposure to high concentrations of TBT (>40 µg/L) immediately prior to the assay. Preincubation of the hemocytes with TBT for 0.5 h reduced the response even further. Similar results were obtained with fish phagocytes exposed to TBT (Wishkovsky et al., in press).

Conclusions

The CL assay provides information regarding the health of an organism because the assay measures changes in phagocytosis, which affects the susceptibility of an individual to pathogenic agents (see Bayne 1983; Feng and Barja 1987). The most attractive features of the CL assay are objectivity, sensitivity, simplicity, and rapidity. To achieve wide use of this tool, further development is necessary and will be achieved with expanded application and standardization of the assay.

Acknowledgments

I thank J. M. Groff, University of California, Davis, and B. S. Roberson, University of Maryland, College Park, for critically reading this manuscript.

References

Allen, R. C. 1975. The role of pH in the chemiluminescent response of the myeloperoxidase-halide-HOOH antimicrobial system. Biochemical and Biophysical Research Communications 63:684–691.

Allen, R. C. 1980. Free-radical production by reticuloendothelial cells. Pages 309–338 *in* A. J. Sbarra and R. R. Strauss, editors. The reticuloendothelial system, a comprehensive treatise, volume 2. Plenum, New York.

Allen, R. C., and L. D. Loose. 1976. Phagocytic activation of a luminol-dependent chemiluminescence in rabbit alveolar and periotoneal macrophages. Biochemical and Biophysical Research Communications 69:245–252.

Allen, R. C., R. L. Stjernholm, and R. H. Steele. 1972. Evidence for the generation of an electronic excitation state(s) in human polymorphonuclear leukocytes and its participation in bactericidal activity. Biochemical and Biophysical Research Communications 47:679–684.

Andersen, B. R., and H. J. Amirault. 1979. Important variables in granulocytes chemiluminescence. Proceedings of the Society for Experimental Biology and Medicine 162:139–145.

Babior, B. M. 1980. The role of oxygen radicals in microbial killing by phagocytes. Pages 339–354 *in* A. J. Sbarra and R. R. Strauss, editors. The reticuloendothelial system, a comprehensive treatise, volume 2. Plenum, New York.

Bayne, C. J. 1983. Molluscan immunobiology. Pages 407–486 *in* A. S. M. Saleuddin and K. M. Wilbur, editors. The Mollusca, volume 5. Academic Press, New York.

Bellavite, P., M. C. Serra, S. Dusi, G. Berton, and M. Chilosi. 1987. The free radical forming system of granulocytes and macrophages: further studies. Advances in the Biosciences 66:161–173.

Brestel, E. P. 1985. Co-oxidation of luminol by hypochlorite and hydrogen peroxide: implications for neutrophil chemiluminescence. Biochemical and Biophysical Research Communications 126:482–488.

Brestel, E. P. 1987. Mechanisms of cellular chemiluminescence. Pages 93–104 *in* K. Van Dyke and V. Castranova, editors. Cellular chemiluminescence, volume 1. CRC Press, Boca Raton, Florida.

Cheson, B. D., R. L. Christensen, R. Sperling, B. E. Kohler, and B. M. Babior. 1976. The origin of the chemiluminescence of phagocytosing granulocytes. Journal of Clinical Investigation 58:789–796.

Cheung, K., A. C. Archibald, and M. F. Robinson. 1985. The origin of chemiluminescence produced by neutrophils stimulated by opsonized zymosan. Journal of Immunology 130:2324–2329.

Dahlgren, C., and O. Stendahl. 1983. Role of myeloperoxidase in luminol-dependent chemiluminescence of polymorphonuclear leukocytes. Infection and Immunity 37:736–741.

DeChatelet, L. R., P. S. Shirley, and R. B. Johnston. 1976. Effect of phorbol myristate acetate on the oxidative metabolism of human polymorphonuclear leukocytes. Blood 47:545–554.

DeChatelet, L. R., and six coauthors. 1982. Mechanism of the luminol-dependent chemiluminescence of human neutrophils. Journal of Immunology 129:1589–1593.

Dikkeboom, R., J. M. G. H. Tijnagel, E. C. Mulder, and W. P. W. Van der Knaap. 1987. Hemocytes of the pond snail *Lymnaea stagnalis* generate reactive forms of oxygen. Journal of Invertebrate Pathology 49:321–331.

Drutz, D. J., and J. Mills. 1982. Immunity and infection. Pages 209–232 *in* D. P. Stites, J. D. Stobo, H. H Fudenberg, and J. V. Wells, editors. Basic and clinical immunology. Lange Medical Publications, Los Altos, California.

Elsasser, M. S., B. S. Roberson, and F. M. Hetrick. 1986. Effects of metals on the chemiluminescent response of rainbow trout (*Salmon gairdneri*) phagocytes. Veterinary Immunology and Immunopathology 12:243–250.

Feng, S. Y., and J. L. Barja. 1987. Cellular and humoral defense mechanisms. Summary. Pages 29–41 *in* W. S. Fisher and A. J. Figueras, editors. Marine bivalve pathology. University of Maryland Sea Grant College Publication UM-SG-TS-87-02, College Park.

Fischer, S., U. Bamberger, B. Reck, and H. Mossmann. 1987. The role of calcium and protein kinase C in the chemiluminescence response of human granulocytes. Pages 165–168 *in* J. Scholmerich, R. Andreesen, A. Kapp, M. Ernst, and W. G. Woods, editors. Bioluminescence and chemiluminescence, new perspectives. Wiley, Chichester, England.

Hatch, G. E., D. E. Gardner, and D. B. Menzel. 1978. Chemiluminescence of phagocytic cells caused by *n*-formyl-methionyl peptides. Journal of Experimental Medicine 147:182–195.

Horan, T. D., D. English, and T. A. McPherson. 1982. Association of neutrophil chemiluminescence with microbicidal activity. Clinical Immunology and Immunopathology 22:259–269.

Jones, H. P., G. Ghali, M. F. Petrone, and J. M. McCord. 1982. Calmodulin-dependent stimulation of the NADPH oxidase of human neutrophils. Biochimica et Biophysica Acta 714:152–156.

Kearns, D. R. 1971. Physical and chemical properties of singlet molecular oxygen. Chemical Reviews 71:395–427.

Klebanoff, S. J. 1975. Antimicrobial systems of the polymorphonuclear leukocyte. Pages 45–60 *in* J. A. Bellanti and D. H. Dayton, editors. The phagocytic cell in host resistance. Raven Press, New York.

Klebanoff, S. J. 1982. Oxygen-dependent cytotoxic mechanisms of phagocytes. Pages 111–161 *in* J. I. Gallin and A. S. Fauci, editors. Advances in host defense mechanisms, volume 1. Raven Press, New York.

Klebanoff, S. J., W. R. Henderson, E. C. Jorg, A. Jorg, R. M. Locksley, and P. G. Ramsey. 1982. Cidal mechanisms and their dependence on hydrogen peroxide. Pages 449–461 *in* M. L. Karnovsky and L. Bolis, editors. Phagocytosis—past and future. Academic Press, New York.

Larson, K. G., F. M. Hetrick, and B. S. Roberson. In press. The effect of environmental pollutants on the chemiluminescence of hemocytes from the American oyster *Crassostrea virginica*. Journal of the Yugoslavian Academy of Science.

O'Flaherty, J. T., and J. Nishihira. 1987. Protein kinase C involvement in neutrophil function. Advances in the Biosciences 66:251–253.

Rosen, H., and S. J. Klebanoff. 1976. Chemiluminescence and superoxide production by myeloperoxidase-deficient leuckocytes. Journal of Clinical Investigation 58:50–60.

Rossi, F., P. Bellavite, and G. Berton. 1982. The respiratory burst in phagocytic leukocytes. Pages 167–191 *in* M. L. Karnovsky and L. Bolis, editors. Phagocytosis—past and future. Academic Press, New York.

Roth, J. A., M. L. Kaeberle, and R. W. Griffith. 1981. Effects of bovine viral diarrhea virus infection on bovine polymorphonuclear leukocytes function. American Journal of Veterinary Research 42:244–250.

Schopf, R. E., B. Lutz, and M. Rehder. 1987. Regulation of phagocyte chemiluminescence by arachidonate metabolism. Pages 173–176 *in* J. Scholmerich, R. Andreesen, A. Kapp, M. Ernst, and W. G. Woods, editors. Bioluminescence and chemiluminescence, new perspectives. Wiley, Chichester, England.

Scott, A. L., and P. H. Klesius. 1981. Chemiluminescence: a novel analysis of phagocytosis in fish. Developments in Biological Standardization 49:243–254.

Scott, A. L., W. A. Rogers, and P. H. Klesius. 1985. Chemiluminescence by peripheral blood phagocytes from channel catfish: function of opsonin and temperature. Developmental and Comparative Immunology 9:241–250.

Sminia, T., W. P. W. van der Knaap, and P. Ede-Lenbosch. 1979. The role of serum factors in phagocytosis of foreign particles by blood cells of the freshwater snail *Lymnaea stagnalis*. Developmental and Comparative Immunology 3:37–44.

Sohnle, P. G., and M. J. Chusid. 1983. The effect of temperature on the chemiluminescence response of neutrophils from rainbow trout and man. Journal of Comparative Pathology 93:493–497.

Stave, J. W., T. M. Cook, and B. S. Roberson. 1987. Chemiluminescent response of striped bass, *Morone saxatilis* (Walbaum), phagocytes to strains of *Yersinia ruckeri*. Journal of Fish Diseases 10:1–10.

Stave, J. W., and B. S. Roberson. 1985. Hydrocortisone suppresses the chemiluminescent response of striped bass phagocytes. Developmental and Comparative Immunology 9:77–84.

Stave, J. W., B. S. Roberson, and F. M. Hetrick. 1983. Chemiluminescence of phagocytic cells isolated from the pronephros of striped bass. Developmental and Comparative Immunology 7:269–276.

Stave, J. W., B. S. Roberson, and F. M. Hetrick. 1984. Factors affecting the chemiluminescent response of fish phagocytes. Journal of Fish Biology 25:197–206.

Stave, J. W., B. S. Roberson, and F. M. Hetrick. 1985. Chemiluminescent responses of striped bass, *Morone saxatilis* (Walbaum), phagocytes to *Vibrio* spp. Journal of Fish Diseases 8:479–483.

Stevens, P., D. J. Winston, and K. Van Dyke. 1978. In vitro evaluation of opsonic and cellular granulocyte function by luminol-dependent chemiluminescence: utility in patients with severe neutropenia and cellular deficiency states. Infection and Immunity 22: 41–51.

Takeshige, K., and S. Minakami. 1981. Involvement of calmodulin in phagocytic respiratory burst of leukocytes. Biochemical and Biophysical Research Communications 99:484–490.

Takeshige, K., and S. Minakami. 1987. Early events and stimulants triggering oxidative metabolism in neutrophils. Pages 113–129 in K. Van Dyke and V. Castranova, editors. Cellular chemiluminescence, volume 1. CRC Press, Boca Raton, Florida.

Tomita, T., E. Blumenstock, and S. Kanegasaki. 1981. Phagocytic and chemiluminescent response of mouse peritoneal macrophages to living and killed Salmonella typhimurium and other bacteria. Infection and Immunity 32:1242–1248.

Trush, M. A., M. E. Wilson, and K. Van Dyke. 1978. The generation of chemiluminescence (CL) by phagocytic cells. Methods in Enzymology 62:462–494.

Van Dyke, K. 1987a. Introduction to cellular chemiluminescence, neutrophils, macrophages, and monocytes. Pages 3–22 in K. Van Dyke and V. Castranova, editors. Cellular chemiluminescence, volume 1. CRC Press, Boca Raton, Florida.

Van Dyke, K. 1987b. Soluble and insoluble activators of neutrophil chemiluminescence. Pages 161–172 in K. Van Dyke and V. Castranova, editors. Cellular chemiluminescence, volume 1. CRC Press, Boca Raton, Florida.

Van Dyke, K., and V. Castranova, editors. 1987. Cellular chemiluminescence, volume 1. CRC Press, Boca Raton, Florida.

Van Dyke, K., C. Van Dyke, D. Peden, M. Matamoros, V. Castranova, and G. Jones. 1981. Preliminary events leading to the production of luminol-dependent chemiluminescence by human granulocytes. Pages 45–53 in M. DeLuca and W. McElroy, editors. Bioluminescence and chemiluminescence. Academic Press, New York.

Warinner, J. E., E. S. Mathews, and B. A. Weeks. 1988. Preliminary investigations of the chemiluminescent response in normal and pollutant-exposed fish. Marine Environmental Research 24:281–284.

Webb, L. S., B. B. Keele, and R. B. Johnston. 1974. Inhibition of phagocytosis-associated chemiluminescence by superoxide dismutase. Infection and Immunity 9:1051–1056.

Welch, W. D. 1980. Correlation between measurements of the luminol-dependent chemiluminescence response and bacterial susceptibility to phagocytosis. Infection and Immunity 30:370–374.

Werb, Z., and I. M. Goldstein. 1982. Phagocytic cells: chemotaxis and effector functions of macrophages and granulocytes. Pages 104–118 in D. P. Stites, J. D. Stobo, H. H. Fudenberg, and J. V. Wells, editors. Basic and clinical immunology. Lange Medical Publications, Los Altos, California.

Wishkovsky, A., E. S. Mathews, and B. A. Weeks. In press. Effect of tributyltin on the chemiluminescent response of phagocytes from three species of estuarine fish. Archives of Environmental Contamination and Toxicology 18.

Wishkovsky, A., B. S. Roberson, and F. M. Hetrick. 1987. In vitro suppression of the phagocytic response of fish macrophages by tetracyclines. Journal of Fish Biology 31 (Supplement A):61–65.

American Fisheries Society Special Publication 18:298–303, 1988

Use of Immunoassays in Haplosporidan Life Cycle Studies

Eugene M. Burreson

Virginia Institute of Marine Science, School of Marine Science
College of William and Mary, Gloucester Point, Virginia 23062, USA

Abstract.—The development of mitigating measures for the major oyster diseases has been hindered by our poor understanding of the life cycles of the pathogens. Evidence from epidemiological studies and transmission experiments suggests that an intermediate host is present in the life cycle of *Haplosporidium* species. Immunoassay is a valuable tool for identifying parasite antigen in an intermediate host, and, because of the potential for stage-specific antigens, assays incorporating polyclonal antibodies may be more effective than assays incorporating monoclonal antibodies. Rabbit antibody against purified spores of *Haplosporidium costale* recognized spores in paraffin sections of oyster tissue, but the antibody did not recognize plasmodia of *H. costale*.

One of the major obstacles to the mitigation of parasite-induced oyster mortality is our poor understanding of the life cycles of the parasites. Four genera contain important disease agents—*Haplosporidium, Marteilia, Bonamia,* and *Perkinsus*—but only species of *Perkinsus* have life cycles that are known (Perkins and Menzel 1966). *Bonamia ostreae* may be transmissible directly (Elston et al. 1986), but the infective stage has not been identified. The lack of information on life cycles has hindered our ability to interpret field observations and has limited many lines of investigation. For example, *Haplosporidium nelsoni* (MSX) moves hundreds of kilometers up Chesapeake Bay in response to increased salinity that results from drought conditions. Because the life cycle and infective stage of *H. nelsoni* are unknown, we do not know how the parasite moves or what stage in the life cycle responds to the change in salinity. Most diseases of oysters have not been transmitted in the laboratory, so it has been impossible to investigate, under controlled conditions, infective dose, defense reactions, pathogenicity, and control measures. Although some management techniques have been developed through field manipulations of diseased oysters, sound management recommendations are also hindered by our poor understanding of the life cycles of the parasites.

A major question is whether parasite transmission between oysters is direct or via an intermediate host. Direct experimental transmission of *Haplosporidium* spp. and *Marteilia* spp. via spores has been unsuccessful and implies that an intermediate host may be involved in the life cycles of these pathogens. Andrews (1984) listed evidence for and against the existence of other hosts in the life cycle of *H. nelsoni*. Intermediate hosts should be distinguished from alternate or reservoir hosts because these terms have been used interchangeably in the literature on oyster diseases. For *H. nelsoni*, an intermediate host is a host, other than oysters, in which some development occurs that is essential for completion of the parasite's life cycle. A reservoir host is any host that serves as a source of *H. nelsoni* from which oysters can become infected. The existence of a reservoir host has been postulated for *H. nelsoni* because spores have only rarely been encountered, yet parasite prevalence is high (Farley 1967; Andrews 1984). Most other species of *Haplosporidium* and those of *Marteilia* sporulate regularly, and a reservoir host is not required to account for the observed prevalence. However, an intermediate host may be required for completion of all haplosporidan life cycles.

Life Cycle Hypotheses

All species of *Haplosporidium* produce spores, and insights into the life cycle can be gained by examining the fate of spores in life cycles of other spore-producing protozoa. Species in the phyla Myxozoa and Microspora and many species in Apicomplexa produce nonmotile spores. In the life cycles that have been elucidated for the species of these phyla, the spore is always eaten by the intermediate or final host, and the sporoplasm emerges from the spore in the gut of the new host. No life cycle is known in which the sporoplasm hatches from the spore in the external environment and exists as a free-living sporoplasm. Thus, it is plausible to hypothesize that spores of *Haplosporidium* spp. are eaten. Because all attempts to infect oysters with spores of *H. nelsoni* and *H. costale* have failed (Andrews 1984), perhaps spores are eaten by an unknown intermediate host.

Four species of *Haplosporidium* are known from the oceanic and estuarine waters of the Chesapeake Bay region. *Haplosporidium costale* in oysters sporulates in connective tissue and causes rapid death of the host. When an oyster dies and gapes, scavengers, such as crabs, amphipods, fishes, and ciliates, feed on the tissue of the oyster, and any of these scavengers could be an intermediate host. Similarly, *Haplosporidium* sp. in the mud crab *Panopeus herbsti* sporulates in all tissues of the host, and the weakened crab is probably ingested by a predator. The oyster parasite *H. nelsoni* sporulates only in digestive diverticula, and spores are probably released from live oysters after localized exfoliation of the diverticula epithelium. *Haplosporidium* sp. from the naval shipworm *Teredo navalis* sporulates in all tissues and can cause death of the host, although spores may be released from live naval shipworms during exfoliation of gill epithelium. Ingestion by scavengers seems unlikely because few organisms could enter the burrow. Spores of *Haplosporidium* spp. in oysters and the naval shipworms probably are ingested by ciliates, small crustaceans, or suspension feeders after the spores leave the host.

Insights into the life cycles of *Haplosporidium* spp. can also be gained by examining the site and morphology of the early infection stages. Evidence is available only for oyster parasites, and it is not clear whether the initial site of infection is the epithelium of the gill or of the gut because infections have been found in only one or the other of these sites. Gut infections imply that the parasite is associated with food whereas gill infections imply that the parasite penetrates as a naked cell from the mantle cavity.

The hypotheses of (1) an intermediate host that ingests spores and (2) infections initiated by a naked cell or by a stage emerging from an oyster food organism are difficult to reconcile. Organisms that ingest spores are unlikely to be ingested by oysters, and organisms ingested by oysters are unlikely to be capable of ingesting spores. An intermediate host that ingests spores and releases a naked-cell stage into the water may be possible, but no such life cycle is known for the spore-producing protozoa. However, the discoveries made by Wolf and Markiw (1984) about the life cycle of *Myxosoma cerebralis* indicate that hypotheses should not be constrained by known life cycles.

Research on the life cycles of *Haplosporidium* spp. should be directed toward determining the fate of spores and identifying the water-borne infective stages. Histological sections of many estuarine organisms have been examined for stages of *H. nelsoni*, but no intermediate stages have been found. This survey type of approach has been hindered by the large number of potential intermediate hosts, by our poor understanding of the normal histology of most estuarine organisms, and possibly by our inability to recognize the parasite even if we encountered it. New techniques developed for the diagnosis of human and veterinary diseases have overcome the latter two problems. Two techniques, nucleic acid probes and immunoassays, hold great promise for the diagnosis of oyster diseases and for the discovery of an intermediate host. Nucleic acid probes are highly specific and have been developed for some human parasites (Wirth et al. 1986), but this technique has not been attempted for oyster parasites. Immunoassay has been successfully used to identify many human and veterinary parasites (Kurstak 1986). A marker is bound to an antibody molecule made against the parasite. The marker may be a fluorescent molecule, an enzyme, colloidal gold particles, or other molecules. When the antibody binds to the parasite, the attached marker can be visualized by several methods. Immunoassays that utilize enzyme markers are most widely used for disease diagnosis; however, the immunogold–silver staining method is gaining usage (Springall et al. 1984; De Mey et al. 1986).

Enzyme Immunoassay Technique

Enzyme immunoassay (EIA) is sensitive for detecting antigen or antibody, rapid, and conservative in the use of reagents. Variations of the technique are used for different objectives and may employ a variety of enzyme–chromogen combinations. In EIA, antibodies detect and distinguish closely related antigens; EIA uses enzymes that accelerate specific chemical reactions and that can be detected by adding appropriate enzyme substrates. Enzyme substrates are utilized that change color during the reaction and may also precipitate from solution. The principle of EIA is the conjugation of an enzyme and antibody; after the immunological reaction occurs, substrate is added and the amount or occurrence of color change is measured (Kurstak 1986).

In the direct EIA (Figure 1A), an enzyme is conjugated to the primary antibody, i.e., the antibody produced in response to a specific parasite. Because antibodies to oyster parasites are not commercially available, the enzyme conjugation for a direct EIA must be done by the investigator.

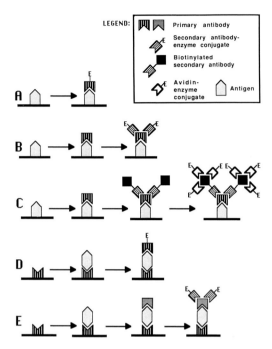

FIGURE 1.—Variations of enzyme immunoassays. **A.** Direct immunoassay. **B.** Indirect immunoassay. **C.** Indirect immunoassay with biotin–avidin bridge. **D.** Direct sandwich immunoassay. **E.** Indirect sandwich immunoassay.

Direct EIA could be used to detect solid-phase parasite antigen. For example, after primary antibody has bound to parasite antigen in a histological section, the addition of an enzyme substrate that precipitates as a colored substance would detect and locate the parasite.

The commercial availability of enzyme-conjugated goat antibodies produced against the immunoglobulin of various other species (e.g., rabbits) and the availability of kits containing required reagents has led to widespread use of indirect EIA. In this variation (Figure 1B), primary antibody is prepared by the investigator and is added to a histological section. If the specific antigen is present in the section, the primary antibody will bind to it. Next, an enzyme-conjugated secondary antibody, made against the immunoglobulin of the species in which the primary antibody was produced, is added, and it will bind to the primary antibody. As in the direct assay, the addition of a substrate that precipitates as a colored substance will detect and locate the parasite. The indirect EIA is more sensitive than the direct EIA because there are more binding sites for the secondary antibody. Although the sensitivity of the indirect

method is about 10 times greater than the direct method (Kurstak 1986), the use of two antibodies in the indirect method increases the possibility of nonspecific binding and false positive readings. Sensitivity can be increased even further by combining a bridge technique with the indirect method. A common bridge is the avidin–biotin complex, which takes advantage of the affinity of avidin for biotin. Commercial kits are available that contain biotin-conjugated antibody and avidin-conjugated enzyme. The assay sequence is primary antibody, biotin-conjugated secondary antibody, and avidin-conjugated enzyme (Figure 1C). The affinity of avidin for biotin results in more bound enzyme and greater sensitivity than in the indirect method without the bridge. This increased sensitivity is illustrated diagrammatically by comparing the amount of enzyme in Figure 1A–C.

A variation called the sandwich EIA may be useful for screening many candidate intermediate hosts for the presence of parasites; however, a positive color change requires more than 10 ng/mL of antigen, an amount perhaps not obtainable from a small intermediate host. In this variation (Figure 1D), primary antibody is adsorbed onto a solid surface, usually the wells of a 96-well microtiter plate. Homogenized tissue from the assay organism is added to the wells and, if parasite antigen is present, the antibody will bind and remove it from solution. The wells are washed, and primary antibody–enzyme conjugate added. The antibody will bind to the parasite antigen present and, if a nonprecipitating substrate is added, the intensity of color in the supernatant will be proportional to the amount of parasite antigen. An indirect sandwich EIA, with a bridge method if desired, can be utilized to increase sensitivity (Figure 1E), but the two primary antibodies must be prepared in different host species or the enzyme-conjugated secondary antibody may bind to the adsorbed primary antibody and yield a false positive result.

Monoclonal versus Polyclonal Antibodies in EIA

Monoclonal and polyclonal antibodies each have advantages and disadvantages for application in EIA (Kurstak 1986; Goding 1987); the choice depends upon the objective of the assay and the availability of facilities and personnel. Antibody specificity is important when EIA is used in the search for an intermediate host. Antibodies recognize epitopes, single antigenic determinants that combine with the antibody paratope. Polyclonal antibodies react with many different epitopes on an antigen; monoclonal antibodies react with a single epitope on an antigen. When the same epitope structure is located on different antigens, a monoclonal antibody cannot distinguish these antigens. Polyclonal antibodies, on the other hand, can distinguish such antigens because their specificity is the result of a unity of hundreds of different clones, and cross-reactions will be random and diluted (Goding 1987). Small changes in the structure of an antigen may have no effect on the binding of polyclonal antibodies because there will always be a subpopulation of paratopes that recognizes some epitopes on the antigen. In contrast, monoclonal antibodies may no longer be able to recognize the changed antigen (Handman and Mitchell 1986), a disadvantage in life cycle research because stage-specific antigens exist in the protozoan groups that have been studied in detail (Handman and Mitchell 1986). For example, monoclonal antibodies that recognize malaria sporozoites in mosquito vectors do not recognize merozoites of the same species in the mammalian host (Yoshida et al. 1980; Zavala et al. 1982), and monoclonal antibodies that recognize merozoites do not recognize sporozoites (Freeman et al. 1980). In contrast, monoclonal antibodies that recognize surface antigens of *Leishmania major* amastigotes and promastigotes reacted with antigens shared by both parasite stages (Alexander and Russell 1985). Monoclonal antibodies have been used effectively in the diagnosis of protozoan diseases (Handman and Mitchell 1986; Wirth et al. 1986); however, because of the possibility of stage-specific antigens, a polyclonal antibody may be better than a monoclonal antibody to detect a stage of an oyster parasite in an intermediate host. Wolf and Markiw (1984) used a polyclonal antibody to identify the intermediate host of the myxozoan *Myxosoma cerebralis*. A monoclonal antibody that is responsive to a single epitope on a parasite in oysters may not react with another stage of the parasite in an intermediate host; however, a mixture of monoclonal antibodies sensitized to various parasite epitopes may be effective. The production, testing, and identification of a monoclonal antibody with high affinity and low cross-reactivity is a long procedure; however, once obtained, the monoclonal antibody can be produced indefinitely, and the unreliability of antiserum production can be avoided.

For the preparation of polyclonal antibodies, a highly purified antigen is needed, and this require-

ment has hindered the development of immunoassays for oyster parasites. Contamination by host tissue will render the antibody useless for diagnosis because the antibody will bind to host tissue and yield false positives. Adsorption of the antibody with host tissue may remove the undesirable antibody, but it may also reduce the activity of the desirable antibody. Contamination with host tissue may be less important if the antibody is to be used to search for parasite antigen in another host species, but generally, the goal should be to obtain the purest antigen. Antigen purity is not as important in monoclonal antibody production because antibodies to host tissue can be discarded, but it is critical that host antigen contaminants are not so abundant that they become immunodominant (Handman and Mitchell 1986). Parasite antigens purified by affinity chromatography with monoclonal antibodies could be used to produce highly specific polyclonal antisera for use in life cycle research (Goding 1987).

Examples of Immunoassays

Research on haplosporidan life cycles at the Virginia Institute of Marine Science has focused on *H. costale* because the organism sporulates regularly every spring and large numbers of spores are present in each gaper (Andrews 1984). Spore-laden oyster tissue was fed to various oyster scavengers, and then the tissues of the scavengers were examined for parasite antigen by EIA. The sporoplasm is the cell that initiates an infection in the next host, so spores must be disrupted to expose the sporoplasm epitopes to the animal in which primary antibodies are to be produced.

Spores of *H. costale* were concentrated first by autodigestion of macerated oyster tissue in large glass beakers. Then, spores were washed three times in sterile seawater and disrupted by shaking at high speed with 0.1-mm glass beads. Spore wall material was not separated from sporoplasm material. Total protein was concentrated by dialysis to 0.145 mg/mL and mixed 1:1 with Freund's complete adjuvant. One milliliter of this suspension was injected intramuscularly into each of three rabbits. After 14 d, spores were disrupted again, and 1.0 mL of spore suspension without adjuvant was injected subcutaneously into each rabbit. The rabbits were bled by cardiac puncture 5 d later, and serum was separated from the clot by incubation for 1 h at room temperature and overnight refrigeration. Serum was adsorbed with homogenized *H. costale*-free oyster tissue, mixed 1:1, overnight at 5°C to remove any possible oyster tissue antibodies. Antigen–antibody complexes were removed by centrifugation, and the serum was filtered, divided into aliquots, and frozen at −21°C. The assay used was an indirect biotin–avidin immunoassay with peroxidase as the enzyme. I used a Vector Laboratories commercial kit that contained biotin-conjugated goat antirabbit immunoglobulin G (IgG) and avidin–peroxidase conjugate. The substrate was hydrogen peroxide with a chromogen that turned blue–black after the reaction was completed; fast green was used as a counterstain. The antigen was *H. costale* in paraffin sections of oyster tissue preserved in 4.0% formaldehyde solution. The assay revealed *H. costale* spores in infected (Figure 2A) but not in uninfected oyster tissue (Figure 2B); reaction with *H. costale* plasmodia was weak. The antibody did not cross-react with spores of *H. nelsoni* or with spores of *Haplosporidium* spp. from the mud crab *Panopeus herbsti* and the naval shipworm. The failure of the antiobdy to recognize plasmodia indicates that epitopes of the spore wall and plasmodia are not similar; this result reinforces the need to isolate plasmodia or sporoplasms as well as spores for antibody production. Roubal and Lester's results (in press) in an immunofluorescent test for *Marteilia sydneyi* were similar: sporonts reacted but the sporoplasm did not. The disruption of *H. costale* spores and the purification of sporoplasm antigen have been hindered to date by an inability to obtain a sufficient number of spores.

For the characterization of *Haplosporidium* sp. of naval shipworms, an immunoassay was developed that used colloidal gold as the marker. Spore purification and primary antibody preparation were similar to that described for *H. costale* except that spores were disrupted in a sonicator and the Ribi adjuvant system was used instead of Freund's adjuvant. A Janssen Life Sciences Auroprobe-LM commercial kit was used that contained 5 nm colloidal gold bound to goat antirabbit IgG. Silver stain reagents were used to enhance the visibility of the colloidal gold. The results of one assay for spores in paraffin sections of naval shipworms are shown in Figure 2C. Strong positive reactions occurred with spores; the background staining with the colloidal gold assay was less than with the enzyme assay.

Immunoassays should increase the chance of delineating the life cycle of *H. nelsoni* or any of the parasites related to it. Assays should be developed for both monoclonal and polyclonal antibodies. Monoclonal antibodies obviate many of the problems of antigen purity that are associated

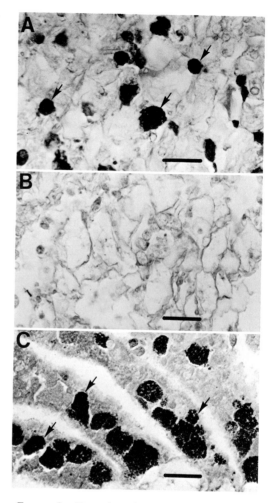

FIGURE 2.—Examples of immunoassay results. **A.** Tissue of oyster infected with *H. costale*. Arrows show positive reaction with sporocysts containing mature spores. Bar = 40 μm. **B.** Uninfected oyster tissue. Bar = 40 μm. **C.** Naval shipworm gills infected with *Haplosporidium* sp. Arrows show positive reaction with sporocysts. Bar = 80 μm.

with the production of polyclonal antibodies, and monoclonals or mixtures of monoclonals will be useful in many aspects of oyster disease research. Polyclonal antibody may be more useful in the search for intermediate hosts because of the potential for stage-specific antigens.

Acknowledgments

I thank J. P. Sypek, Tufts University School of Medicine, and M. Elizabeth Robinson, Virginia Institute of Marine Science, for use of the immunogold assay micrograph. This research was supported in part by the National Sea Grant College

Program, under grant NA85AA-D-SG016, project R/MP-2, through the Virginia Sea Grant College Program. This is contribution 1445 of the Virginia Institute of Marine Science.

References

Alexander, J., and D. G. Russell. 1985. Parasite antigens, their role in protection, diagnosis and escape: the leishmaniases. Current Topics in Microbiology and Immunology 120:43–67.

Andrews, J. D. 1984. Epizootiology of diseases of oysters (*Crassostrea virginica*), and parasites of associated organisms in eastern North America. Helgoländer Meeresuntersuchungen 37:149–166.

De Mey, J., G. W. Hacker, M. De Waele, and D. R. Springall. 1986. Gold probes in light microscopy. Pages 71–88 in J. M. Polak and S. Van Noorden, editors. Immunocytochemistry: practical applications in pathology and biology, 2nd edition. Wright-PSG, Bristol, England.

Elston, R. A., C. A. Farley, and M. L. Kent. 1986. Occurrence and significance of bonamiasis in European flat oysters *Ostrea edulis* in North America. Diseases of Aquatic Organisms 2:49–54.

Farley, C. A. 1967. A proposed life cycle of *Minchinia nelsoni* (Haplosporida, Haplosporidiidae) in the American oyster *Crassostrea virginica*. Journal of Protozoology 14:616–625.

Freeman, R. R., A. J. Trejdosiewicz, and G. A. M. Cross. 1980. Protective monoclonal antibodies recognizing stage-specific merozoite antigens of a rodent malaria parasite. Nature (London) 284:366–368.

Goding, J. W. 1987. Monoclonal antibodies: principles and practice. Academic Press, New York.

Handman, E., and G. F. Mitchell. 1986. Monoclonal antibodies in the study of parasites and host–parasite relationships. Pages 113.1–113.15 in D. M. Weir, editor. Handbook of experimental immunology, volume 4. Blackwell, Boston.

Kurstak, E. 1986. Enzyme immunodiagnosis. Academic Press, Orlando, Florida.

Perkins, F. O., and R. W. Menzel. 1966. Morphological and cultural studies of a motile stage in the life cycle of *Dermocystidium marinum*. Proceedings National Shellfisheries Association 56:23–30.

Roubal, F. R., and R. J. G. Lester. In press. Development of an immunofluorescent test for *Marteilia sydneyi*, agent of QX disease in the Sydney rock oyster, *Saccostrea commercialis*. International Journal for Parasitology 19.

Springall, D. R., G. W. Hacker, L. Grimelius, and J. M. Polak. 1984. The potential of the immunogold–silver staining method for paraffin sections. Histochemistry 81:603–608.

Wirth, D. F., W. O. Rogers, R. Barker, Jr., H. Dourado, L. Suesebang, and B. Albuquerque. 1986. Leishmaniasis and malaria: new tools for epidemiologic analysis. Science (Washington, D.C.) 234:975–979.

Wolf, K., and M. E. Markiw. 1984. Biology contravenes taxonomy in the Myxozoa: new discoveries show alternation of invertebrate and vertebrate hosts. Science (Washington, D.C.) 225:1449–1452.

Yoshida, N., R. S. Nussenzweig, P. Potocnjak, V. Nussenzweig, and M. Aikawa. 1980. Hybridoma produces protective antibodies directed against the sporozoite stage of malaria parasite. Science (Washington, D.C.) 207:71–73.

Zavala, F., R. W. Gwadz, F. H. Collins, R. S. Nussenzweig, and V. Nussenzweig. 1982. Monoclonal antibodies to circumsporozoite proteins identify the species of malaria parasite in infected mosquitoes. Nature (London) 299:737–738.

American Fisheries Society Special Publication 18:304–310, 1988

Monoclonal Antibodies: A Tool for Molluscan Pathology

Eric Mialhe, Viviane Boulo, and Henri Grizel

Institut Français de Recherche pour l'Exploitation de la Mer
Laboratoire de Pathologie et de Génétique des Invertébrés Marins
17390 La Tremblade, France

Hervé Rogier and Françis Paolucci

Laboratoire d'Immunodiagnostics, Sanofi, 34082 Montpellier, France

Abstract.—Hybridoma technology is reviewed, and the characteristics of monoclonal antibodies are compared with those of polyclonal antibodies. The contribution of monoclonal antibodies to molluscan pathology is developed with special emphasis on their use as diagnostic tools. The results of studies with monoclonal antibodies prepared against the protozoan oyster pathogen *Bonamia ostreae* are briefly described.

The development of hybridoma technology as elaborated by Kohler and Milstein (1975) has made an impact in many fields of biological research such as immunology, biochemistry, and pathology (Yelton and Scharff 1981; Krakauer 1985; Seiler et al. 1985). In pathology, mouse monoclonal antibodies have been used in diagnosis (Van Der Auwera 1987) and therapy (Blythman et al. 1981; Frankel 1985). In this article, we briefly review hybridoma technology. We compare the properties of monoclonal and polyclonal antibodies, and we consider the prospects for use of monoclonal antibodies in molluscan pathology, especially for the diagnosis of infectious diseases.

Hybridoma Technology and Production of Monoclonal Antibodies

The principle of hybridoma technology is the continuation of the nonproliferative line of antibody-producing lymphocytes by fusing the lymphocytes with a tumor cell line (myeloma cells; Figure 1). Hybrid cells, called hybridomas, are obtained which retain both the ability of individual lymphocytes to secrete antibody and the ability of myeloma cells to grow without limit. Thus, the homogenous antibody derived from a single clone of hybridomas is called a monoclonal antibody (MAB). The properties of the lymphocyte can also be retained by infecting the lymphocyte with a transforming virus or by transfecting it with tumorigenic DNA (Schonherr and Houwink 1984). The production of MAB has been reviewed by Kennett et al. (1980), Goding (1983), Pau et al. (1983), Schonherr and Houwink (1984), and Paolucci et al. (1986). The different steps in the production are shown in Figure 2.

Immunization, Preparation of Cells, and Fusion

A mouse is immunized by successively injecting it with antigenic preparations (Figure 2). Purified antigen is not necessary, but the hybridization yield is partly conditioned by the level of sensitization of the animal. After the last injection, lymphocytes isolated from the spleen are fused with myeloma cells either by chemical treatment with polyethylene glycol (Paolucci et al. 1986) or by electrical treatment (Vienken and Zimmermann 1985).

Selection of Hybridomas

Because the yield of stable hybridomas from parent cells is low, about 10^6–10^7, the nonfused myeloma cells deficient in hypoxanthine guanine phosphoribosyltransferase (HGPRT−) must be kept from overgrowth. The HGPRT-deficient cells cannot grow in a medium containing hypoxanthine aminopterin thymidine, so this medium is used to select the HGPRT+ hybridomas after fusion.

Selection of Hybridomas Producing Specific Antibody

The hybridomas must be screened as early as possible to distinguish and eliminate those producing nonspecific antibody from those producing specific antibody. The screening technique must be rapid, simple, and suitable for the large number of hybridomas grown in the wells of microculture plates. The techniques used most frequently are radioimmunoassay (RIA), enzyme-linked immunosorbent assay (ELISA), and immunofluorescence (IF). Screening is a key step in lymphocyte hybridization, and its success depends entirely on the availability of purified antigen.

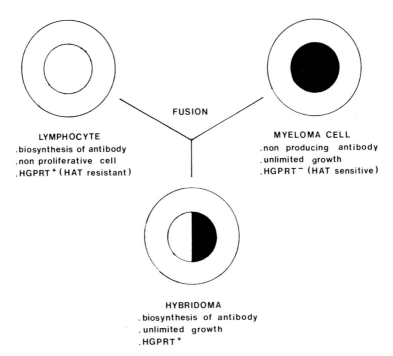

FUSION

LYMPHOCYTE
.biosynthesis of antibody
.non proliferative cell
.HGPRT⁺(HAT resistant)

MYELOMA CELL
.non producing antibody
.unlimited growth
.HGPRT⁻(HAT sensitive)

HYBRIDOMA
.biosynthesis of antibody
.unlimited growth
.HGPRT⁺

FIGURE 1.—Principle of preparation of hybridomas. HGPRT = hypoxanthine guanine phosphoribosyltransferase, HAT = hypoxanthine aminopterin thymidine.

Cloning of Hybridomas Producing Specific Antibody

As soon as positive hybridomas (i.e., those producing specific antibodies) are identified (Figure 2), they must be cloned to reduce the risk of overgrowth by negative cells (i.e., those producing nonspecific antibody). The cells can be cloned in soft agar, with an electronic cell sorter, or by limiting dilutions. The last method is used most frequently; cells are successively diluted to a point at which, statistically, there is less than one viable cell per microculture well. The capacity of the cloned cells to produce specific antibody is then assayed.

Production of Monoclonal Antibodies and Cryopreservation

Because the hybridomas are descended from a line of tumor cells, they can be grown indefinitely in culture and they can produce monoclonal antibodies in vitro or in vivo. Usually, the cloned hybridomas are grown in the ascites fluid of pristane-pretreated mice: On average, 3 mL of fluid or 3–30 mg of antibody per mouse are obtained. Monoclonal antibodies are purified by affinity chromatography on immobilized protein A, which selects immunoglobulins according to isotype. Cryopreservation of hybridomas is an essential safeguard against loss of valuable hy-bridoma lines. Cells that are cryoprotected with dimethyl sulfoxide (7.5%) in fetal calf serum can be stored in liquid nitrogen.

Comparison of Polyclonal and Monoclonal Antibodies

Polyclonal antisera obtained from immunized animals are characterized by a heterogeneity of antigen-specific immunoglobulins, a low titer of these specific antibodies, and variability between serum batches (Figure 3). Nevertheless, the use of highly purified antigens for immunization and in immune adsorption techniques has led to the development of specific and sensitive test systems, for example, indirect immunofluorescence (Boulo et al., in press). Polyclonal antibodies give rise to difficult problems in the development of quantitative and reproducible immunodiagnostic methods.

Because of their specificity, unlimited availability, and homogeneity (Figure 3), monoclonal antibodies can be standardized for use in highly sensitive and specific immunoassays, especially for detecting small amounts of infectious agents in clinical specimens. Also, they may be valuable in detecting antigenic variation between different stages or strains of parasites, which would be a more difficult procedure with polyclonal antibodies.

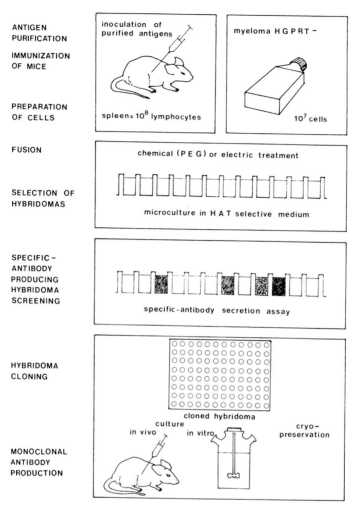

ANTIGEN
PURIFICATION

inoculation of
purified antigens

myeloma H G P R T −

IMMUNIZATION
OF MICE

PREPARATION
OF CELLS

spleen= 10^8 lymphocytes

10^7 cells

FUSION

chemical (P E G) or electric treatment

SELECTION OF
HYBRIDOMAS

microculture in H A T selective medium

SPECIFIC −
ANTIBODY
PRODUCING
HYBRIDOMA
SCREENING

specific-antibody secretion assay

HYBRIDOMA
CLONING

cloned hybridoma

culture

in vivo   in vitro   cryopreservation

MONOCLONAL
ANTIBODY
PRODUCTION

FIGURE 2.—Steps in hybridoma technology. HGPRT = hypoxanthine guanine phosphoribosyltransferase; PEG = polyethylene glycol; HAT = hypoxanthine aminopterin thymidine.

Monoclonal Antibodies for Diagnosis of Molluscan Pathogens

Heretofore, infectious diseases of molluscs have been diagnosed from histological preparations. Although parasites and procaryotes can be detected and identified by this technique, the method has several limitations and disadvantages, especially in epidemiological surveys. The preparation and observation of specimens are time-consuming. The availability of personnel and material is probably insufficient for useful disease prophylaxis. Moreover, it is difficult to precisely quantify infections by this procedure, and the method cannot be used to diagnose viral infections of molluscs (Johnson 1984; Elston and Wilkinson 1985).

Alternative methods used in human and veterinary pathology include immunoassays based on specific antigen–antibody reactions. Monoclonal antibodies are especially suitable for detecting antigens in epidemiological studies of parasitic (Ungar et al. 1985; Wirtz et al. 1985), procaryotic (Holley et al. 1984; Kotani and McGarrity 1985; Morris and Ivanyi 1985), or viral (Beards et al. 1984; Monath et al. 1986) diseases.

Because invertebrates have no demonstrable humoral immune response, immunodiagnosis depends on the utilization of specific antibodies to reveal the presence of infectious agents in them. Among molluscs, the barrier to the development of such technology has been the inability to prepare purified parasite suspensions to immunize

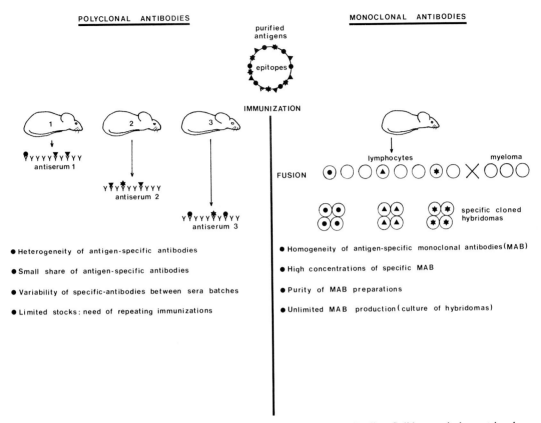

FIGURE 3.—Comparison of properties of polyclonal and monoclonal antibodies. Solid star, circle, or triangle = antigenic determinant; star, circle, or triangle + Y = immunoglobulin specific for antigenic determinant; circled star, circle, or triangle = lymphocyte secreting immunoglobulin specific for antigenic determinant.

mice and to assay hybridomas. However, the recent purification of two pathogens of oysters (Mialhe et al. 1985; Mialhe et al. 1988a) has overcome this barrier.

Pathogens can be detected by three types of immunoassays, direct, indirect, and direct sandwich (Figure 4). The different solid-phase immunoassay systems, IF, RIA, and ELISA, depend on the labeling element that is conjugated to the antibody. Immunofluorescence is mainly a qualitative and sensitive technique adapted for analyzing a few specimens (Nairn 1976), whereas RIA and ELISA are suitable for detecting and quantifying pathogens in many specimens. However, radioactive isotopes with short half-life radiation hazards limit the use of RIA to specially authorized laboratories. The sensitivity of ELISA (Voller et al. 1976; Voller et al. 1978; Yolken 1982) relies on the conversion of many substrate molecules by a single molecule of enzyme–antibody conjugate. These conjugates are stable and can be stored frozen. Microtiter-plate colorimeters are available to rapidly measure many samples. Substrates that give rise to colored and precipitated reaction products (Dao 1985; Turner 1986) are well adapted for simple assay without any instruments. Consequently, ELISA, especially the direct type (Figure 4), constitutes a good immunodiagnostic method for the quick, easy, and precise determination of the percentage and rate of infection of diseases in molluscs.

Monoclonal Antibodies against
Bonamia ostreae

In several European countries, the breeding of the edible oyster *Ostrea edulis* is adversely affected by the intrahemocytic parasite, *Bonamia ostrea* (Ascetospora). In consideration of the limits of the histological diagnostic method and the recent purification of this pathogen (Mialhe et al. 1988a), it became important to produce monoclonal antibodies for immunodiagnosis. Some brief

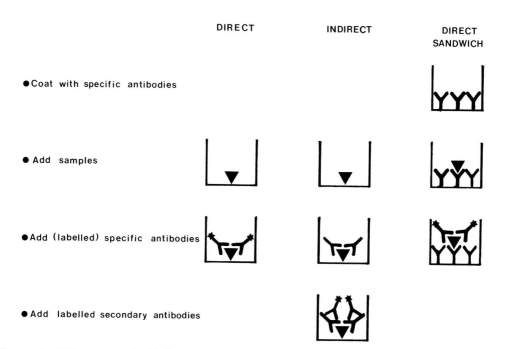

FIGURE 4.—Different types of solid phase immunoassays. Y = immunoglobulin; triangle = antigenic determinant; triangle + Y = immunoglobulin specific for antigenic determinant; Y + star = immunoglobulin labeled with an enzyme or radioisotope.

results are presented below to illustrate the applicability of monoclonal antibody technology to molluscan pathology. The details of the research will soon appear in print (Boulo et al., in press; Rogier et al., in press).

About 700 hybridomas were obtained from a fusion of lymphocytes of a hyperimmunized Balb/c mouse with myeloma cells. Eight hybridomas were then selected from these on the basis of their production of antibody specific for *B. ostreae*. The epitope specificity of these eight hybridomas was defined precisely by an inhibition RIA test and by IF antibody pattern analysis. Monoclonal antibodies 20B2 and 15C2, specific for cytoplasmic membrane epitopes, were retained for diagnosing *B. ostreae* by indirect IF in three clinical studies. The results were related to those diagnosed histologically. In light infections, when only a few *Bonamia* cells were observed on a smear, the IF test with MAB permitted quicker detection of *B. ostreae* because it could be observed at lower magnification.

The two radiolabeled monoclonal antibodies differentiate between hemolymph of diseased oysters from that of healthy oysters (Figure 5). On this basis, we are now developing an enzymatic immunoassay for detecting *Bonamia* disease.

Discussion

The economic impact of infectious diseases on bivalve culture and the absence of both resistant races and antiparasitic treatment indicate how necessary preventive measures are to insure the continuity of bivalve culture. Such measures depend mainly on the development of quantitative

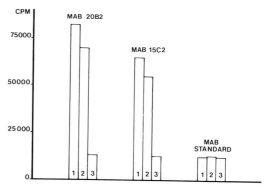

FIGURE 5.—Specificity analysis of monoclonal antibodies specific to *Bonamia ostreae* by direct radioimmunoassay. CPM = counts/min; MAB = monoclonal antibody. Bar 1 = purified *B. ostraeae* cells immobilized on nitrocellulose; bar 2 = parasitized oyster hemolymph immobilized on nitrocellulose; bar 3 = healthy oyster hemolymph immobilized on nitrocellulose.

immunodiagnostic methods such as ELISA, which is quick, reliable, and applicable by all investigators concerned with a specific disease. Monoclonal antibody methods provide a desirable alternative to polyclonal immunosera techniques. Our first results with *B. ostreae* suggest opportunities for other investigations into monoclonal antibodies against the major identified pathogens of economically important bivalves.

The value of monoclonal antibodies for fundamental research in molluscan pathology must also be noted. Monoclonal antibodies constitute new and powerful tools with which to study the antigenic variation of parasites during development or from different geographical areas (Mialhe et al. 1988b). Also, such antibodies are suitable reagents for discerning the role of receptors implicated in the recognition process between host cell and parasite. Finally, collaboration between molluscan pathologists and specialists in hybridoma technology is essential to quickly elaborate prophylactic strategies on an international scale.

References

Beards, G. M., and eight coauthors. 1984. Enzyme-linked immunosorbent assays based on polyclonal and monoclonal antibodies for rotavirus detection. Journal of Clinical Microbiology 19:248–254.

Blythman, H. E., and six coauthors. 1981. Immunotoxins: hybrid molecules of monoclonal antibodies and toxin subunits specifically kill tumour cells. Nature (London) 290:145–146.

Boulo, V., H. Rogier, F. Paolucci, E. Mialhe, and H. Grizel. In press. Immunodiagnostic of *Bonamia ostreae* (Ascetospora) parasite of *Ostrea edulis* and subcellular identification of epitopes by monoclonal antibodies. Journal of Fish Diseases.

Dao, M. L. 1985. An improved method of antigen detection on nitrocellulose. In situ staining of alkaline phosphatase conjugated antibody. Journal of Immunological Methods 82:225–231.

Elston, R. A., and M. T. Wilkinson. 1985. Pathology, management and diagnosis of oyster velar virus disease (OVVD). Aquaculture 48:189–210.

Frankel, A. E. 1985. Antibody–toxin hybrids: a clinical review of their use. Journal of Biological Response Modifiers 4:437–446.

Goding, J. W. 1983. Monoclonal antibodies: principles and practice. Academic Press, New York.

Holley, D. L., S. D. Allen, and B. B. Barnett. 1984. Enzyme-linked immunosorbent assay, using monoclonal antibody, to detect enterotoxic *Escherichia coli* K99 antigen in feces of dairy calves. American Journal of Veterinary Research 45:2613–2616.

Johnson, P. T. 1984. Viral diseases of marine invertebrates. Helgoländer Meeresuntersuchungen 37: 65–98.

Kennett, R. H., T. J. McKearn, and K. B. Bechtol. 1980. Monoclonal antibodies. Hybridomas: a new dimension in biological analysis. Plenum, New York.

Kohler, G., and C. Milstein. 1975. Continuous cultures of fused cells secreting antibodies of predefined specificity. Nature (London) 256:495–497.

Kotani, H., and G. J. McGarrity. 1985. Rapid and simple identification of mycoplasmas by immunobinding. Journal of Immunological Methods 85: 257–267.

Krakauer, H. 1985. Clinical applications of monoclonal antibodies. European Journal of Clinical Microbiology 4:1–9.

Mialhe, E., E. Bachère, D. Chagot, and H. Grizel. 1988a. Isolation and purification of the protozoan *Bonamia ostreae* (Pichot and coll. 1979), a parasite affecting the flat oyster *Ostrea edulis* L. Aquaculture 71:293–299.

Mialhe, E., E. Bachère, C. Le Bec, and H. Grizel. 1985. Isolement et purification de *Marteilia* (Protozoa: Ascetospora) parasites des bivalves marins. Comptes Rendus de l'Académie des Sciences, Série III, Sciences de la Vie 301:137–142.

Mialhe, E., and seven coauthors. 1988b. Serological analysis of *Bonamia* in *Ostrea edulis* and *Tiostrea lutaria* using polyclonal and monoclonal antibodies. Aquatic Living Resources 1:1–3. (Gauthier Villars, Montrouge Cedex, France.)

Monath, T. P., and six coauthors. 1986. Sensitive and specific monoclonal immunoassay for detecting yellow fever virus in laboratory and clinical specimens. Journal of Clinical Microbiology 23: 129–134.

Morris, J. A., and J. Ivanyi. 1985. Immunoassays of field isolates of *Mycobacterium bovis* and other mycobacteria by use of monoclonal antibodies. Journal of Medical Microbiology 19:367–373.

Nairn, R. C. 1976. Fluorescent protein tracing. Churchill-Livingstone, London.

Paolucci, F., B. Basuyaux, C. Clavies, M. Piechaczyk, and B. Pau. 1986. Lymphocyte hybridization: basic principles and selection of monoclonal antibodies directed against hormones. Serono Symposia Publications from Raven Press 30:3–13.

Pau, B., P. Poncelet, and P. Gros. 1983. L'hybridation lymphocytaire: principe et potentiel d'application en biologie clinique. Revue Française des Laboratoires 115:5–16.

Rogier, H., and seven coauthors. In press. Monoclonal antibodies against *Bonamia ostreae* (Ascetospora) parasite of the flat oyster *Ostrea edulis*. Hybridoma.

Schonherr, O. T., and E. H. Houwink. 1984. Antibody engineering, a strategy for the development of monoclonal antibodies. Antonie van Leeuwenhoek Journal of Microbiology 50:597–623.

Seiler, F. R., and seven coauthors. 1985. Monoclonal antibodies: their chemistry, functions and possible uses. Angewandte Chemie International Edition in English 24:139–160.

Turner, B. M. 1986. Use of alkaline phosphatase-conjugated antibodies for detection of protein antigens on nitrocellulose filters. Methods in Enzymology 121: 848–855.

Ungar, B. L. P., R. H. Yolken, and T. C. Quinn. 1985.

Use of monoclonal antibody in an enzyme immunoassay for the detection of *Entamoeba histolytica*
in fecal specimens. American Journal of Tropical
Medicine and Hygiene 34:465–472.

Van Der Auwera, P. 1987. Rapid diagnosis of infectious
diseases. Internal Clinic Products Review 1:12–27.

Vienken, J., and U. Zimmermann. 1985. An improved
electrofusion technique for production of mouse
hybridoma cells. FEBS (Federation of European
Biochemical Societies) Letters 182:278–280.

Voller, A., A. Barlett, and D. E. Bidwell. 1976. Enzyme
immunoassays for parasitic diseases. Transactions
of the Royal Society of Tropical Medicine and
Hygiene 70:98–106.

Voller, A., A. Barlett, and D. E. Bidwell. 1978. Enzyme
immunoassays with special reference to ELISA

technique. Journal of Clinical Pathology (London)
31:507–520.

Wirtz, R. A., T. R. Burkot, R. G. Andre, R. Rosen-
berg, W. E. Collins, and D. R. Roberts. 1985. Identification of *Plasmodium vivax* sporozoites in mosquitoes using an enzyme-linked immunosorbent assay. American Journal of Tropical Medicine and
Hygiene 34:1048–1054.

Yelton, D. E., and M. D. Scharff. 1981. Monoclonal
antibodies: a powerful new tool in biology and
medicine. Annual Review of Biochemistry 50:
657–680.

Yolken, R. H. 1982. Enzyme immunoassays for the
detection of infectious antigens in body fluids. Current limitations and future prospects. Reviews of
Infectious Diseases 4:35–68.

American Fisheries Society Special Publication 18:311–315, 1988

Bivalve Mollusc Cell Culture

Eric Mialhe, Viviane Boulo, and Henri Grizel

Institut Français de Recherche pour l'Exploitation de la Mer
Laboratoire de Pathologie et de Génétique des Invertébrés Marins
17390 La Tremblade, France

Abstract.—The historical background of bivalve cell culture and its potential uses in molluscan pathology are reviewed. Then, primary cultures for the study of some host–pathogen interactions at the cellular level are considered. Next, alternative methods to classical cell cultures are examined that may lead to the production of bivalve cell lines.

Bivalve breeding, an economically important element of aquaculture, is becoming increasingly subject to the hazards of disease. Infectious diseases affect the species of bivalves bred throughout the world and are caused by intracellular pathogenic agents, viruses, microorganisms, or parasites. The diseases develop into epizootics that are responsible for considerable economic loss.

The study of these diseases must go beyond description to establish methods for prevention and therapy. Given the results obtained in human and veterinary pathology, the achievement of such measures depends partly on the development of bivalve cell cultures that permit in vitro cultivation of the intracellular pathogenic agents. Such cell cultures could help us to determine and understand the host–pathogen relationship at the cellular and molecular level (Trager 1983; Agabian and Eisen 1985). Practically, with the attainment of in vitro production of parasites, experimental reproduction of the disease could lead to the selection of resistant animals. The cultures would also constitute a diagnostic tool adapted to revealing fastidious pathogens such as viruses (Fields 1985). In addition, cell cultures would permit in vitro screening of chemical substances as candidate antiparasitic agents (Campbell and Rew 1986).

Cell cultures of marine molluscs date from 1960s and were inspired by the establishment of several insect cell lines (Vago 1972). In these cultures of molluscs, bivalve tissues were maintained in simple culture media to which vertebrate serum was added (Chardonnet and Peres 1963; Tripp 1963; Perkins and Menzel 1964; Li et al. 1966; Tripp et al. 1966; Vago 1972). The cells survived only a few days or a few weeks in spite of claims that cell lines had been established (Vago and Chastang 1960; Doutko 1967).

More recently, as a result of several economically catastrophic viral and parasitic epizootics

(Sindermann 1976; Meuriot and Grizel 1984; Grizel 1985), research on the culture of bivalve cells was resumed (Cousserans 1975; Brewster and Nicholson 1979; Stephens and Hetrick 1979a, 1979b; Hetrick et al. 1981). These experiments gave rise to improved techniques for the establishment of primary cultures, especially of embryonic and larval cells, but cell lines were not established. This failure underscored certain difficulties: the need for practical methods of regularly obtaining aseptic tissues, which are essential for the attainment of primary cultures (Millar and Scott 1967), and the lack of physiological and biochemical data needed to devise suitable culture media.

Nonetheless, the establishment of an embryonic cell line of the bloodfluke planorb *Biomphalaria glabrata* (Hansen 1976; Bayne et al. 1978) supported the feasibility of experiments in cell cultures of bivalves. This success and the understanding that cell systems are essential for progress in bivalve pathology indicate the need for further research in this area. Short-term research should focus on primary cultures and their direct use for in vitro study of intracellular pathogens; longer-term work should address methods for establishing stable cell lines. These areas of research are considered in this article. The technical aspects of cell culture can be found in more specialized publications (Malek and Cheng 1974; Bayne 1976; Jakoby and Pastan 1979).

Primary Cultures

Primary cultures are those initiated either from explants (from which the cells migrate) or from cells that have been isolated by mechanical or enzymatic dissociation of tissues. Theoretically, any cell type may be placed in culture, but its survival in vitro varies greatly according to its nature and degree of differentiation.

The major difficulty in establishing primary

cultures rests at the preliminary decontamination of the tissues. The majority of bivalve organs are in direct contact with the environment, which is contaminated by various types of microorganisms and protozoans. Some relatively efficient protocols for decontamination have been developed for adult tissues (Stephens and Hetrick 1979a, 1979b). The presence of numerous intrinsic contaminants impedes the attainment of aseptic tissues from the digestive gland. For larval tissues, the most reliable method is in vitro fertilization with gametes removed aseptically.

Primary cultures should be maintained in culture media whose composition is based on the physicochemical characteristics of bivalve hemolymph. Biochemical data on the hemolymph are limited to concentrations of free amino acids (Bishop et al. 1983), total protein and sugar contents (De Zwaan 1983), and certain inorganic constituents (Deane and O'Brien 1979). These data, however, vary greatly between species, between individuals, and according to the physiological condition of each animal. On the other hand, certain physicochemical criteria, such as osmotic pressure and pH, are better defined (Burton 1983). After these characteristics are taken into account, certain components used classically in cultures of animal cells, are added to the medium. These include compounds such as nucleotide precursors, vitamins, and growth factors. Other additives may include fetal calf serum, yeast hydrolysates, and vitamin solutions. In some experiments, media have been supplemented with homologous hemolymph (Boulo, unpublished data). An important feature of molluscan cells seems to be a low nutritive requirement, because primary cultures generally show longer survival in media diluted with sterile seawater.

Cell culture work in our laboratory with the edible oyster *Ostrea edulis* leads us to the following observations.

(1) Cells taken from paleal and palpeal tissues are of two types: ciliated cells that lose their cilia after a few days of culture and then degenerate rapidly; and smaller, spherical, less-differentiated cells that remain in suspension and survive for several weeks without showing mitotic figures.

(2) Primary cultures of larval cells are distinctly more promising. In particular, cultures established from veliger larvae are at the optimal stage of differentiation. At an earlier stage, the cells are too charged with vitellus and are very fragile. At a later stage, the larval tissues are too differentiated and newly formed shells constitute a major obstacle to tissue dissociation. Mechanical dissociation of young larval tissues is preferable to enzymatic dissociation because it is less stressful for the cells. Mechanical dissociation regularly leads to a large degree of adherence of explants and much cellular migration. Cell activity remains apparent for several weeks as contractile elements cause a rythmic movement in certain explants. Nonetheless, as of the sixth week of culture, the cells progressively change and become suspended without any observable cellular multiplication.

(3) Primary cultures of hemocytes have been easy to establish because of the low risk of contamination during removal of the hemolymph and the ease with which the promial vein can be punctured. Hemocytes in culture regularly survive 6 weeks.

Thus, primary cultures already constitute cellular systems that permit a short-term approach to the culture and in vitro study of pathogenic agents. Primary cultures of embryonic and larval tissues of the Pacific oyster *Crassostrea gigas* could be used, for example, to study the iridovirus responsible for oyster velar disease (Elston and Wilkinson 1985). Such a procedure has already been used to cultivate insect viruses and rickettsias in vitro (Vago and Quiot 1982).

Primary hemocyte cultures can be easily established throughout the year and maintained for periods regularly lasting 4–6 weeks. They constitute a select system for cultivating viral or protozoan pathogens, which develop in the hemocytes of different species of bivalves. The isolation and purification of *Bonamia ostreae* (Mialhe et al. 1988) has enabled us to experimentally infect primary hemocyte cultures and undertake the in vitro study of the relationships between this protozoan and the hemocytes of both the naturally susceptible edible oyster and the experimentally resistant Pacific oyster. This model has already made possible the description of the initial stages of parasite development, and has provided an approach to the mechanisms of recognition, penetration, and survival in the hemocyte. This model is comparable to those in human and veterinary parasitology (Trager 1983) for parasites for which there is no experimental cell line (e.g., *Plasmodium* spp., which develop in erythrocytes).

Cell Lines

Subsequent to cell multiplication and at the time of the first subculture, primary culture becomes a cell line. A cell line thus corresponds to a population of cells whose capacity for multiplication is unlimited. Hence, development of an empirical and uncertain nature is the origin of numerous cell lines (Hink 1980), including the only cell line of a mollusc, the bloodfluke planorb (Hansen 1976). Generally, primary cultures are established from tumors, embryonic and larval tissues, or connective tissues. The different tissue types have a low degree of differentiation, and mitotic activity is characteristic of the tissue of origin. The principal advantage of cell lines over primary cultures lies in their capacity for multiplication. Thus, the ability to obtain a cell line from host tissue, if it is susceptible to infection with a given pathogenic agent, will greatly facilitate in vitro studies of that pathogen.

The experience acquired in working with animal cell lines confers a special a priori interest in culturing embryonic and larval cells of bivalves. This was also the choice of Hetrick et al. (1981), who used embryos of the eastern oyster *Crassostrea virginica* at the 16- and 32-cell stage. Our work with embryos of edible oysters has led us to select young veliger stages for the reasons already given. Also, larval production in a hatchery can be performed throughout the year, a major advantage for conducting continuous experimentation and for optimizing culture conditions.

The infrequent observation of mitosis in primary cultures suggests the absence of mitogenic factors in the culture medium. This underscores the need to test various supplements to the medium, such as homologous hemolymph or extracts from cerebroganglions of the common Atlantic slipper snail *Crepidula fornicata*; the latter contains a nonspecific mitotic factor (Le Gall et al. 1987). Tests could also be conducted with extracts of genital glands, which may prove favorable to cell multiplication (Anonymous 1985).

It is also important to undertake experiments in tumoral transformations, either in vivo or in vitro. Tumors occur naturally in bivalves (Mix 1986; Peters 1988, this volume), and some knowledge has been acquired about chemical carcinogenesis (Heidelberger 1975; Hecker et al. 1982; Huberman and Barr 1985; Huberman and Jones 1985). Chemical carcinogenesis is, at present, considered a two-stage process in which sequential events, initiation and promotion, lead to the appearance of tumors (Hecker et al. 1982). Initiation presumably involves an irreversible "mutational event of a tumor gene." This event is caused by a carcinogen capable of binding covalently to DNA and forming DNA adducts. Some chemicals bind directly (nitrosamides) and others only after cellular conversion to a chemically reactive form (polycyclic aromatic hydrocarbons, aromatic amines, nitrosamines, aflatoxins). The second stage of the carcinogenic process is tumor promotion (Diamond et al. 1980). Tumor promoters are devoid of mutagenic activity but enhance cell transformation. They may exert their promotional effect by causing the expression of mutated tumor genes similar to gene expression during cell differentiation. Some experiments with molluscs have been reported that involved benzo[a]pyrene, 3-methylcholanthrene, and a direct-acting mutagen, N-methyl-N'-nitro-N-nitrosoguanidine (Hetrick et al. 1981). The negative results observed in vivo and in vitro could be explained by the absence of tumor promoters. With initiator and promoter chemicals, some positive experimental results have been reported in vivo for a planarian (Hall et al. 1986) and in vitro for human and rodent primary cultures (Huberman and Jones 1985). These experiments have stimulated our own research with edible and Pacific oysters now in progress in our laboratory.

Finally, it will be useful for molluscan pathologists concerned with cell culture to take into account research on oncogenes (Bishop 1983; Bradshaw 1986; Garrett 1986; Klein and Klein 1986), because neoplastic transformation (Ratner et al. 1985) of vertebrate primary cells was achieved by transfection with active oncogenes. Oncogenes are widely conserved evolutionarily from yeasts to insects to mammals (Shilo and Weinberg 1981; Jenkins et al. 1984).

References

Agabian, N., and H. Eisen. 1985. Molecular biology of host–parasite interactions, UCLA (University of California Los Angeles) Symposia on Molecular and Cellular Biology, New Series 13.

Anonymous. 1985. Artificial pearls with cultivated cells. Technocrat 18(4).

Bayne, C. J. 1976. Culture of molluscan organs: a review. Pages 61–74 *in* K. Maramorosch, editor. Invertebrate tissue culture: research applications. Academic Press, New York.

Bayne, C. J., A. Owczarzak, and J. R. Allen. 1978. Molluscan (*Biomphalaria*) cell line: serology, karyotype, behavioral and enzyme electrophoretic characterization. Journal of Invertebrate Pathology 32:35–39.

Bishop, J. M. 1983. Cellular oncogenes and retroviruses. Annual Review of Biochemistry 52:301–354.

Bishop, S. H., L. L. Ellis, and J. M. Burcham. 1983. Amino acid metabolism in molluscs. Pages 244–328 *in* K. M. Wilbur and C. M. Yonge, editors. The Mollusca, volume 1. Academic Press, New York.

Bradshaw, T. K. 1986. Cell transformation: the role of

oncogenes and growth factors. Mutagenesis 1:91–97.

Brewster, F., and B. L. Nicholson. 1979. *In vitro* maintenance of amoebocytes from American oyster (*Crassostrea virginica*). Journal of the Fisheries Research Board of Canada 36:461–467.

Burton, R. F. 1983. Ionic regulation and water balance. Pages 291–352 *in* A. S. M. Saleuddin and K. M. Wilbur, editors. The Mollusca, volume 5. Academic Press, New York.

Campbell, W. C., and R. S. Rew. 1986. Chemotherapy of parasitic diseases. Plenum, New York.

Chardonnet, Y., and G. Peres. 1963. Essai de culture de cellules provenant de mollusque: *Mytilus galloprovincialis*. Comptes Rendus des Séances de la Société de Biologie et de ses Filiales 157:1593–1595.

Cousserans, F. 1975. Recherche sur la culture de cellules de mollusques marins et sur l'emploi de ces sytèmes cellulaires en pathologie marine. Doctoral dissertation. Université des Sciences et Techniques du Languedoc, Montpellier, France.

Deane, E. M., and R. W. O'Brien. 1979. Composition of the haemolymph of *Tridacna maxima* (Mollusca: Bivalvia). Comparative Biochemistry and Physiology A, Comparative Physiology 66:339–341.

De Zwaan, A. 1983. Carbohydrate catabolism in bivalves. Pages 138–169 *in* K. M. Wilbur and C. M. Yonge, editors. The Mollusca, volume 1. Academic Press, New York.

Diamond, L., T. G. O'Brien, and W. Baird. 1980. Tumor promoters and the mechanism of tumor promotion. Advances in Cancer Research 32:1–74.

Doutko, Y. P. 1967. Obtention d'une lignée de cellules transplantables d'un foie de *Mytilus galloprovincialis* L. Tsitologiya i Genetika 1:61–64.

Elston, R. A., and M. T. Wilkinson. 1985. Pathology, management and diagnosis of oyster velar virus disease (OVVD). Aquaculture 48:189–210.

Fields, B. N. 1985. Virology. Raven Press, New York.

Garrett, C. T. 1986. Oncogenes. Clinica Chimica Acta 156:1–40.

Grizel, H. 1985. Etude des récentes épizooties de l'huître plate *Ostrea edulis* L. et de leur impact sur l'ostreiculture bretonne. Doctoral dissertation. Université des Sciences et Techniques du Languedoc, Montpellier, France.

Hall, F., M. Morita, and J. B. Best. 1986. Neoplastic transformation in the planarian. I—cocarcinogenesis and histopathology. Journal of Experimental Zoology 240:211–227.

Hansen, E. L. 1976. A cell line from embryos of *Biomphalaria glabrata* (Pulmonata): establishment and characteristics. Pages 75–99 *in* K. Maramorosch, editor. Invertebrate tissue culture: research applications. Academic Press, New York.

Hecker, E., N. E. Fusegenic, W. Kunz, F. Marks, and H. W. Thielmann. 1982. Carcinogenesis: a comprehensive survey, volume 7. Raven Press, New York.

Heidelberger, C. 1975. Chemical carcinogenesis. Annual Review of Biochemistry 44:79–121.

Hetrick, F. M., E. Stephens, N. Lomax, and K. Lutrell. 1981. Attempts to develop a marine molluscan cell line. University of Maryland, Sea Grant College Program Technical Report UM-SG-TS-81-06, College Park.

Hink, W. F. 1980. The 1979 compilation of invertebrate cell lines and culture media. Pages 553–578 *in* E. Kurstak and K. Maramorosch, editors. Invertebrate systems in vitro. Elsevier, Amsterdam.

Huberman, E., and S. H. Barr. 1985. Carcinogenesis: a comprehensive survey, volume 10. Raven Press, New York.

Huberman, E., and C. A. Jones. 1985. The control of mutagenesis and cell differentiation in cultured human and rodents cells by chemicals that initiate or promote tumor formation. Pages 77–100 *in* A. D. Woodhead and C. J. Shellabarger, editors. Assessment of risk from low-level exposure to radiation and chemicals. Plenum, New York.

Jakoby, W. B., and I. H. Pastan. 1979. Methods in enzymology, volume 58. Academic Press, New York.

Jenkins, J. R., K. Rudge, and G. A. Lurrie. 1984. Cellular immortalization by a cDNA clone encoding the transformation associated phosphoprotein, p53. Nature (London) 312:651–654.

Klein, G., and E. Klein. 1986. Conditioned tumorigenicity of activated oncogenes. Cancer Research 46:3211–3224.

Le Gall, S., C. Feral, C. Lengronne, and M. Porchet. 1987. Partial purification of the neuroendocrine mitogenic factor in the mollusc *Crepidula fornicata* L. Comparative Biochemistry and Physiology B, Comparative Biochemistry 86:393–396.

Li, M. F., J. E. Stewart, and R. E. Drinnan. 1966. In vitro cultivation of cells of the oyster, *Crassostrea virginica*. Journal of the Fisheries Research Board of Canada 23:595–599.

Malek, E. A., and T. C. Cheng. 1974. Tissue and organ culture. Pages 242–263 *in* E. A. Malek and T. C. Cheng, editors. Medical and economic malacology. Academic Press, New York.

Meuriot, E., and H. Grizel. 1984. Note sur l'impact économique des maladies de l'huître plate en Bretagne. Institut Scientifique et Technique des Pêches Maritimes Rapport Technique 12, Nantes, France.

Mialhe, E., E., Bachère, D. Chagot, and H. Grizel. 1988. Isolation and purification of the protozoan *Bonamia ostreae* (Pichot and coll. 1979), a parasite affecting the flat oyster *Ostrea edulis* L. Aquaculture 71:293–299.

Millar, R. H., and J. M. Scott. 1967. Bacteria-free culture of oyster larvae. Nature (London) 216:1139–1140.

Mix, M. C. 1986. Cancerous diseases in aquatic animals and their association with environmental pollutants: a critical literature review. Marine Environmental Research 20:1–141.

Perkins, F. O., and R. W. Menzel. 1964. Maintenance of oyster cells *in vitro*, Nature (London) 204:1106–1107.

Peters, E. C. 1988. Recent investigations on the disseminated sarcomas of marine bivalve molluscs. American Fisheries Society Special Publication 18:74–92.

Ratner, L., S. F. Josephs, and F. Wong Staal. 1985. Oncogenes: their role in neoplastic transformation. Annual Review of Microbiology 29:419–449.

Shilo, B. Z., and R. A. Weinberg. 1981. DNA sequences homologous to vertebrate oncogenes are conserved in *Drosophila melanogaster*. Proceedings of National Academy of Sciences of the USA 78:6789–6792.

Sindermann, C. J. 1976. Oyster mortalities and their control. Pages 349–361 *in* Advances in aquaculture. FAO Technical conference on aquaculture, Kyoto, Japan. Fishing News Books, Farnham, England.

Stephens, E. B., and F. M. Hetrick. 1979a. Cultivation of granular amoebocytes from the American oyster. Pages 991–992 *in* Tissue Culture Association Manual, volume 5. Tissue Culture Association, Rockville, Maryland.

Stephens, E. B., and F. M. Hetrick. 1979b. Decontamination of American oyster tissues for cell and organ culture. Pages 1029–1031 *in* Tissue Culture Association Manual, volume 5. Tissue Culture Association, Rockville, Maryland.

Trager, W. 1983. In vitro growth of parasites. Pages 39–51 *in* J. Guardiola, L. Luzzatto, and W. Trager, editors. Molecular biology of parasites. Raven Press, New York.

Tripp, M. R. 1963. Cellular response of molluscs. Annals of the New York Academy of Sciences 113:467–474.

Tripp, M. R., L. A. Bisignani, and M. T. Kenny. 1966. Oyster amoebocytes *in vitro*. Journal of Invertebrate Pathology 8:137–140.

Vago, C. 1972. Invertebrate cell and organ culture in invertebrate pathology. Pages 245–278 *in* C. Vago, editor. Invertebrate tissue culture, volume 2. Academic Press, New York.

Vago, C., and S. Chastang. 1960. Culture de tissus d'huîtres. Comptes Rendus de l'Académie des Sciences, Série III, Sciences de la Vie 250:2751.

Vago, C., and J. M. Quiot. 1982. Infection *in vitro* d'embryons d'invertébrés: un principe d'étude de pathogénèse intracellulaire. Comptes Rendus de l'Académie des Sciences, Série III, Sciences de la Vie. 295:461.